Principles of Solidification

Martin Eden Glicksman

Principles of Solidification

An Introduction to Modern Casting and Crystal Growth Concepts

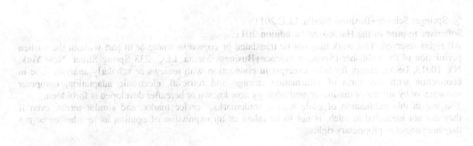

 Springer

Martin Eden Glicksman
Materials Science and Engineering Department
University of Florida
Gainesville, FL 32611-6400, USA
mglic@mse.ufl.edu

ISBN 978-1-4899-8185-1   ISBN 978-1-4419-7344-3 (eBook)
DOI 10.1007/978-1-4419-7344-3
Springer New York Dordrecht Heidelberg London

Printed on acid-free paper

Springer is part of Springer Science+Business Media (www.springer.com)

To my wife, Lucinda

Preface

> How many have truly benefited from my classes I cannot
> judge. But I have the consolation that one person has learned
> much thereby, and that is me.
>
> *Ludwig Boltzmann, Vienna, 1905 [1]*

This book is a distillation based on more than three decades of teaching university courses and providing professional seminars on solidification, casting and welding of metals, and crystal growth. The courses themselves were offered for many years at Rensselaer Polytechnic Institute and, most recently, at the University of Florida. Their purpose, consistently, was to present to a variety of engineering and science students a logical progression of the essential elements of materials science relevant to molten phases and processes leading to their crystallization. This text provides a comprehensive survey of scientific and engineering fundamentals for understanding crystallization processes, dwelling especially on applications to pure materials, metallic alloys, oxides, semiconductors, and polyphase systems. The didactic approach adhered to derives in large measure from the author's personal lecture notes that were prepared annually for one-semester courses populated by advanced undergraduate and graduate students in materials science, chemical and mechanical engineering, geology and physics. This book is designed for teaching this group, as well as for professionals interested in specific topics or exposure to integrated aspects of solidification and crystal growth.

The development of most chapters included herein favors, where appropriate, a quantitative approach, augmented by scientific descriptions anchored in materials science and condensed matter physics. Familiarity with the calculus, ordinary and partial differential equations, and at least introductory chemical thermodynamics is assumed, all of which will prove helpful to most readers. Due care has been taken to provide the reader comprehensive, logical developments supported by citations of pertinent research papers and helpful reviews. Although the literature on solidification and crystal growth still lacks a consistent, canonical terminology, I have endeavored, with some personal biases, to apply best practices throughout, as I observed their application by other researchers and thoughtful authors attentive to this field.

The text's present author, in various ways, has also attempted to mirror the approach and logical structure so successfully used in Professor Bruce Chalmers's highly regarded monograph by the same title, published almost 50 years ago [2], while Dr. Chalmers was Gordon Mackay Professor of Metallurgy at Harvard University. I used Professor Chalmers's book when first teaching the subject of solidification and crystal growth; his book was reprinted for my classes in later years through the gracious permission of his estate. That work remained our background text, as I expanded some topics using my personal notes to up-date them or to introduce new ones. The present book, as already mentioned, resulted from those serial teaching efforts, through steady improvements based on feedback from students and colleagues, and progress in the field. This book is written with the express desire that it recaptures at least some of the spirit and purpose originally stated by Bruce Chalmers, namely,

> ... to provide a critical review of the state of knowledge and understanding of the process of solidification, defined for this purpose as the discontinuous change of state from liquid to crystalline solid.

Professor Chalmers's personal charm, his many research contributions, his unique technical and expository style, and, of course, his book itself, were mainstays in the early development of solidification science. It is the author's further hope that the current monograph will continue to assist students and practicing scientists and engineers in obtaining a firmer grasp of this interesting and important field, which has continued its development at a brisk pace ever since the publication of Professor Chalmers's, *Principles of Solidification*.

I owe thanks to former students for their feedback each year that I taught this subject, which collectively helped shape the present form and content of this volume. My deepest appreciation extends to my dear friend Professor Markus Rettenmayr, Friedrich Schiller University, Jena, Germany, who provided numerous useful comments, alternative perspectives, and detailed suggestions for improving the clarity and consistency of the book, and who was willing to read critically an early draft of this work; to my friend and collaborator Professor Paulo Rios, Universidad Federal de Volta Redonda, Brazil, for his encouragement, discussions, and unflagging interest in my writing this book; and to my friend and colleague Professor Diran Apelian, Worcester Polytechnic Institute, who successfully located out-of-print reference materials that proved so useful.

Support provided to the author through the Florida twenty-first Century Scholar Program, and the warm encouragement and interest always received from Professor Kevin Jones, Chairman, Department of Materials Science & Engineering, and from the College of Engineering, University of Florida, are gratefully acknowledged and were much appreciated during many months of writing.

Finally, my heartfelt appreciation and love extend to my wife, Lucinda, whose understanding, forbearance and tolerance of the lengthy and often intrusive process of book writing, along with her provisioning countless nourishing snacks and meals that kept me going through the writing of this work.

Gainesville, FL   Martin Eden Glicksman

References

1. E. Broda, *Ludwig Boltzmann, Man · Physicist · Philosopher*, Ox Bow Press, Woodbridge, CT, 1983, p. 102.
2. B. Chalmers, *Principles of Solidification*, Wiley, New York, NY, 1964.

Contents

Part I
Introductory Aspects

Chapter 1
Crystals and Melts

1.1 Crystals

A crystal is defined as:

> A homogenous solid formed by a repeating, three-dimensional pattern of atoms, ions, or molecules and having fixed distances between constituent parts.

Crystalline phases consist of regular (periodic) space-filling assemblages of atoms, ions, or molecules that, more or less, compactly fill 3-dimensional space at an average mass density, ρ_s. Crystalline phases are subject to the rigorous constraint imposed by long-range order (LRO) and satisfy the spatial invariance determined by the lattice's point-group symmetry. That is, LRO assures the presence of special symmetries and provides long-range correlations of the particle positions, even over macroscopic distances. Thermodynamic equilibrium selects the particular crystalline phase that will minimize a system's free energy at some fixed pressure for all temperatures below the phase diagram's co-existence curve with the melt phase. Crystalline phases, as defined here, also broadly constitute many of the common 'engineering materials', including metals and alloys, crystalline ceramics, semiconductors, and many important polymeric systems.

A key issue to be discussed in detail in this initial chapter is that the underlying lattice structures present in crystalline phases exhibit extremely long relaxation times for local atomic rearrangements—a special situation that permits crystalline solids to support significant shearing stresses almost indefinitely at all temperatures up to their melting point. The elastic resistance of crystalline materials to internal shear stresses contrasts with the prompt flow and stress relaxation induced by even small non-hydrostatic stresses applied to their melts. The high-temperature mechanical response of crystalline solids may be described roughly by their enormous viscosities ($> 10^{15}$ poise), or, equivalently, by extremely small values of their viscoelastic creep rates. Such inelastic behavior in solids at high temperatures can occur even in otherwise near-perfect crystals, because of the presence of lattice vacancies. Lattice vacancies are always present as a dilute population of mobile equilibrium point defects that develop at all temperatures above 0 K. Their influence on a crystal's self-diffusion rate and mechanical creep behavior usually becomes appreciable only at temperatures above about half the absolute melting point, where

M.E. Glicksman, *Principles of Solidification*, DOI 10.1007/978-1-4419-7344-3_1,
© Springer Science+Business Media, LLC 2011

self-diffusion becomes a significant atomic process [1]. Nevertheless, single crystals retain impressively high resistances to creeping flow virtually right up to their melting points or solidus temperatures.

1.2 Melts

Melts are condensed fluids; specifically, they are the liquid phases derived by congruently, or non-congruently, fusing a crystalline solid. Melts exhibit a definite volume at any fixed temperature and pressure, and are derived by progressively melting a crystalline phase at a discontinuous crystal-melt or solid-liquid interface. Both of these terms are used interchangeably in this book.

Melts, moreover, are highly mobile assemblies of atoms or molecules aggregated at an average mass density, ρ_ℓ, which has a value comparable to (usually within $\pm5\%$) the mass density for their conjugate crystal, ρ_s. Melts, in sharp contrast with either gases and crystalline solids, exhibit short-range atomic order (SRO), thereby minimizing the energy of small clusters of atoms present in their structure. The loss of long-range spatial order in melts, which increases over increasing length scales,[1] allows them to maximize their entropy, and thereby minimize their free energy at temperatures above the co-existence line with the crystal. Thus, melts persist in thermodynamic equilibrium at fixed pressures over a range of temperatures above their fusion point, but beneath their boiling point. Except for important, but complicated, spatial correlations developed among locally interacting clusters in a melt, the atoms and molecules comprising a typical 'simple' melt are located essentially at random positions, and, consequently, lack perceptible long-range order over distances exceeding just a few nanometers. Melts, as do most fluids, will not support steady, non-hydrostatic stress states of any magnitude, but instead flow viscously (irreversibly) under the application of a shear stress, or pressure gradient. The lack of any long-range lattice structure in melts is also responsible for their characteristic short relaxation times for local atomic or molecular rearrangements. Although melts do not contain discernible vacancies, they have copious 'free volume', which is mutably distributed throughout their more tightly bound atomic or molecular clusters. Evidence supports this free volume distribution through melts as the key structural feature responsible for allowing orders-of-magnitude higher self-diffusion and interdiffusion rates than are possible in well-ordered crystalline phases and, of course, their readiness to flow [1, 4].

---

[1] So-called 'liquid crystals' provide the main exception to this rule, by exhibiting long-range order, in one or two spatial directions, and being rapidly responsive to applied electric fields that align their molecules. These substances comprise a large class of mostly organic compounds [2], and have important technical applications in electronic displays for TVs, cell phones, and many other devices. Their solidification behaviors are unusual—and extremely interesting—but lie beyond the scope intended for this book. Readers interested in liquid crystals should see the editorial by Luckhurst and Samulski [3].

1.2.1 Viscosity of Melts

Although incapable of sustaining applied shear stresses over long time scales, melts do support hydrostatic stresses indefinitely. Melts flow easily to assume the shape of a container or mold, and thereby quickly relax any internal non-hydrostatic stress. Most low-viscosity melts will relax the deviatoric stress components in a few milliseconds or even microseconds. Their mechanical behavior stands in stark contrast with that exhibited by crystalline solids, which exhibit both a fixed volume at a given temperature and pressure and a fixed shape. Indeed, it is likely that the observation of these remarkable qualitative changes in mechanical (flow) behavior that occur upon melting and freezing stimulated mankind over the centuries to develop an impressive range of technical processes involving solid-melt processing:

1. *Casting*: A molten alloy may be poured or cast into a mold and solidified into a beautiful (and/or useful) shaped object. Diverse casting methods were developed by perceptive skilled artisans over the millennia. The ability to cast materials such as bronze, gold, and silver into jewelry and functional forms is still used by archeologists and historians as a measure of the sophistication of early human societies. Today, casting still constitutes the basis of many efficient, advanced manufacturing processes for the production of a multitude of objects made from metals, alloys, polymers, glasses, and some ceramics.
2. *Welding and Soldering*: Metallic solids can be heated locally, and the resulting (or added) melt allowed to cool and freeze to form a strong bond joining them. Welding, brazing, and soldering processes, although used to fashion gold objects throughout history, have been employed more broadly throughout the industrialized world for centuries, and currently provide the most effective methods for forming reliable mechanical and electrical connections between metallic materials. Such processes have the virtue of being highly automated, such as the robotic welding of automotive frames, or wave soldering of microelectronic circuits.
3. *Crystal growth*: Some melts can be transformed slowly, through controlled freezing, into a single crystal of arbitrary shape, size, and atomic-scale perfection. Crystal growth processes for technological applications began late in the nineteenth century, and today form the cornerstone of virtually all modern semiconductor electronics, photonics, and single crystal high-temperature alloy components for high-temperature turbomachinery. Variants of these crystal growth processes also provide society a source of artificial gems and gemstone simulants.

1.2.1.1 Dynamic Shear Viscosity

As mentioned above, melts flow easily to assume the shape of their containers, primarily because they lack a lattice structure, are devoid of long-range order, and have an abundance of atomically distributed free volume. Free volume within the atomic structure of a melt permits relatively easy thermally-activated motions of the component atoms or molecules, and local relaxations that relieve nonhydrostatic (deviatoric) stresses. In fact, the largest, most easily sensed physical change that occurs upon melting a crystalline solid is its sudden and remarkably large drop in shear viscosity, η—a decrease of some 15–18 orders-of-magnitude! Moreover, the time-scale of stress relaxation in an incompressible melt depends proportionately on its shear viscosity,[2] defined as

$$\eta \equiv \frac{\tau}{\mu} = \frac{\tau}{\frac{\partial U_x}{\partial y}}. \tag{1.1}$$

Here τ represents the local shear stress, and μ is the time rate of shear strain occurring in the melt. As indicated on the RHS of Eq. (1.1), $\partial U_x/\partial y$ is the gradient in the y-direction of the fluid velocity component in the x-direction, U_x. The y-gradient is taken normal to the flow direction, x, where x and y are Cartesian coordinates. Shear viscosities of melts usually decrease exponentially with the absolute temperature. Shear, or dynamic viscosity, is a transport property that is closely related to a melt's self-diffusion coefficient, because viscous flow involves the erratic interactions and thermally activated motions of molecules or atoms moving past one another, as does molecular diffusion. Under isothermal conditions, the shear viscosity of most simple melts increases slowly—perhaps by up to a factor of 10, or so—for even large increases in pressure, circa 10^4 bar. The dynamic viscosity variation with pressure found in melts contrasts strongly with that found for other fluids, such as the permanent gases, which are highly compressible fluids that exhibit viscosities that are nearly independent of the pressure.

1.2.1.2 Kinematic Viscosity

Another useful measure of viscosity in melts is the 'kinematic' viscosity.[3] The kinematic viscosity, or momentum diffusivity, ν, as it is sometimes termed in fluid mechanics, is defined as the ratio of a fluid's shear viscosity to its mass density, and directly compares a melt's internal viscous forces with its inertial forces, namely,

$$\nu \equiv \frac{\eta}{\rho}. \tag{1.2}$$

Note, that the physical units of the kinematic viscosity, as defined in Eq. (1.2), are identical to those for the mass diffusion coefficient, D [m^2/s], although ν *falls* exponentially with the temperature, whereas D *rises* exponentially with the temperature [1]. The kinematic viscosity, ν, is used to describe the frictional response of a melt to long-range convective motions driven by pressure gradients, whereas the diffusivity, D, provides the spontaneous mixing response of a melt's molecular or

---

[2] The CGS unit of a melt's dynamic shear viscosity, η, is the poise [P], named for J.L.M. Poiseuille, where 1[P]=1[g/cm·s]=1[dyne·s/cm^2]. The System International (SI) unit is the Pascal-second [Pa·s], where 1[Pa·s]=10[P].

[3] The kinematic viscosity, ν, has CGS units of stokes, [St], after G.G. Stokes, where 1[St]=1[cm^2/s], and has SI units of [m^2/s].

atomic components in the presence of a concentration gradient. Both quantities are important in many solidification and crystal growth processes.

Dynamic shear viscosities for a variety of solids (crystals and glasses) and fluids (melts and gases) were compared by D. Turnbull on the basis of their respective 'homologous' temperatures [5]. The homologous temperature was chosen as a convenient dimensionless measure of the energy of a material's interatomic or intermolecular bonding relative to the material's average thermal energy. Turnbull chose the homologous temperature $\tau \equiv k_B T / \Delta H_v$, where ΔH_v is the heat of vaporization—a convenient measure of interatomic bond strength—T is absolute temperature, and $k_B = 1.3806 \times 10^{-23}$ J/K is Boltzmann's constant. Turnbull's comparison of the viscous properties of glasses, crystals, and melts using the homologous temperature is highly illuminating.

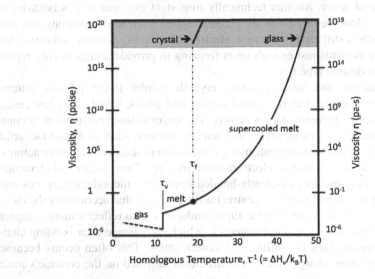

Fig. 1.1 Turnbull's correlation of dynamic shear viscosity and reciprocal homologous temperature, $\tau^{-1} \equiv \Delta H_v / k_B T$. Upon cooling a melt, either a discontinuous jump in viscosity occurs (*vertical dash-dot line*) at its homologous freezing point ($\tau_f \approx 0.05$), or a steady increase develops as the melt is supercooled toward its glass transition temperature near $\tau \approx 0.022$. The sudden change in shear viscosity upon melting and freezing can amount to more than 18 orders-of-magnitude

Figure 1.1 illustrates how the dynamic shear viscosity, η, (plotted on a log-scale) varies with the reciprocal of the homologous temperature, τ^{-1}. Figure 1.1 also clearly demonstrates that an enormous discontinuity occurs in the shear viscosity upon melting a crystalline solid at $\tau = \tau_f$, (at the black dot, where a change in viscosity occurs of about 18 orders-of-magnitude)! Instead, if phase change is avoided, a steady *continuous* transition can occurs in the melt's viscosity as a melt is cooled below its freezing point, or as a glassy solid is heated and steadily softens. The latter continuous path produces ultimately a melt that steadily becomes less and less viscous. Finally, the melt discontinuously vaporizes to an even less viscous gas phase at the homologous boiling point, $\tau = \tau_v \approx 0.08$.

Remarkably, and despite these impressive differences in their shear viscosities, a solid and its corresponding melt often exhibit similar mass densities at the same temperature and pressure. For example, the structural transitions associated with melting typical metallic alloys show relative mass density changes of only a few percent, typically 1–5%, and seldom exceeding 10%. As discussed in Section 1.6.2, these small mass density changes on melting are associated with the presence of highly labile free volume that accounts for the fluidity of melts.

Many crystalline solids melt into slightly less dense liquids, and do so because the loss of LRO leads to a less efficient *average* packing of space and, usually, to an increase in the molar volume of their melts. Some crystalline solids, however, form denser melts, the most common example of which is H_2O ice, where hydrogen bonding plays major roles in both rhombohedral crystalline ice and in the molecular structure of water. Another technically important example of this behavior is type metal—a low-melting point alloy compounded from lead, antimony, tin, and copper—which is still used in the (non-electronic) specialty printing industry. This metallic alloy expands just enough upon freezing to provide a high-fidelity reproduction of the desired type font.

Semiconductors and some ceramics crystals exhibit larger relative volume changes upon freezing than do typical metals and alloys, which, in a few cases, even exceed a 10% increase in mass density. The larger volume contractions accompanying the melting of many crystalline semiconductors, such as Si and Ge, arise from electronic transitions accompanying destruction of their crystalline structures, viz., the release of copious free electrons upon melting. Thus, melting of elemental semiconductors transforms covalently-bonded intrinsic semiconductor crystals into highly conductive liquid metals. The structural transitions that accompany the melting of ceramics, such as metal-oxide superconductors, also reflect a more complex variety of atomic defects in these materials, which can change their bonding characteristics between their crystalline and molten states. This often occurs because of local electrostatic charge balance requirements imposed on the ceramic's ionic components.

The significant conclusion reached here is that despite only rather modest differences in their intermolecular distances and bond interaction strengths, crystals and their melts can exhibit enormous changes in certain mechanical properties, including especially the shear and kinematic viscosities. These drastic changes in flow response enable the use of ingot and casting processes as key steps in the industrial production of primary metals (both ferrous and non-ferrous), in polymer and ceramic melt processing, and in the growth of bulk single crystals.

1.3 Comparison of Crystalline and Melt Structures

Structural differences between crystals and their melts may be compared conveniently, albeit in a somewhat limited manner, on the basis of their radial distribution functions (RDFs). RDFs provide quantitative information on the nature of the average environment of atoms or molecules in a phase at small length scales.

1.3.1 RDFs and Local Atomic Densities

RDFs, as conventional structural measures, can be determined only with considerable difficulty by using advanced experimental methods of X-ray, electron, or neutron diffraction. An RDF, after extraction by mathematical methods from diffraction data, provides a limited measure of the short-range atomic or molecular order in a material. The RDF for a material may be reported in several variants as outline below.

Although methods for measuring RDFs are involved and somewhat technical, their detailed experimental and analytical procedures will be found in standard texts on X-ray diffraction methods [6]. Their description is limited here to their elementary definitions, in an effort to clarify what they reveal about the underlying differences between crystals and their corresponding melt phases. The following local densities may be defined:

1. $N(r)$ is the *total* time-averaged number of atoms expected within a small region defined by the volume, $V(r) = (4\pi/3)r^3$, enclosed by a sphere of radius r, centered about the location $\mathbf{r} = 0$, chosen within a material at the center of any arbitrary 'reference' atom.
2. $n(r)$ is the additional number of atoms per unit volume counted in a thin spherical shell located at a distance $r + dr$ from the origin, $\mathbf{r} = 0$. Thus, $n(r)$ is related to the probability of finding the centers of other atoms at a specific distance, r, from the center of an average atom. The relationships among the total number of atoms contained in a spherical region of radius r, $N(r)$, their total volume, $V(r)$, and their radial number density, $n(r)$, are expressed by the following integrals:

$$N(r) = \int_0^r 4\pi r'^2 n(r') dr'. \tag{1.3}$$

and, of course,

$$V(r) = \int_0^r 4\pi r'^2 dr'. \tag{1.4}$$

Comparing Eq. (1.3) with Eq. (1.4) one sees that

$$N(r) = \langle n(r) \rangle \cdot V(r), \tag{1.5}$$

where $\langle n(r) \rangle$ is the spatial average of the atom number density over the radial distance r.

3. $\rho(r)$ is the (local) radial mass density defined as

$$\rho(r) \equiv n(r)\frac{A}{N_A}, \tag{1.6}$$

where A is the atomic or molecular weight, and N_A is Avogadro's number, 6.0221×10^{23} [mole^{-1}]. One may recognize, using the definition in Eq. (1.6) of the radial mass density, $\rho(r)$, that the conventional, or bulk mass density of a molten phase, ρ_ℓ, is the macroscopic density limit reached at a sufficiently large distance from any reference atom. Thus, the ordinary (Archimedean) mass density of a melt, ρ_ℓ, is defined as the limit

$$\rho_\ell \equiv \lim_{r \to \infty} n(r) \frac{A}{N_A}, \qquad (1.7)$$

where ρ_ℓ denotes the macroscopic mass density—the usual property listed in handbooks, with which we are all familiar.

1.3.2 Radial Distribution Functions (RDFs)

X-ray and neutron diffraction scattered intensity data of liquids may be reduced via the Zernicke and Prins analysis [7] to report the local radial mass and number densities by employing radial distribution functions, or RDFs. Specifically, the commonly used RDFs are $4\pi r^2 \rho(r)$ or, equivalently, $4\pi r^2 n(r)$, and its scaled version, $W(r)$. The scaled RDF, $W(r)$, is defined as the *dimensionless* mass or number density,

$$W(r) = \frac{\rho(r)}{\rho_\ell} = \frac{n(r)}{n_\ell}, \qquad (1.8)$$

where n_ℓ is the average number of atoms per unit volume in the melt. $W(r)$ approaches unity at large distances from the arbitrary origin, $\mathbf{r} = 0$, because at relatively large distances from a reference atom $\rho(r) \to \rho_\ell$, and $n(r) rightarrow n_\ell$. The RDFs defined as $4\pi r^2 \rho(r)$ and $4\pi r^2 n(r)$ describe local density functions, bearing units of mass, or atoms, per unit radial distance from an arbitrary reference point. RDFs of liquids usually oscillate at small distances from the reference atom, exposing useful short-range structural information, and then gradually approach the smoothly rising 'parabolas' for the average mass, or number density, i.e., $4\pi r^2 \rho_\ell$ and $4\pi r^2 n_\ell$, respectively.

1.4 Structure of Melts

1.4.1 RDFs in Monatomic Melts

Elemental sodium—a BCC crystalline phase at 1 bar pressure—melts at 97.8 C to form a typical monatomic liquid metal. The RDF of molten sodium, plotted as the function $4\pi r^2 \rho(r)$ in Fig. 1.2, was determined from x-ray diffraction experiments [8] via application of the Zernicke-Prins integral transform procedure [7]. In

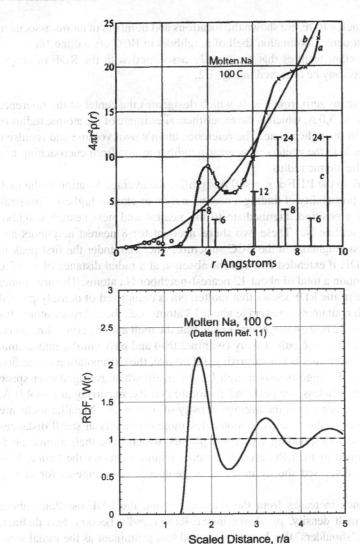

Fig. 1.2 *Top*: Radial distribution function (curve *a*) for molten Na at 1 atm. pressure and 100 C. These data [6] show that the atomic radius of Na in the molten state is $\approx$ 3 Å, locating where the RDF lifts sharply away from the *r*-axis. Nearest-neighbor Na atoms reside at an average location of $r \approx 4$ Å from the center of any reference atom, which is intermediate to the first and second coordination shells in the crystalline phase, shown here as vertical bars. The RDF oscillates significantly about the mean radial density, $4\pi r^2 \rho_\ell$ (curve *b*), for distances out to about 4 atomic radii ($\approx$ 12 Å), and then converges towards its bulk RDF value. *Bottom*: The $W(r/a) \equiv \rho(r)/\rho_\ell$ for molten Na at 100 C, determined using modern instrumentation [9]. Here, the distance variable, *r*, is scaled by sodium's atomic radius based on its atomic volume, $a \approx 2.5$ Å. The data for both RDF's are remarkably consistent

addition to the melt's RDF, are shown the locations and numbers of atoms associated
with each subsequent coordination shell of neighbors in BCC crystalline Na.

Several important features that are usually associated with the RDF of *simple*
monatomic melts may be observed in Fig. 1.2.

1. The RDF remains zero from $r = 0$, which designates the center of the reference
   Na atom, to $r = 3.0$ Å, which is the experimental estimate of the atomic radius of
   metallic Na in its molten state. The reference atom's own volume and repulsive
   potential preclude the centers of nearest-neighbor atoms from encroaching any
   closer than the atomic radius.
2. The first peak in the RDF at $r \approx 4.0$ Å signifies the average location in the melt
   at which the probability of finding a nearest-neighbor atom is highest. This peak
   is located at a position intermediate to the nearest and next-nearest neighbor
   shells in crystalline Na. These two shells account for 8 nearest neighbors and
   6 next-nearest neighbors in the BCC structure. The area under the first peak in
   the melt's RDF, if extended down to the abscissa at a radial distance of $r \approx 5.0$
   Å, would contain a total of about 12 nearest-neighbor Na atoms. The area under
   the first peak of the RDF shows that molten Na is comprised of densely-packed
   clusters, each containing on average about 13 atoms, i.e., the reference atom plus
   its statistically 12 nearest neighbors. Given that the melt and the crystalline phase
   exhibit nearly the same bulk density (within 2.0%) and have similar interatomic
   forces, this observation is not surprising. Of course, the extrapolation of the first
   peak in the RDF suggests only a high *local* density when averaged over space
   and time. Nevertheless, the peak radial atomic density, occurring at $r \approx 4.0$ Å,
   is almost 60% higher than the average density of molten Na. Metallic melts are
   often able to pack their atoms or molecules more efficiently at small distances
   than can their corresponding crystalline phases, which have their atomic positions constrained by the LRO crystallographic requirements for the lattice. This
   point will be discussed shortly in terms of geometric requirements for atomic
   packing.
3. As the distance increases from the reference atom, the RDF oscillates about
   the average melt density, ρ_ℓ. Peaks in the RDF rapidly become less distinct,
   appearing as 'shoulders' that become less and less prominent as the radial sampling distance increases. Beyond about 3 atomic or molecular radii, the RDFs of
   most melts start approaching their average-density parabola, $4\pi r^2 \rho_\ell$, exhibiting
   weaker and weaker departures from its bulk density.

1.4.2 RDFs in Molecular Melts

Consider the RDF of molten elemental phosphorus, P_4. This fluid provides an
example of a *molecular* melt consisting of randomly oriented tetrahedrally-bonded
phosphorus atoms. The P_4 tetrahedra are packed in space in a dense, but essentially disordered, manner. The P_4 melt phase results from the equilibrium melting
at 317 K and 1 bar pressure of the complex cubic α-crystalline form of the element.

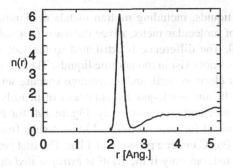

Fig. 1.3 Radial number density, $n(r)$, for molten molecular P_4. Here the radial number density of P atoms is plotted against the distance, r, from the center of any reference atom. The major peak falls to zero at approximately $r = 2.6$ Å. The area under this peak is $n=3$ atoms, and if combined with the reference atom centered at $r = 0$, the peak represents one P_4 molecule. Subsequent oscillations in the $n(r)$ RDF, although not uniquely interpretable, do approximate the location of next-nearest neighbors and provide additional information on the cluster geometry and structural correlations occurring in the molten state. Data from reference [10]

The RDF, now in the form of the number density of P atoms, $n(r)$, is shown plotted in Fig. 1.3. The RDF rises quickly, peaks, falls to zero, and then rises more slowly again in a monotonic manner, oscillating only slightly, and gradually approaching the bulk macroscopic density, ρ_ℓ, of this molten element. The sharp initial peak in the $n(r)$ curve for molecular P_4 is anchored at zero at both ends. This closed peak defines an area on this RDF plot that equals 3 atoms, precisely. The closed peak in this RDF unambiguously expresses the fact that the melt is composed of distinct tetrahedral P_4 molecular units. Each P-atom in the melt always correlates with its well-defined 'cage' formed by its 3 nearest neighbors, which, including the reference atom, defines the P_4 molecular units present in the fluid. Thus, with any random P-atom taken as the origin, the radial density function for molten P_4 falls to zero at the spatially averaged distance of the molecular radius. Clearly, the short-range tetrahedral grouping of all the molecules imparts an average character-istic radial distance about the reference atom at which the local radial density must again statistically approximate zero. This might seem peculiar at first glance, but, surprisingly, at any distance between **r** = 0 (at the center of the reference atom) and its atomic radius, $r \approx 1.1$ Å, and at a distance just beyond the location $r \approx 3.2$ Å of its cage of 3 neighboring atoms, the *local* atomic density of molten P_4 vanishes. Its density immediately starts rising again at radial distances exceeding the reference atom's shell of second-nearest neighbors. The structural interpretation of subsequent shoulders and peaks on the RDF is, however, not nearly as straightforward as the interpretation of its first 'closed' peak, which is well defined by P_4's molecular geometry. Indeed, the uncertainty about more distant peaks and valleys proves to be a stringent limitation in using RDF's to interpret the structure of melts. That is, at distances larger than the first shell of neighbors, an atomistic interpretation, or spatial visualization, of the RDF becomes progressively more ambiguous, as to how much each subsequent 'shell' of neighbors actually contributes to the average local melt density.

RDFs of simpler liquids, including molten metals or liquified noble gases, by contrast with those of molecular melts, never decrease to a value of zero for any radial distance $r > 0$. The difference is attributed to the fact that a well-defined 'molecular radius' does not exist in monatomic liquids. There is only a mean atomic radius set by an outer electron shell, and an average spacing set by the interatomic potential. Most metallic and noble-gas melts consist of densely-packed individual atoms that fill space efficiently, but randomly. Figure 1.4, for example, shows the $W(r)$ function determined for pure molten gold that results from melting the FCC crystalline phase at 1,063 C under a pressure of 1 bar. The first peak of molten gold's RDF may be interpreted, but only if the RDF is *extrapolated* down to zero atomic density, (see shaded region in Fig. 1.4). The extrapolated peak may be thought of, roughly speaking, as approximating the first 'coordination shell' of gold atoms surrounding the origin, which is chosen as the center of any arbitrary gold atom. The first peak in the $W(r)$ expresses strong, but localized, atom correlations occurring over short distances in the melt. Subsequent peaks, observable at larger radial distances in the $W(r)$ measurements, are not as easily interpreted. Overlaps from neighboring contributions to the $W(r)$ again make the unique interpretation of the radial density distribution highly uncertain.

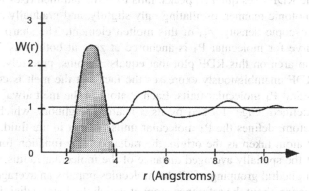

Fig. 1.4 Radial distribution function for molten gold. The RDF is the dimensionless $W(r) \equiv \rho(r)/\rho_\ell$. In the case of Au, as in other liquid metals, the first peak in the RDF never decreases to zero radial density. Molten gold consists of densely packed Au nuclei freely moving in a 'sea' of free electrons. No fixed molecular configurations occur; instead the atoms associate briefly to form icosahedral arrangements consisting of polytetrahedral groupings. The extrapolated (*gray*) area under the first peak suggests an *average* initial coordination number of $\langle Z \rangle \approx 11$ nearest neighbors. Again, a precise interpretation of the subsequent peaks and valleys in the RDF of molten Au is not possible

1.5 Theoretical Estimates of RDFs

Theories of randomly packed melt structures are based on mathematical ideas developed by Coxeter [11]. Bernal [12, 13], a crystallographer, then constructed the first detailed structural model that yielded realistic RDF's of melts and other polytetrahedral phases having well-developed SRO, but without LRO. Other investigators [14–16], using Bernal's approach of 'random dense-packing' of equal-size hard-spheres, carried his structural analysis of melt structures and amorphous phases even further. More recently the physics of Bernal's random dense model has been investigated as a 'universal' state of hard-sphere fluids [17].

Bernal's model of a melt consisted of a physically constructed network containing spheres of equal size, but packed so well that they everywhere tended to maximize their *local* density without suffering 'crystallization'. Bernal found that a dense, but random, arrangement of equal-size spheres develops an extended three-dimensional structure lacking long-range order. Consequently, Bernal postulated that simple melts and some amorphous phases seem to maximize their SRO and minimize their LRO simultaneously. Bernal's insights on the liquid and amorphous solid states have proven to be fundamentally correct as theoretical models became more quantitative and sophisticated [16, 18], as they became based on realistic interatomic potentials and quantum mechanics. A comparison of Bernal's RDF with that found experimentally for liquid argon is shown in Fig. 1.5.

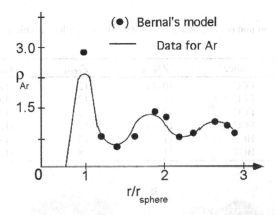

Fig. 1.5 Comparison of the theoretical mass density prediction from Bernal's random dense-packing model for liquids and amorphous solids with an RDF measured experimentally for liquid argon. Liquid argon may be considered as condensed spherical Ar atoms. Such atoms interact through an approximately 'hard-sphere' interatomic potential that has an attractive interaction up to contact, at which point strong repulsion starts. As such, molten argon realistically portrays the behavior of a monatomic 'hard sphere' liquid with dense-random packing

1.5.1 Atomic Coordination in Melts

An average coordination number, $\langle Z \rangle$, in a melt with atomic radius r_a, may be defined using its RDF, $n(r)$. This RDF equals the number of atoms per unit volume included by the counting integral taken over the radial distance between the center of any reference atom, $r = 0$, and the centers, $r = 2r_a$, of all atoms contacting the reference atom.

$$\langle Z \rangle = \int_0^{2r_a} 4\pi r^2 n(r) dr. \tag{1.9}$$

The coordination number, $\langle Z \rangle$, given in Eq. (1.9), equals the *average* number of closest neighbors surrounding an arbitrary atom in the melt. The quantity $\langle Z \rangle$ is analogous to the coordination number found in crystals, Z, which is based on the lattice's geometrically well-defined 'shell' of nearest neighbors. It is interesting to note that in many melt phases, especially those derived from close-packed crystals, the average coordination number $\langle Z \rangle \approx 11$. This average coordination number for metallic melts is typically lower, by one, than the corresponding coordination number for the close-packed (FCC and HCP) crystalline solids, which have $Z_{cryst} = 12$. See Table 1.1 and Fig. 1.6 for a comparison of the coordination numbers, atomic radii, and local packings found in metallic crystals and their melts. Also, when solids melt from more 'open', non-close-packed crystal structures, such as BCT-Sn, they frequently form melts exhibiting a $\langle Z \rangle$-value that is higher than Z in the crystalline state. Thus, the level of SRO upon melting non-close packed crystals actually

Table 1.1 Coordination numbers, Z and $\langle Z \rangle$, for crystals and melts. Atomic radii, (melt/crystal) are in Ångstroms

Metal	Lattice	$\langle Z \rangle_{melt}$	Z_{cryst}	r_{atom} (Å)
Al	FCC	10–11	12	(1.49/1.43)
Au	FCC	11	12	(1.43/1.43)
Pb	FCC	12	12	(1.86/1.75)
Ar	FCC	10–11	12	(1.98/1.92)
Zn	HCP	11	12	(1.47/1.40)
Sn	BCT	11	8	(1.60/1.53)

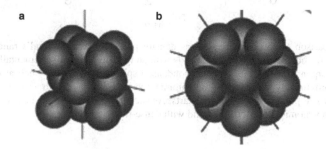

a   **b**

Fig. 1.6 Locally dense packings: (**a**) The *left* cluster represents a close-packed (FCC) crystalline structure, dominated by octahedral atomic arrangements. Such clusters, and a similar one for HCP crystals, yield a local packing density of $PD = 0.7405$. (**b**) The *right* cluster represents a 13-atom icosahedral dense packing, exhibiting polytetrahedral order (PTO). This structure consists of densely packed tetrahedra with pentagonal (icosahedral) symmetry. It exhibits 5 distinctive 'gaps' surrounding the cental atom, and is incapable of being extended to form a lattice with LRO. Surprisingly, the local packing density of this 'non-crystalline' arrangement, $PD = 0.7796$, is higher than that for close-packed crystalline structures, although both structures have identical coordination numbers, $Z_{FCC} = Z_{PTO} = 12$

increases, and so does the average coordination in such a melt, $\langle Z \rangle$, although all vestiges of LRO would, of course, be destroyed.

The mean, or bulk mass density, however, can either increase or decrease during melting. Clearly the *net* changes occurring in bulk density upon melting a crystal depend both on the short-range atomic interactions that result in the melt, and on the fact that the LRO of the solid state is destroyed. The destruction of LRO implies that some type of 'looseness', in the form of free volume, is distributed throughout a melt's atomic structure. In addition, the density changes that occur upon melting also depend on electronic state transitions that influence both the local packing and the effective atomic radii. Significant electronic transitions occur when melting semimetals, for example Bi, Sb, and As, as well as semiconductors, such as elemental Si and Ge that revert upon melting—both structurally and electronically—to liquid metals via major band-structure changes. That is, the open diamond cubic (DC) structures of the covalent semiconductors 'collapse' upon fusion, releasing their paired electrons to form denser metallic melts with a higher $\langle Z \rangle$-value and a much higher density of 'free' electrons. Because of the sharp electronic band transition from crystalline semiconductor to liquid metal, both Ge and Si also exhibit unusually large ($\approx 10\%$) density increases upon melting.

1.5.2 Atomic Packing

Recent theoretical models of melts, based on molecular dynamic simulations, support the idea that tetrahedral and polytetrahedral clusters are most abundant. In crystals, on the other hand, which have LRO and lattices exhibiting translational symmetry, one finds a preponderance of octahedral as well as tetrahedral atomic arrangements.

As discussed in Section 1.5, the first quantitative model of tetrahedral clustering in melts was proposed in 1960 by Bernal, who showed that if one attempts to fill space with spheres of equal size, without benefit of LRO from crystallization, then, what is termed 'random dense-packing' develops. Random dense-packing in melts involves the onset of *local* polytetrahedral arrangements of the atoms. The particles forming such melts are densely packed locally as icosahedral clusters consisting of 13 atoms arranged as 20 tetrahedra. See Fig. 1.6. Icosahedral clustering provides the central atom within a cluster with a coordination shell consisting of 12 nearest neighbors, which matches the coordination number in close-packed structures such as FCC and HCP. What is unexpected, however, is that in case of liquids and glasses the local packing density, PD, which is the fraction of space occupied by atoms in icosahedral clusters, is remarkably high, viz., $PD \approx 0.7796$. For the case of perfect close-packed crystals a *lower* packing density is achieved, $PD \approx 0.7405$! The close-packed crystalline case, however, allows the atomic packing density to be uniform over length scales extending from the unit cell (nanometers) out to macroscopic distances, whereas icosahedral order in melts is strictly local, and not extensible to macroscopic distances. If the 'free volume' distribution needed to extend a random dense liquid to larger length scales is included, then the average atomic packing density falls considerably to $PD \approx 0.636$, which, as explained, is smaller than the packing density achieved in close-packed crystal structures, and even about 10% below that for more 'open' structures such as BCC, for which the atomic packing density is reduced to $PD \approx 0.680$.

1.5.3 Bonding in Liquids

Interatomic bonds in liquids are similar to those found in the corresponding crystalline state. This conclusion merely depends on the facts that melts and their crystals have comparable average atomic densities and, consequently, exhibit comparable interatomic distances, and experience nearly identical interatomic potentials setting the forces bonding the atoms. Except for slight higher-order differences in bond energies established by both neighbors and next-nearest neighbors, crystals and their melts are bonded similarly. Thus, one finds that the remarkable, often dramatic, difference between the properties of melts and crystals depends not on their respective interatomic forces, but rather on the detailed spatial distribution among nearest and next nearest neighbors—i.e., on their detailed packing arrangements.

In crystals, for example, the coordination number or near-neighbor bonding is fixed by symmetry for all equivalent sites in the crystal. A more complicated situation prevails in melts, which lack a lattice structure. In liquids a bond-number counting operation, similar to that used for crystals, provides a useful procedure called the 'Voronoi tessellation' or 'Dirichlet division'. One chooses an arbitrary reference atom, and connects its center to the centers of all its nearest neighbors. Then one bisects the connecting lines between atomic centers, and constructs a normal plane at every bisection. The normal planes are then extended to enclose the reference atom in a closed polyhedral shell called the Voronoi (or Dirichlet) polyhedron. The number of edges per face, n, on the faces of the Voronoi polyhedra in liquids equals the local bond number, ζ, which, in turn, equals the number of tetrahedra packed about the bond line normal to that face. One samples the bond number counting for a representative number of atoms in the structure and calculates the bonding probability function, $\zeta(n)$.

Ichikawa [19], Bennett [20], and Finney [21], independently determined bonding numbers for dense, randomly-packed spheres, both by using experiment and computer simulation. The results of their studies, see Fig. 1.6, shows that a characteristic distribution of polytetrahedral packing occurs. Finney found that icosahedral packing (5 tetrahedra surrounding a bond line) is the most prevalent. The maximum coordination number in such a dense-random arrangement is also surprisingly high, $Z_{max} \approx 13.394$, and well above that ever achieved in crystalline matter [22, 23]. Thus, melts and other simple liquids manage to pack their atoms or molecules extremely efficiently at small scales (nanometers), but because they lack LRO they require a distribution of free volume to extend their random structures to macroscopic dimensions. The combination of dense-random packing with distributed free volume provides the fluidity and viscous behavior that are so well known in melts, yet these random structures retain values of their internal energy and specific heat that are similar to those found in their crystalline state. Not surprisingly, perhaps, the thermodynamic property that most distinguishes the crystalline state from the molten state is the molar entropy. The high entropy of melts, compared to their crystals at similar temperatures and pressures, is linked directly to the polytetrahedral SRO of liquids, in contrast to the octahedral LRO found in crystals.

1.5.4 Influence of Temperature on RDFs

Most melts are thermodynamically stable over a surprisingly wide range of temperatures. Specifically, many molten phases exhibit a wide temperature separation (even at moderate pressures near 1 bar) between their melting point, T_m, and their boiling point, T_v. At its melting point, a melt has partial molar free energies of each component that equal those for the corresponding equilibrium crystalline phase, and at higher temperatures, specifically at the boiling point, a melt has partial molar free energies of its components that are equal to those for the corresponding gas or vapor phase. Some interesting changes also occur in a melt's atomic structure, as measured through its RDF, with increasing temperature at constant pressure. The specific case of the RDF for pure metallic mercury was studied in some detail through a series of X-ray investigations at increasing temperatures between the freezing point and boiling point. The RDF's shown in Fig. 1.7 for molten Hg exhibit several interesting features caused by the change of temperature:

- Some of the peaks in mercury's RDF at low temperatures (near its freezing point) correspond, but only approximately, to the radial distances where the sharp Bragg peaks of diffracted X-ray intensity occur in the crystal's own RDF. In this limited sense—not to be taken too literally—the melt exhibits at relatively low temperatures some 'crystal-like' SRO. Mercury, however, crystallizes in a non-close-packed crystal structure, so its crystalline RDF corresponds only crudely with the melt's RDF. All one may really ascertain from the melt's RDF at low temperatures is that the nearest-neighbor distances in the crystal and the average interatomic separation in the melt are comparable.
- At relatively high temperatures—nearer mercury's normal boiling point of 356.6 C, at 10 MPa—the oscillations appearing in its RDF near the melting point tend to die out and display much smaller local variations from the average, smooth RDF of mercury's vapor phase. At these higher temperatures the RDF of liquid Hg is, in some sense, becoming 'gas-like' in character, and liquid Hg is approaching the behavior of its vapor phase.
- At and above mercury's critical point, which occurs at much higher pressures (167 MPa) and temperatures (1,478 C), molten Hg becomes indistinguishable from that of its supercritical vapor [24]. Above its critical point, the melt and its vapor are identical phases with indistinguishable RDF's.

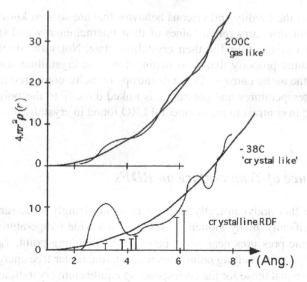

Fig. 1.7 Radial distribution functions for mercury. Two RDF's are shown: The upper RDF is obtained at 200.0 C, which closer to the boiling point of Hg (356.6 C) at a pressure of 1 bar (0.1 MPa). At this temperature and pressure the RDF oscillates rather weakly about the smooth mass density parabola, $4\pi r^2 \rho_\ell$, that represents the average radial density for the high-temperature 'gas-like' phase. By contrast, at a temperature of −38.0 C, just above the freezing point of pure Hg, −38.8 C, the lower RDF exhibits a considerably more pronounced oscillatory structure over comparable radial distances. Specifically, one sees larger peaks and valleys, and a greater departure from the average mass density parabola representing the average liquid-state radial mass density function near the freezing point. Bragg diffraction peaks for crystalline Hg are placed along the lower abscissa for reference. At most, only a vague correlation exists between the peaks in the low-temperature RDF for the molten-state and the sharp Bragg peaks obtained from the crystalline phase

1.6 Theories of Melting and the Liquid State

1.6.1 Lattice Vibrational Instability

Interest in developing physically-based models of the melting transition of solids arose early in the twentieth century by F. Lindemann [25], who postulated that fusion of a crystal results when the amplitude of atomic thermal vibrations reaches some critical fraction, $f^\star$, of the equilibrium lattice spacing, a_0. The value of $f^\star = 1/2$ was selected arbitrarily as that fraction of the interatom spacing at which neighboring hard-sphere atoms could 'collide' when vibrating in opposition. Such collisions, presumably, would lead to instabilities of the crystalline lattice. Other early attempts to describe the onset of the melting transition were developed by Lennard-Jones, who used order-disorder arguments based on kinetic gas theory, i.e., determining conditions for LRO→SRO, as the physical criterion for lattice fusion. A recent thorough review by Rettenmayr [26] covers theories and models proposed to explain the onset of melting in crystals.

In more recent times improvements were made on chemically-based models of the liquid and amorphous solid state [27–29]. These works were comprehensively reviewed in the context of the growing interest in metallic glasses by Ramachandrarao et al. [30]. These models of the amorphous solid state, however, remain of limited interest in explaining fundamentals of solidification and melting, and will not be discussed further. A comparative review of Bernal's structural model and Eyring's hole model of the molten state, plus several other formulations of liquid-state theory, are included in [31].

1.6.2 Hole Theories of Liquids

Structural models of liquids, based on atomistic-level modeling, were proposed in the mid-1930s by Frenkel, a Russian physicist. Frenkel postulated that the 1–10% volume change upon melting from the crystalline state was accounted for by the spontaneous introduction of vast numbers of mobile 'gaps' or 'cracks' among the atoms [32]. Frenkel termed these empty spaces 'holes', to distinguish them clearly from much larger atom-sized lattice monovacancies, or the geometrically defined interstices of any crystalline lattice.

Altar [33] and Fürth [34] developed models for the molten state that consisted of adding high concentrations of monovacancy-sized holes that were distributed through a continuum, or 'framework', consisting of particles placed at random at the appropriate melt density. These holes were allowed four degrees of freedom: three translational modes, plus a 'breathing' or dilatational mode, where the volume of a hole can fluctuate, or even disappear by evaporation of particles into it. Fürth's model has been applied, with only limited success, to estimate the specific heats, viscosities, thermal conductivities, and other liquid-state properties [35].

In 1941, Eyring et al. applied chemical rate theory arguments and developed a rudimentary hole theory of liquids [36, 37] that agrees with some of the basic thermodynamic characteristics of the melting-freezing process, and is consistent with accepted structural features of melts already discussed in Section 1.4. Eyring suggested that starting with a perfect crystalline solid, and then heating it to higher and higher temperatures, subsequent melting or sublimation events would involve a sequence of *average* structural and bonding changes occurring within the original defect-free crystal. These include:

1. Vaporization from the melt, which involves completely removing an atom by pulling apart, on average, all 11 interatomic bonds, each of energy U_v, relative to vacuum. These bonds correspond to those comprising the $\langle Z \rangle \approx 11$ nearest-neighbor bonds surrounding (on average) every atom or molecule in the melt. The total energy change associated with rupturing these bonds and removing the atom to form a gas phase from the melt is the energy of vaporization, $\Delta H_v \approx 11 U_v$.
2. Melting, according to Eyring, involves accumulating, on average, one 'hole', equivalent to one missing atom, for each of the crystal's unit cells, thereby

reducing the coordination number in the case of close-packed crystals from $Z = 12$ to that of a melt, $\langle Z \rangle \approx 11$. The addition of such a 'hole', or monovacancy[4]—to use more modern solid-state terminology—corresponds roughly to eliminating 1 bond pair of the crystal's binding energy, U_f, relative to its molten state, so that $\Delta H_f \approx U_f$.

Eyring reasoned that if these average bonding and structural changes on fusion were approximately correct from a thermodynamic perspective, then it follows that

$$\frac{\Delta H_v}{\Delta H_f} \approx 11 \frac{U_v}{U_f}. \tag{1.10}$$

The absolute boiling temperature, T_v, and absolute fusion temperature, T_m, scale individually in direct proportion with their corresponding total change in binding energies per atom, namely $11U_v$ and U_f. This seemingly straightforward result underscores the fact that temperature provides a direct measure of a system's average thermal energy. By using Eyring's ratio, Eq. (1.10), these phase-change enthalpies may also be compared approximately to their transition temperatures as

$$\frac{\Delta H_v}{\Delta H_f} \approx 11 \frac{T_v}{T_m}. \tag{1.11}$$

Now, there exist two well-known classical entropy-change correlations established empirically from large collections of thermo-chemical measurements concerning all of the quantities appearing in Eq. (1.11), specifically, we cite the ratios,

$$\Delta S_v = \frac{\Delta H_v}{T_v} \approx 21 \left[\frac{\text{cal}}{\text{mol} \cdot \text{K}} \right], \tag{1.12}$$

known as Trouton's rule, and

$$\Delta S_f = \frac{\Delta H_f}{T_m} \approx 2 \left[\frac{\text{cal}}{\text{mol} \cdot \text{K}} \right], \tag{1.13}$$

known as Richard's rule. Trouton's and Richard's rules, Eqs. (1.12) and (1.13), respectively, may be combined to demonstrate that the ratio of binding energy

---

[4] It is known, however, that most metal crystals, even at their melting points, develop only dilute populations of monovacancies and divacancies. Monovacancy concentrations in metals close to their melting point, for example, seldom exceed a few missing lattice atoms per 10^3–10^4 lattice sites, which are concentrations far below that of $1/(\langle Z \rangle + 1)$, which is equivalent to one missing atom per cluster of $(\langle Z \rangle + 1)$ atoms in the melt. One should note that Eyring's 'holes' are similar to those described by Fürth, whereas Frenkel's 'holes' seem to be entirely different entities. Nevertheless, as a comparative measure, the average coordination number of monatomic melts, is, arguably, about one fewer than its crystalline conjugate. See again Table 1.1. for a comparison of local coordination numbers for metals.

changes for the phase-change processes of vaporization and fusion, as estimated from thermodynamic data, is approximately

$$\frac{\Delta H_v}{\Delta H_f} \approx 10.5 \frac{T_v}{T_m}. \tag{1.14}$$

Equation (1.14), based on calorimetric measurements made on many solid-melt systems, corresponds reasonably with Eyring's 'hole' model of melts, Eq. (1.11), as is demonstrated by the data comparison displayed between the fourth and fifth columns in Table 1.2. This hole model is certainly overly simplified—indeed wrong in some aspects—especially in view of modern descriptions of melt structures and crystals based on X-ray and neutron diffraction analyses and our understanding of condensed matter based on molecular dynamics. Nonetheless, Eyring's model captures several useful structural correlations and roughly agrees with some well-known 'rules of thumb' among the thermophysical properties of molten phases. As many scientists would agree, even a 'wrong' theory is better than having none.

Table 1.2 Comparison of thermodynamic quantities for melting and boiling

$[T_m$ and T_b in K; ΔS_f in $\frac{cal}{mol-K}]$

Metal	T_m	T_b	$\frac{\Delta H_v}{\Delta H_f}$	$11\frac{T_b}{T_m}$	ΔS_f
Al	933	2, 753	27.8	32.5	2.75
Au	1, 336	3, 223	26.7	26.4	2.21
Cu	1, 356	2, 848	23.4	23.2	2.30
Fe	1, 809	3, 343	22.4	20.3	2.00
Zn	693	1, 180	16.0	18.6	2.55
Cd	594	1, 038	15.6	19.3	2.46
Mg	923	1, 376	15.4	16.3	2.32

1.7 Summary

The subject of solidification and crystal growth was introduced via several comparisons between the crystalline and liquid states. First, the phenomenology of the enormous viscosity changes occurring between solids and melts was discussed, accounting for the long history over which mankind used casting methods to produce utilitarian and artistic objects of gold, silver, and bronze.

Distinctions were drawn on the bases of long-range and short-range ordering, which distinguish crystalline solids from their melts. Crystals can be classified on the basis of their lattices and point-group symmetries, with their molecular structure described well by an unique unit cell that captures all the essential symmetries. Melts, despite their similar mass density and bond strengths, lack long-range order, and must be described statistically via radial distribution functions (RDFs). RDFs are shown to capture local short-range two-point correlations that exist statistically within a melt, but cannot provide details about longer range interactions among the atoms or molecules. A few approaches to calculate RDFs for melts based on physical modeling of random-dense structures are described.

Finally, several theories of crystalline fusion are described briefly, each of which have some utility in estimating liquid-state properties. These approaches include Lindemann's catastrophic lattice disruption theory, and the Frenkel-Eyring 'hole' theory, the latter describing distributions of free volume in melts. Hole theories do account nicely for the typical 1–10% relative volume change that occurs when forming a molten phase from the crystalline solid state, and they are consistent with several thermodynamic empirical rules established for the melting transition.

References

1. M.E. Glicksman, *Diffusion in Solids: Field Theory, Solid-State Principles and Applications*, Wiley Interscience Publishers, New York, NY, 2000.
2. M. Che, W. Grellmann and S. Seidler, *Appl. Polym. Sci.*, **64** (1997) 1079.
3. G.R. Luckhurst and E.T. Samulski, *Liq. Cryst.*, **1** (1986) 1.
4. F.C. Frank, *Proc. R. Soc. Lond.*, **215A** (1952) 43.
5. D. Turnbull, *J. Chem. Phys.*, **20** (1952) 411.
6. B.E. Warren, *X-Ray Diffraction*, Dover Publications, Inc., New York, NY, 1990.
7. F. Zernicke and J.A. Prins, *Z. Physik*, **41** (1927) 184.
8. L.P. Tarasov and B.E. Warren, *J. Chem. Phys.*, **4** (1936) 236.
9. M. Ishitobi and J. Chihara, *J. Phys. Condens. Matter*, **4** (1992) 3679.
10. T. Morishita, *Phys. Rev. Lett.*, **87** (2001) 105701-1.
11. H.S.M. Coxeter, *Ill. J. Math.*, **2** (1958) 726.
12. J.D. Bernal, *Nature*, **185** (1960) 68.
13. J.D. Bernal, *Proc. R. Soc. Lond.*, **280A** (1964) 43.
14. G.D. Scott, *Nature*, **188** (1960) 908.
15. M.F. Ashby, F. Spaepen and S. Williams, *Acta Metall.*, **78** (1978) 1647.
16. J.L. Finney, *Proc. R. Soc. Lond.*, **319A** (1970) 497.
17. R.D. Kamien and A.J. Liu, *Phys. Rev. Lett.* **99** 155501.
18. G.S. Cargill III, *Solid State Phys.*, F. Seitz and D. Turnbull, Eds., **30**, Academic Press, New York, 1975, 227.
19. T. Ichikawa, *Phys. Stat. Solidi*, **A19** (1973) 707.
20. C.H. Bennett, *J. Appl. Phys.* **43** (1972) 2727.
21. J.L. Finney, *Mater Sci. Eng.*, **23** (1976) 199.
22. D.R. Nelson, *Phys. Rev.*, **B28** (1983) 5515.
23. D.R. Nelson and F. Spaepen, *Solid State Phys.*, **42**, Academic Press, New York, 1989, p. 1.
24. Y. Kajihara et al., *J. Non-Cryst. Solids*, **312–314** (2002) 489.
25. F.A. Lindemann, *Physik. Z.*, **11** (1910) 609.
26. M. Rettenmayr, *Int. Mater. Rev.*, **54** (2009) 1.
27. J.H. Hildebrand, *J. Chem. Phys.*, **15** (1947) 225.
28. J. Huggins, *J. Phys. Chem.*, **52** (1948) 248.
29. M.H. Cohen and D. Turnbull, *J. Chem. Phys.*, **31** (1959) 1164.
30. P. Ramachandrarao, B. Cantor and R.W. Cahn, *J. Mater. Sci.*, **12** (1977) 2488.
31. D. Tabor, Gases, Liquids, and Solids, and Other States of Matter, 3rd Ed., Cambridge University Press, Cambridge, UK, 1991.
32. J. Frenkel, *Kinetic Theory of Liquids*, Dover Publications, Inc., New York, NY, 1955, p. 176.

33. T. Altar, *J. Chem. Phys.*, **5** (1941) 577.
34. R. Fürth, *Proc. Camb. Phil. Soc.*, **37** (1941) 252.
35. F.C. Auluck and D.S. Kothari, *Nature*, **3895** (1944) 777.
36. S. Glasstone, K.J. Laidler and H. Eyring, *Theory of Rate Processes*, McGraw-Hill, New York, NY, 1941, 477.
37. N. Hirai and H. Eyring, *J. Appl. Phys.*, **29** (1958) 810.

Chapter 2
Thermodynamics of Crystal-Melt Phase Change

2.1 Enthalpy: Heat and Work Exchanges in Solid-Liquid Transformations

The enthalpy, or heat content, provides a useful thermodynamic state function [1] when dealing with the energetic changes encountered in many types of solidification and melting processes, especially when they are carried out under constant pressure (isobaric) conditions. The enthalpy, $H(S, P, N)$, is an extensive thermodynamic property of a phase, derived as the Legendre transform [2] of the internal energy, $U(S, V, N)$, already used in Chapter 1.[1] In the particular case of the enthalpy, however, pressure, P, replaces volume, V, as one of the independent variables, and S and N remain the system's molar entropy and mass, respectively. In this instance, adding the pressure-volume product, PV, to the internal energy, U, provides the Legendre transform of the internal energy called enthalpy, or heat content, $H(S, P, N)$. Consequently, the molar enthalpies of a pure solid phase, s, and its conjugate melt phase, ℓ, are found as,

$$H_i = U_i + PV_i \ (i = s, \ell).$$ (2.1)

The enthalpy provides a direct measure of the energy increase of a system as heat is added at constant pressure. Consequently, the enthalpy is often the best choice for a thermodynamic state variable when tracking energy changes during solidification processes involving isobaric heat transfer. The total differential of the function defined in Eq. (2.1) is

$$dH_i = dU_i + PdV_i + V_i dP,$$ (2.2)

where we note that the last term on the RHS of Eq. (2.2) vanishes for changes occurring at constant pressure, where $dP = 0$.

---

[1] See Appendix A for details to derive any of the thermodynamic potentials from the internal energy function.

M.E. Glicksman, *Principles of Solidification*, DOI 10.1007/978-1-4419-7344-3_2,
© Springer Science+Business Media, LLC 2011

Now, the infinitesimal change in internal energy, dU_i, for a single-phase system of fixed mass receiving or releasing heat at constant pressure consists of differential changes in the heat, Q, and work, W. As is expressed by the 1st law of thermodynamics,

$$dU = \hat{d}Q + \hat{d}W. \tag{2.3}$$

The symbol, $\hat{d}$, in Eq. (2.3) is used to denote an 'inexact', or path-dependent, differential operator.[2] If the work involved with a melting or solidification process is restricted to ordinary mechanical pressure-volume work, then the RHS of Eq. (2.3) may be written as

$$dU_i = TdS_i - PdV_i, \tag{2.4}$$

where the differential of the heat added or subtracted via 'reversible' heat transfer from the system's surroundings at temperature, T, is given by Clausius's entropy formula,

$$\hat{d}Q_{rev} = TdS_i \ (i = s, \ell). \tag{2.5}$$

Here dS_i is the differential change of the entropy of phase i. The molar, or gram-atomic entropy of a phase, $S_i(U_i, V_i)$, as indicated, is a thermodynamic state function of the extensive variables for an Avogadro's number of particles ($\approx 6.0221 \times 10^{23}$), i.e., for a unit molar, or gram-atomic mass. Reversible heat transfer, wherein heat energy is transferred to or from the surroundings of a thermodynamic system under the action of a *negligibly small* thermal gradient, provides the classical method for calculating entropy changes by integration of Eq. (2.5). The more general statement for a real irreversible processes is the so-called Clausius inequality,

$$\Delta S_{tot} \geq 0. \tag{2.6}$$

Equation (2.6) constitutes one of thermodynamic's most important theoretical constructs, defining the *total* change of the entropy as always unidirectional, i.e., becoming more positive. This is a form of the 2nd law of thermodynamics, the deeper implication of which is that real processes spontaneously run in one direction, and in one direction only. Unless the transfer of heat energy to or from a phase, and all the attendant internal changes proceed infinitesimally slowly, the total entropy change (including unavoidable frictional or viscous processes) will be positive. All real processes, such as melting and solidification, of course, are always to some extent

---

[2] Its purpose here is to remind the reader that the amount of heat and work added to a system both depend on the particular path, or process, taken between arbitrary starting and ending thermodynamic states. Heat and work are *not* state functions, but their combined changes equal that of the internal energy, which is a state function.

'irreversible', and invariably involve the presence of entropy-increasing dissipation. All real processes, without exception, require that the *total* entropy change of the system and its environment increases, i.e., $\Delta S \geq 0$. For an excellent introductory discussion on thermodynamic reversibility and irreversibility and the second law see [3].

Substituting Eq. (2.4) into Eq. (2.2) yields the well-known differential expression connecting the entropy change at constant pressure, dS_i of a phase ($i = s, \ell$) with its corresponding change in enthalpy, dH_i.

$$dH_i = TdS_i. \ (P = \text{const.}) \tag{2.7}$$

At constant pressure, which is the environment often encountered in ordinary solidification and melting processes, one may integrate Eq. (2.7) between a pair of states A and B and obtain

$$\Delta H_i = \int_A^B TdS_i = \Delta Q. \ (P = \text{const.}) \tag{2.8}$$

Equation (2.8) shows formally that enthalpy changes are exactly equal to the *reversible* heat exchanges between a phase and its environment, the latter chosen in this instance to maintain the system's pressure constant. That is, the heat energy, ΔQ, transferred reversibly under steady pressure, by heating or cooling a phase, or by transforming it from solid to liquid, or vice versa, is equal to the changes in the system's enthalpy function.

Equation (2.8) also demonstrates an extremely useful property of the enthalpy function that allows one to sketch an enthalpy-temperature phase diagram for a pure substance, held at fixed pressure, exhibiting an equilibrium melting (or freezing) point, T_{eq}.

The enthalpy-temperature phase diagram, Fig. 2.1, shows that the equilibrium liquid and solid occupy distinct, non-overlapping regions within the H-T plane, with P fixed. Specifically, either solid or liquid exists as the preferred phase anywhere along their individual existence lines, each located, respectively, below and above the thermodynamic transition point, T_{eq}. A discontinuity in the enthalpy function always occurs at T_{eq} between the two-phase existence lines. This enthalpy 'jump' locates the unique combination of pressure and temperature at which the substance can exist in either, or both, phase states, depending on the system's total heat content, or enthalpy. The enthalpy jump acts as a tie-line passing through a mixed-state, where both phases can coexist in equilibrium, in proportions that depend on the system's overall enthalpy level. The enthalpy of the mixed (solid plus liquid) state may therefore vary between $H_s(T_{eq})$, for 100% solid, up to $H_\ell(T_{eq})$ for 100% liquid. Enthalpy values intermediate between H_s and H_ℓ correspond to a mole of solid-plus-liquid, with the lever rule and rule of mixtures controlling the exact proportions of these phases at equilibrium. More importantly, a two-phase mixture produced by solidification (even of a pure substance) represents, in actuality, a *solidification microstructure*. Solid–liquid microstructures consist of these

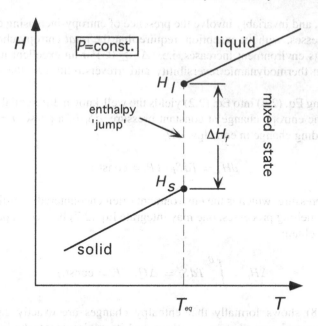

Fig. 2.1 Enthalpy-temperature phase diagram for a pure substance. The enthalpy jump, ΔH_f, represents the difference in heat content per unit mass of liquid and solid at their equilibrium melting-freezing point, T_{eq}. The difference in heat content is proportional to the entropy difference between the well-ordered crystalline solid state having low entropy, and the much less orderly (amorphous) liquid state having high entropy

phases in intimate, temporal contact, on length scales established by kinetic and other extrinsic factors, all of which reside beyond the scope of classical thermodynamics. The reader will find the kinetics of microstructure formation caused by different solidification processes discussed in considerable detail in later chapters, after various preliminary concepts are developed, including thermodynamics, heat and mass transfer, interfacial phenomena, and morphological stability.

In the case of isobaric equilibrium melting, the enthalpy-temperature phase diagram also shows that the jump in the heat content of the coexisting phases at T_{eq} defines an important property of the phase transformation itself, namely, the gram-atomic, or molar latent heat of fusion, ΔH_f. This enthalpy difference between crystal and melt, per unit molar mass of substance, is indicated in Fig. 2.1 as

$$\Delta H_f = H_\ell(T_{eq}) - H_s(T_{eq}) \geq 0. \tag{2.9}$$

The inequality indicated on the RHS of Eq. (2.9), consistent with Fig. 2.1, states that upon melting, the jump in heat content is positive. Indeed, common experience shows that most materials do melt upon heating.[3] There are, however, a few

---

[3] The less common occurrence of melting upon cooling occurs in the case of metatectic systems [4].

exceptions to this rule that are thermodynamically permissible: namely, that some materials undergo 'retrograde melting' when cooled. This peculiar exception to normal melting behavior occurs under unusual circumstances where the entropy of the solid phase is greater than that of its melt at the same temperature and pressure. This can occur in quantum systems, such as solid–liquid He at cryogenic temperatures, and in some classical binary alloy systems at elevated temperatures, where a solid phase will spontaneously melt on cooling. Generally, of course, this is not the case, and the latent heat of fusion, ΔH_f, is normally considered to be a positive quantity.

Now, if Eq. (2.1) is substituted into the righthand side of Eq. (2.9), one finds that

$$\Delta H_f = U_\ell - U_s + PV_\ell - PV_s, \tag{2.10}$$

or

$$\Delta H_f = \Delta U_f + P\Delta V_f, \tag{2.11}$$

where $\Delta V_f \equiv V_\ell - V_s$ is the volume change on melting a mole of material, and $\Delta U_f \equiv U_\ell - U_s$ is the internal energy change per mole of material melted at the equilibrium melting point.

2.2 Comparison of Terms in ΔH_f

Equation (2.11) shows that the enthalpy change for the melting of a pure substance consists of two terms: (1) the internal energy change for the transformation, ΔU_f, plus (2) any pressure-volume work, $P\Delta V_f$, done by (+), or received from (−), the system's surroundings, which is taken as standard sea-level atmosphere, where $P = 0.101$ MPa.

A typical case to be considered here is that of a gram-atomic weight of pure crystalline copper melting gradually at 1,084.6 C (in an equilibrium, or reversible manner) at normal atmospheric pressure. The contributions of the energy terms appearing in Eq. (2.11) for this melting transition are as follows:

1. $(V_\ell - V_s)/V_s = 0.05$, and $V_s = 10.0 \frac{cm^3}{g \cdot atom}$.

2. $\Delta V_f = 0.5 \frac{cm^3}{g \cdot atom} = 5.0 \times 10^{-4} \frac{liters}{g \cdot atom}$.

3. $P \cdot \Delta V_f = 0.101 \cdot \left(5.0 \times 10^{-4} \frac{liter \cdot MPa}{g \cdot atom}\right) \times \left(\frac{1.98}{8.31 \times 10^{-3}} \frac{cal}{liter \cdot MPa}\right)$
   $= 1.2 \times 10^{-2} \frac{cal}{g \cdot atom}$.

4. $\Delta H_f = \Delta U_f + P\Delta V_f = 3 \times 10^3 \frac{cal}{g \cdot atom}$.

Clearly, the pressure-volume work, $P\Delta V_f$, indicated in steps (3) and (4), is negligible compared to the internal energy change ΔU_f. This exercise suggests that the enthalpy change for the melting or solidification of a typical metal at atmospheric pressure is well-approximated by just the change in its internal energy alone. Thus, one may conclude that for most ordinary solidification processes, $\Delta H_f \approx \Delta U_f$, which is a result that is accurate under the stated conditions to about 4 ppm! This example also serves to show that the enthalpy (heat content) changes for common solidification and melting processes, i.e., those occurring at or near atmospheric pressure, primarily reflect the large entropy changes occurring during the solid$\leftrightarrow$melt transformation. The work produced by the volume change acting against the isobaric surroundings is indeed extremely small, and may be safely neglected in most situations.[4]

2.3 Co-existence of Solids and Melts

The co-existence of solid and liquid phases under varying thermodynamic conditions presents a classic example of heterophase equilibria, defined by Gibbs as requiring three fundamental conditions:

1. Equality of temperature, or thermal equilibrium, so $T_s = T_\ell$.
2. Equality of pressure, or mechanical equilibrium, so that $P_s = P_\ell$.
3. Equality of chemical potentials, or chemical equilibrium, so $\mu_s = \mu_\ell$.

The internal energies of these conjugate phases are defined as,

$$U_i(S_i, V_i, N_i) = TS_i - PV_i + \mu_i N_i, \quad (i = s, \ell), \tag{2.12}$$

where S_i, V_i, and N_i are the total entropies, volumes, and molar masses of the phases. The Gibbs potentials for a fixed mass of N moles, of either solid or melt, are given as,

$$G_i(T, P, N_i) = U_i(S_i, V_i, N_i) - TS_i + PV_i, \quad (i = s, \ell). \tag{2.13}$$

Co-existence of a pure, one-component (unary) crystal and its conjugate melt, expressed through the combined first and second laws, requires unique combinations of temperature and pressure that allow the Gibbs potential per mole, or chemical potential, of the phases to match. Thus, it is required that at thermodynamic equilibrium, Gibbs's heterophase conditions hold, namely

---

[4] The pressure-volume term, $P\Delta V_f$, is normally considered as being negligible compared to the internal energy change, excepting when the pressure during solidification exceeds several kilobars. Such conditions do arise in geological solidification and melting of the Earth's mantle rocks, and are even encountered in some important engineering processes carried out at elevated pressure, such as hot isostatic pressing (HIP) of partially molten alloys, squeeze casting, and high-pressure liquid-phase sintering. Generally, however, for melting and freezing under moderate pressures, enthalpy changes alone provide a convenient and reasonably accurate approximation of the internal energy changes.

$$\frac{G_s}{N} = \frac{U_s - TS_s + PV_s}{N} = \frac{G_\ell}{N} = \frac{U_\ell - TS_\ell + PV_\ell}{N}. \tag{2.14}$$

Thermodynamic co-existence of crystal and melt at combinations of temperature and pressure are equivalent to the isothermal, isobaric (reversible) melting of a mole, or gram-atomic weight of crystal, or, similarly, the (reversible) freezing of a mole of melt. These transitions must result in a zero change in the system's Gibbs potential. That is, for reversibly melting a mole of crystal,

$$\frac{G_\ell - G_s}{N} = 0 = \Delta U_f - T\Delta S_f + P\Delta V_f, \tag{2.15}$$

where $\Delta U_f = (U_\ell - U_s)/N$, $\Delta S_f = (S_\ell - S_s)/N$, and $\Delta V_f = (V_\ell - V_s)/N$, are, respectively, the changes in internal energy, entropy, and volume upon melting one mole of crystal.

Of interest here is determining the condition for solid–liquid phase co-existence where changes in temperature and pressure are allowed, subject to the equality of the chemical potentials, per the Gibbs criterion. The differential form of Eq. (2.15) is

$$0 = d(U_f) - d(T\Delta S_f) + d(P\Delta V_f). \tag{2.16}$$

Substituting the individual total differentials into Eq. (2.16) gives

$$0 = Td(\Delta S_f) - Pd(\Delta V_f) - Td(\Delta S_f) - \Delta S_f dT + Pd(\Delta V_f) + \Delta V_f dP, \tag{2.17}$$

which after cancellation of terms yields the condition for phase co-existence,

$$0 = -\Delta S_f dT + \Delta V_f dP. \tag{2.18}$$

Equation (2.18) is the Clausius–Clapeyron equation, often written as

$$\left(\frac{dP}{dT}\right)_\mu = \frac{\Delta S_f}{\Delta V_f} = \left(\frac{\Delta H_f}{\Delta V_f}\right)\frac{1}{T_{eq}}. \tag{2.19}$$

The curve along which a pure crystal and its melt maintain equal chemical potentials may be found by integrating Eq. (2.19) from an initial state of co-existence, a, to another state of co-existence, b.

$$\int_{Pa}^{Pb} dP = \int_{T_{eq}^a}^{T_{eq}^b} \left(\frac{\Delta H_f}{\Delta V_f}\right)\frac{dT_{eq}}{T_{eq}}, \tag{2.20}$$

which yields the form of the solid–liquid co-existence curve, namely

$$P^b = P^a + \left(\frac{\Delta H_f}{\Delta V_f}\right) \ln \frac{T_{eq}^b}{T_{eq}^a}. \tag{2.21}$$

Equation (2.21) is valid for moderate temperature and pressure changes over which the latent heat of fusion and the volume change on fusion remain reasonably constant. Plots of Eq. (2.21) are shown in Fig. 2.2 for two elements: Pb, an FCC metal, the volume of which expands about 6% on melting, and Si, a diamond-cubic (DC) semiconductor that contracts nearly 10% on melting. The influence of pressure on the melting point of these elements is opposite, because of the sign reversal in their respective volume changes on transforming.

The rather large pressure changes required to shift the melting point of materials just a few K usually makes pressure control impractical for improving the operation of crystal growth processes. High-pressure crystal growth, however, has been employed to elevate the freezing point of certain materials, such as selenium, in order to lower its melt viscosity and augment the rate of crystal growth.

2.3.1 Solid-Melt Critical Point

The question has arisen: Could the temperature and pressure be elevated sufficiently to cause a crystal and its co-existing conjugate melt to reach a critical point? Many examples exist, of course, of liquids and vapors having their co-existence curves terminate at critical points. Water and CO_2 are two important examples of liquids exhibiting critical points, and having widespread industrial applications above criticality.

Radial distribution functions for liquid mercury were discussed in Chapter 1, and certainly showed a 'gas-like' tendency at high temperatures. But to achieve criticality, two phases must have their molar volume- and entropy-differences *vanish*. These circumstances appear highly unlikely to occur in the case of solids and their melts, as these phases fundamentally differ in their long-range order parameters, whereas liquids and their vapors, below criticality, merely differ in their short-range order parameters. P.W. Bridgman, the Nobel prize winning high-pressure physicist, provides an informed opinion about the above question concerning crystal-melt criticality [5]:

> ... unless there is some reversal of the trend in the present experimental range at pressures beyond those now attainable, there is no reason to think that the melting [co-existence] curve can end in a critical point.

2.4 Thermal Supercooling and Metastability

With reference to the enthalpy-temperature, or H-T, diagram shown in Fig. 2.3, consider the process of gradually supercooling, or 'undercooling', a mole of pure melt phase from the initial state at its equilibrium point, T_{eq}, to its spontaneous

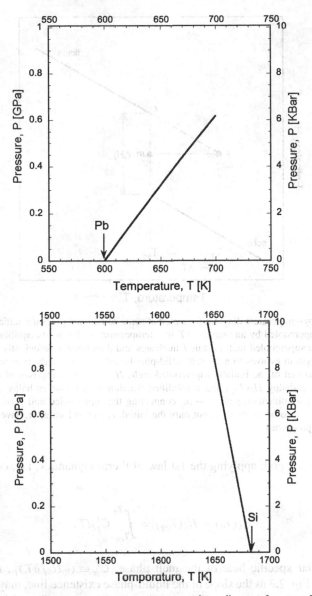

Fig. 2.2 Co-existence lines on a temperature-pressure phase diagram for crystalline Pb (*upper plot*) and Si (*lower plot*) with their conjugate liquid phases, according to the Clausius–Clapeyron equation, Eq. (2.20). The *vertical arrows* indicate the equilibrium melting points of these elements at ordinary atmospheric pressure. As the pressure increases, the melting point of Pb rises, whereas that of Si decreases. The axes of both plots cover equivalent ranges of temperature and pressure

nucleation temperature, T_0. The enthalpy change for the liquid cooling from its equilibrium melting point, state ℓ, to its (metastable) supercooled condition, state i, is $H_\ell(T_0) - H_\ell(T_{eq})$. This enthalpy change may be calculated using the standard

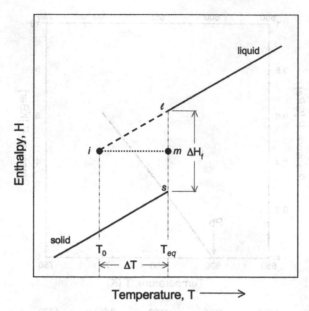

Fig. 2.3 Enthalpy-temperature diagram for a melt at some some fixed pressure, initially in state ℓ, which is then supercooled by an amount ΔT to a temperature T_0 below its equilibrium freezing point, T_{eq}. If the supercooled melt in state i nucleates and then freezes sufficiently quickly (adiabatically) the system evolves to a 'mixed' solid-plus-liquid state, m, with its average enthalpy, $H_{s+\ell}$ identical to that of the initially supercooled melt, $H_\ell(T_0)$, but comprised of a liquid fraction with molar enthalpy $H_\ell(T_{eq})$ and a solidified fraction with molar enthalpy $H_s(T_{eq})$. The (non-equilibrium) solidification path, $i \to m$, connecting the supercooled and partially solidified states is constrained to be isenthalpic, but only the initial, i, and end state, m, have well-defined thermodynamic properties

heat-exchange integral, applying the 1st law of thermodynamics, i.e., conservation of energy,

$$H_\ell(T_0) - H_\ell(T_{eq}) = \int_{T_{eq}}^{T_0} C_p^\ell dT. \tag{2.22}$$

Here, the molar specific heat of the melt phase, $C_p^\ell \equiv (\partial H_\ell/\partial T)_p$, represented graphically in Fig. 2.3 as the slope of the liquid-phase existence line, may be considered to be approximately constant for small or moderate amounts of supercooling (typically less than about 100 K). The heat-exchange integral, Eq. (2.22), may be simplified with this approximation and integrated as

$$H_\ell(T_0) = H_\ell(T_{eq}) + C_p^\ell(T_0 - T_{eq}). \tag{2.23}$$

Now, if the supercooled melt, in state i, is allowed to freeze spontaneously, and the transformation achieves a rate sufficiently fast relative to the time required for long-range heat exchanges with the melt's surroundings, then an overall *adiabatic* condition is established between the system and its surroundings. The solidifying system's rapidly (partially) solidified state, m, is constrained to remain at the same enthalpy level, $H_m(T_{eq})$, as the initially supercooled melt, namely, $H_\ell(T_0) = H_i$. This type of partial freezing can occur in metals in just a fraction of a second, and is termed adiabatic solidification. Adiabatic solidification results in the production of a well-defined mass fraction, X_s, of solid from the initially supercooled liquid. The average enthalpy of the resultant two-phase state, (solid plus liquid coexisting at their equilibrium temperature, T_{eq}) is equal to the initial enthalpy of the super-cooled melt, as no heat has been exchanged with the surroundings. The resultant mass fraction solidified, X_s may be expressed through the enthalpy conservation condition as

$$H_i = H_m = X_s H_s(T_{eq}) + (1 - X_s) H_\ell(T_{eq}). \tag{2.24}$$

The RHS of Eq. (2.24) is an expression for the average enthalpy of the mixed state, m. If one equates the RHS of Eqs. (2.23) and (2.24) and solves for the mass fraction frozen adiabatically, X_s, one obtains the result

$$X_s = \left(\frac{C_p^\ell}{\Delta H_f}\right) \Delta T, \tag{2.25}$$

where in Eq. (2.25) the last term, $\Delta T = T_{eq} - T_0$, designates the initial melt supercooling, and the definition for the molar latent heat of fusion is $\Delta H_f \equiv H_\ell(T_{eq}) - H_s(T_{eq})$.

2.4.1 Characteristic Melt Supercooling

The mass fraction solidifying adiabatically from a supercooled pure melt, X_s, is referred to in the heat transfer literature as the thermal Stefan number, St, or dimensionless supercooling. Note X_s (or, equivalently, St) is equal to the dimensionless ratio of the initial supercooling of the melt (in Kelvins) to a property of the system called the 'characteristic' supercooling,

$$\Delta T_{char} \equiv \Delta H_f / C_p^\ell. \tag{2.26}$$

The characteristic supercooling, also expressed in Kelvins, is a thermodynamic property of the melt. Values for the characteristic supercooling of different materials vary widely, however, covering the range of ca. 10–500 K. If a melt were subjected to a supercooling equal to or greater than the melt's characteristic supercooling, $\Delta H_f / C_p^\ell$, then $St \geq 1$, and the entire melt mass would be capable of solidifying adiabatically, causing $X_s \to 1$. This is the only situation—and one seldom either achieved, or especially desired, in practice—for which external heat transfer is not required to complete the solidification process. This extreme condition, and its usually attendant high-speed freezing, have been observed in the laboratory for small (gram-sized) quantities of many metallic and ceramic materials [6]. By contrast, however, most industrial-scale solidification processes are carried out at levels of supercooling that seldom exceed several Kelvins, and, thus, typically represent only a small fraction of a melt's characteristic supercooling, $\Delta H_f / C_p^\ell$. Consequently, in castings, ingots, and welds, as examples of standard industrial solidification processes, the Stefan number for spontaneous freezing is usually small ($St \ll 1$), so little solid is produced promptly and adiabatically following nucleation, and, therefore, external, long-range heat transfer would normally be required to complete the solidification process. The fact that heat must be removed from the crystallizing system and transferred to its environment to allow the major solidification reactions to proceed to completion suggests immediately that heat transfer, unavoidably, is an essential engineering issue in the design of practical solidification and crystal growth processes. Rudiments of solidification heat transfer—both macroscopic and microscopic—will be presented and discussed in subsequent chapters.

Finally, it is worth noting that the subject area of supercooled melts, just touched upon in this section from the energetic standpoint, also extends to a wide range of other interesting materials phenomena. Supercooled (metastable) melts, for example, are relevant to such diverse topics as network (silicate) glass forming systems, phase separating melts, amorphous alloys, crystal growth, atmospheric processes, biological and geological systems, etc. [7, 8].

2.5 Free Energy Changes During Freezing and Melting

The ability for any phase transformation to progress spontaneously requires that the system's free energy decreases, or, at a minimum, remains stationary for thermodynamically reversible processes. This statement derives from the 2nd law of thermodynamics [1]. Specifically, for systems held both at constant temperature, T, and pressure, P, one chooses the Gibbs potential, or free energy, $G(T, P, N)$, as the function testing a solidifying system's stability. The choice of the Gibbs potential is that its functional minima define thermodynamic equilibrium states—both stable and metastable. The Gibbs potential (See Appendix A) is formally defined as the Legendre transform of the system's internal energy, $U(S, V, N)$, where temperature, T, replaces entropy, S, and pressure, P, replaces volume, V, as independent variables [2]. For a gram-atomic, or molar, mass of a unary (single component) system

$$G(T, P, N = 1) \equiv U + PV - TS = H - TS. \qquad (2.27)$$

2.5.1 Reversible Solidification

The gradual freezing, or melting, of a pure material at its melting temperature, T_m, and at a fixed pressure, P, provides a good approximation of a thermodynamically reversible transformation. For reversible melting or freezing, $\Delta G = 0$. Equation (2.27), when applied to the liquid-solid phase change restricted to *reversible* conditions, shows that

$$\Delta H_f = T_m \Delta S_f, \tag{2.28}$$

where the molar entropy of fusion, defined at some pressure, P, is just the difference in molar entropies between the liquid and crystalline states,

$$\Delta S_f \equiv S_\ell(T_m) - S_s(T_m). \tag{2.29}$$

The molar entropy change, ΔS_f, as now defined in Eq. (2.29), is almost always a positive quantity when the solidification-melting reaction proceeds in the direction $s \to \ell$. Thus, ΔS_f represents the (normally positive) entropy change associated with the large decrease in long-range atomic order (LRO) as the melting transition proceeds. This sudden decrease in long-range atomic order occurs, as already described in Chapter 1, as a crystal melts to form its conjugate liquid phase at, or close to, its normal thermodynamic melting point. The decrease in LRO increases the system's entropy in direct proportion to the mass of solid melted. The accompanying entropy increase during melting under equilibrium is 'supplied' by *reversible* heat transfers delivered from the system's surroundings. The fact that heat must be added continually from its boundaries so that melting advances over time, again suggests that heat transfer remains a major technical issue in the design of practical melting and solidification processes.

2.5.2 Irreversible Solidification

A particularly simple example of irreversible solidification, where $\Delta G \neq 0$, is that of the partial adiabatic freezing of a supercooled melt, as shown in the enthalpy diagram, Fig. 2.3. A consequence of irreversibility is that $\Delta G < 0$ and that $\Delta S_{tot} > 0$. The term ΔS_{tot} represents the summed entropy changes for the solidifying system plus its surroundings. The 2nd Law of Thermodynamics requires that operating any real (i.e., irreversible) process, where frictional resistance and dissipation are present *at any length scale*, will always increase the total entropy of the universe. Here the word 'universe' may be taken as the crystal-melt system plus its local surroundings, which may be accepting or supplying heat and or work during the irreversible transformation. A formal, and succinct statement of this principle is provided by Clausius's inequality, which states: for any cyclic process or path[5] involving heat exchanges (enthalpy changes) between a system and its surroundings, the closed

---

[5] A 'cyclic' process is a sequence of connected thermodynamic steps, such as heat transfer, work exchanges, and transformations, all of which eventually returns the system and its surroundings to their initial states.

line integral of the irreversible heat exchanged, divided by the absolute temperature, along any segment of the cycle must be less than the line integral taken along the same path for reversible heat exchange. Thus,

$$\oint \frac{\hat{d}Q}{T} \leq \oint \frac{\hat{d}Q_{rev}}{T}. \tag{2.30}$$

For segments of the cyclic pathway that are reversible, the RHS of Eq. (2.30) equals the change in the classical entropy for the system, or minus the same entropy change for the surroundings with which the reversible heat exchanges occur.

$$\Delta S_{sys} = -\Delta S_{surr} = \oint \frac{\hat{d}Q_{rev}}{T}. \tag{2.31}$$

The path integral of reversible heat exchanges thus encompasses a series of state function changes. For a cyclic path, the starting point equals the end point, and this line integral for reversible processes vanishes. If, however, irreversible steps are anywhere involved, such as in the rapid solidification of a supercooled melt, then the value of the integral on the LHS of Eq. (2.30) will always be *less* than that of the one traversing the reversible cycle.

Now, as mentioned, entropy, S, like enthalpy, H, is a thermodynamic state function. So, even if a cyclic thermodynamic path involves some irreversibility, it follows that for *any* segment of the cycle connecting two states, say the initially supercooled melt, i, and the partially frozen mixture, m, shown in Fig. 2.3, the entropy change between those states must be identical. This remains true whether or not these state changes are accessed via either reversible or irreversible processes. Thus, *irrespective of path reversibility*,

$$\Delta S_{irrev}[i \rightarrow m] = \Delta S_{rev}[i \rightarrow m]. \tag{2.32}$$

If, however, the process connecting states i and m is a real one, and dissipation, or friction, is present, one expects some additional entropy to be produced within the system that does not involve any heat exchange with the surroundings. Indeed, the spontaneous fast freezing of a metastable, supercooled melt involves complex internal mesoscopic heat transfer (occurring on a scale of millimeters or micrometers) under extremely steep thermal gradients, and the process entails rapid molecular attachment and growth of the interphase interfaces. Interface growth and propagation requires irreversible atomic motions (occurring over microscopic dimensions of fractions of a nanometer). The additional entropy production, ΔS_{prod}, incurred by free energy dissipation is indistinguishable from, and added to, any entropy changes directly caused by heat exchanges with the surroundings. So for any irreversible segment it follows that,

$$\Delta S_{irrev} = \int \frac{\hat{d}Q}{T} + \Delta S_{prod}. \tag{2.33}$$

In view of Eqs. (2.32) and (2.31), one sees that the Clausius inequality, Eq. (2.30), must be obeyed, as $\Delta S_{prod} > 0$, and the integral of the heat exchange divided by the absolute temperature for an irreversible cycle will always be smaller than that for a reversible cycle. In Section 2.7 we will develop the details of such a comparison with an irreversible solidification process.

2.6 Principal Types of Binary Alloy Solidification

Many of the theoretical models, examples, and experiments, discussed throughout this book are directed at so-called 'engineering' materials. These are often, but not exclusively, metals and alloys that are melted, cast, soldered and welded in the course of their production, fabrication, and application. The microstructures, solidification responses, and properties of numerous commercial casting alloys have been studied intensively, and reported in detailed microstructure atlases [9], usually organized broadly under categories based on the main elemental component (e.g., copper alloys, aluminum alloys, cast irons and steels). These microstructure atlases are invaluable for their practical guidance on alloy selection and subsequent processing, but the underlying scientific principles explaining the origin and formation of their cast microstructures are seldom included. In addition to ordinary casting alloys, solidification principles are also applicable to the production and control of sophisticated high-temperature alloy castings, composed of large, shaped, superalloy single crystals for use in critical applications, such as jet aircraft engines and large land-based turbo-generators. The fundamental approaches to be described here also apply to other important classes of modern materials [10], such as the preparation of semiconductors [11, 12], superconductors [13], and opto-electronic materials [14]. Nevertheless, some of the best confirmation of the theories and explanations to be presented in later chapters are for binary alloys that often form the starting point for the development of practical engineering materials, the economic importance of which to modern society is nothing short of extraordinary.

The following discussion is based upon a generic binary phase diagram [15] that captures the major phase relationships applicable to some of the typical alloys to be treated in detail in later chapters.

1. (A & B) are the pure components: The components of a binary alloy locate the ends of the diagram (See Fig. 2.4.), and typically solidify from their respective supercooled melts as pure A- or B-dendritic (branched) crystals, or, as a slowly grown, smooth solidification front, leading to nearly featureless monocrystals. Pure materials provide limited applications, aside perhaps from some major importance as metallic electrical conductors and semiconductors. Pure materials tend to be too weak mechanically for most other engineering applications. They do afford, however, the simplest examples for studying the solidification process.
2. (SS) solid solution alloys: Such dilute alloys typically solidify as α- or β-phase solid-solution 'primary' dendrites, along with small amounts of non-equilibrium phases that eventually appear within the cast microstructure. Binary solid

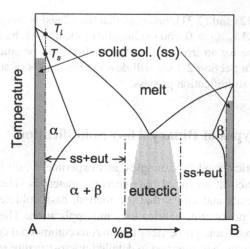

Fig. 2.4 Binary phase diagram, displaying some typical compositional zones that exhibit distinctive solidification behaviors. The curves T_ℓ and T_s indicate the liquidus and solidus boundaries, which tract the temperatures where equilibrium solidification starts, T_ℓ, and completes, T_s, at a given composition, $\%B$, and temperature. The shaded 'terminal' regions, located near the pure components A and B, are labeled 'ss', which denotes the limited composition zones for obtaining single-phase microstructures of either of the two solid-solution phases, α or β. The gray area, labeled 'eutectic', located around the mid-composition range, indicates where the eutectic microstructural constituent dominates during freezing. The two regions, 'ss+eut', suggest compositions over which the microstructure consists of primary phase and some eutectic microconstituent. Adapted from [15]

solutions are a step closer to practical casting alloys, which usually contain at least three, or more, controlled elemental components. More complex alloys cast for specialty purposes, such as tool steels, stainless steels, aero-structural alloys, or superalloys for high-temperature applications may contain over a dozen controlled components!

3. $(SS + Eu)$ solid solution primary phase + eutectic: More solute-rich alloys initially solidify as either α- or β-dendrites, depending on composition, accompanied by a significant amount of α/β eutectic micro-constituent, appearing later in the overall freezing process. The eutectic constituent is often distributed throughout the interdendritic spaces of each grain in a casting. This is probably the most common situation encountered in practical casting alloys.

4. (Eu) Eutectic (polyphase): Binary eutectics solidify by an isothermal α/β eutectic reaction. Many eutectics are characterized by relatively short-range diffusion paths among α, β, and the eutectic melt from which they crystallize in close proximity. Eutectic binary alloys form the basis for common solders and brazing alloys, as well as cast irons and type metal. Polyphase solidification, as occurs in eutectic alloys, involves complex interfacial and diffusion processes among the three phases, and will be discussed in more detail in Chapter 16.

The reader should note at this point that the major distinction among solid-solution dendritic alloys and eutectics is primarily the scale at which the solidification structure forms. Eutectics tend to develop fine-scale microstructures (ca. 1–10 μ m), whereas dendritic structures exhibit relatively coarse morphologies (ca. 10–100 μ m). Every alloy type yields a range of cast structures that depends sensitively on the alloy's chemical composition, initial superheating of the melt, solidification geometry and mold-wall characteristics, supercooling and cooling rate, plus 'modifiers' that may be added as grain refiners or to control gas porosity.

2.7 Illustrative Examples

One gram-atom of pure aluminum is melted and held at its equilibrium temperature, T_{eq}, at atmospheric pressure, P. This melt, held in a well-insulated inert crucible, is then supercooled to a lower temperature, T_0, nucleated, and then (partially) solidified adiabatically. The remainder of the melt, after partial adiabatic freezing, is subsequently totally re-solidified at its equilibrium temperature, T_{eq}, by slowly withdrawing heat from the container. The following thermo-chemical data apply:

- Gram-atomic mass Al = 26.98 g
- $T_{eq} = 933.2$ K
- $\Delta T = T_{eq} - T_0 = 60$ K
- $\Delta H_f = 10.47 \frac{kJ}{\text{gatom}}$
- $\Delta T_{char} = \Delta H_f / C_p^\ell = 375.5$ K
- $C_p^\ell = 28.74 \frac{J}{\text{g-atom·K}}$
- $S_s(T_{eq}) = 59.18 \frac{J}{\text{g-atom·K}}$
- $S_\ell(T_{eq}) = 71.13 \frac{J}{\text{g-atom·K}}$

1. *Once nucleated, how many grams of crystalline Al freeze adiabatically from the initially supercooled melt?*
   When supercooled by 60 K, molten Al can nucleate and spontaneously freeze, forming a mass fraction of solid phase, X_s. Spontaneous solidification of Al at this large supercooling is rapid, so partial freezing occurs under virtually adiabatic conditions. The latent heat released is trapped within the insulated system, quickly raising the temperature of the solid–liquid mixture to its equilibrium melting point, T_{eq}, at which time further solidification ceases. The mass fraction of Al crystals, X_s, solidifying to the end-point of adiabatic freezing may be found using energy conservation, by applying the 'lever rule' to the metastable extension of the enthalpy-temperature, H-T, phase diagram. See again, Fig. 2.3 to follow these geometric constructions:

$$X_s = \frac{\ell m}{\ell s}, \tag{2.34}$$

where $\ell m/im = C_p^\ell$, $im = T_{eq} - T_0$, and $\ell s = \Delta H_f$. Substituting all these quantities into the RHS of Eq. (2.34) yields the mass fraction of solid Al

$$X_s = \frac{T_{eq} - T_0}{\Delta H_f/C_p^\ell} = \frac{\Delta T}{\Delta T_{char}} = 0.16. \qquad (2.35)$$

The mass of Al, M_s, solidified adiabatically is therefore 0.16×26.98 g $= M_s = 4.3$ g. This leaves a mass of untransformed liquid Al, $M_\ell = 22.68$ g, with the system adiabatically self-heated back to its equilibrium temperature, $T_{eq} = 933.2$ K. The crystals and melt produced by partial adiabatic freezing are usually found in the form of a fine mixture, or slurry of dendrites, called a 'mush'. The scale, morphological characteristics, and kinetic behavior of rapidly formed solidification microstructures comprise interesting important topics, and are dealt with in Chapters 13 and 17.

2. *Determine the entropy change for the partial adiabatic freezing described in (1).* First, with reference to Fig. 2.5, the *reversible* entropy change per g-atom Al solidified slowly at its equilibrium point, T_{eq}, is $-\Delta S_f = S_s(T_{eq}) - S_\ell(T_{eq}) = -11.95$ [J/g-atom·K]. Next, the entropy per g-atom of the initially supercooled melt, $S_\ell(T_{eq} - 60$ K$)$, may be found by calculating the Clausius heat exchange integral, Eq.(2.31), for the liquid phase cooled slowly and reversibly from its equilibrium temperature, T_{eq}, into its supercooled state at $T_{eq} - 60$ K. One obtains

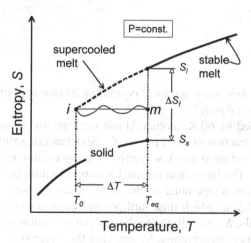

Fig. 2.5 Entropy-temperature, S-T, diagram for a pure melt. The entropies for the stable crystalline solid and liquid phases are shown as *solid curves*. The entropy for the metastable supercooled melt is the *dashed curve* extending below T_{eq} of the stable melt. The wiggly curve connecting the entropy for the initially supercooled melt, and the resultant mixed-state of solid plus melt at T_{eq} suggests some unknown non-equilibrium pathway for rapid adiabatic solidification. Irreversible pathways cannot be displayed on the S-T equilibrium phase diagram

$$S_\ell(T_{eq} - 60\text{ K}) = S_\ell(T_{eq}) + \int_{T_{eq}}^{T_{eq}-60\text{ K}} C_p^\ell \frac{dT}{T}. \qquad (2.36)$$

Substituting values into Eq.(2.36) from the list of thermochemical data for Al yields the entropy per mole of molten Al cooled 60 K below its equilibrium temperature,

$$S_\ell(T_{eq} - 60\text{ K}) = 71.13 + 28.74 \times \left(\ln \frac{933.2\text{ K} - 60\text{ K}}{933.2\text{ K}} \right)$$

$$= 69.22 \left[\frac{\text{J}}{\text{g} - \text{atom} \cdot \text{K}} \right],$$

As might be expected, supercooled (metastable) molten Al is found to have a lower entropy per unit mass than does (stable) molten Al at its melting point.

The entropy of the mixed state (crystal plus melt) formed immediately after partial adiabatic freezing may be determined by using the rule of mixtures, recognizing that the mass fractions of solid and melt are also equivalent to their atom fractions in the partially solidified mixture, and that entropy—a thermodynamic state function—is independent of the path taken between the initial (supercooled) and final (partially solidified) states. Thus,

$$S_{mixed} = S_s(T_{eq}) \cdot X_s + S_\ell(T_{eq}) \cdot X_\ell, \qquad (2.37)$$

and so

$$S_{mixed} = (59.18)(0.16) + (71.13)(0.84) = 69.22 \left[\frac{\text{J}}{\text{g} - \text{atom} \cdot \text{K}} \right],$$

Comparison of Eq. (2.36) and Eq. (2.37) shows that the molar entropy change caused by adiabatically freezing the supercooled melt to a mixed state containing about 16% crystalline Al at T_{eq} is exactly zero. As heat flow during *adiabatic* freezing is precluded either to or from the system's external environment, one may conclude that the environment itself does not experience any entropy change during the partial freezing process.

Any entropy production attending the irreversible solidification cycle will therefore depend on certain (unknown) kinetic details of the non-equilibrium (irreversible) phenomena that accompanied the rapid solidification process. Nonequilibrium solidification and crystal growth phenomena will be discussed in Chapter 17.

The cyclic value of Clausius's heat exchange integral may be evaluated for adiabatic solidification by starting with the melt at its equilibrium temperature, T_{eq}, and then (1) reversibly cooling it to its initial metastable state at a supercooling $\Delta T = T_{eq} - T_0$; (2) allowing adiabatic solidification to occur ($\Delta Q = 0$); (3) then reversibly remelting the fraction of solid, X_s, and, thereby, returning the system to its starting condition at T_{eq}.

$$\oint \frac{\hat{d}Q}{T} = \int_{T_{eq}}^{T_0} \frac{C_p^{\ell}dT}{T} + 0 + X_s \frac{\Delta H_f}{T_{eq}}. \tag{2.38}$$

Evaluating the integral on the RHS of Eq. (2.38), substituting Eq. (2.35) for the coefficient X_s, and then canceling terms yields

$$\oint \frac{\hat{d}Q}{T} = C_P^{\ell} \ln\left(\frac{T_0}{T_{eq}}\right) + C_P^{\ell} - C_P^{\ell} \frac{T_0}{T_{eq}}. \tag{2.39}$$

Division of both sides of Eq. (2.39) by the common factor, C_P^{ℓ}, provides a dimensionless form of the cyclic heat integral, namely,

$$\mathcal{I}(\rho) = \ln \rho + 1 - \rho, \tag{2.40}$$

where $\mathcal{I}(\rho)$ is the dimensionless heat exchange integral, and $\rho \equiv T_0/T_{eq}$ is the ratio of the nucleation temperature, T_0, to the equilibrium temperature. A plot of Eq. (2.40) is shown in Fig. 2.6.

3. *Find the additional entropy change for isothermally (reversibly) freezing the remaining molten Al at its melting temperature.*

The remaining 22.7 g of molten Al can be solidified by gradually transferring the required amount of latent heat to the system's surrounding at the equilibrium

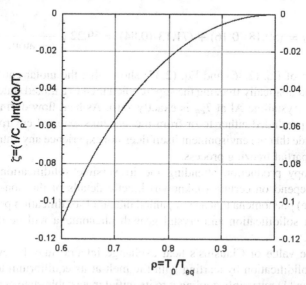

Fig. 2.6 Plot of Eq. (2.40), showing the cyclic heat exchange integral for irreversible adiabatic solidification. Note that the function, ζ, is always negative, in agreement with Clausius's inequality [1], Eq.(2.30), confirming that the cyclic heat exchanged between a solidifying system and its surroundings is *always* less than that for a reversible cycle, where Clausius's integral vanishes

temperature, T_{eq}. Slow, isothermal solidification under negligible thermal gradients approximates reversible heat transfer from the surroundings.
Thus, the Clausius inequality may also be stated as,

$$\Delta S \geq \oint \frac{\partial Q_{irr}}{T}. \tag{2.41}$$

The reversible entropy change for the remaining melt to be solidified at its equilibrium temperature is

$$\Delta S_{rev} = \frac{(59.18 - 71.13)\left[\frac{J}{g-\text{atom}\cdot K}\right] \times 22.7\,[g]}{26.98\left[\frac{g}{g-\text{atom}}\right]} = -10.05\left[J\cdot K^{-1}\right], \tag{2.42}$$

The system's surroundings that receive the latent heat released from solidifying the remaining melt of would, of course, increase its entropy by the same, or greater amount, thus, satisfying Clausius's inequality, Eq. (2.41). Were this last portion of the solidification cycle (slow freezing) truly reversible, its added entropy change for the system plus its surroundings would be zero.

4. *What is the enthalpy change for partial adiabatic freezing?*
   The enthalpy, a thermodynamic function of state, can also be used to characterize adiabatic freezing. The molar enthalpy change may be found simply as, $\Delta H_{tot} = H_{final} - H_{initial}$. The final system enthalpy, H_{final}, may be found again using the rule of mixtures, namely

$$H_{final} = H_\ell(T_m)(0.84) + H_s(T_m)(0.16), \tag{2.43}$$

so that

$$H_{final} = H_\ell(T_m)(0.84) + (H_\ell(T_m) - \Delta H_f)(0.16), \tag{2.44}$$

and

$$H_{final} = H_\ell(T_m) - 1.68\left[\frac{kJ}{g-\text{atom}}\right]. \tag{2.45}$$

The initial enthalpy of the supercooled melt can be calculated using the heat integral, remembering that C_p^ℓ is approximately constant over moderate temperature changes.

$$H_{initial} = H_\ell(T_m) + \int_{T_m}^{T_m - 60} C_p^\ell dT \approx H_\ell(T_m) - 28.74\left[\frac{J}{g-\text{atom}}\right] \times 60. \tag{2.46}$$

so

$$H_{initial} = H_\ell(T_m) - 1.72 \left[\frac{kJ}{g - atom}\right].\qquad(2.47)$$

Subtracting the results found in Eq. (2.47) from that found in Eq. (2.45) shows that the total change in enthalpy, within the precision of these calculations, is $\Delta H_{tot} = H_{final} - H_{initial} \approx 0$. A zero net change in the total enthalpy is the result expected for any 'isenthalpic' process, such as adiabatic solidification. (See again Fig. 2.3.) In essence, the partial solidification process merely 'rearranged' the internal distribution of the system's enthalpy (heat content), but does not alter its total value. We shall discover later that cast alloy microstructures evolve through the spatial redistribution of both enthalpy and chemical components. An advantage of using the enthalpy function, H, to follow energy changes during solidification, rather than using the classical entropy, S, is that the influence of the (constant pressure) environment is automatically accounted in the calculations.

5. *Estimate the Gibbs free energy change for the spontaneous partial solidification of the supercooled melt.*

   The initial state of the system, prior to its partial freezing, is metastable melt phase supercooled 60 K. See the Gibbs potential versus temperature plot, Fig. 2.7. The Gibbs free energy per mole of the solid and melt are equal at the mixture's final equilibrium temperature, T_{eq}, and pressure, P. The Gibbs free energy change per mole solidified may be found as the difference between the

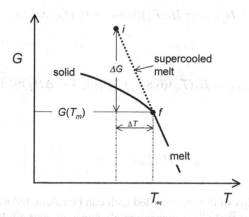

Fig. 2.7 Gibbs free energy versus temperature. The stable phases are shown as *solid curves* that intersect at the equilibrium melting temperature, T_{eq}.. The Gibbs energy for the metastable melt, which is supercooled an amount ΔT, is represented as the *dotted extension* of the stable melt curve below T_{eq}. The initial supercooled melt has its free energy at point i, and the final state of the partially solidified two-phase mixture is shown at point f. The total change in Gibbs free energy for spontaneous partial freezing is shown as the vertical drop, ΔG, from states $i \rightarrow f$ The non-equilibrium path actually connecting states i and f cannot be represented on this diagram

free energy in the final (partially solidified) state, f, and the initial (fully molten) supercooled state, i.

$$\Delta G = G_{final} - G_{initial} = G(T_{eq}) - G_\ell(T_{eq} - \Delta T). \qquad (2.48)$$

The change in free energy for supercooling the melt from state $f \rightarrow i$ is given as

$$G(T_{eq}) - G_\ell(T_{eq} - \Delta T) = \int_{T_{eq}-\Delta T}^{T_{eq}} S_\ell dT \approx S_\ell(T_{eq}) \cdot \Delta T, \qquad (2.49)$$

where the approximation is employed of considering the entropy of the melt to remain constant. Using the thermochemical data provided, the Gibbs free energy change per mole of supercooled melt that spontaneously solidifies (states $i \rightarrow f$) is given by $-S_\ell(T_{eq}) \cdot \Delta T$, and thus,

$$\Delta G = -71.13 \left[\frac{J}{g - atom \cdot K} \right] \times 60\ K = -4.27 \left[\frac{kJ}{g - atom} \right], \qquad (2.50)$$

which, indeed, represents a relatively large release of free energy. Although the irreversible path connecting the initial supercooled melt, point i, and the system's final mixed-phase state, point f, cannot be traced directly on the free energy diagram, the net free energy change between these states is easily found. A negative value of the Gibbs energy change is *always* required for spontaneous (thermodynamically permissible) physical processes such as adiabatic solidification.

6. *Show that for (reversible) isothermal freezing of the rest of the melt, controlled through extremely slow external heat transport, the Gibbs free energy change $\Delta G = 0$.*
The Gibbs function was defined earlier in Eq. (2.7) as $G = H - TS$. For an isothermal, isobaric process, such as the slow solidification of a pure material at its equilibrium temperature, the change per gram-atom of the Gibbs potential is

$$\Delta G = \Delta H_f - T \Delta S_f. \qquad (2.51)$$

Substituting Eq. (2.8) for the enthalpy of fusion, ΔH_f, in Eq. (2.51), which is a valid relationship for isobaric, isothermal, reversible phase changes, proves that $\Delta G = 0$. This result for slow equilibrium (reversible) solidification contrasts with our earlier finding in exercise (5) for spontaneous solidification, which always results in a negative free energy change.

The key difference demonstrated here between *reversible* and *irreversible* solidification processes is that in the latter case at least some free energy would be consumed by unavoidable kinetic processes attending the rapid solidification reaction. Non-equilibrium processes in rapid solidification are discussed in later chapters, and include transport of latent heat, diffusion of solutes (in the case of alloys), viscous flows in the melt, solid–liquid interface growth and motion, and, finally, formation of non-equilibrium crystalline defects generated within the solid phase as a crystal grows rapidly. These defects include trapped excess lattice vacancies and solutes, reduced crystallographic order, dislocation networks, mosaic structures, and grain boundaries; any or all may result when a supercooled melt nucleates spontaneously and crystals grow rapidly.

2.8 Summary

This chapter introduces the notions of heat and work associated with solid–liquid transformations at constant pressure. The first law of thermodynamics is used to discuss the magnitudes of these contributions to the enthalpy, or heat content function, showing that conventional melting and freezing processes carried out near ambient pressures have negligible pressure-volume terms compared to their internal energy changes as regarding the magnitude of the latent heat of transformation. The internal energy change itself reflects primarily the entropy change associated with the loss, or gain, of long-range order upon melting, or freezing, respectively. Major exceptions to this conclusion occur in cases of melting and freezing at kilobar pressure levels, which are indeed common in geological solidification of mantle rocks, and in some industrial high-pressure liquid metal sintering processes.

The coexistence conditions for crystals and their conjugate melt phases are developed using the so-called combined first and second laws of thermodynamics. The Gibbs criterion for heterophase equilibrium is introduced, requiring that mechanical, thermal, and chemical equilibria all occur simultaneously. The Clausius–Clapeyron equation is then derived to specify the allowed combinations of temperature and pressure that equate the chemical potentials of a crystal and its melt, and thus permitting solid–liquid co-existence to occur.

Next, the important concepts of phase metastability and thermal supercooling are introduced, using the simple example of a pure melt, sub-cooled below its thermodynamic freezing point, and then allowed to freeze adiabatically. These concepts occur multiple times in various contexts as subsequent chapters cover kinetic topics where supercooling and metastability arise as factors. Associated with these concepts are the foundational ideas of reversible and irreversible processes. The Clausius inequality is used to explain the precise thermodynamic criterion for distinguishing between reversible and irreversible processes. To make those abstractions more concrete, detailed calculations are used as examples of the entropy production and free energy dissipation that are encountered during irreversible freezing.

Finally, binary alloy solidification is introduced by discussing a few generic types of solidification alloys: (1) unary, or one-component freezing; (2) binary solid-solution alloys; (3) solid-solution plus eutectic microconstitutents; and (4) polyphase solidification, such as eutectics and peritectics. Some illustrative thermochemical calculations are also provided covering the simpler case of freezing pure aluminum both adiabatically and isothermally, where the cyclic heat exchange is explored as a measure of irreversibility, and Gibbs potential changes are calculated to distinguish reversible and irreversible solidification clearly.

References

1. W.J. Moore, *Physical Chemistry*, 2nd Ed., Prentice-Hall, Inc., Englewood Cliffs, NJ, 1955.
2. H.B. Callen, *Thermodynamics and an Introduction to Thermostatics*, 2nd Ed., Wiley, New York, NY, 1985.
3. R.T. DeHoff, *Thermodynamics in Materials Science*, Chap. 3, McGraw-Hill, Inc. New York, NY, 1993.
4. M. Stier and M. Rettenmayr, *J. Cryst. Growth*, **311** (2008) 137.
5. P.W. Bridgman, *Physics of High Pressure*, Bell, London, 1949.
6. 'Undercooled metallic melts: properties, solidification and metastable phases', *Proceedings of the NATO Advanced Research Workshop*, D.M. Herlach, I. Egry, P. Baeri and F. Spaepen, Eds., II Ciocco, Italy, 1993.
7. J.T. Fourkas et al., *Supercooled Liquids*, ACS Symposium Series 676, American Chemical Society, Washington, DC, 1997.
8. *Schmelze, Erstarrung und Grenzflächen*, I. Egry, P.R. Sahm and T. Volkmann, Eds., Vieweg Verlag, Braunschweig, 1999.
9. L. Backerud, C. Guocai and J. Tamminen, *Solidification Characteristics of Aluminium Alloys*, **2**, American Foundry Society, Des Plaines, IL, 1990.
10. *Solidification Science and Processing*, I. Ohnaka and D.M. Stefanescu Eds., TMS, Warrendale, PA, 1996.
11. *Crystal Growth*, C.H.L. Goodman, Ed., **1**, Plenum Press, New York, NY, 1978.
12. *Crystal Growth*, C.H.L. Goodman, Ed., **2**, Plenum Press, New York, NY, 1978.
13. Y. Yamada, *J. Mater. Res.*, **10** (1995) 1601.
14. N.B. Singh et al., *J. Cryst. Growth*, **250** (2003) 107.
15. W. Kurz and D.J. Fisher, *Fundamentals of Solidification*, Chap. 1, 3rd Ed., Trans Tech Publications, Aedermannsdorf, 1989.

Chapter 3
Thermal Concepts in Solidification

3.1 Near-Equilibrium Freezing

Casting, welding, and crystal growth processes all depend sensitively on the rates of heat extraction used during solidification. There are several levels at which one may examine these processes, but starting with the simplest geometry and description of heat flows to and from the solid-melt interface from the solidification environment draws attention to basic relationships between those rates of heat transfer and corresponding process response.

3.1.1 Preliminary Definitions

Begin by considering a small quantity of molten mass that is undergoing a transition from liquid to solid phase. See Fig. 3.1 defining the system under consideration (solid plus liquid regions) and the surroundings, at the system's boundary, S.

The mass of the thermodynamic system is considered to be small, as one may assume at this point that the liquid, ℓ, and solid, s, phases remain in thermal equilibrium at all times. Several additional simplifying assumptions are also required for developing a rudimentary thermal analysis of near-equilibrium solidification:

1. The pressure, P, remains constant throughout the system.
2. The molar, or gram-atomic, specific heats of the two phases are constant, and are assumed to be equal to each other, so

$$c_p^\ell = c_p^s = c_p \equiv \frac{\partial H}{\partial T} \qquad (3.1)$$

3. In addition, the molar volumes, Ω_ℓ and Ω_s, and densities, ρ_ℓ and ρ_s, of the phases are also assumed to be constant and equal, so

$$\Omega_\ell = \Omega_s = \Omega, \qquad (3.2)$$

and

$$\rho_\ell = \rho_s = \rho. \qquad (3.3)$$

M.E. Glicksman, *Principles of Solidification*, DOI 10.1007/978-1-4419-7344-3_3,
© Springer Science+Business Media, LLC 2011

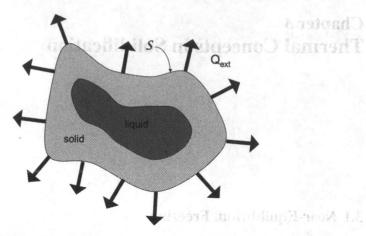

Fig. 3.1 Small region of solid and liquid undergoing near-equilibrium, or 'quasi-static' freezing. The solid–liquid interface is shown as irregular, and the external boundary, S, is subject to the outward heat flow, Q_{ext}, to its surroundings

4. The temperature within each phase remains constant and equal, and so thermal gradients within the phase regions may be neglected. Thus,

$$T_\ell = T_s = T. \tag{3.4}$$

5. The chemical compositions of the solid and liquid phases, C_s, and C_ℓ, are uniform, but *not* equal.
6. Consistent with the first and fourth assumptions for equilibrium solidification the molar enthalpy of the melt, H_ℓ, and that of the solid, H_s, are everywhere constant within those phases. Thus, their difference, which defines the latent heat of fusion, $\Delta H_f \equiv H_\ell - H_s$, is also a constant.

Now consider slowly traveling along an enthalpy 'path' represented on the molar enthalpy versus temperature diagram, Fig. 3.2. The solidification path of interest begins at a relatively high temperature, T_ℓ, in the all-liquid state, and ends at a lower temperature, T_s, in the all-solid state. These starting and ending temperatures could be thought of as designating, respectively, the 'liquidus' and 'solidus' temperatures, defined on the system's phase diagram, but that assumption is not necesssary at this point in our discussion. See again the generic binary phase diagram Fig. 2.4.

The dashed metastable extensions added to Fig. 3.2 suggest, respectively, the tendency of melts to supercool appreciably prior to freezing, and the corresponding possibility of achieving some superheating of the solid prior to the onset of melting.[1]

---

[1] The ease of supercooling melts, and the difficulty of superheating crystals might, at first glance, seem like a paradoxical asymmetry of nature. Indeed the acquisition of metastable states for both phases involves Gibbs potential changes that are nearly identical for every degree of supercooling or superheating achieved. Melts easily supercool several 100 K, whereas their crystals seldom

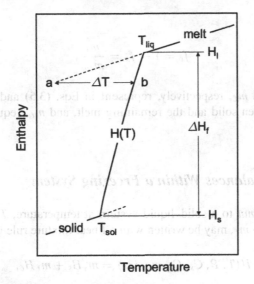

Fig. 3.2 Molar enthalpies of the solid, H_s, up to the solidus, T_{sol}, and for the melt, H_ℓ, above the liquidus, T_{liq}. Mixed phases occur between those limits, with the curve $H(T)$, the average enthalpy, plotted as a function of the temperature. Slow, quasi-static melting and freezing occur between T_{liq} and T_{sol} along the steeply decreasing enthalpy function for the mixed state. The liquid phase can be supercooled along the *dashed* metastable extension below T_{liq}, whereas the crystal would be superheated along the *dashed* metastable extension above T_{sol}. Adiabatic solidification of supercooled melt occurs along the horizontal (isoenthalpic) path a to b. Superheating of the crystalline solid above T_s, is seldom ever observed

The curve $H(T)$ connecting the solid-phase enthalpy line with the melt-phase enthalpy line represents the 'mixed-state' enthalpy encountered along the solidification path. The mixed state comprises distinct regions of solid and melt of constantly changing proportions as the temperature changes and the pressure remains fixed.

At each instant of time, the fractions of solid and liquid, f_s and f_ℓ, respectively, are given by

$$f_s = \frac{m_s}{m_s + m_\ell} = \frac{m_s}{m_{tot}}, \tag{3.5}$$

---

exhibit superheating of more than a small fraction of 1 K. The origin of this marked asymmetry lies in part with their relative difficulty of nucleating the conjugate phase. Melts require relatively large supercoolings to overcome the kinetic barrier for nucleation of the solid phase, whereas crystals contain a variety of free and internal surfaces and dislocation line defects that effectively reduce the superheating needed to nucleate their liquid phase. Other important factors contributing to this surprising asymmetry are discussed by Rettenmayr [1] in his review on this subject: they include the huge differences in the self-diffusion rates in crystals and their melts at the same temperature, and non-reciprocal influences of the solid–liquid interface on the ease of allowing the melting/crystallization transitions.

and

$$f_\ell = 1 - f_s = \frac{m_\ell}{m_{tot}}. \tag{3.6}$$

The terms m_s and m_ℓ, respectively, represent in Eqs. (3.5) and (3.6) the molar masses of the frozen solid and the remaining melt, and m_{tot} equals the system's total mass.

3.1.2 Energy Balances Within a Freezing System

The enthalpy *internal* to a solid–liquid system at temperature, T, pressure P, of mass, $m_{tot} = m_s + m_\ell$, may be written with a linear mixture rule as

$$H(T, P, C_s, C_\ell, m_s, m_\ell) = m_s H_s + m_\ell H_\ell, \tag{3.7}$$

The molar enthalpy of the isothermal, isobaric mixed-phase system with phase compositions, C_i ($i = s, \ell$), becomes

$$H_{int} = \frac{H(T, P, C_i, m_i)}{m_{tot}} = f_s H_s(T, P, C_s) + f_\ell H_\ell(T, P, C_\ell). \tag{3.8}$$

Unless otherwise specified, we now suppress, for the sake of brevity, the explicit, but cumbersome, notation used in Eq. (3.8), indicating the implied functional dependences of the system's enthalpy functions on temperature, phase composition, and pressure. Thus, using Eq. (3.6), regrouping the terms in Eq. (3.8), and inserting the definition of the molar heat of fusion, ΔH_f, yields

$$H_{int} = H_\ell - f_s \Delta H_f. \tag{3.9}$$

Differentiate Eq. (3.9) with respect to time to obtain

$$\dot{H}_{int} = \dot{H}_\ell - \dot{f}_s \Delta H_f, \tag{3.10}$$

where the over-dot notation implies total differentiation with respect to time. Applying the chain differentiation rule to the first term on the RHS of Eq. (3.10) yields the expected linear form

$$\dot{H}_\ell = \frac{\partial H_\ell}{\partial T} \frac{dT}{dt} = c_p \dot{T}. \tag{3.11}$$

Substituting Eq. (3.11) into Eq. (3.10) and solving for the time-rate of temperature change, $\dot{T}$, yields

$$\dot{T} = \frac{\dot{H}_{int}}{c_p} + \frac{\Delta H_f}{c_p} \dot{f}_s. \tag{3.12}$$

Equation (3.12) shows that the time-rate of change of the system's temperature (or equivalently the slope of the cooling or heating curve) generally consists of two terms: (1) the rate of change of the system's molar enthalpy, $\dot{H}_{int}$, and (2) the rate of change of the mass fraction solidified, $\dot{f}_s$. The first of these arises from any enthalpy changes to which the phases are subjected as the temperature changes, whereas the second term arises from the changing amounts of solid and melt accompanying freezing or melting. Additional terms accounting for changes in pressure or composition could be easily added to Eq. (3.12). For the purposes of the present discussion on energy balances during solidification, however, we limit the enthalpy changes shown in Eq. (3.12) to those caused by changes in the system's temperature.

3.1.3 Energy Conservation

The time rate of enthalpy change caused by heat exchanged with the surroundings and entering the system through its *external* boundary, S, as shown in Fig. 3.2, may be written as the surface area integral of the normal component of the heat flux vector, $\dot{\mathbf{q}} \cdot \hat{\mathbf{n}}$, where $\hat{\mathbf{n}}$ denotes the local unit normal anywhere along the interface, namely

$$\dot{H}_{ext} = \oint_S \dot{\mathbf{q}} \cdot \hat{\mathbf{n}} dA. \tag{3.13}$$

The integrand of Eq. (3.13) represents the normal component of the heat flux vector to or from at any point on S. The convention used in Eq. (3.13) is that if the vector flux points away from the system, it is considered to be negative, whereas if the heat flux vector points inward it is positive. The integral of the flux may be replaced by its area-weighted average over the closed system, namely

$$\oint_S \dot{\mathbf{q}} \cdot \hat{\mathbf{n}} dA = \langle \dot{q}_n \rangle S. \tag{3.14}$$

Conservation of energy may be expressed for isobaric changes occurring at fixed phase compositions within and external to the system as the enthalpy balance

$$\dot{H}_{ext} = m_{tot} \dot{H}_{int}, \tag{3.15}$$

where $\dot{H}_{int}$ is the rate of enthalpy release within the body per unit mass. If Eq. (3.15) is solved for $\dot{H}_{int}$, which is then substituted back into Eq. (3.12), one obtains the expressions,

$$\dot{T} = \frac{\langle \dot{q}_n \rangle\, S}{m_{tot}\, c_p} + \frac{\Delta H_f\, \dot{f}_s}{c_p} = \left(\frac{\Omega}{c_p}\right)\left(\frac{S}{V}\right)\langle \dot{q}_n \rangle + \left(\frac{\Delta H_f}{c_p}\right)\dot{f}_s. \qquad (3.16)$$

The first equality in Eqs. (3.16) is a mass-referenced expression for the rate of temperature change, whereas the second equality is volume-referenced. The terms in the second equality are grouped to expose clearly all the important materials and system parameters affecting the temperature response:

1. Volumetric specific heat, c_p/Ω.
2. Surface-to-volume ratio, S/V.
3. Characteristic temperature, $T_{char} \equiv \Delta H_f/c_p$.

Note again that the rate of temperature change depends on two distinct physical processes: (i) the addition, or loss, of heat energy at the system's external boundary, (ii) the rate of phase transformation (melting or freezing) occurring within the system, with the latter expressed as the time rate of solid fraction change.

Now consider several important cases of solidification with differing types of energy exchanges with its surroundings to which Eqs. (3.16) may be applied.

3.1.3.1 Adiabatic Recalescence

An adiabatic condition precludes all heat flow into or out of the solid–liquid system from its surroundings, so that $\langle \dot{q}_n \rangle = 0$. The melt, being initially supercooled, is metastable relative to the solid. Once nucleated, metastability allows partial freezing to progress spontaneously, and thus $\dot{f}_s \geq 0$. See again the dashed extension of the liquid's enthalpy function in Fig. 3.2. Inserting these conditions, namely that $\langle \dot{q}_n \rangle = 0$ and $\dot{f}_s \geq 0$, into Eq. (3.16) shows that $\dot{T} = (\Delta H_f/c_p)\dot{f}_s$. A supercooled system, isolated from its surroundings, is initially at the enthalpy-temperature coordinate a indicated on Fig. 3.2. When induced to freeze adiabatically, the system spontaneously self-heats (recalesces) from its initially molten metastable state to a mixed state of equal enthalpy at the *isenthalpic* coordinate b. The system's temperature rises by an amount ΔT as the fraction solidified, $f_s(t)$, increases to its final mixed-phase state, $0 < f_s(\infty) \leq 1$, the final value for which depends both on the initial supercooling, ΔT, and the material's characteristic temperature.

3.1.3.2 Cooling by External Heat Transfer

A typical casting or weldment cools by conductive and radiation exchanges with its surroundings. Figure 3.2 shows the cooling path through enthalpy-temperature space as the system moves from $T(a)$ towards $T(b)$, so $\dot{T} \leq 0$. Consequently, the initially molten system gradually freezes and $f_s(t) \to 1$. This common solidification scenario requires that the average outward energy flow from the solidifying system is described by Eq. (3.16).

$$\langle \dot{q}_n \rangle = -\left(\frac{\Delta H_f}{\Omega}\right)\left(\frac{V}{S}\right)\dot{f}_s + \left(\frac{c_p}{\Omega}\right)\left(\frac{V}{S}\right)\dot{T} < 0. \qquad (3.17)$$

Under the conditions of external energy exchange specified here, both terms in Eq. (3.17) become negative. If one substitutes the chain-rule expression that $\dot{f_s} = (df_s/dT)\dot{T}$ into the cooling rate, Eq. (3.16) for the system, another interesting result obtains, which was first pointed out by Kurz and Fisher in their original monograph [2] for the case of quasi-static freezing,

$$\dot{T} = \frac{\left(\frac{\Omega}{c_p}\right)\left(\frac{S}{V}\right)\langle \dot{q_n} \rangle}{1 - \frac{\Delta H_f}{c_p}\frac{df_s}{dT}}. \tag{3.18}$$

The denominator of Eq. (3.18) contains the differential coefficient df_s/dT, which is negative for most materials; thus, the denominator is almost always positive. The numerator of Eq. (3.18) is negative for net heat transferred to the surroundings. Thus, in contrast to the earlier result found for the case of adiabatic freezing, where the temperature of a supercooled melt (already discussed in Section 2.3) *rises* with increased formation of solid, the internal temperature of a solidifying system usually *falls* as solidification advances through the process of external removal of thermal energy. A few unusual exceptions to these results occur in binary alloy systems with phase diagrams that exhibit 'retrograde melting', where $df_s/dT > 0$. Alloys exhibiting retrograde melting actually melt on cooling, solidify on heating, and exhibit negative values for ΔH_f! This peculiar behavior arises from strong non-idealities in the alloy's solution thermodynamics. Although interesting, we shall ignore the case of retrograde solidification behavior at this point in our discussion of quasi-static freezing.

3.1.3.3 Invariant Freezing

Invariant freezing and melting occur in cases such as the melting and freezing of pure materials, eutectics, monotectics, peritectics, etc.—all of which represent solidification reactions in alloy systems that can transform quasi-statically at *constant temperature*. Thus, the cooling and heating rates during invariant freezing or melting, respectively, are zero. The average rate of heat loss for invariant solidification, where $\dot{T} = 0$, as predicted from Eq. (3.17) is

$$\langle \dot{q_n} \rangle = -\left(\frac{\Delta H_f}{\Omega}\right)\left(\frac{V}{S}\right)\dot{f_s}. \tag{3.19}$$

Note Eq. (3.19) shows that in the special cases of invariant solidification and melting, where $\dot{T} = 0$, the rate of energy loss is proportional to the freezing rate, $\dot{f_s}$. Equation (3.19) also shows that for alloys subject to invariant melting and freezing, the temperature coefficient of solidification, $df_s/dT \to -\infty$, a result that is consistent with the observation that melting and freezing can progress despite the fact that $\dot{T} = 0$.

3.2 Energy Balances at Crystal-Melt Interfaces

In general, obtaining a finite heat flux during solid–liquid phase change, $\langle \dot{q}_n \rangle$, requires that there are temperatures gradients in one or both of the phases adjacent to the interface where phase change is occurring. It is easiest, and perhaps most instructive, to consider several one-dimensional situations that deal with energy balances at moving solidification fronts. Most of the subsequent analytical development can be focussed on one-dimensional problems. Specifically, one-dimensional problems, such as depicted in Fig. 3.3, are advantageous as they permit simplification of the mathematical description of the location and geometrical form of the crystal-melt interface. In addition, the relevant boundary conditions are imposed along easily visualized planes, which also simplify the required mathematical description without sacrificing any of the essential physics.

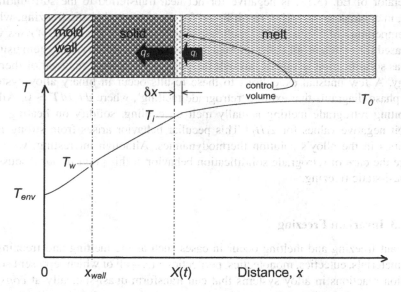

Fig. 3.3 Temperature distributions in a one-dimensional bar mold undergoing solidification. The internal solid-mold interface at x_{wall} and the *moving* crystal-melt interface, $X(t)$, require specification of boundary conditions establishing their behavior. Data are also needed to specify the initial melt temperature, T_0, and the exterior mold wall temperature, T_{env}, set by the environment. The temperatures T_w and T_i satisfy continuity of the interior thermal fields at the mold-solid and solid-melt interfaces, respectively. The *black arrows* suggest the direction and magnitudes of the heat fluxes, $\dot{q}_\ell$ in the melt, and $\dot{q}_s$ in the solid. The motion of the solid-melt interface toward $+x$ generates latent heat that increases the magnitude of $\dot{q}_s$ leaving the control volume

3.2.1 Stefan's Interface Energy Balance

A *local* interfacial energy balance may be developed on a small mass of liquid, dm_ℓ of unit cross-sectional area and thickness dx that changes phase during a brief increment of freezing time, dt. This energy balance was first formulated late in the nineteenth century for heat transfer/phase-change problems by the Slovene physicist Jozef Stefan. Dynamical phase-change problems are still called 'Stefan problems', and the local energy balance at a moving phase boundary is referred to as the 'Stefan condition'. This energetic balance involves the 'flow' of thermal energy per unit area towards or away from a moving interface, accounting for any latent heat released or absorbed at the interface. Specifically, with reference to Fig. 3.3, over a small time increment dt, the Stefan condition requires that the heat fluxes adjacent to the solid–liquid interface, $\dot{q}_\ell$, in the melt, and $\dot{q}_s$ in the solid, plus any latent heat released by the *displacement* of the interface, dx, all sum to zero. The Stefan condition, in fact, is a statement of *local* energy conservation. It recognizes that an interface is capable of releasing or absorbing energy, but not storing it. The Stefan energy balance may be written as,

$$(\dot{q}_s - \dot{q}_\ell)\,dt + \frac{\Delta H_f}{\Omega}\,dx = 0, \qquad (3.20)$$

where $\dot{q}_s$ and $-\dot{q}_\ell$ represent, respectively, the heat flux leaving the left-hand face of a small control volume surrounding the interface and the heat flux entering the right-hand face of the control volume. The second term on the LHS of Eq. (3.20) represents the latent heat produced from the motion of the interface. This term consists of the enthalpy of freezing per unit volume (ΔH_f divided by the system's molar volume, Ω) multiplied by the small volume of material solidified over the time increment, viz., $dx \times$ unit area.

Equation (3.20) may be rewritten in differential rate form using the interface tracking function $X(t)$ as,

$$\dot{q}_s - \dot{q}_\ell = -\rho\,\Delta H_f \frac{d}{dt}X(t), \qquad (3.21)$$

where $\rho = \Omega^{-1}$, the mass density, is substituted for the reciprocal of the gram-atomic or molar volume. The effect of the small density change upon $\ell \to s$ transformation, already discussed in detail in Section 1.3, can easily be incorporated into the Stefan balance; however, for the present discussion these density differences will be ignored, as the results are essentially unaffected by the small mass density changes that accompany melting and solidification.

It now remains to express the differential rate expression in Eq. (3.21) in terms of the temperature gradients that extend throughout the solid-melt system. Gradients in the temperature directly reflect any spatial variations in the temperature fields, $T_s(x, t)$, for the solid, or, $T_\ell(x, t)$, for the melt. Discussion of these temperature fields and their associated gradients and heat fluxes must be set aside momentarily until the general laws of heat conduction are delineated.

3.2.2 Fourier's Law of Heat Conduction

Joseph Fourier propounded the laws of heat conduction and developed the first analytic theory of heat in continuous bodies, his first discourses on these subjects given to the French Academy of Sciences being published in 1807-1808 [3]. His laws are based both on his quantitative measurements of conductive heat transfer through a solid material, and on his prodigious mathematical skills. The experimental observations were based on an iron plate steadily transferring heat through a uniform thickness, d, measured between its two broad surfaces, which were maintained at different uniform temperatures, T_h and T_c, where $T_h > T_c$, as is sketched in Fig. 3.4.

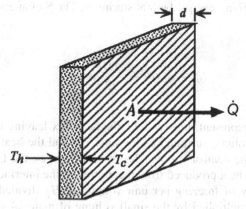

Fig. 3.4 The heat conducted per unit time, $\dot{Q}$, through a flat plate of cross-sectional area, A, and thickness d, subject to a temperature gradient, $(T_h - T_c)/d$. Joseph Fourier used such an arrangement to formulate his laws of conduction heat flow

Fourier noted that the rates of heat transfer per unit area through the plate, $\dot{Q}$, often referred to figuratively as 'heat flow'—although nothing actually flows—were proportional to the temperature difference established across the plate, $T_h - T_c$, and were inversely proportional to its thickness, d. His observations may be expressed mathematically as the linear response function,

$$\dot{Q} = kA\frac{(T_h - T_c)}{d},$$ (3.22)

where k, a property specific to the plate's material, is the constant of proportionality, and bears the modern units of W/m-K. The more technically suitable term for this type of thermal energy transfer is 'heat conduction'. The constant of proportionality, k, that appears in Eq. (3.22) is the thermal conductivity. Fourier's experimental observation may be generalized to a small region in a material supporting an infinitesimal temperature difference in some direction, x. The orientation of the coordinate x within the material is chosen along its direction of maximum temperature change with distance:

$$\frac{\dot{Q}}{A} \equiv \dot{q}_n = -k\frac{dT}{d|\mathbf{r}|}, \qquad (3.23)$$

where $\dot{Q}/A$ is the heat flux, $\dot{q}_n$, at the point under consideration, and the coordinate $\mathbf{r}$, chosen as positive in the direction of maximum energy transport and decreasing temperature, requires a minus sign to appears with the corresponding temperature gradient. It should be apparent that, in general, the heat flux, or heat transfer rate per unit area, is a vector quantity, having both magnitude and direction.

In Cartesian (x, y, z) coordinates, Fourier's law may be written compactly in any dimensionality using standard vector notation,

$$\dot{\mathbf{q}} = -k\nabla\mathbf{T} = -k\left(\frac{\partial T}{\partial x}\mathbf{e}_x + \frac{\partial T}{\partial y}\mathbf{e}_y + \frac{\partial T}{\partial z}\mathbf{e}_z\right), \qquad (3.24)$$

where $\mathbf{e}_x$, $\mathbf{e}_y$, and $\mathbf{e}_z$ are Cartesian unit vectors. The heat flux can be written for *anisotropic* materials subjected to a thermal gradient, $\nabla\mathbf{T}$, as the tensor contraction

$$\dot{\mathbf{q}} = -[\mathbf{k_{ij}}] \cdot \nabla\mathbf{T}, \qquad (3.25)$$

where $[\mathbf{k_{ij}}]$ is the material's thermal conductivity matrix.

In solidification and melting problems, in particular, the tensorial aspects of the thermal conductivity are usually unimportant, excepting perhaps in the analysis of thermally-induced strains in solidified anisotropic (non-cubic) crystalline bodies and complex composite castings. By contrast, the inclusion of the temperature dependence of the thermal conductivity, $k(T)$, is important when performing accurate heat transfer calculations for alloys that solidify over wide temperature ranges, perhaps encompassing as much as several hundred K. The major point here, however, is that the conduction heat flux response, $\dot{q}$, to a temperature gradient is controlled entirely by the material's thermal conductivity.

If a temperature gradient is imposed on different materials a wide range of fluxes can develop. For some common materials as disparate in their thermal conductivities as are 'rock wool' fiber insulation and metallic copper, the flux responses can vary by factors as great as 10^4. In fact, a few recently-developed solid materials exhibit even greater extremes in their thermal conductivities. For example, diamond films grown from isotopically pure C^{12} display hundreds of times the conductivity of metallic copper, whereas silica aerogels can exhibit thermal conductivities as low as one-tenth that of the best ordinary solid insulators. Thus, thermal flux responses in solids can vary by factors greater than one million.

The preferred units for the thermal conductivity under System International (SI) are [W/m $-$ K], or [W/m^2 $-$ K/m]. Other conductivity units appear in the heat transfer literature and in engineering handbooks, but preferred engineering practice is to adhere to System International (SI) units.

3.2.3 Thermal Conduction

Thermal conduction in solids and fluids is the result of quantum phenomena, where by thermal energy is transported by phonons (quantized atomic vibrations) and 'free' electrons (especially in metals). The contributions from these transport mechanisms are lumped into a phenomenological parameter termed the thermal conductivity, k, which connects the heat flux to the thermal gradient, as shown in Eq. (3.25). Figure 3.5 demonstrates the 5 order-of-magnitude range of thermal conductivities encountered in various states of matter.

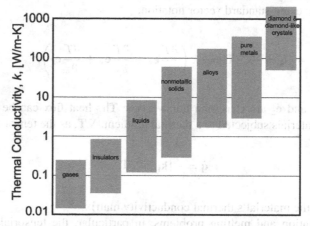

Fig. 3.5 The approximate range of thermal conductivities for different states of matter. Gases, amorphous insulators, such as glass and liquids, have low conductivities, whereas metals generally have high conductivities. A few covalent crystals, such as natural diamond, C^{12}, and SiC, exhibit unusually high thermal conductivities as non-metals. The heat conduction mechanisms vary from molecular diffusion for gases and liquids, free electron transport in the case of metals, to phonon processes in insulator crystals and polymers

Thermal conductivity remains fundamentally as the constant of proportionality in Fourier's law of heat conduction, Eq. (3.24). Conductivity, at the most basic level, reflects fundamental quantum processes that occur at molecular scales in solids. However, one need not be concerned about the detailed atomic or molecular events responsible for the diffusion and transport of thermal energy. Atomic-scale mechanisms of thermal conduction are now well explained using basic theories of solid-state physics and statistical mechanics. It is well understood, for example, that the excellent conductivity of metals is caused by their high density of nearly 'free' electrons, which efficiently carry thermal energy past the lattice-bound nuclei to distant points. The collective vibrational modes of the ionic nuclei comprising the lattice of metallic crystals—described by physicists as a phonon spectrum—actually contribute only a small component of a metal's thermal conductivity. By contrast, most polymers, and many non-metallic solids, such as silicate glasses, contain virtually no free electrons, and depend instead on molecular vibrations (phonon motion) for their thermal conduction. Finally, fluids—both gases and liquids—exhibit more complex intermolecular interactions, such as translation and other collective molecular modes to transfer energy from hot regions to cooler regions. Fluids, in addition, including both gases and condensed molten phases encountered in solidification and crystal growth, are subject to buoyancy-induced (gravity-mediated) and forced (mechanically mediated) convection. Convection provides an independent mechanism for long-range energy transport based on macroscopic fluid dynamical motions occurring on length scales that are much larger compared to intermolecular distances and phonon wavelengths relevant to solid-state conduction.

3.2.4 Newton's Law of Cooling: Heat Transfer Coefficients

Internal heat conduction, as described in the previous two sections, is not the sole mechanism needed for energy transport during solidification: radiation and environmental convection provide two other essential mechanisms that link the energy transfered from solidifying bodies to their surroundings.

Newton's law of cooling expresses phenomenologically a linearization simplifying extremely complex, coupled heat transfer phenomena associated with radiation and convection into the environment adjacent to a heated solid surface, such as a casting. The heat transfer situation shown in Fig. 3.6 portrays a solidifying system in contact with a convecting fluid environment. The fluid develops a thermal boundary layer, across which the temperature drops from the value T_w toward the environmental temperature, T_∞. The redistribution of the thermal energy convected

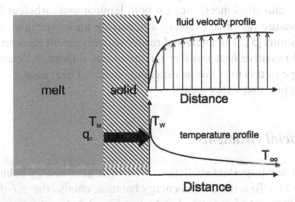

Fig. 3.6 Heat released at the solid–liquid interface at temperature T_m flows through the freezing solid region toward the exterior boundary at temperature T_w. The environment far from the solid/fluid boundary is fixed at temperature T_∞. A typical fluid velocity profile is shown above the temperature sketch, with the hydrodynamic 'no-slip' condition, $V = 0$, imposed on the fluid contacting the exterior boundary. The component of the fluid velocity parallel to this boundary rises as a sublinear power of the distance away from the solid boundary. The fluid velocity increases approximately parabolically adjacent to the solid surface. The precise fluid velocity profile depends on many details, including the solid surface roughness, and whether or not the fluid flow remains laminar or becomes turbulent

and or radiated into such an ambient environment may be written as the following linearized response:

$$\dot{q}_n = h(T_w - T_\infty),\tag{3.26}$$

where the heat transfer coefficient, h, used in Eq. (3.26), unlike the thermal conductivity, k, already discussed in Sections 3.2.2 and 3.2.3, is *not* an unique property of the material, and even bears different units [W/m^2-K]. In fact, Eq. (3.26) really constitutes the definition of h, which embodies all of the following factors that influence the complicated processes of transferring thermal energy across a material boundary into an adjacent fluid dynamical environment:

1. Thermophysical properties of the fluid environment.
2. Surface roughness and geometry.
3. Fluid flow type.
4. Spectral characteristics of the radiation surface.
5. Radiative heat transfer (linearized).
6. Total temperature differences.

The concept of a heat transfer coefficient, h, or 'heat conductance' provides a useful engineering approach to characterize such difficult-to-describe thermal interaction processes, including radiative and convective heat transfer across air gaps at mold walls, or the heat loss from a solidifying continuous casting subject to radiation and external spray cooling. Heat transfer coefficients describing such processes are, at their core, *empirical*, although engineers have learned how to correlate these coefficients with a number of important process and control variables using the physics of heat transfer, plus fluid mechanics for both laminar and turbulent flows [4], and for high-temperature radiative exchanges occurring under specular, blackbody, or graybody conditions [5]. The interested reader should consult these monographs on convective and radiative heat transfer for numerous important details that extend beyond the scope of this text. For an excellent survey of heat transfer topics applied to solidification processes also see [6].

3.2.5 Interfacial Gradients

Now consider a few important subtleties of the interfacial energy balance, or Stefan balance, Eq. (3.21). To reiterate this energy balance, choose the $+x$-direction from left to right in the control volume sketched in Fig. 3.3. For the temperature fields discussed in Section 3.2.1, the interfacial heat flux, $\dot{q}_s$, represents a negative energy contribution to the control volume, whereas $\dot{q}_\ell$, the heat conducted from the hotter melt, provides a positive energy contribution. The interfacial speed, $d\mathrm{X}/dt$, is positive for motion directed toward $+x$ as shown during solidification. Fourier's law, Eq. (3.24), may now be inserted in the LHS of Eq. (3.21). Retaining consistency of the algebraic signs for each energy contribution to the control volume yields an expression for the gradients acting on each side of the moving solid–liquid interface:

$$k_\ell \left(\frac{dT_\ell}{dx}\right)_{x=X^-} - k_s \left(\frac{dT_s}{dx}\right)_{x=X^+} + \rho \Delta H_f \frac{dX}{dt} = 0. \qquad (3.27)$$

The cumbersome subscript notation $x = X^-$ and $x = X^+$ used in Eq. (3.27) just serves to remind the reader that the thermal fields—specifically their gradients— adjoining a moving interface are locally *discontinuous*. Moreover, these gradients are 'one-sided' and have 'jumps' in their values as the interface, located at the plane $X(t)$, is crossed. The solid-side of the interface in Fig. 3.3 is denoted X^-, whereas the liquid-side of the interface is denoted X^+. The very concept of distinguishable 'sides' to a solid–liquid interface represented by a moving mathematical plane, $X(t)$, seems artificial at first, indeed even strange, but a number of important mathematical and numerical models[2] of solid–liquid interfaces employ this idea effectively. The reader ought to recognize, even at this early stage in the development of the subject, that a 'sharp' interface description, i.e., one limited to a mathematical plane of zero thickness is absurd, considering that atoms and molecules have finite sizes. Thus, one should accept the notion that solid–liquid interfaces possess some thickness, albeit just a few molecular diameters, but further elaboration of this particular point will be deferred.

The determination of the interface tracking function $X(t)$ requires finding the explicit forms for the temperature fields $T_s(x, t)$ and $T_\ell(x, t)$. Once these fields are known, the two thermal gradients can be evaluated, and then the interfacial equation for $X(t)$ may (at least in principle) be solved. Unfortunately, this mathematical procedure can be difficult, and exact (analytical) solutions for $T(x, t)$, and hence for the interface motion, $X(t)$, are available only for a few geometrically simple cases. This limitation holds even for one-dimensional, invariant freezing, which is the case depicted in Fig. 3.3. Numerical models of impressive sophistication are now available to solve solidification problems accurately in complex geometries, as is often encountered in shape castings. These computer codes are available to foundries and die casting shops as proprietary commercial software packages.

In succeeding Chapters some simplifying assumptions about the internal release of latent heat and the motions of the solid–liquid interface will be explored that yield approximate but useful heat transfer results. More importantly, these approximations are chosen to illustrate the combinations of individual thermo-physical parameters that influence engineering Stefan problems, as well as to provide deeper insights into more complex aspects of solidification, such as including solute diffusion in the case of freezing alloys, and multiphase interactions in the case of eutectic systems.

---

[2] Several basic mathematical descriptions of heat and solute release at moving interfaces, such as Greens functions and Heaviside graphs, use the notion of discontinuous gradients acting 'across' an interface of zero thickness. These methods are also incorporated into modern computer algorithms used to simulate interfacial gradients within complex casting and crystal growth problems.

3.3 Summary

Thermal phenomena, heat conduction and radiation transport are of major importance in most melting and freezing processes, due to the required addition, or release, of both sensible and latent heat for a system to achieve melting or freezing, respectively. The enthalpy functions for crystals and their melts are discussed to clarify both the equilibrium co-existence of these phases, and, more importantly, their tendency, from an energetic standpoint, to become metastable and transform spontaneously. Once formed, a solid-melt interface can change phase at rates determined through the conditions imposed by heat transfer. Heat conduction and the temperature fields that develop within a freezing system satisfy both Fourier's law, and energy conservation. Rates of freezing are both material- and system-dependent, as they are influenced by factors such as the thermal conductivity—a material-specific property—as well as by the rates of external heat transfer from the solidifying system to its environment. Heat transfer coefficients, which account for combined energy transport processes, including radiative, evaporative, and convective heat transfer from a hot casting surface, linearize and 'lump' all essential energy transport needed to achieve overall control of complex solidification processes.

References

1. M. Rettenmayr, *Int. Mater. Rev.*, **54** (2009) 1.
2. W. Kurz and D.J. Fisher, *Fundamentals of Solidification*, Hemisphere Press, Aedermannsdorf, 1980.
3. J. Fourier, *Bull. des Sci.*, (1808) 112.
4. A. Bejan, *Convection Heat Transfer*, Wiley, New York, NY (1984).
5. D.P. DeWitt and G.D. Nutter, *Theory and Practice of Radiation Thermometry*, Wiley, New York, NY (1988).
6. D.R. Poirier and E.J. Poirier, *Heat Transfer Fundamentals for Metal Casting*, 2nd Ed., The Minerals Metals & Materials Society, Warrendale, PA, 1994, 9.

Chapter 4
Solidification of Pure Materials

4.1 Quasi-static Theory

4.1.1 Introduction

With the preliminaries of the structure, thermodynamics, and energy balances encountered in crystal-melt systems and their transformations discussed in earlier chapters, one may now consider additional details of the progressive freezing of a pure melt phase with some initial thickness L, for a slab casting, or with a radius, R_0, for solidification of a circular right cylinder. These representative casting geometries serve to elucidate some important findings from heat transfer theory and help describe interfacial motion that quantify the kinetic behavior of freezing events occurring on macroscopic scales.

Prior to the start of solidification at $t = 0$, a pure (one-component) melt was poured into an insulating mold at a uniform temperature just slightly above the material's melting temperature, T_m. A sketch of this configuration for a slab casting is shown in Fig. 4.1. Convective cooling of the solidifying system is allowed to take place at the external heat transfer surface, $x = 0$, separating the solidifying system from its ambient environment, the temperature of which at large distances to the left, $x \ll 0$, is T_∞. The cooling to the environment is characterized by a heat transfer coefficient, h, the concept for which was developed earlier in Section 3.2.4.

The temperature distribution within the solidified layer at any future instant of time is assumed to be linear. That is, the temperature field acts as if the solidifying layer were *static*, and was 'passively' conducting the latent heat from the solid–liquid interface, at $x = \hat{x}$, to the heat transfer surface to the environment, at $x = 0$. Stated somewhat differently, at any given time, $t \geq 0$, the solidified portion of this slab casting has sufficient time to develop its steady-state temperature field established between the solid–liquid interface, at temperature T_m, and the heat transfer surface, which is maintained at temperature T_w. Later in this analysis it will be shown that this assumed quasi-static temperature distribution provides a reasonable approximation of the true temperature field, provided that the product of the solid's specific heat, c_p, and its total temperature difference, $T_m - T_w$, is small compared

M.E. Glicksman, *Principles of Solidification*, DOI 10.1007/978-1-4419-7344-3_4,
© Springer Science+Business Media, LLC 2011

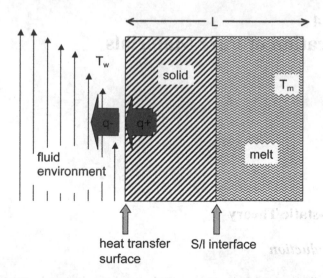

Fig. 4.1 Configuration of a partially frozen 'slab' casting of thickness L. The instantaneous position of the solid–liquid interface, always at its equilibrium melting temperature, T_m, denotes the extent of solidification after a time, t. The heat transfer surface maintains a steady temperature, T_w, and physically separates the casting from its environment as suggested on the sketch. Heat fluxes leaving the casting are shown as entering the environment at a temperature T_w, and then convected away in a fluid moving parallel to the solid boundary toward some far-field temperature, T_∞, not shown in this figure

to the latent heat of solidification, and that radiative heat transfer remains negligible compared to heat transfer via conduction. The detailed temperature distributions in the three zones of the slab casting—the melt, the solid, and the environment—are each shown schematically in Fig. 4.2.

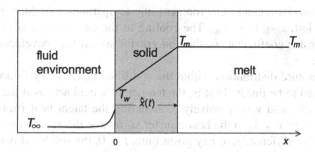

Fig. 4.2 Temperature distributions within the solidifying slab casting of a pure material of total thickness L. See sketch in Fig. 4.1. The coordinate system fixes the left boundary between the fluid environment and the solid at $x = 0$, and locates the extent of solidification that has occurred at time t by the position of the solid-melt interface at $x = \hat{x}(t) \leq L$. The temperature distribution through the melt remains uniform at $T(x, t) = T_m$, which is a constant, whereas the temperature falls in a linearly manner through the growing solid from T_m to T_w. The temperature then drops precipitously as heat enters the flowing fluid that represents the environment

4.1.2 Analysis of 1-Dimensional Freezing

Consider the highly simplified example of freezing in one spatial dimension, x, of a pure melt at its melting point. The quasi-static temperature distribution across the thickening slab of solid, $T_s(x, t)$, is assumed to develop as a fully relaxed linear field, providing that certain conditions, to be discussed shortly, are satisfied.

$$T_s(x, t) = T_w + \frac{(T_m - T_w)}{\hat{x}(t)} x, \ (x \geq 0) \tag{4.1}$$

for which the thermal gradient anywhere within the solid at time t is given by the linear form,

$$\frac{\partial T_s}{\partial x} = \frac{(T_m - T_w)}{\hat{x}(t)}. \tag{4.2}$$

The heat flux, $\dot{q}^{(+)}$, arriving at $x = 0$, the heat transfer surface, is given by Fourier's law of heat flow, Eq. (3.23), in accord with the quasi-static temperature gradient specified in Eq. (4.2). The heat flux reaching the environment from the freezing solid is

$$\dot{q}^+ = -k_s \frac{\partial T_s}{\partial x} = -k_s \frac{(T_m - T_w)}{\hat{x}(t)}. \tag{4.3}$$

The heat flux, $\dot{q}^-$, leaving the slab casting at $x = 0$, via heat transfer to the environment, is found using Newton's law of cooling, Eq. (3.26),

$$\dot{q}^- = -h(T_w - T_\infty). \tag{4.4}$$

It is required that $\dot{q}^+ = \dot{q}^- = \dot{q}_{wall}$ to assure a conserved, smoothly changing transfer of energy from the melt, through the solidified slab, and into the environment. If the as yet unknown wall temperature, T_w, is eliminated from Eqs. (4.3) and (4.4), the quasi-static heat flux may be found after a few steps of algebra as

$$\dot{q}_{wall} = \frac{T_m - T_\infty}{\frac{1}{h} + \frac{\hat{x}(t)}{k_s}}. \tag{4.5}$$

An analysis of Eq. (4.5) shows that the heat flux at the wall, $x = 0$, has two distinguished limits:

1. *Thin solid layer; highly conducting material; $\hat{x}(t)/k_s << 1/h$*: The wall heat flux formula, Eq. (4.5), reduces to $\dot{q}_{wall} = h(T_m - T_\infty)$. This indicates that when $\hat{x}(t)$ is small, after short freezing times, the solidification rate, as measured by the rate of latent heat extracted by the environment, is determined solely by the heat transfer coefficient, h. The thickness of the solidified layer, at this stage in the process, has little influence on the rate of freezing. That is, freezing rates

are initially limited by the system's *external* rate of heat transfer, and that rate is proportional to the difference in temperature between the melting point of the solidifying material and the environmental temperature, T_∞.

2. *Thick solid layer; poorly conducting material;* $\hat{x}(t)/k_s >> 1/h$: The heat flux formula now reduces to $\dot{q}_{wall} = k_s(T_m - T_\infty)/\hat{x}(t)$. This result shows that the rate of solidification is determined by the internal resistance to heat transfer through the frozen solid, as limited by the thermal gradient established within that solid. As the solid layer thickens further, $\hat{x}(t) \rightarrow L$, the gradient falls further, and the rate of freezing declines.

Next, if Eq. (4.5) is substituted into the Stefan interfacial energy balance, Eq. (3.2.1), one finds the relationship between the interface speed, $d\hat{x}/dt$, and the conducted heat flux,

$$\rho \Delta H_f \frac{d\hat{x}}{dt} = \frac{T_m - T_\infty}{\frac{1}{h} + \frac{\hat{x}(t)}{k_s}}. \tag{4.6}$$

Several algebraic steps must be performed on Eq. (4.6), including multiplying both sides of this equation by the ratio of the specific heat, c_p, to the thermal conductivity of the solid, k_s, and also multiplying through by the square of the slab's thickness, L^2. These steps when carried out yield a complicated looking expression, namely,

$$\frac{\rho c_p}{k_s} \left(\frac{L^2}{L}\right) \frac{d\hat{x}}{dt} = \frac{c_p}{\Delta H_f} \frac{(T_m - T_\infty)}{(\frac{k_s}{h} + \hat{x}(t))\frac{1}{L}}. \tag{4.7}$$

The following dimensionless numbers and thermal properties may be chosen. These groupings, which are commonly used in heat transfer theory, also prove extremely useful for subsequent applications and further discussion of Eq. (4.7):

1. Thermal diffusivity: $\alpha \equiv k_s c_p/\rho$.
2. Dimensionless position: $f_s \equiv \hat{x}(t)/L$ (fraction solid).
3. Dimensionless time: $Fo \equiv \alpha/L^2 t$ (Fourier number).
4. Dimensionless heat transfer coefficient: $Bi \equiv hL/k_s$ (Biot number).
5. Dimensionless temperature difference: $St \equiv (T_m - T_\infty)/(\Delta H_f/c_p)$ (Stefan number).
6. Alternative dimensionless time: $\tau \equiv St \cdot Fo$.

Equation (4.7) may now be recast into two convenient forms using the dimensionless parameters defined above.

$$\frac{df_s}{dFo} = \frac{St}{Bi^{-1} + f_s}, \tag{4.8}$$

and

$$\frac{df_s}{d\tau} = \left(Bi^{-1} + f_s\right)^{-1}. \tag{4.9}$$

Now the variables in Eq. (4.9), f_s and τ, may be separated, and the expression integrated. One applies the initial condition that the freezing slab starts out as all melt, so that $f_\ell = 1$, and $f_s = 1 - f_\ell = 0$ when $\tau = 0$.

$$\int_0^{f_s} \left(Bi^{-1} + f_s\right) df_s = \int_0^{\tau} d\tau. \tag{4.10}$$

Carrying out the integrations indicated in Eq. (4.10) yields the dimensionless relation between the extent of solidification and time, namely,

$$\frac{f_s}{Bi} + \frac{f_s^2}{2} = \tau. \tag{4.11}$$

It is convenient to solve Eq. (4.11) for the independent variable, f_s, which gives the desired relationship describing the solidification of a slab of a pure molten substance poured initially at its melting point. Application of the quadratic formula to Eq. (4.11) provides a more useful expression for estimating the quasi-static freezing of a slab of liquid phase at its melting temperature,

$$f_s = \sqrt{\frac{1}{Bi^2} + 2\tau} - \frac{1}{Bi}. \tag{4.12}$$

Equation (4.12) is plotted in Fig. 4.3 for a range of Biot numbers. Here the fraction solid, f_s, is shown as a function of dimensionless freezing time, τ, with the Biot number, Bi, selected as the appropriate heat transfer parameter. As the casting's Biot number decreases, less and less solidification is achieved for a given *dimensionless* time period, τ. Complete solidification is expected in dimensionless times less than unity for Biot numbers greater than 2. Figure 4.3 suggests that slab castings solidified at small Biot numbers (including thin-walled castings, small environmental heat transfer coefficients, or high internal thermal conductivities) solidify relatively slowly compared with slabs with large Biot numbers. The shorter dimensionless freezing time shown in Fig. 4.3 must, however, be carefully interpreted as a Fourier number, which itself is inversely proportional to the square of the slab thickness, L. Thus, for a given Fourier number, the complete freezing of a thin slab takes a much smaller dimensional time than it would for a thick slab. Moreover, because $\tau \equiv St \cdot Fo$, the 'real time' to freeze a slab of any specified thickness, thermal conductivity, or heat transfer coefficient, remains inversely proportional to the Stefan number, or total temperature difference provided during solidification. Clearly, the intuitive interpretation of freezing characteristics are best made after one performs an appropriate return to *dimensional* parameters of real time, temperature, and distance.

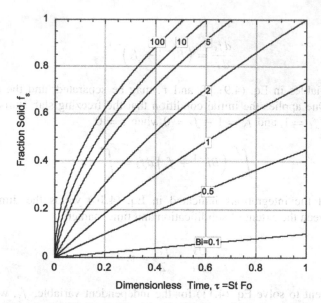

Fig. 4.3 Plot of the relationship, Eq. (4.12), giving the fraction solidified as a function of the dimensionless freezing time, τ, for various Biot numbers. At small Biot numbers the fraction solidified progresses slowly and is nearly linear in dimensionless time, whereas at large Biot number the fraction frozen increases rapidly, and is distinctly non-linear in dimensionless time

The dimensionless time to complete the process of solidification, τ_f, is also of interest, and may be determined easily from Eq. (4.12) by setting $f_s = 1$. Inserting this condition into Eq. (4.12) yields the result

$$\tau_f = \frac{1}{2} + \frac{1}{Bi}. \tag{4.13}$$

If the dimensionless time in Eq. (4.12) is multiplied and divided by the right-hand and left-hand sides of Eq. (4.13), a new expression may be created giving the fraction solidified as a function of the fraction of the total dimensionless freezing time.

$$f_s = \sqrt{\frac{1}{Bi^2} + \left(1 + \frac{2}{Bi}\right)\frac{\tau}{\tau_f}}. \tag{4.14}$$

Note, as concluded just above, large Biot numbers tend to decrease the dimensionless time needed to complete the solidification process.[1] However, even in the limit of infinite Biot numbers, $Bi \to \infty$, the *minimum* dimensionless time to complete the solidification process, τ_f, never becomes smaller than $1/2$. This result

[1] Extensions using fast-convergent analytic methods to solve transient heat conduction problems at high Biot numbers in finite solid bodies, including plates, cylinders, and spheres, were developed recently by Ostrogorsky and Mikic [1].

formally corresponds to the case of 'ideal' heat transfer with negligible thermal resistance at the external heat transfer surface. See again the casting configuration sketched in Fig. 4.1. When heat transfer to the environment is ideal, then the dimensional time, t_f, needed to freeze a pure material slab of thickness L may be determined from the definition of alternative dimensionless freezing time, $\tau_f = St \cdot Fo$, so

$$\tau_f = \frac{1}{2} = St \cdot Fo = \frac{T_m - T_\infty}{(\Delta H_f / c_p)} \frac{\alpha}{L^2} t_f. \tag{4.15}$$

Solving Eq. (4.15) for t_f shows that

$$t_f = \frac{L^2 \Delta H_f}{2 \alpha c_p (T_m - T_\infty)}. \tag{4.16}$$

As is observed in the behavior of Eq. (4.16), the dimensional freezing time, t_f, for one-dimensional quasi-static solidification increases quadratically with the thickness of the casting, L, and decreases with the temperature difference across the solid. Kinetic relationships between time and length scales, such as just shown, are called 'scaling laws', which provide engineers approximate, but easily-applied guidance for developing casting processes. Scaling laws also supply important insights as to the fundamental origins of empirically-based 'rules-of-thumb' that traditional foundrymen have known from practice and observation. Today, engineers still rely on scaling laws, but routinely improve and refine them with critical experiments, or use more sophisticated computer foundry codes that are capable of augmenting the estimates for solidification behavior in more complex casting shapes and compositions. Much more discussion and use of scaling laws in solidification will be provided on a number of topics covered later in this book.

The reader is reminded that formulas such as Eq. (4.16) retain their validity over an enormous range of time and length scales. Such scaling laws apply equally to engineering the solidification of tiny solder joints in microelectronics, with length scales as small as micrometers, and over corresponding time scales of fractions of a second, as it would to geological melting and freezing problems of sea ice, which occur on a scale of many kilometers, and over time periods measured in centuries, or perhaps millennia.

4.1.3 Quasi-static Results

Heat transfer occurring under quasi-static conditions for the freezing of a slab of pure material may be determined directly provided that the location of the solid–liquid interface, $\hat{x}(t)$, is known explicitly. The position of the solid–liquid interface may be found by solving Eq. (3.27), if the interface gradients are known, or, alternatively, by combining Eqs. (4.6) and (4.7) given above, if the interface speed,

$\hat{\dot{x}}(t)$, is known. Specifically, the steady heat flux passing through the solid into its surroundings is

$$q_s = \frac{(T_m - T_\infty)}{\frac{1}{h} + \frac{\hat{x}(t)}{k_s}}. \tag{4.17}$$

The initial heat flux occurs at the onset of solidification, at $\tau = 0$, when $\hat{x}(t) = 0$. Inserting $\hat{x}(t) = 0$ into Eq. (4.17) yields the initial heat flux, $\dot{q}(0)$ as

$$\dot{q}(0) = h(T_m - T_\infty). \tag{4.18}$$

If one divides both sides of Eq. (4.17) by the respective sides of Eq. (4.18) one obtains

$$\frac{\dot{q}_s}{\dot{q}(0)} = \frac{1}{1 + Bi \cdot f_s}. \tag{4.19}$$

Substituting for f_s in Eq. (4.19), using Eq. (4.8), and then carrying through several algebraic steps, yields the quasi-static heat-flux ratio at any arbitrary dimensionless time, τ.

$$\frac{\dot{q}_s}{\dot{q}(0)} = \frac{1}{\sqrt{1 + 2Bi^2\tau}}. \tag{4.20}$$

If the dimensionless time is multiplied and divided by the LHS and the RHS of Eq. (4.13), respectively, one obtains the desired result

$$\frac{\dot{q}_s}{\dot{q}(0)} = \frac{1}{\sqrt{1 + (2Bi + Bi^2)\frac{\tau}{\tau_f}}}. \tag{4.21}$$

Figure 4.4 shows f_s, Eq. (4.14), and the heat flux ratio, Eq. (4.21), plotted together for various Biot numbers, versus the fraction of the *total* dimensionless freezing time, τ_f. These graphs indicate that a solidifying system characterized by a high Biot number exhibits relatively non-uniform releases of latent heat over time, with a correspondingly rapid increase in f_s. By contrast, during solidification of a system with a small Biot number, one would observe a slower increase in f_s, accompanied by more uniform latent heat release throughout the entire course of freezing.

4.1.4 Solidification of a Circular Cylinder

The quasi-static approximation developed in prior sections on one-dimensional solidification of a slab of pure material will now be applied to the case of two-dimensional freezing of melt contained in a long, right circular cylinder. This

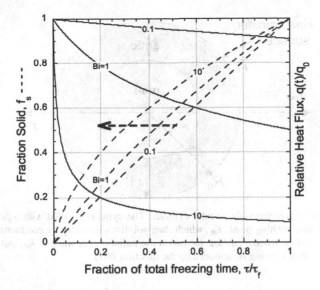

Fig. 4.4 Plot of the relationships, Eqs. (4.14) and (4.21) against the fractional (dimensionless) freezing time, τ/τ_f. *Dashed arrow* directs use of the left-hand ordinate for the *dashed curves*, whereas *solid curves* are to be read against the right-hand ordinate scale. Small Biot numbers allow a uniform release of latent heat during the entire freezing process, right ordinate scale, whereas large Biot numbers cause non-uniform heat release. The fraction solidified at small Biot numbers, $Bi \ll 1$, left ordinate scale, is nearly proportional to the elapsed freezing time, but becomes progressively more non-linear for larger Biot numbers, $Bi \geq 1$

geometry is studied as it is descriptive of large commercial ingots, water-cooled consumable arc melting molds, and the output shape of continuous strand-casting machines that directly produce cast products such as solders.

The significant change from unidirectional freezing is that the internal thermal resistance to heat transfer develops over time across a thickening annular ring of frozen solid of thickness $H(t)$. See sketch of this system in Fig. 4.5. Again, one uses Fourier's law of heat conduction, now expressed in cylindrical coordinates, r, θ, z, to determine the thermal field, gradients, and heat fluxes. A latent heat flux, $\dot{q}_r$, is released in radial directions uniformly over the angular variable, θ, from the quasi stationary cylindrical solid–liquid interface, located at the radial position $\hat{r}(t) = R_0 - H(t)$. The quasi-static outward flow of heat per unit length along z, passing through any arbitrary cylindrical surface located within the solid annulus, $R_0 - H(t) \leq r \leq R_0$, may be written as

$$\dot{q}_r = -2\pi k_s r \frac{\partial T}{\partial r}. \tag{4.22}$$

If one separates the (independent) spatial variable, r, from the (dependent) temperature variable, T, shown in Eq. (4.22), the result may be easily integrated as

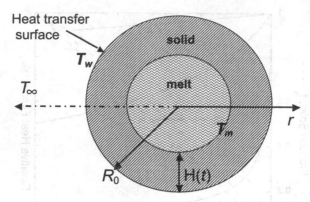

Fig. 4.5 Sketch of a freezing right circular cylinder. The cylinder is filled with a pure melt just above its equilibrium melting point, T_m, which then solidifies inwardly by conducting the latent heat released to the heat transfer surface, located at the radial coordinate $r = R_0$, and at temperature, T_w. The extent of freezing is measured by the function $H(t)$

$$\dot{q}_r \int_{R_0-H(t)}^{R_0} \frac{dr}{r} = -2\pi k_s \int_{T_m}^{T_w} dT. \tag{4.23}$$

Carrying out the integrations indicated in Eq. (4.23) yields an expression for the quasi-static radial heat flux

$$\dot{q}_r = \frac{2\pi k_s}{\ln\left(\frac{R_0}{R_0-H(t)}\right)} (T_m - T_w). \tag{4.24}$$

Equation (4.24) shows that the total heat current through a cylindrical casting is proportional to the temperature drop across the solid annular cylinder, with its 'internal' heat transfer coefficient, or conductance, equal to the denominator in Eq. (4.24). The thermal resistance associated with internal heat conduction is just the inverse of this conductance. Insofar as thermal resistances are additive—as are electrical resistors in a series circuit—one may simply add additional thermal resistance from the external Newtonian cooling, and then balance the heat generated at the solid–liquid interface with the heat flow through the entire system, now consisting of the solidifying casting plus its environment. The thermal balance for the cylindrical casting in the fluid environment under discussion is,

$$\frac{k_s(T_m - T_\infty)}{k_s/R_0 h + \ln\left(\frac{R_0}{R_0-H(t)}\right)} = \rho \Delta H_f (R_0 - H(t))\frac{dH(t)}{dt}. \tag{4.25}$$

The following quantities are useful for introduction into Eq. (4.25):

1. The ratio of the frozen layer to cylinder radius: $\zeta \equiv H(t)/R_0$.
2. Stefan number: $St \equiv c_p(T_m - T_\infty)/\Delta H_f = (T_m - T_\infty)/T_{char}$.

3. Thermal diffusivity: $\alpha \equiv k_s/\rho c_p$.
4. Biot number: $Bi \equiv hR_0/k_s$.
5. Fourier number: $Fo \equiv \alpha t/R_o^2$.
6. Dimensionless freezing time: $\tau \equiv St \cdot Fo$.

Several straighforward algebraic steps must be performed to yield the desired relationship between the dimensionless freezing time, τ, and the frozen thickness ratio, ζ, for the cylinder.

$$\tau = \frac{1}{Bi}\left(\zeta - \frac{\zeta^2}{2}\right) + \frac{1}{4}\left(2(\zeta - 1)^2 \ln(1 - \zeta) - \zeta(\zeta - 2)\right). \qquad (4.26)$$

Figure 4.6 shows a semi-logarithmic plot of Eq. (4.26), with the abscissa selected as the dimensionless freezing time, $\tau = St \cdot Fo$, (the independent variable) and the ordinate as $\zeta = H(\tau)/R_0$, the frozen thickness ratio. Curves are plotted for a range of Biot numbers. Again, one notes that inward freezing cylinders, similar to the case of slabs treated earlier, tend to solidify in shorter dimensionless times at larger Biot numbers. As indicated in this plot, little change in behavior occurs as the Biot number exceeds 10. Although Fig. 4.6 captures a relatively wide range of solidification behaviors, the clearest interpretation of the freezing process occurs in dimensional units of time and distance. Dimensional units are needed, ultimately, to clarify and remove the coupled influence that the cylinder radius, R_0, has on both the Biot and Fourier numbers.

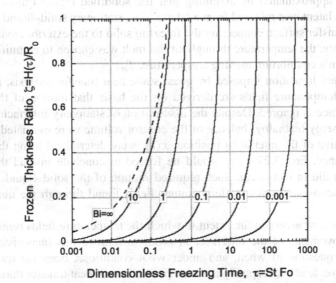

Fig. 4.6 Progress of solidification in right-circular cylinders, tracked by the function $\zeta = H(\tau)/R_0$, at various Biot numbers, versus dimensionless freezing time (log scale). In a cylindrical geometry, the solidification behavior would change only slightly for Biot numbers above about 10

A word of caution is injected at this point regarding application of the quasi-static approximation to freezing and melting problems. The extent of freezing under quasi-static conditions is always predicted theoretically to be less than it actually occurs under more realistic conditions where interfacial motions couple with the thermal field and steepen the gradients. Moreover, it is known that if the Biot number exceeds certain well-defined limits, then predictions based on the quasi-static approximation can fail utterly, and predict physically implausible results. The interested reader is referred to [2] for an in-depth discussion on this particular limitation. Further comparisons will be provided in the next section concerning the relationship of quasi-static approaches versus more accurate moving boundary methods.

4.2 Moving Boundary Analysis

4.2.1 Introduction

In prior sections of this chapter we considered two simple examples of quasi-static heat transfer during solidification. The one-dimensional progressive freezing of a pure melt was described (1) in the form of a slab casting of thickness L, and (2) as a long, right circular cylinder of radius R_0, undergoing two-dimensional radial solidification. The rates of heat transfer for these basic casting problems were determined using quasi-static temperature fields. In brief, the quasi-static temperature fields were approximated by assuming that the solidified region transferred, via conduction, latent heat released from a 'static', i.e., stationary solid–liquid interface to a heat transfer surface connecting the freezing solid to the exterior environment. For simplicity, the temperature throughout the melt was chosen to be uniform, and at the system's equilibrium melting temperature, T_m.

A stringent limitation imposed by quasi-static heat transfer analysis, however, is that the temperature fields are derived on the basis that motion of the solid–liquid interface is ignored. Despite the assumption of stationary interfacial behavior, if the energy (enthalpy) balance of the control volume were evaluated, and the time derivative of the interfacial position, $\hat{x}(t)$, were determined using the Stefan energy balance, Eq. (3.27), one would be forced to conclude instead that interface motion did in fact occur. Such required motion of the solid–liquid interface is clearly inconsistent with the temperature fields found through the quasi-static analysis.

One must now ascertain the extent to which the temperature fields remain 'coupled' to movements of the interface, in order to remove this inconsistency and answer the question of when, and under what conditions, does the quasi-static approximation lead to acceptably accurate predictions of heat transfer during solidification. Stated slightly differently, the quasi-static mathematical approximation results in a mathematical linearization of the (non-linear) influences of the coupling between the temperature fields and the necessary motions of the solid–liquid interface that are responsible for the release of latent heat. A quantitative evaluation of the accuracy of quasi-static methods used in specific solidification problems requires that a more general analysis be available as a basis for comparison. Such an analysis, can be found using Neumann's 'moving boundary' solution, which reduces to the simpler quasi-static form under well-defined mathematical conditions. Specifically, it is found that Neumann's moving boundary analysis does indeed approach the quasi-static form, provided that the Stefan number responsible for the solidification is sufficiently small. Specifically, a moving boundary begins to 'act' quasi-statically if, and only if $St \ll 1$. Lastly, several important examples will be discussed involving unidirectional freezing against a mold wall, or solidification of a casting aided by the action of a cooling environment. These examples will be discussed as extensions to the basic Neumann moving boundary problem.

4.2.2 Conduction-Advection

Solidification and crystal growth are processes that involve both thermal conduction through the melt and freezing solid, plus advective heat flow caused by any motions of the phases relative to the coordinate system. Heat conducted and advected through homogeneous regions is described by a form of the energy equation that predicts the development of the temperature distribution. Each region has specified thermal properties including: the thermal conductivity, mass density, radiative transmittance, and specific heat. Solidification, where one region, the melt, transforms to the solid, also adds a source term, viz., the latent heat.

The conduction-advection equation is based on an enthalpy balance, to which any appropriate constitutive laws of heat generation or transport may be added, e.g., conduction, radiation, convection, internal friction, viscous dissipation. General cases become complex, so we limit the description to a simple example.

The constitutive equation needed here for a basic moving-boundary solidification problems is again Fourier's law of heat conduction, but now in moving coordinates, where advected energy must be included. A control volume relates the rate of heat conduction into and out to the material's local temperature gradients, plus the rate of energy advected in and out by the relative motion of matter passing through it. Figure 4.7 shows a sketch of a conductive-advective control volume needed to determine the enthalpy balance at an arbitrary point within the liquid or solid phases. The centroid of this control volume is located at any general Cartesian coordinate (x, y, z), and given rectilinear edge lengths Δx, Δy, Δz, respectively. Its volume is $\Delta V = \Delta x \times \Delta y \times \Delta z$.

The flux component, $q_x(x - \Delta x/2, y, z)$, entering the left-hand face of the control volume is slightly different from the magnitude of the flux component at the control volume's centroid, $q_x(x, y, z)$, as is also the flux component, $q_x(x + \Delta x/2, y, z)$, leaving the right-hand face. These differences also arise independently for flux components pointed in each of the other two orthogonal Cartesian directions, as the flux (and, ultimately, the temperature) are both considered continuous, differentiable functions of space and time.

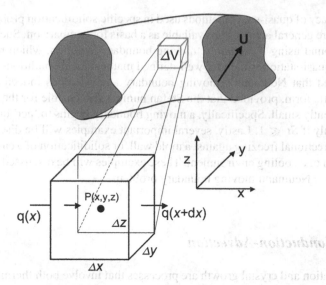

Fig. 4.7 Control volume used to develop the standard energy equation in a continuum moving w.r.t. Cartesian axes, x, y, z at a uniform relative velocity **U**. The heat fluxes parallel to the x-axis, entering and leaving a control volume that is centered at point $P(x, y, z)$, are shown in detail. An arbitrary heat flux would also consist of flux components acting parallel to the y- and z-axes

The enthalpy balance contributed just by fluxes in the x-direction is

$$q_x\left(x - \frac{\Delta x}{2}\right)\Delta y\Delta z - q_x\left(x + \frac{\Delta x}{2}\right)\Delta y\Delta z + \rho_i H_i U_i\left(x - \frac{\Delta x}{2}\right)\Delta y\Delta z$$

$$-\rho_i H_i U_i\left(x + \frac{\Delta x}{2}\right)\Delta y\Delta z = \rho_i \frac{\partial H_i}{\partial t}\Delta x\Delta y\Delta z, \quad (i = s, \ell). \qquad (4.27)$$

The left-hand side of Eq. (4.27) contains the energy rates of the conduction heat flux component, q_x, and any *advected sensible heat* entering and leaving the control volume via the x-component of the relative velocity, **U**, between the coordinate system and the material. The macro-motion of the region occurs as rigid-body motion (advection) at a constant velocity, **U**, relative to the coordinate axes. The assumption of uniform phase motion may be replaced by the equivalent requirement that each phase, i, acts incompressibly, so that their mass densities, ρ_i may be taken as constants. The small density differences between the crystal and its melt is ignored.

Actually, the most restrictive assumption that will be encountered in formulating the enthalpy balance and then solving for the thermal field is assuming that the individual thermal conductivities of the phases are constants. The analytical solutions to the energy equation to be developed also impose additional constraints on the allowed solidification geometry, so their primary value is affording quick compre-hension of thermal transport during freezing and melting.[2]

With few exceptions, the thermal conductivity of materials depends on tempera-ture, and sometimes sensitively. Allowing the individual conductivities of the solid and liquid to remain functions of these variables, rather than selecting an average value for each, introduces the mathematical difficulty of non-linearity, a complica-tion that shall be avoided at present. Accurate solutions to the energy equation that do account for temperature-dependent thermal properties may be obtained numeri-cally with standard finite difference computer codes that solve the non-linear PDEs.

Consider the RHS of Eq. (4.27). One sees that the net energy conducted and advected through the faces of the control volume that are normal to the x-axis equals its fraction of the total time rate of change of the enthalpy within the control volume. This statement is the central physical argument of energy conservation in the conduction-advection control volume. It's just a restatement of the 1st law of thermodynamics that energy in the forms of heat and work may be transformed but not destroyed. Identical energy balances, of course, may be formulated for conduction and advection occurring parallel to the remaining Cartesian directions, y and z.

Now each term on the LHS of Eq. (4.27) may be expanded in a Taylor series. Each series is truncated beyond the linear term, and the control volume is reduced to infinitesimal dimensions, so $\Delta x \to dx$, $\Delta y \to dy$, and $\Delta z \to dz$:

$$\left(q_x(x) - \frac{\partial q_x}{\partial x}\frac{dx}{2} \right) \cdot dydz - \left(q_x(x) + \frac{\partial q_x}{\partial x}\frac{dx}{2} \right) \cdot dydz$$

$$+ \left(\rho_i U_x H_i(x) + \rho_i U_x \frac{\partial H_i}{\partial x}\frac{dx}{2} \right) \cdot dydz - \left(\rho_i U_x H_i(x) - \rho_i U_x \frac{\partial H_i}{\partial x}\frac{dx}{2} \right) \cdot dydz$$

$$= \rho_i \frac{\partial H_i}{\partial t} dxdydz. \qquad (4.28)$$

Gathering terms in Eq. (4.28) and canceling yields the differential enthalpy balance at the point (x, y, z) associated with energy exchanges just along the x-axis,

$$-\frac{\partial q_x}{\partial x} + \rho_i U_i \frac{\partial H_i}{\partial x} = \rho_i \frac{\partial H_i}{\partial t}. \qquad (4.29)$$

---

[2] The differences between using numerical simulations and analytical solutions are that the latter quickly give the analyst or engineer an intuitive grasp of a system's behavior over as wide a range of the process parameters as required. Numerical simulations provide enormous flexibility and breadth in their application, and become indispensable when dealing with complexities introduced by variable thermal properties and complicated casting geometries. Computer simulations, how-ever, still pose considerable interpretive challenges when one explores wide regions of a multi-parameter space that many solidification problems require. Clearly, each methodology—numeric and analytic—can assist the casting engineer.

Fourier's law of heat conduction, discussed in Section 4.1, is restated here to describe this constitutive law, but just for heat flow parallel to the x-axis. The component of the vector flux, $\mathbf{q}$, which is parallel to the x-axis, is

$$\mathbf{q} \cdot \mathbf{e}_x = q_x = -k_i \frac{\partial T}{\partial x}. \tag{4.30}$$

The molar enthalpy function, or heat content of a phase i, at constant pressure, $H_i(T)$, was defined in Chapter 2 as

$$H_i(T) = H_i(T_0) + \int_{T_0}^{T} c_p^i dT. \tag{4.31}$$

Here T_0 is some arbitrary reference temperature, and c_p^i is the molar specific heat at constant pressure of phase i. The partial derivatives with respect to time and space of the enthalpy of a phase follows immediately from Eq. (4.31) as

$$\frac{\partial H_i}{\partial t} = c_p^i \frac{\partial T}{\partial t} \quad (i = s, \ell), \tag{4.32}$$

and

$$\frac{\partial H_i}{\partial x} = c_p^i \frac{\partial T}{\partial x} \quad (i = s, \ell). \tag{4.33}$$

Substituting the relations shown in Eqs. (4.30), (4.32), and (4.33) into Eq. (4.29) yields the partial energy balance for conduction and advection through control volume faces that are normal to the x-axis,

$$k_i \frac{\partial^2 T}{\partial x^2} + \rho_i U_x c_p^i \frac{\partial T}{\partial x} = \rho_i c_p^i \frac{\partial T}{\partial t} \quad (x \text{ contribution only}). \tag{4.34}$$

Enthalpy conservation equations comparable to Eq. (4.34) may be derived identically for the remaining faces of the cubical control volume element that are normal to the y-axis and the z-axis. If all three independent energy conservation equations are summed, one obtains the conduction-advection equation in a phase i, moving at a uniform speed, $\mathbf{U}$, with respect to fixed Cartesian coordinates, namely,

$$k_i \left(\frac{\partial^2 T}{\partial x^2} + \frac{\partial^2 T}{\partial y^2} + \frac{\partial^2 T}{\partial z^2} \right) + \rho_i c_p^i \left(U_x \frac{\partial T}{\partial x} + U_y \frac{\partial T}{\partial y} + U_z \frac{\partial T}{\partial z} \right) = \rho_i c_p^i \frac{\partial T}{\partial t}. \tag{4.35}$$

This result is a linear, 2nd-order, partial differential equation (PDE) that is easily generalized to other coordinate systems using standard notation of the vector calculus, namely:

$$k_i \nabla^2 T + \rho_i c_p^i \mathbf{U} \cdot \nabla T = \rho_i c_p^i \frac{\partial T}{\partial t}, \qquad (4.36)$$

where ∇^2 is the Laplacian operator, and $\mathbf{U} \cdot \nabla T$ is the inner product between the phase's translational velocity and the temperature gradient. If both sides of Eq. (4.36) are divided through by the volumetric specific heat, $\rho_i c_p^i$, one obtains the standard 'conduction-advection' form of the energy equation,

$$\frac{k_i}{\rho_i c_p^i} \nabla^2 T + \mathbf{U} \cdot \nabla T = \frac{\partial T}{\partial t}. \qquad (4.37)$$

Equation (4.37) may be rearranged slightly, by recalling that the thermal diffusivity of the ith phase is defined as $\alpha_i \equiv k_i / \rho_i c_p^i, (i = s, \ell)$. Thus,

$$\frac{\partial T_i}{\partial t} = \alpha_i \nabla^2 T_i + \mathbf{U} \cdot \nabla T_i \ (i = s, \ell). \qquad (4.38)$$

Further manipulations of Eq. (4.38) will be reserved for specific problems for which solutions are desired. However, it should be noted that any mathematical solution to Eq. (4.38) requires specification of the parameters α_i and $\mathbf{U}$ for each phase, plus the imposition of two boundary or initial conditions.

Six commonly encountered conditions found in casting and crystal growth problems are enumerated next:

1. *Imposed surface temperature at $x = 0$: $T(0, t) = T_w$.*
2. *Steady imposed heat flux at $x = 0$: $-k \left(\frac{\partial T}{\partial x} \right)_{x=0} = q = constant$.*
3. *Adiabatic (insulated) surface at $x = 0$: $\left(\frac{\partial T}{\partial x} \right)_{x=0} = 0$.*
4. *Convective heat transfer at $x = 0$ to an environment: $k \left(\frac{\partial T}{\partial x} \right)_{x=0} = h[T(0, t) - T_{envir}]$.*
5. *Equilibrium temperature maintained at the solid–liquid interface, $\hat{x}(t)$: $T(\hat{x}, t) = T_m$.*
6. *'Natural' boundary conditions*: So-called 'natural' boundary conditions are frequently invoked at the boundaries of a phase to insure continuity of both the temperature field and that of the heat flux.

4.2.3 Neumann's Solution: Semi-infinite Freezing

Only a few analytical solutions are available for solidification problems in one spatial dimension, and most are restricted to infinite or semi-infinite phase regions [2–5]. Indeed, finite-domain solidification problems pose challenges in all spatial dimensions, and as they usually require series solutions one must employ numerical methods. See, for example, the discussions on the analysis of phase-change problems provided in standard heat transfer compendia such as provided in references [6] and [7].

Today, fortunately, with the wide availability of efficient digital computing, the paucity of analytical solutions for phase-change or moving boundary problems no longer presents an impediment for engineers. As an historical note, F. Neumann is credited with developing this mathematical solution, which originally appeared as part of his lectures on mathematical physics given in 1860. His *exact* moving boundary solution is worthwhile reviewing, as it compactly demonstrates the major features encountered in conduction-limited solidification problems, and may be compared with the approximate quasi-static solutions discussed earlier in this chapter.

The region undergoing solidification is sketched in Fig. 4.8, where the starting liquid phase—in this instance a pure melt—occupies the entire positive half-space, $x > 0$. The melt is static, at constant pressure, and initially has a uniform temperature designated T_0, which is above the material's equilibrium freezing point, T_m. Freezing is initiated at the origin, $x = 0$, at time $t = 0$, by suddenly lowering the temperature there to a fixed value, $T_w < T_m$. A thickening region of crystalline solid forms in time as the solid-melt interface at $\hat{x}(t)$, travels toward the right.

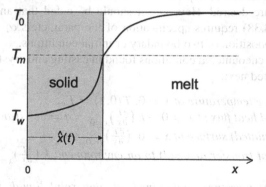

Fig. 4.8 Configuration of solid and liquid regions, and their temperature fields, treated in Neumann's 1-dimensional, moving boundary solution. Neumann's solution yields the extent of freezing, $\hat{x}(t)$, for a semi-infinite bar of pure melt poured initially at a uniform temperature above its freezing temperature, so $T_0 \geq T_m$

Two governing energy equations are require in the solution this moving boundary problem: one valid in the solid, $x < \hat{x}(t)$, the other in the melt region, $x > \hat{x}(t)$, respectively:

$$\frac{\partial^2 T_s}{\partial x^2} = \frac{1}{\alpha_s} \frac{\partial T_s}{\partial t}, \quad x < \hat{x}(t), \tag{4.39}$$

and

$$\frac{\partial^2 T_\ell}{\partial x^2} = \frac{1}{\alpha_\ell} \frac{\partial T_\ell}{\partial t}, \quad x > \hat{x}(t). \tag{4.40}$$

Equations (4.39) and (4.40) comprise a pair of coupled PDEs, each of second order, indicating that their combined solution will require a total of four initial and boundary conditions.

The boundary and interfacial conditions appropriate to the solution of Neumann's moving boundary problem are:

1. A prescribed temperature is set at the exterior left-hand boundary, $T(0, t) = T_w$.
2. The initially prescribed temperature of the melt, which holds far from the solid-melt interface for all time is $T(\infty, t) = T_0$.
3. The temperatures on either side of the moving solid-melt interface are equal and fixed at the equilibrium melting point, and thus $T_s(\hat{x}, t) = T_\ell(\hat{x}, t) = T_m$.
4. Stefan's condition—a local energy balance for the solid–liquid interface—always holds at $\hat{x}(t)$, so that

$$
k_\ell \left(\frac{\partial T_\ell}{\partial x} \right)_{x=\hat{x}} + \rho_s \Delta H_f \frac{d\hat{x}}{dt} - k_s \left(\frac{\partial T_s}{\partial x} \right)_{x=\hat{x}} = 0.
$$

Linear heat-flow solutions to Eqs. (4.39) and (4.40), subject to the conditions enumerated above, may be found using standard mathematical techniques, such as Laplace transforms. The interested reader can refer to [8] for full details on deriving this particular solution.

Neumann's moving boundary solution is expressed conveniently in terms of two *similarity* variables, $\xi_\ell = x/\sqrt{4\alpha_\ell t}$ and $\xi_s = x/\sqrt{4\alpha_s t}$, for the liquid ($x > \hat{x}$) and solid ($x < \hat{x}$) regions, respectively. Remarkably, similarity variables capture the spatial and temporal behavior of the temperature fields in terms of certain functional combinations of the ordinary space variable, x, and time, t. The thermal fields found from solutions to Eqs. (4.39) and (4.40) that are of interest here include for the solid,

$$
\frac{T_s(x, t) - T_m}{T_w - T_m} = 1 + A \, \mathrm{erf}(\xi_s), \quad (0 \le x < \hat{x}(t)) \tag{4.41}
$$

and for the liquid,

$$
\frac{T_\ell(x, t) - T_m}{T_0 - T_m} = 1 + B \, \mathrm{erfc}(\xi_\ell), \quad (x > \hat{x}(t)). \tag{4.42}
$$

The error function, erf(z), and its complement, erfc(z), appearing on the RHS of Eqs. (4.41) and (4.42), respectively, are standard tabulated functions defined as follows:

$$
\mathrm{erf}(z) \equiv \frac{2}{\sqrt{\pi}} \int_z^\infty e^{-u^2} \, du, \tag{4.43}
$$

and

$$
\mathrm{erfc}(z) \equiv 1 - \mathrm{erf}(z). \tag{4.44}
$$

Further details on the properties of the error function and its derivative, their use in calculating Neumann-type solutions, some approximations for the functions defined in Eqs. (4.43) and (4.44), and an extensive tabulation of their values may be found in [8]. Today, however, engineers have at their disposal a variety of mathematical packages available on workstations and PCs that deal routinely with the evaluation of the error function.[3]

One may evaluate the temperature fields that are solutions to the energy equations in the melt and solid given as Eqs. (4.42) and (4.41), respectively, by applying the constant temperature condition that holds at any time, $t > 0$, on the moving solid-melt interface, $x = \hat{x}(t)$. (See boundary condition (3), listed above.) Inserting this condition sets the left-hand sides of Eqs. (4.42) and (4.41) equal to zero, so that

$$-\frac{1}{A} = \text{erf}\left(\frac{\hat{x}}{\sqrt{4\alpha_s t}}\right),$$ (4.45)

and

$$-\frac{1}{B} = \text{erfc}\left(\frac{\hat{x}}{\sqrt{4\alpha_\ell t}}\right).$$ (4.46)

Equations (4.45) and (4.46) clearly remain valid for *all* times during freezing, because the left-hand sides are constants. We conclude, therefore, that when $x = \hat{x}(t)$, the similarity variables appearing as the arguments of the error function and the error function complement solutions are themselves constants. Thus, for $\hat{\xi}_s$ and $\hat{\xi}_\ell$ to be constants as $\hat{x}$ increases with time, it is also required that

$$\frac{\hat{x}(t)}{\sqrt{4\alpha_s t}} = \lambda_s,$$ (4.47)

and

$$\frac{\hat{x}(t)}{\sqrt{4\alpha_\ell t}} = \lambda_\ell,$$ (4.48)

where λ_s and λ_ℓ denote the 'freezing constants' or moving boundary solution characteristics for this problem. Inspection of Eqs. (4.47) and (4.48) shows that for them to hold at any arbitrary time, t, the solid-melt interface must move outward on the x-axis proportional to the square-root of time. This result is not at all surprising, inasmuch as most processes that are kinetically dominated by diffusive phenomena, such as thermal conduction and chemical diffusion, often evolve as the square-root

---

[3] This chapter, despite its focus on classical analytic solutions for casting and crystal growth problems, is also included to encourage the use of mathematics software and computers to explore more deeply the details and nuances of heat flow during solidification.

of time [6, 8]. The characteristic of square-root dependences in solidification, with
a few notable exceptions, will be encountered elsewhere in this book.

In fact, Eqs. (4.47) and (4.48) may be generalized slightly,

$$\hat{x}(t) = \lambda_i \sqrt{4\alpha_i t}, \quad (i = s, l), \qquad (4.49)$$

where the solution characteristics, λ_i, are related as

$$\frac{\lambda_\ell}{\lambda_s} = \sqrt{\frac{\alpha_s}{\alpha_\ell}}. \qquad (4.50)$$

An expression for the interfacial speed, $v = d\hat{x}/dt$, which slows inversely with
the square-root of freezing time, may be found immediately by differentiating
Eq. (4.49),

$$v = \frac{d\hat{x}}{dt} = \lambda_i \sqrt{\frac{\alpha_i}{t}}. \qquad (4.51)$$

To determine the values of the solution characteristics one applies Stefan's interfacial energy balance (condition (4), listed above) by inserting the interface speed,
the time derivative on the RHS of Eq. (4.51), and the temperature gradients at the
interface, the latter of which may be found by partially differentiating the temperature fields, Eqs. (4.42) and (4.41), with respect to x. This gives the exact Stefan
energy balance at the moving interface as,

$$\rho_s \Delta H_f \lambda_s \sqrt{\frac{\alpha_s}{t}} + k_\ell (T_0 - T_{eq}) \frac{Be^{-\left(\frac{\alpha_s}{\alpha_\ell}\right)\lambda_s^2}}{\sqrt{\pi \alpha_\ell t}} - k_s (T_{eq} - T_w) \frac{Ae^{-\lambda_s^2}}{\sqrt{\pi \alpha_s t}} = 0. \qquad (4.52)$$

The unknown constants A and B, introduced with the solutions for the temperature
fields developed within the solid and melt, Eqs. (4.42) and (4.41), may now be found
by substituting Eq. (4.47) into Eq. (4.45), and Eq. (4.48) into Eq. (4.46), along with
using Eq. (4.50). Following these steps yields expressions for the two field constants,

$$A = \frac{-1}{\text{erf}(\lambda_s)}, \qquad (4.53)$$

and

$$B = \frac{-1}{\text{erfc}(\lambda_\ell)} = \frac{-1}{\text{erfc}\left(\lambda_s \sqrt{\frac{\alpha_s}{\alpha_\ell}}\right)}. \qquad (4.54)$$

Finally, if these expressions for the field constants A and B are substituted back
into the Stefan energy balance, Eq. (4.52), one obtains an awkward transcendental
equation that defines the freezing constant, λ_s,

$$\rho_s \Delta H_f \sqrt{\alpha_s} \; \lambda_s + \frac{k_\ell (T_0 - T_m) e^{-\frac{\alpha_s}{\alpha_\ell} \lambda_s^2}}{\sqrt{\pi \alpha_\ell} \; \text{erfc}\left(\lambda_s \sqrt{\frac{\alpha_s}{\alpha_\ell}}\right)} = \frac{k_s (T_m - T_w) e^{-\lambda_s^2}}{\sqrt{\pi \alpha_s} \; \text{erf}(\lambda_s)}. \tag{4.55}$$

Equation (4.55) can be evaluated numerically to calculate the freezing constant, λ_s, or 'solution characteristic', as it is designated in the heat transfer literature on solidification and crystal growth. The solution characteristic for the current problem is a function of the parameters T_0 and T_w, and depends on six thermodynamic and transport constants. The satisfaction of obtaining Neumann's analytical solution for such a simple freezing problem seems, at first sight, diminished by discovering that its characteristics, λ_i ($i = s, \ell$), are functions of no less than eight parameters!

Nonetheless, by resorting to graphing methods, and with carefully guided choices for the system parameters and field variables, plus applying additional assumptions that simplify the Neumann problem even further, one still can learn a great deal about heat transfer during one-dimensional freezing, without becoming 'lost' in an 8-parameter space. Here is one noteworthy simplification:

4.2.3.1 No Superheating of Melt

The unidirectional (one-dimensional) solidification of a pure melt being considered here can occur under the limitation that there was no initial superheat applied to the melt phase. That is, the temperature of the melt is set initially uniformly at its melting point, $T_0 = T_m$, and remains so for the duration of solidification.

The temperature field in the melt, $T_\ell(x, t)$, in this special case, becomes trivial, as all the space and time dependences of the thermal field in the melt, $x \geq \hat{x}$, are eliminated. Inasmuch as $T_\ell(x, t) = T_m$, except for the progressive motion of the solid-melt interface that continually shifts the location of the starting coordinate, $\hat{x}(t)$, denoting the molten region. Note also that the second term on the RHS of Eq. (4.55) vanishes without any superheat. Removing the spatial and temporal dependences of the temperature field within the melt leads to the simpler result,

$$\sqrt{\pi} \lambda_s e^{\lambda_s^2} \text{erf}(\lambda_s) = \frac{k_s (T_m - T_w)}{\rho_s \Delta H_f \alpha_s} = \frac{(T_m - T_w)}{\Delta T_{char}}. \tag{4.56}$$

The second equality in Eq. (4.56) is the ratio of the total temperature difference across the solid that drives freezing to the system's characteristic temperature. This ratio defines the system's Stefan number, St, which yields the final transcendental equation for the solution characteristic of Neumann's moving boundary problem when lacking superheat in the melt,

$$\sqrt{\pi} \lambda_s e^{\lambda_s^2} \text{erf}(\lambda_s) = St. \tag{4.57}$$

The predictions using Eq. (4.57) can now be compared with the development in Section 4.1 of the *quasi-static* approximation of heat transfer through a slab casting, with the melt also initially poured at its melting temperature. The earlier quasi-static approximation derived for that problem as Eq. (4.12) was

$$f_s = \frac{\hat{x}}{L} = \sqrt{\frac{1}{Bi^2} + 2\tau} - \frac{1}{Bi}. \tag{4.58}$$

If the Biot number in Eq. (4.58) is allowed to approach infinity ($Bi \to \infty$) then the Newtonian cooling condition specified at its external environment is replaced instead by a fixed-temperature boundary condition at $x = 0$, which is identical to that used in Neumann's problem without superheat. Specifically, imposing an infinite Biot number allows 'perfect' heat transfer to occur at the casting wall, so $T(0, t) = T_w$. With this simplification, Eq. (4.58) reduces to

$$\frac{\hat{x}}{L} = \sqrt{2\tau} = \sqrt{2St \cdot Fo}. \tag{4.59}$$

Inserting the definition of the Fourier number for the dimensionless time into Eq. (4.59) yields

$$\frac{\hat{x}}{L} = \sqrt{2St} \sqrt{\frac{\alpha_s t}{L^2}}, \tag{4.60}$$

which when rearranged slightly gives the result for quasi-static freezing with ideal heat transfer to the environment,

$$\frac{\hat{x}}{\sqrt{4\alpha_s t}} = \lambda_s = \sqrt{\frac{St}{2}}. \tag{4.61}$$

Equation (4.61) may now be inverted to provide the quasi-static approximate solution relating the characteristic, λ_s, and the Stefan number,

$$2\lambda_s^2 = St. \tag{4.62}$$

The solutions to Neumann's moving boundary problem, Eq. (4.57), and the solution derived using the quasi-static approximation, Eq. (4.61), are plotted together in Figs. 4.9 and 4.10. These graphs show that the simpler quasi-static result may be considered accurate, provided that the Stefan number in the solid is 'sufficiently small'. Specifically, these graphs show that better than 1% agreement is achieved between the quasi-static approximation and moving boundary theory provided that the temperature difference established across the solid, expressed as a Stefan number, is less than ca. 0.03. Of course, as the Stefan number increase toward unity, or larger, the discrepancies between these theories increase rapidly, and the quasi-static approximation quickly becomes an inadequate description of the heat transfer during unidirectional freezing.

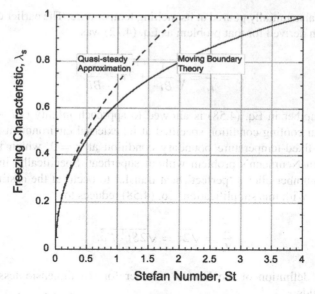

Fig. 4.9 Linear plots of the freezing characteristic, λ_s, versus Stefan number, St, for the quasi-static approximation and the solution for Neumann's unidirectional moving boundary problem. The disparity between the predictions of moving boundary theory and the quasi-static approximation decreases rapidly as the Stefan number becomes small ($St \ll 1$). The quasi-static approximation overestimates the freezing characteristic, and therefore underestimates the time needed for solidification to occur

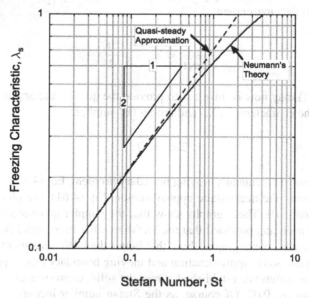

Fig. 4.10 Heat transfer solutions for 1-dimensional solidification plotted on log-log coordinates. These plots clearly show that the quasi-static approximation and Neumann's solution agree in the asymptotic limit $St \to 0$, where the freezing characteristic, λ_s varies as $\sqrt{St/2}$

As the value of the characteristic temperature, $\Delta H_f / c_p^s$, varies from one material to another, it becomes necessary to obtain an estimate for what temperature difference set across a solidifying solid slab constitutes a 'large' or a 'small' freezing Stefan number. Metals typical have $\Delta T_{char} \approx 100s$ K, whereas many organic crystals have $\Delta T_{char} \approx 10s$ K. Thus, the judgment of whether or not the quasi-static approximation suffices in heat transfer calculations for a given slab casting or melt-grown crystal depends on the material being solidified as well as on the temperature difference imposed on the solid. When dealing with relatively large Stefan numbers, which occurs in cases of metals with high melting points, the need for Neumann's more complicated solution is justified.

4.2.3.2 Casting Speed and the Stefan Number

Here the melt, a liquid metal, is initially held at its melting temperature, $T_m = 1,000$ C, and has a characteristic temperature $\Delta T_{char} = \Delta H_f / c_p^s = 300$ K. In addition, the left-hand wall of the freezing casting is maintained at a relatively low constant temperature of 100 C by a continuous spray of water. The Stefan number for the temperature difference across the freezing solid is St = (1000K−100K)/300K = 3, which is considered a large value. The graph of the temperature fields that develop within this solidifying slab casting are shown in Fig. 4.11 without superheat, and with enhanced environmental cooling. Moreover, the freezing time associated with this plot is selected when solidification is half complete, so $f_s = 0.5$.

The temperature fields in both phases, T_s, and T_ℓ, are plotted together with the quasi-static approximation in Fig. 4.11. The quasi-static approximation for this temperature distribution is based on Eq. (4.12), developed in Section 4.1, whereas the Neumann temperature distribution is predicted using Eq. (4.57). Comparison of the curves in Fig. 4.11 shows that the thermal gradients predicted by Neumann's moving boundary formula are smaller than those predicted by the quasi-static approximation. This is consistent with the fact that the quasi-static solidification speeds are predicted to be faster than those from Neumann's solution, and the times to attain total solidification are less.

To take this example a step further, apply the formula for the fraction solidified derived using the quasi-static approximation with $Bi \to \infty$, Eq. (4.59), now written in the form,

$$f_s = \sqrt{2St \cdot Fo},$$
(4.63)

and solve Eq. (4.63) for the Fourier number (dimensionless time) when the fraction solid is $f_s = 0.5$, and the Stefan number, $St = 3$.

$$Fo = \frac{f_s^2}{2St} = \frac{(0.5)^2}{2 \times 3} = 4.17 \times 10^{-2}.$$
(4.64)

The equivalent calculation with Neumann's formula for the freezing time to reach the same casting configuration as shown in Fig. 4.11 starts with the definition of the solution characteristic given in Eq. (4.50), $\lambda_s \equiv \hat{x} / \sqrt{4\alpha_s t}$, with solidification half

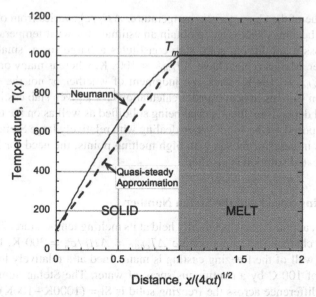

Fig. 4.11 Temperature fields that develop in a metal casting when half solid, so $f_s = 0.5$. The fields are plotted against the non-dimensional similarity variable $x/(4\alpha_s t)^{1/2}$. The Stefan number during freezing is $St = 3.0$, based on the ratio of the temperature difference across the freezing solid, 900 K, and the system's characteristic temperature, $T_{char} = 300$ K. The slopes of these temperature fields represent (scaled) thermal gradients within the casting. The gradients in the solid adjacent to the advancing solid-melt interface, based on Neumann's moving boundary solution, are smaller than those predicted with quasi-static theory. Neumann's solution, therefore, predicts a longer time for total solidification to occur, i.e., $f_s \to 1$, than that estimated with the quasi-static approximation. Quasi-static theory would be accurate enough for many engineering purposes provided that $St \ll 1$. That requirement, however, is violated in this particular example where large temperature differences occur across the slab

completed, i.e., setting $f_s = 0.5$. Multiplying and dividing by the system's arbitrary total thickness, L, gives the prediction,

$$\lambda_s = \frac{\hat{x}L}{L\sqrt{4\alpha_s t}} = \frac{f_s}{2\sqrt{Fo}}. \tag{4.65}$$

Solve Eq. (4.65) for the Fourier number to achieve the condition $f_s = 0.5$, gives

$$Fo = \frac{f_s^2}{4(\lambda_s)^2} = \frac{(0.5)^2}{4 \cdot (0.92)^2} = 7.38 \times 10^{-2}, \tag{4.66}$$

where the value for λ_s in the denominator of Eq. (4.66) is conveniently obtained from Eq. (4.57) using $St = 3.0$ and the graphical solutions presented in Figs. 4.9 or 4.10. Comparison of the Fourier numbers calculated from Eqs. (4.64) and (4.66) shows that the quasi-static estimate of the time to freeze half this casting is actually a poor one—the approximate result being only about 56% of the freezing time predicted with the more accurate Neumann formula.

4.2.3.3 Related Casting Problems

Readers interested in other closely related application areas for moving boundary problems, including numerical solutions for alloy solidification, heat transfer to phase change materials, low-speed dip forming for cladding, semiconductor fabrication, and freezing in permafrost should consult [3].

This chapter concludes with mention of two commonly encountered unidirectional casting processes that relate closely to Neumann's basic unidirectional moving boundary problem.

1. *Solidification against an infinitely thick cold wall.* Thick-walled molds can absorb the heat of fusion released during solidification at the contact surface, portrayed in Fig. 4.12 as 'perfect' heat transfer, $h \to \infty$. The kinetics of the process depend on both properties within the casting and thermal properties of the wall. Figure 4.12 shows this thermal configuration, with the cold mold temperature chosen initially at T_{env}.

2. *Solidification on a cold wall with thermal contact resistance.* Most contact surfaces between a mold wall and the solidified solid show thermal resistance. The resistance can be caused by a thin gas film that develops from lack of wetting between the melt and the mold wall, or from subsequent thermal contraction of the slab as the system cools. The thermal impedance of a contact surface can become sufficiently severe to exhibit a sudden temperature 'jump', as is sketched in Fig. 4.13. The complexities of mold wall air gaps are best captured by using an overall heat transfer coefficient, h that accounts in an average way for the combined effects of convective, conductive, and radiative heat transfer processes. The interested reader should consult the numerous monographs on heat transfer in casting [4, 5, 9–12].

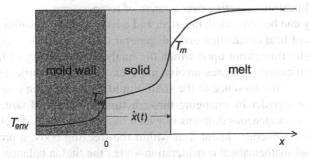

Fig. 4.12 Configuration of the melt and freezing solid phase during solidification against a thick mold wall that conducts the latent heat of fusion with high efficiency toward its external environment at T_{env}. If the heat transfer between the solid and the mold is nearly perfect, so $h \to \infty$, the temperature at the right-hand mold wall, T_w, matches with the thermal distribution established in the solid. Similarly, the heat transfer through the solid-melt interface is also close to ideal, so the temperatures in the solid and melt match at their mutual melting temperature, T_m. Significant exceptions to this situation can occur during rapid solidification processing, which are discussed in Chapter 17

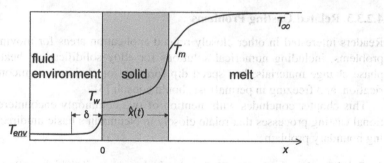

Fig. 4.13 Temperature distributions in a slab casting cooled by a fluid flowing past the solid surface at $x = 0$. A thermal boundary layer of thickness Δ is shown, depicting the distance over which the temperature drops suddenly between the hot casting and the cooler environment. An extremely steep fall, or jump in temperature can develop between the boundary of the hot solid at $x = 0$ and the contacting fluid environment, which may nucleate surface boiling and form a transient vapor film. By contrast, the heat transfer between the advancing solid and its melt at $\hat{x}(t)$ remains nearly ideal, so the temperature fields in the solid and melt match at $T(\hat{x}) = T_m$. The Neumann moving boundary solution, Eq. (4.65), also applies to such an externally cooled case. The thickness of the solid, $\hat{x}(t)$, can be determined providing that the overall heat transfer coefficient of the solid-environment boundary is known

4.3 Summary

The transfer of heat energy by various mechanisms during casting and crystal growth comprise the main topics of this chapter. Pure substances serve as model systems for understanding enthalpy balances and energy redistribution mechanisms accompanying solidification. Treating heat flow in pure materials also avoids the attendant complications of solute rejection or incorporation at the solid-melt interface, and solidification occurring over a range of temperatures.

Heat energy can be conducted, radiated, and advected through solids and liquids. Fourier's laws of heat conduction and the general time-dependent energy equation form the analytic framework upon which the mathematical theory of heat transfer rests. All solidification processes involve important, and often intricate heat transfer issues, because the presence of the solid–liquid transformation complicates the analysis of heat transfer by changing through time the physical size, shape, and position of the homogeneous domains in which the thermal fields develop. In addition, solidification liberates latent heat within the freezing body, a process which demands special mathematical consideration—viz., the Stefan balance—applied at the moving interface.

Quasi-static heat transfer theory essentially ignores the direct influences of the moving solid-melt interface on local temperature gradients, and allows the time-dependent temperature fields to 'relax' into simpler states. Quasi-static theory tends to underestimate how long it takes to solidify a body, an error that increases as the Stefan number increases. The full influences of the solid-melt interface on the thermal fields are captured in Neumann's moving boundary theory, which is analytically manageable for a few solidification geometries and boundary conditions admitting error-function type solutions to the energy equation. Finally the effects of the external environment, such as mold walls and convective fluid surroundings may be gathered as 'lumped' heat transfer coefficients, which include the mechanisms of radiation and convection, thereby completing the transfer of heat from the casting to the environment.

References

1. A.G. Ostrogorsky and B.B. Mikic, *Heat Mass Transfer*, DOI 10.1007/s00231-008-0438-9, Springer, New York, NY, 2008.
2. V. Alexiades and A.D. Solomon, *Mathematical Modeling of Melting and Freezing Processes*, Chap. 3, Hemisphere Publishing Corporation, Washington, DC, 1993.
3. *Moving Boundary Problems*, D.G. Wilson, A.D. Solomon and P.T. Boggs, Eds., Academic Press, New York, NY, 1978.
4. V.J. Lunardini, *Heat Transfer with Freezing and Thawing*, Elsevier, Amsterdam, 1991.
5. D. Poulikakos, *Conduction Heat Transfer*, Chaps. 9 and 10, Prentice Hall, Englewood Cliffs, NJ, 1994.
6. H.S. Carslaw and J.C. Jaeger, *Conduction of Heat in Solids*, 2nd Ed., Clarendon Press, Oxford, 1984.
7. J. Crank, *Free and Moving Boundary Problems*, Clarendon Press, Oxford, 1984.
8. M.E. Glicksman, *Diffusion in Solids: Field Theory, Solid-State from an and Applications*, Wiley Interscience Publishers, New York, NY, 2000.
9. G.H. Geiger and D.R. Poirier, *Transport Phenomena in Metallurgy*, Addison-Wesley, Reading, MA, 1973.
10. F.P. Incropera and D.P. DeWitt, *Fundamentals of Heat and Mass Transfer*, 3rd Ed., Wiley, New York, NY, 1990.
11. S. Kou, *Transport Phenomena and Materials Processing*, Wiley, New York, NY, 1996.
12. D.R. Poirier and E.J. Poirier, *Heat Transfer Fundamentals for Metal Casting*, 2nd Ed., The Minerals Metals & Materials Society, Warrendale, PA, 1994, 9.

Part II
Macrosegregation

Chapter 5
Solute Mass Balances: Macrosegregation

5.1 Local and Global Interfacial Equilibrium

Solute rejection and its redistribution during freezing and melting are basic topics introduced by the Dutch physical chemist H.W.B. Roozeboom more than a century ago [1, 2] to explain the influence of phase equilibria on the solidification behavior of alloys. Solute redistribution occurs at a moving solid–liquid interface when limited—or so-called *local*—thermodynamic equilibrium prevails, often aided by transport processes (diffusion and convective mixing) acting primarily within the melt phase ahead of the advancing interface. In this chapter several important settings are explored involving the rejection and redistribution of solutes under differing conditions of solidification. Specifically, during alloy freezing it will be shown that the condition of local thermodynamic equilibrium usually holds, whereas total—or so-called *global*—equilibrium seldom, if ever, occurs. Nevertheless, both local and global equilibria represent useful idealizations as thermodynamic limits, between which operate most solidification and single crystal growth processes encountered in practice. Local equilibria at solid–liquid interfaces are also exploited in several solidification-based ultrapurification methods, such as repeated unidirectional freezing and multipass zone melting, both of which processes will be explored in detail in this chapter.

5.1.1 Introduction

Solute partitioning between phases occurs by virtue of the fact that at any temperature and pressure at which solid and liquid co-exist, according to the equilibrium phase diagram, compositional 'tie lines' occur at any co-existence temperature. Tie lines connect unique, equilibrium, solid-phase compositions, $\hat{C}_s(T)$, with their respective interfacial liquid-phase compositions, $\hat{C}_\ell(T)$. Tie lines, therefore, graphically represent the temperature-dependent concentration 'jumps' expected at equilibrium between conjugate phase compositions. These jumps, or composition discontinuities, equalize the activities and chemical potentials of the components shared across the interface between a crystal and its molten phase. It is not

M.E. Glicksman, *Principles of Solidification*, DOI 10.1007/978-1-4419-7344-3_5,
© Springer Science+Business Media, LLC 2011

surprising, of course, that phases as different as are crystals and melts co-existing in thermodynamic equilibrium, and joined in physical contact at their common interface, must also differ in their chemical compositions. One may argue qualitatively that the large differences between the local atomic or molecular structure of unary (one component) solids and liquids, as already discussed in Chapter 1, would certainly justify distinctly different partition of an added component to achieve mutual saturation with respect to that component.

More specifically, the lack of long-range order (LRO) in melts usually—but not always—allows a greater uptake, or retention, of most added components, denoted as 'solutes', or accidentally added components, denoted as 'impurities', before saturation occurs. The crystalline solid, which has a high degree of LRO, simply provides fewer ways to accommodate solute atoms and impurities within its well-ordered lattice structure. Thus, for many alloy systems, the solubility of a component at any temperature (and fixed pressure) is greater in the melt than in the crystalline state. Some notable exceptions to this qualitative statement do occur, and their occurrences are not unusual. For example, where two elements happen to exhibit complete mutual solid-solubility, such as occurs in nearly ideal systems such as Ag-Au, the component (Ag) with the lower melting point *must* accept a higher solubility of the other component (Au) in its crystalline form, and a lower solubility in its liquid phase. The solidus and liquidus temperatures of Ag-rich solid solution therefore *both* rise monotonically with increasing solute content. Other exceptions are caused by the opposite thermodynamic situation where unusually strong interactions exist between the solute atoms and their surrounding lattice atoms in the solid. These phase relationships and the corresponding relative solubility of other components in their crystals and melts are quantified in Section 5.1.2.

5.1.1.1 Local Equilibrium

Fundamental conditions for *local* thermodynamic equilibrium to exist between crystalline solids and their melts may, of course, be stipulated irrespective of *any* structural or atomistic details of the solid and liquid phases. These fundamental conditions require that the following equalities hold during local equilibrium among a specially selected subset of the intensive thermodynamic properties of the solid and liquid phases in an N-component alloy:

1. *Thermal equilibrium:* $T_s = T_\ell$.
2. *Mechanical equilibrium:* $P_s = P_\ell$.
3. *Chemical equilibrium:* $\mu_i^s = \mu_i^\ell, (i = 1, 2, \cdots N)$.

Conditions 1 and 2 listed above for thermal and mechanical equilibria were already introduced in Chapters 2 and 4. The third condition, chemical equilibrium, requires equality of the component chemical potentials for each phase. When satisfied across a solid–liquid interface,[1] equality of the chemical potential for any component, i, leads to unequal concentration, or saturation, of that i^{th} component in the phases, viz., $\hat{C}_s(T, P) \neq \hat{C}_\ell(T, P)$.

The 'hat' symbols overlying these concentration functions serve to remind the reader that the thermodynamic equilibrium under discussion here is *local*, and therefore limited to phase regions in extremely close proximity to the interface. Generally, spatial differentiation of 'hatted' (local) concentration functions is *not* permitted. Additional statements or explanations are provided when dealing with such local (discontinuous) concentrations where spatial differentiation is contemplated.

As the concepts of macrosegregation and, later, microsegregation, are more fully developed in this and later chapters, and then applied to various solidification processes, distinctions will be drawn between local and global states of thermodynamic equilibrium. In addition, the same 'hat' notation is used elsewhere in this book to help identify occurrences of different idealizations for approximating thermodynamic conditions during freezing. Finally, where the mathematical 'hat' symbol is applied to tracking functions such as $\hat{x}(t)$, it designates the instantaneous position of a moving solid–liquid interface, which is generally a *continuous* function of time, and therefore differentiable.

5.1.2 The Distribution Coefficient, k_0

A fundamental property of any solid–liquid binary system is its equilibrium distribution coefficient, k_0. This number represents a key shared property of both phases, and appears in a great number of mathematical expressions involving the redistribution of solute between solids and melts undergoing freezing or melting. In the solidification and crystal growth literature the quantity k_0 is variously referred to as the distribution coefficient, chemical segregation coefficient, and the partition coefficient. All remain in usage, but we shall use the term *equilibrium* distribution coefficient.

Where the subscript is either dropped, or altered from the symbol, k_0, some change is implied in its value, usually caused by kinetic phenomena, such as fluid flow occurring near the interface, or by rapid interface motion that precludes local equilibrium. Changes in the distribution coefficient from the equilibrium value may also reflect extrinsic interactions affecting the interface, induced by application of an electric, magnetic, gravitational, or stress field. The equilibrium distribution coefficient, k_0, depends only on the interface temperature and pressure, provided that

---

[1] The chemical potentials in an N-component phase are the intensive thermodynamic variables associated with the phase's Gibbs free energy. Formally, for each component, $i = 1, 2, 3, \ldots N$, the chemical potential is $\mu_i = (\partial G/\partial N_i)_{T,P,N_{j \neq i}}$. See also Appendix A for additional details of this definition. The related partial molal Gibbs free energies of the components forming a phase diagram are found by reading the intersections of the common tangent extended to the pure component edges of the diagram. The partial molal Gibbs energy equals the chemical potentials of the components relative to the reference state for which the phase diagram is constructed [3].

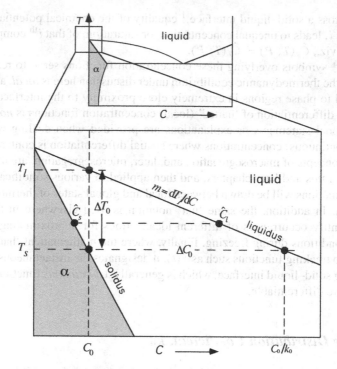

Fig. 5.1 *Upper*: Binary phase diagram (pressure fixed) showing the α solid solution on the left-hand side; *Lower*: Detail of the terminal corner of this phase diagram defining the equilibrium distribution coefficient, k_0, for a dilute alloy with overall composition C_0. Several other important phase diagram-derived parameters are also displayed: including the slopes of the solidus and liquidus, the alloy's overall freezing range, ΔT_0, and the initial tie-line jump, ΔC_0, for the first melt to form upon heating

the interface remains quiescent, or quasi-static.[2] As is indicated on Fig. 5.1, k_0 is defined directly from the system's phase diagram. Its value, moreover, is just the tie-line ratio of the equilibrium concentration of the solid, $\hat{C}_s$, to that of the liquid phase, $\hat{C}_\ell$, at the specified temperature and pressure, namely,

$$k_0 \equiv \frac{\hat{C}_s(T, P)}{\hat{C}_\ell(T, P)}. \tag{5.1}$$

---

[2] The term 'quasi-static' was already introduced in Chapter 4 to denote the effects of moderate interface speeds on the thermal fields during freezing of a pure substance. Here, however, one deals with alloys, where chemical equilibrium occurs in addition to thermal and mechanical equilibria. Chemical equilibrium normally develops far more slowly than do thermal or mechanical equilibria. A quantitative distinction between 'moderate rates of interface motion' versus 'rapid solidification' is provided later in Chapter 17. The judgment made on the rapidity of the interface for specific solidification processes is based on the time scales required for short-range diffusive motions of solute atoms to allow adequate transit between solid and liquid.

The dependence of the distribution coefficient on the interface temperature and pressure, indicated on the RHS of Eq. (5.1), is shown here only to remind the reader that k_0, in general, is *not* a constant. Indeed, in most alloy systems k_0 exhibits substantial dependences on the temperature and composition of the melt, as the slopes of the solidus, $m_s = \partial T / \partial \hat{C}_s$, and liquidus, $m_\ell = \partial T / \partial \hat{C}_\ell$, can not remain constant with composition. It is through the constitutive relations provided by the phase diagram via the solidus and liquidus curves that the indicated temperature dependence results. In Section 5.7 the importance of these dependences on temperature and other variables will be demonstrated in practical applications of materials purification.

As shown in Fig. 5.1 for the case of a binary alloy, the numerator and denominator of Eq. (5.1) represent, respectively, the equilibrium, or 'tie-line', compositions of the solid and liquid phases co-existing in local equilibrium at the solid–liquid interface at the temperature, compositions, and pressure of interest. Local equilibrium implies that components A and B, present on both sides of the solid–liquid interface—but in different concentrations—exchange sufficiently by short-range diffusive fluctuations across the phase boundary as to maintain equality of their chemical potentials. Equality of these chemical potentials may be assumed to hold, irrespective of any moderate motion of the interface. Thus, the validity of the definition of k_0 in Eq. (5.1) *during* net solidification or melting rests on the assumption of local thermodynamic equilibrium. The equality of the chemical potentials is expected to be limited to just a small spatial region where the phases remain in intimate molecular contact, communicating via frequent diffusive fluctuations. The spatial extent of such an interfacial locale, where local equilibrium occurs, is severely limited by the time-scale needed for such exchanges to occur when the interface moves. Classical diffusion theory stipulates mathematical relationships linking the spatial and temporal scales required for local equilibrium during solidification or melting. These relationships will be explored in some detail later in Chapter 17, where local equilibrium is precluded during rapid solidification processing. To reiterate, for a binary alloy undergoing freezing or melting, the specification of local equilibrium requires two additional thermodynamic statements beyond the stipulation of equality of temperature and pressure:

$$\mu_A^s = \mu_A^\ell, \tag{5.2}$$

and

$$\mu_B^s = \mu_B^\ell. \tag{5.3}$$

The imposition of local chemical equilibrium through Eqs. (5.2) and (5.3) also provides a critically important physical assumption used in many kinetic solidification theories, some of which will be discussed in this and subsequent chapters.

Finally, local thermodynamic equilibrium limits the interface during melting and freezing to a narrow region—usually just a few atomic diameters—in order to guarantee the sufficiency of diffusive molecular interchanges and fluctuations. Thin transition zones (<10 nm) develop between many metallic alloys and their melts, attributed to their extreme differences in local structure and packing. Moreover, these same interfaces often exhibit remarkably high mobilities, moving easily when even small net thermodynamic forces are available. Crystal-melt interfaces in other materials, such as semiconductors, ceramics, and polymers, can develop much larger kinetic resistances to motion, and often require greater departures from local equilibrium to move them at all. Interfacial mobility, the kinetic concept that quantifies the ease of such interfacial motions under net driving forces, will be explored in detail in later chapters.

The equilibrium distribution coefficient in many solidification models is assumed to remain constant throughout the entire process of solidification. This assumption either requires that the solidus and liquidus are straight lines on the phase diagram, so that their individual slopes, m_s and m_ℓ, respectively, remain independent of the temperature and composition, or that by some rare coincidence their composition ratio at different temperatures happens to hold steady. As mentioned earlier, this cannot be true across the phase diagram. Consequently, such a simplifying assumption will usually remain valid for dilute alloys, and for solidus–liquidus lines bordering some eutectic and peritectic reactions. However, even where the liquidus and solidus appear markedly curved, Eq. (5.1) may still be applied point-wise, i.e., using the tie line at each temperature where solid and melt coexist. Lastly, the small (but practically important) volume changes accompanying solidification and melting will be ignored, which is tantamount to assuming that $\rho_s = \rho_\ell$. That implicit assumption avoids unnecessarily complicating the mathematical expressions to be derived subsequently for predicting alloy macrosegregation and impurity removal using the processes of melting and freezing. Fortunately, the volume changes that accompany melting and freezing induce only minor changes in the quantitative predictions of chemical macrosegregation during solidification and melting.

5.2 Solute Rejection at the Solid–Liquid Interface

5.2.1 Interfacial Solute Balance

Consider the solid-melt configuration shown partially solidified in Fig. 5.2. One assumes that the melt phase remains well mixed throughout by either rapid liquid-state diffusion or by long-range forced or natural convective flows. Therefore, the composition field in the liquid, $C_\ell(x)$, is taken everywhere as uniform, and fixed at the value $\hat{C}_\ell(T, P)$. Note that this assumption of 'perfect mixing', combined with a constant equilibrium distribution coefficient, k_0-value, also fixes the interfacial composition of the solid phase, $\hat{C}_s(T, P)$. As local chemical equilibrium is tacitly assumed to hold, the interface composition ratio is given by Eq. (5.1). Thermal and mechanical equilibria are also assumed throughout the system, so the phase compositions must also remain consistent with the solidifying system's uniform temperature and pressure. Beyond the fact that the system's temperature is spatially uniform, the temperature must fall over time as solidification progresses, following

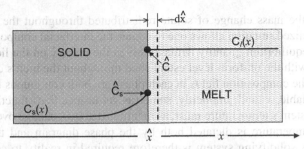

Fig. 5.2 Schematic of local equilibrium at a solid–liquid interface during freezing of an alloy. The solid–liquid interface at $\hat{x}$ is slowly advancing to the right consuming the uniformly mixed binary melt at concentration $C_\ell(x)$ by transforming it to a crystalline solid of varying composition, $C_s(x)$. Local equilibrium at the solid–liquid interface requires the 'tie line' concentration jump, $\hat{C}_\ell - \hat{C}_s$, where the compositions indicate the concentration of the minor component, or solute, B. As suggested in this schematic diagram, the compositions solidifying from an alloy melt form a continuous spatial field, $C(x)$, to be discussed below as either macro- or microsegregation

the downward course of the interface equilibrium temperature, $\hat{T}(\hat{C}_i)$ $(i = s, \ell)$, as dictated by two-phase equilibrium in the binary phase diagram. See again Fig. 5.1.

The chemical diffusivity of solids, even at temperatures as high as their solidus, remains orders-of-magnitude smaller than those for their melts at the same temperature [4, 5]. Thus, we will assume at this point that solute diffusion in the solid phase may be totally ignored by approximating the solute diffusivity in the solid state as $D_s = 0$. However, either in cases of extremely slow solidification, such as might occur on long geological time scales in the slow growth of certain minerals, or, especially, for solidification occurring over small length scales, as found in microsegregation,[3] and in electronic doping of semiconductors for microcircuits, the latter of which involves atom motions over nanoscopic length scales, solid-state diffusion must be included for accuracy. Discussions of solid-state and liquid-state diffusion during solidification or crystal growth will be deferred until Section 5.4. This topic also arises in Chapters 6 and 14, where solid-state diffusion is shown to influence macrosegregation, microsegregation, and interface stability.

The solute mass distribution of component B is sketched in Fig. 5.2. Here the solute concentrations are shown as increasing toward the right in the solid phase ($x \leq \hat{x}$), implying that $k_0 < 1$, and remaining level in the liquid ($x \geq \hat{x}$), which is mixed perfectly throughout. The overall mass balance for an infinitesimal increment of solidification, $d\hat{x}$, into such a system may be written as,

$$dm_s + dm_\ell = 0. \tag{5.4}$$

The mass differentials, dm_s and dm_ℓ on the left-hand side of Eq. (5.4) are, respectively, the mass change of solute, B, rejected during an increment of additional

---

[3] See Chapter 14 for a detailed discussion of small-scale diffusion during dendritic microsegregation.

freezing, and the mass change of solute redistributed throughout the well-mixed melt. This idealized situation allows one to equate the interfacial composition of the melt, $\hat{C}_\ell$, (an equilibrium quantity defined only at the point $\hat{x}$ on the liquid side of the interface) with the uniform level established throughout the melt's composition field, $C_\ell(x)$. The composition fields in each phase are both continuous functions of the spatial variable, x, and, implicitly, some (as yet unspecified) function of time, because the system's temperature must, as mentioned above, change over time. The decrease in temperature is dictated both by the phase diagram and the imposed cooling rate. A solidifying system is therefore required in reality to cool through and beyond its total freezing range, ΔT_0 to complete the freezing process. See again Fig. 5.1.

The interfacial compositions, $\hat{C}_\ell(T)$, for the liquid and $\hat{C}_s(T)$ for the solid at an arbitrary instant of time remain linked via local tie-line equilibria through the phase diagram's liquidus and solidus, namely

$$\hat{C}_\ell(T) = C_\ell(x) \ \ (x \geq \hat{x}), \tag{5.5}$$

and

$$\hat{C}_s(T) = k_0 \hat{C}_\ell(T), \tag{5.6}$$

where Eq. (5.6) is merely a restatement of the definition for k_0 given by Eq. (5.1). Substituting these interfacial compositions into the differential solute balance, Eq. (5.4), yields the solute mass balance for an infinitesimal increment of solidification, now expressed in terms of both local and field composition variables that indicate the concentrations of B, and the fractional phase volumes, f_s and f_ℓ,

$$\left[\hat{C}_s(T) - \hat{C}_\ell(T)\right] df_s + f_\ell \, dC_\ell = 0. \tag{5.7}$$

Factoring out $\hat{C}_\ell(T)$ from the first term on the LHS of Eq. (5.7), and applying conservation of volume, i.e., $f_s + f_\ell = 1$, gives

$$\hat{C}_\ell(T)(k_0 - 1)df_s = (f_s - 1)dC_\ell. \tag{5.8}$$

Equation (5.5), a mass balance based on *perfect* mixing in the melt, justifies replacing the local interfacial composition, $\hat{C}_\ell(T)$, on the left-hand side of Eq. (5.8) by the continuous (and constant) field variable, $C_\ell(f_s)$.

Now, the variables appearing in Eq. (5.8) can be separated, and the differential solute mass balance solved by integration,

$$\int_0^{f_s} \frac{df_s}{1 - f_s} = \int_{C_0}^{C_\ell} \frac{dC_\ell}{(1 - k_0)C_\ell}. \tag{5.9}$$

These elementary integrals are evaluated as

$$\ln\left(\frac{1}{1-f_s}\right) = \frac{1}{k_0-1}\ln\left(\frac{C_0}{C_\ell}\right). \tag{5.10}$$

Taking anti-logarithms of Eq. (5.10) and rearranging the terms provides the chemical macrosegregation relationship for a solidifying bar of alloy melt in terms of the fraction solid, f_s, namely,

$$\frac{C_\ell}{C_0} = (1-f_s)^{k_0-1}. \tag{5.11}$$

The equivalent macrosegregation relationship for the variation of the solid's composition with fraction solidified may be found by inserting the definition of the distribution coefficient, k_0, given by Eq. (5.1), into the result for the liquid phase, Eq. (5.11).

$$\frac{\hat{C}_s}{C_0} = k_0(1-f_s)^{k_0-1}. \tag{5.12}$$

If, as already assumed in the mass balance, all solid-state diffusion is precluded both during and after solidification, it then follows that at each instant, as the interface at location $\hat{x}$ cools to some temperature, T, between the liquidus and the solidus, that $\hat{C}_s(T) = C_s(\hat{x})$. The lack of any solute transport occurring within the solid permits writing Eq. (5.12)—a local interface equation—as the chemical macrosegregation equation for the frozen solid, namely,

$$\frac{C_s}{C_0} = k_0(1-f_s)^{(k_0-1)}. \tag{5.13}$$

5.3 Gulliver–Scheil Macrosegregation Theory

The companion pair of equations just derived by using a differential solute mass balance, Eqs. (5.11) and (5.13), is now called the Gulliver–Scheil segregation law [6–9]. G.H. Gulliver [10], an English metallurgist teaching engineering in Scotland, published these equations by considering mass balances for a sequence of infinitesimal crystallization stages, each occurring at a slightly reduced temperature. The incrementally added solid, of composition C_s, is separated from the remaining fraction of liquid at each stage, and the temperature is lowered slightly to continue the freezing process. Gulliver's original macrosegregation result was derived in the awkward form of an infinite product, and, unfortunately, was overlooked for many years, although he claimed, correctly, but without adding the needed proof, that it reduced to Eq. (5.11) and Eq. (5.12). Over the next 20 years, German investigators E. Scheuer [9], and then E. Scheil [6], independently 'discovered' Gulliver's macroseg-regation law by applying the more familiar differential equation approach shown above. Their famous (identical) results underscore clearly the critically important point that the equilibrium phase diagram, along with its associated equilibrium 'lever rule', provide misleading predictions for how molten alloys solidify under most practical conditions, where *global* equilibria seldom, if ever, occurs. The Gulliver–Scheil equations represent the correct limiting behavior to solute redistri-bution occurring in a freezing alloy, where mixing is totally effective within the melt and insignificant within the solid. The lever law, by contrast, provides the opposing and unrealistic limit for the redistribution of solute during freezing under full, i.e., global thermodynamic equilibrium.

5.3.1 Determination of the Distribution Coefficient

Most alloy systems, as explained qualitatively in Section 5.1.1, exhibit values for the equilibrium distribution coefficient, k_0, that are less than unity, but some also have $k_0 > 1$. The special case of $k_0 = 1$ is always precluded on thermodynamic grounds. Were the condition $k_0 = 1$ allowed by a phase diagram—and it never is—it would imply that the tie-lines defining the interfacial composition jump, $\hat{C}_\ell - \hat{C}_s$, would shrink to zero length, so that the single-phase liquid region would directly con-tact the single-phase solid region, without benefit of any intervening mixed-phase region. That situation is not permitted thermodynamically in any alloy.

The equilibrium distribution coefficient, k_0, can, of course, be found easily at any temperature for a binary alloy provided that accurate thermodynamic data are available. If the phase diagram is available, all that is needed are the solid–liquid tie lines over the alloy's freezing range. A practical alternative to finding k_0 by thermodynamic means is measuring the *effective* distribution coefficient, k_{eff}, using a solidification experiment. The advantages of measuring k_{eff} rather than k_0 is that the complicated influences of melt convection, interface stability, and freezing speed can be collectively ascertained by applying the Gulliver–Scheil analysis to segre-gation data obtained from simple mold geometries, such as freezing in a vertical cylinder-, or slab-mold, or cast as a horizontal rod or bar. These mold geometries allow the fraction solidified to be measured easily. If the solute concentration, C_s relative to the initial concentration, C_0, is determined for a few values of the fraction solidified, f_s, the data can be plotted log-log using Eq. (5.13), with k_{eff} substituted for k_0, specifically,

$$\log_{10} \frac{C_s}{C_0} = \log_{10} k_{eff} + \left(k_{eff} - 1\right) \log_{10} \left(1 - f_s\right). \tag{5.14}$$

As Fig. 5.3 demonstrates, one may extract independent estimates of k_{eff} from exper-imental segregation data by using both the slopes and ordinate intercepts.

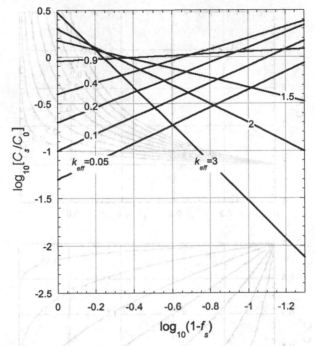

Fig. 5.3 Plot of Eq. (5.14) to allow estimates of effective distribution coefficients, k_{eff}, to be extracted from experimental data of solute concentration versus fraction solidified, f_s. Ordinate is the measured relative concentration, $\log_{10}(C_s/C_0)$, and the abscissa is the $\log_{10}(1 - f_s)$. Both the ordinate intercepts of the curves and their slopes provide independent estimates for k_{eff}

Later, in Chapter 17, it will be demonstrated that the solute distribution coefficient can depend on both the melt composition and the freezing rate, especially if that rate becomes sufficiently high to inhibit attainment of *local* equilibrium. Under the extreme kinetic conditions met during high-speed solidification, the distribution coefficient in an alloy can be forced to approach unity, and 'partitionless' freezing ($k_{eff} = 1$) can, in fact, be realized.[4]

Figures 5.4 and 5.5 show plots of Gulliver–Scheil segregation curves, Eqs. (5.11) and (5.13), respectively, for the remaining liquid fraction, and the solidified solid fraction. These macrosegregation curves do not account for any restrictions to solute macrosegregation caused by solubility limits imposed by the phase diagram, or for the appearance at lower temperatures of any polyphase reactions, such as eutec-

---

[4] Again, it is important to keep in mind that in order for the solid phase and liquid phase to solidify with *identical* composition, one needs to achieve extremely rapid solidification conditions that operate far from thermodynamic equilibrium. Achieving such extreme non-equilibrium conditions in both the laboratory and at industrial scales by rapid solidification or, avoiding solidification altogether and forming rapidly quenched metallic glasses directly from the melt, have become technically feasible over the past 30 years by using a variety of clever processing methods, a few of which will be discussed later in Chapter 17.

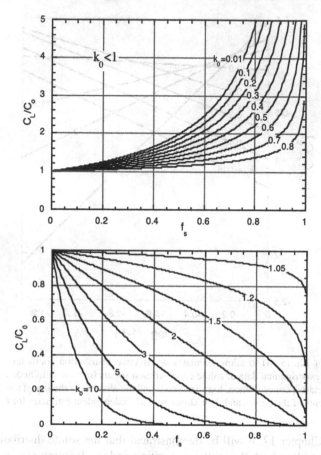

Fig. 5.4 *Upper:* Gulliver–Scheil macrosegregation curves for the liquid phase in alloys with $k_0 < 1$. The composition ordinate is the composition of the remaining liquid phase, C_ℓ, divided by its starting composition, C_0. As the fraction solidified, f_s, approaches unity, the liquid compositions rise rapidly, and eventually will exceed any physically meaningful concentration level. The phase solubilities given in the phase diagram must be used to truncate predictions based on these macrosegregation curves. *Lower:* Gulliver–Scheil macrosegregation curves for the liquid phase in alloys with $k_0 > 1$. As the fraction solidified, f_s, approaches unity, the liquid compositions for $k_0 > 1$ fall continuously and approach zero. Singularities as $f_s \rightarrow 1$ in the upper set of curves for $k_0 < 1$ do not develop when $k_0 > 1$

tics or peritectics. Such complications of the actual solidification process remain as important factors that must be added independently to interpret properly binary alloy segregation curves. A few of these complications will be discussed once certain additional physical phenomena, such as solid-state diffusion, melt convection, and the nature of secondary solidification reactions, are introduced into the solutal mass balance.

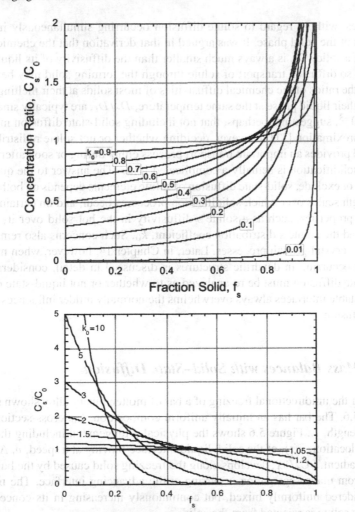

Fig. 5.5 *Upper:* Gulliver–Scheil macrosegregation curves for the solid in alloys with $k_0 < 1$. The composition ordinate is the composition ratio of the solid phase, C_s, divided by the alloy's (uniform) starting composition, C_0. As the fraction solidified, f_s, approaches unity, concentrations of solute in the solid rise rapidly, and eventually exceed physically meaningful concentration levels. The rapid rise in solid composition toward the termination of freezing is delayed toward f_s approaching unity, both for small values of k_0, and for values of $k_0 \rightarrow 1$. *Lower:* Gulliver–Scheil macrosegregation curves for the solid with $k_0 > 1$. The solid compositions for alloys with $k_0 > 1$ fall and approach zero as the fraction solidified, f_s, increases

5.4 Macrosegregation with Solid–State Diffusion

5.4.1 Solid-State Diffusion

The Gulliver–Scheil macrosegregation equations [6, 7, 10], derived and discussed in Sections 5.2 and 5.3, only considered solute mass balances at moving solid–liquid interfaces, without regard to solute diffusion occurring simultaneously in the hot portions of the solid phase. It was argued in that derivation that the chemical diffusivity of a solid, D_s, is always much smaller than the diffusivity of its liquid phase, D_ℓ, and so diffusion transport of solute through the forming solid may be ignored. In fact, the ratios of the chemical diffusivities of most solids at their melting point to that for their liquid phase at the same temperature, D_s/D_ℓ, are typically smaller than about 10^{-3}, suggesting, perhaps, that not including solid-state diffusion might be a safe approximation [11]. However, deciding whether or not solute redistribution in the solid provides an important transport process either during or soon after completion of solidification is actually a complicated issue. The answer to the question to include, or exclude, solid-state diffusion of solute elements depends on both the time and length scales over which solidification processes occur, and on certain system-specific properties, such as a solute's diffusivity in the hot solid over its freezing range, and the solute's distribution coefficient, k_0. Such concerns also remain valid for many crystal growth processes. Later, in Chapter 14, however, when microsegregation occurring in dendritic structures is discussed in detail, considerations of solid-state diffusion must be re-analyzed as to whether or not liquid-state diffusion near unstable interfaces always overwhelms the normally milder influences of solid-state diffusion.

5.4.2 Mass Balances with Solid–State Diffusion

Consider the unidirectional freezing of a bar of molten binary alloy shown sketched in Fig. 5.6. The bar has an initially uniform composition, C_0, cross-sectional area, A, and length, L. Figure 5.6 shows the physical arrangement, including the instantaneous location, $\hat{x}(t)$, of the solid–liquid interface moving at a speed, v. A composition gradient, $\partial C_s/\partial x$, develops along the freezing solid caused by the build-up of solute from macrosegregation occurring at the advancing interface. The melt may be considered uniformly mixed, but continuously increasing in its concentration, $C_\ell(t)$, as solute is rejected from the solid.

Choose a control volume of cross-sectional area A, differential thickness, dx, that is co-located at the moving interface coordinate, $\hat{x}(t)$. The local solute mass balance affecting the control volume now consists of three contributions:

1. Solute mass, dm_1, which is rejected by macrosegregation into the melt to allow the infinitesimal transformation of liquid-to-solid as the interface moves from $\hat{x}(t) \rightarrow \hat{x}(t + dt)$:

$$dm_1 = -A(\hat{C}_s - \hat{C}_\ell)d\hat{x}. \tag{5.15}$$

2. Solute mass, dm_2, that 'back-diffuses' out of the left-hand face of the control volume into the already-formed, adjacent hot solid. Fick's 1st law of diffusion [11] is invoked to approximate the mass flux flowing down the concentration gradient, i.e., towards the previously-solidified solid:

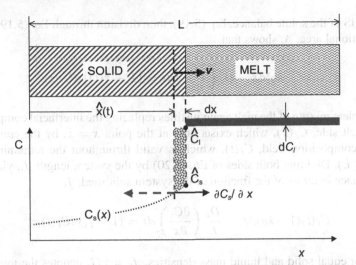

Fig. 5.6 Schematic of an advancing solid–liquid interface during freezing of an alloy melt in the form of a bar of length L. The solid–liquid interface has reached the position $\hat{x}(t)$, and is traveling to the right at a speed, v, steadily consuming the uniformly-mixed melt of concentration C_ℓ. The equilibrium jump in concentration at the interface is depicted as $\hat{C}_\ell - \hat{C}_s$. The concentration field that develops in the solidifying solid, $C_s(x)$, is shown as increasing with distance and time, thus forming a concentration gradient, dC_s/dx. The hot solid, subject to its own concentration gradient, responds by sending a flux of solute down the concentration gradient (*dashed left-pointing arrow*)

$$dm_2 = -AD_s \left(\frac{\partial C_s}{\partial x}\right)_{\hat{x}(t)} dt. \tag{5.16}$$

3. Rejected solute mass dm_3 is uniformly dispersed throughout *all* the remaining molten phase. This additional infusion of solute raises the solute concentration in the melt by an infinitesimal amount $dC_l(t)$:

$$dm_3 = A\left[L - \hat{x}(t)\right] dC_\ell(t). \tag{5.17}$$

The mass balance for the solute after an infinitesimal volume increment of freezing, $Ad\hat{x}$, is given by the conservation statement,

$$dm_1 + dm_2 = dm_3. \tag{5.18}$$

Substituting Eqs. (5.15), (5.16), and (5.17) into Eq. (5.18) yields the solutal mass balance for an infinitesimal advance of the solid–liquid interface.

$$-A(\hat{C}_s - \hat{C}_\ell)d\hat{x} - AD_s \left(\frac{\partial C_s}{\partial x}\right)_{\hat{x}} dt = A\left[L - \hat{x}\right] dC_\ell(t). \tag{5.19}$$

If the local equilibrium distribution coefficient connecting the ratio of the solid and liquid phase tie-line compositions, $k_0 = \hat{C}_s/\hat{C}_\ell$, is substituted into the first term on the LHS of the solute balance, Eq. (5.19), then division through Eq. (5.19) by the cross-sectional area, A, shows that

$$\hat{C}_\ell(1 - k_0)d\hat{x} - D_s\left(\frac{\partial C_s}{\partial x}\right)_{\hat{x}}dt = [L - \hat{x}(t)]dC_\ell(t). \qquad (5.20)$$

The complete mixing of the melt again justifies replacing the interfacial composition on the melt side, $\hat{C}_\ell(\hat{x})$, which exists only at the point $x = \hat{x}$, by the *continuous* uniform composition field, $C_\ell(t)$, which is valid throughout the remaining melt ($\hat{x} \leq x \leq L$). Dividing both sides of Eq. (5.20) by the system length, L, yields the mass balance in terms of the fraction of the system solidified, f_s.

$$C_\ell(t)(1 - k_0)df_s - \frac{D_s}{L}\left(\frac{\partial C_s}{\partial x}\right)_{\hat{x}}dt = (1 - f_s)dC_\ell(t), \qquad (5.21)$$

where, for equal solid and liquid mass densities, $f_s = \hat{x}/L$ denotes the mass fraction of the molten bar already solidified when the solid–liquid interface reaches the location $x = \hat{x}(t)$.

The composition gradient developed by the macrosegregation process in the hot solid located just to the left of the solid–liquid interface can be related to the rate of change of liquid composition with the fraction solidified, $\partial C_l/\partial f_s$. This differential coefficient can be developed in three steps:

1. Applying chain differentiation: $\partial C_\ell(t)/\partial x = (\partial\hat{C}_\ell(t)/\partial f_s)(\partial f_s/\partial\hat{x})$;
2. Recognizing that $\partial f_s/\partial\hat{x} = 1/L$;
3. Using local equilibrium: $C_\ell(t) = \hat{C}_\ell(t) = \hat{C}_s(t)/k_0$;

If these steps are inserted into Eq. (5.21), one obtains an estimate for the solute concentration gradient that develops in the melt adjacent to the moving solid–liquid interface, $\hat{x}(t)$.

$$\left(\frac{\partial C_s}{\partial x}\right)_{\hat{x}(t)} = k_0\frac{\partial\hat{C}_\ell}{\partial f_s}\left(\frac{\partial f_s}{\partial\hat{x}}\right) = \frac{k_0}{L}\frac{\partial C_\ell}{\partial f_s}. \qquad (5.22)$$

Equation (5.22) may now be substituted back into Eq. (5.21). After several additional algebraic rearrangements it yields the desired differential solute mass balance including the effect of back-diffusion through the already solidified solid,

$$(1 - k_0)C_\ell\frac{\partial f_s}{\partial t} - \left(\frac{D_s k_0}{L^2}\right)\frac{\partial C_\ell}{\partial f_s} = (1 - f_s)\frac{dC_\ell}{dt}. \qquad (5.23)$$

5.4.3 Time Dependence of Fraction Solidified

If the directional solidification of the alloy shown in Fig. 5.6 follows so-called parabolic time kinetics, which is generally consistent with heat transfer-limited freezing, (See for example either the quasi-static heat transfer approximation developed in Chapter 4, or Neumann's moving boundary solution to the energy equation, solved in Section 4.2.) then f_s may be expressed as an explicit parabolic function of the physical solidification time, t, or its dimensionless form, τ, namely,

$$f_s(t) = \sqrt{\frac{t}{t_f}} = \sqrt{2\tau}. \tag{5.24}$$

This result can be easily found from the quasi-static approximation by taking the limit of large Biot numbers appropriate to the case of solidification limited only by thermal conduction, i.e., heat transfer unimpeded by significant external (environmental) thermal resistance. In Eq. (5.24), τ equals the dimensionless solidification time, defined as $t/2t_f$. Here t_f is the physical time at which solidification is completed, which corresponds to the situation where f_s equals unity, and when $\tau \to 0.5$. Differentiation of Eq. (5.24) yields the dimensionless freezing rate for ideal heat conduction,

$$\frac{\partial f_s}{\partial \tau} = \frac{1}{\sqrt{2\tau}} = \frac{1}{f_s}, \tag{5.25}$$

which when inserted back into the first term in the solute mass balance, Eq. (5.23), gives

$$\frac{(1 - k_0)}{2f_s t_f} C_\ell - \left(\frac{D_s k_0}{L^2}\right) \frac{\partial C_\ell}{\partial f_s} = (1 - f_s)\frac{dC_\ell}{dt}. \tag{5.26}$$

The RHS of Eq. (5.26) can be modified by twice applying the chain differentiation rule,

$$\frac{dC_\ell}{dt} = \left(\frac{\partial C_\ell}{\partial f_s}\right)\left(\frac{\partial f_s}{\partial \tau}\right)\left(\frac{d\tau}{dt}\right), \tag{5.27}$$

Differentiating the parabolic kinetic freezing relationship, Eq. (5.24), provides the differential relationship $d\tau/dt = 1/2t_f$ that can be substituted into the RHS of Eq. (5.27) to form the ordinary differential equation,

$$\frac{(1 - k_0)}{2f_s t_f} C_\ell - \left(\frac{D_s k_0}{L^2}\right) \frac{\partial C_\ell}{\partial f_s} = \frac{(1 - f_s)}{2f_s t_f}\frac{dC_\ell}{df_s}. \tag{5.28}$$

After several additional steps of algebra to separate the dependent and independent variables, C_ℓ and f_s, respectively, one obtains the separated form of Eq. (5.28), which is the desired solute mass balance with solid-state back-diffusion through the hot solid,

$$\frac{1}{1-k_0}\left(\frac{dC_\ell}{C_\ell}\right) = \frac{df_s}{1 - f_s\left(1 - \frac{2k_0 D_s t_f}{L^2}\right)}. \tag{5.29}$$

Equation (5.29) can be integrated easily using the initial system conditions for the solutal mass balance: namely, $C_\ell = C_0$, when $f_s = 0$. This initial condition yields the particular integrals,

$$\frac{1}{1-k_0}\int_{C_0}^{C_\ell}\left(\frac{dC_\ell}{C_\ell}\right) = \int_0^{f_s}\frac{df_s}{1 - f_s\left(1 - \frac{2k_0 D_s t_f}{L^2}\right)}. \tag{5.30}$$

5.4.4 Solutal Fourier Number

By grouping appropriate terms in Eq. (5.30), a solutal Fourier number, Fo_s, may be defined that provides a convenient dimensionless measure of solidification time with solid-state diffusion, namely,

$$Fo_s \equiv \frac{D_s t_f}{L^2}. \tag{5.31}$$

If the definition for Fo_s, given by Eq. (5.31), is substituted back into the denominator on the RHS of Eq. (5.30), and the definite integrals are each evaluated by forward integration, one obtains Brody and Flemings' modification of the Gulliver–Scheil macrosegregation equations [12, 13].

The Brody–Flemings model, developed in the mid-1960s, permitted the first estimates of the amount of chemical segregation that is expected in the melt when some solid-state diffusion occurs during solidification. Specifically, by evaluating the integrals in Eq. (5.30) one obtains the Brody–Flemings solution, Eqs. (5.32) and (5.33), which show that the extent of diffusion-modified macrosegregation depends on three independent dimensionless factors:

(1) the distribution coefficient, k_0, obtained from the system's phase diagram;
(2) the degree of solidification, f_s; and
(3) the time taken to accomplish that solidification, Fo_s.

$$\frac{C_l}{C_0} = [1 - f_s(1 - 2k_0 Fo_s)]^{\frac{k_0-1}{1-2k_0 Fo_s}}. \tag{5.32}$$

The companion equation for estimating the amount of macrosegregation that develops in the solid in the presence of back-diffusion is easily obtained by multiplying both sides of Eq. (5.32) by the distribution coefficient, k_0, to yield

$$\frac{C_s}{C_0} = k_0[1 - f_s(1 - 2k_0 Fo_s)]^{\frac{k_0-1}{1-2k_0 Fo_s}}. \tag{5.33}$$

The pair of macrosegregation formulas, Eqs. (5.32) and (5.33), actually include only a first approximation for estimating the amount of solute transported through the adjacent solid. An additional but unstated approximation in formulating these equations is that although diffusion transport in the hot solid obeys Fick's 1st law, the solution does *not* obey the time-dependent diffusion equation, or Fick's 2nd law [11]. That limitation in the Brody–Flemings macrosegregation model is, however, not substantial, because in many practical solidification problems the Fourier number setting the time scale during which solidification occurs remains small. With reference again to Fig. 5.6, we note that as the Fourier number gets larger, solute diffusing away from the moving solid–liquid interface at $x = \hat{x}(t)$ eventually reaches the system's left-hand boundary at $x = 0$. Diffusing solute atoms, in such a circumstance, that reach the left boundary would be 'reflected' back into the hot solid toward the solid–liquid interface. This so-called 'diffusion-reflection' effect— which is well-known in many solid-state diffusion problems [11]—acts to reduce the concentration gradient near the interface, and, consequently would reduce the amount of back diffusion. Thus, the Brody–Flemings results, Eqs. (5.32) and (5.33), tend to overestimate the amount of solid-state diffusion that actually occurred during freezing, particularly in cases of 'slow' solidification, where the solutal Fourier number is large.

The reader should note that in order for the solutal Fourier number to be considered large enough to cause appreciable error in the mass balance, the dimensionless group $2k_0 Fo_s$, which appears in Eqs. (5.32) and (5.33), must be comparable to unity. Stated somewhat differently, if the time scale measured with the solutal Fourier number is negligibly small, i.e., $Fo_s \ll 1/2k_0$, then the Brody–Flemings model properly reduces to the original Gulliver–Scheil macrosegregation model, the results for which were discussed in Section 5.3. When $Fo_s \approx 1/2k_0$, or larger, then the Brody–Flemings model becomes increasingly affected by the occurrence of 'diffusion reflection' of the solute. These mathematical requirements impose restrictions on the solidification length and time scales over which the Brody–Flemings mass balance for *macrosegregation* remains valid.[5] Specifically, for chemical *macrosegregation* occurring in cases where the system's length, L, might be large (e.g., on the order of several centimeters), the solutal Fourier number would usually be negligible, excepting instances where the solidification time scale, t_f, stretched out to enormous (geological) magnitudes. The criteria needed for judging the acceptable ranges of compatible length and time scales used in this theoretical model will be revisited later, when developing the subject of chemical *microsegregation* in dendritic microstructures, as dealt with in Chapter 14.

---

[5] When Brody–Flemings theory is applied to cases of *microsegregation*, increasing the solutal Fourier number may have other compensating effects, such as allowing phase coarsening to develop, which can increase the effective length scale over which diffusion must operate. Thus, where there is a microstructure that can respond to being held at elevated temperature, the situation regarding diffusion time and length scales becomes more complicated.

5.5 Limits of the Brody–Flemings Solute Balance

Two limits may be distinguished from Eq. (5.32) that are of interest here:

1. *No back-diffusion* ($Fo \to 0$): The fractional composition occurring throughout the well-mixed liquid fraction ($f_\ell = 1 - f_s$) remaining is

$$\frac{C_\ell}{C_0} = (1 - f_s)^{k_0-1}, \qquad (5.34)$$

and the companion equation for the solid phase's fractional composition instantaneously freezing out is

$$\frac{C_s}{C_0} = k_0(1 - f_s)^{k_0-1}. \qquad (5.35)$$

Equations (5.34) and (5.35), as mentioned above, recapture the classical Gulliver–Scheil macrosegregation equations [Cf. Eqs. (16.11) and (16.12)] derived earlier that neglect solid-state diffusion during solidification.

2. *Equilibrium freezing* ($Fo \to 1/2$): If the solutal Fourier number gets sufficiently large, in this case approaching $1/2$, then Eq. (5.32) reduces to

$$\frac{C_l}{C_0} = \frac{1}{1 - (1 - k_0)f_s}, \qquad (5.36)$$

with the companion equation for the solid phase composition freezing out,

$$\frac{C_s}{C_0} = \frac{k_0}{1 - (1 - k_0)f_s}. \qquad (5.37)$$

Equations (5.36) and (5.37) are formally equivalent to the equilibrium lever rule that one normally applies to phase diagrams. The lever rule determines the mass fractions and phase constitutions expected at a given temperature, pressure, and overall composition, when *global* equilibrium of the bulk phases prevails. Global equilibrium, however, seldom, if ever, actually occurs under any practical solidification conditions. Specifically, the following limits would be expected under global equilibrium solidification of a melt of initial composition C_0:

1. For the first solid formed on cooling and nucleating the melt, $f_s \approx 0$, with $C_\ell = C_0$, and $C_s = k_0 C_0$.
2. For the last solid frozen out, $f_s \to 1$, with $C_l = C_0/k_0$, and $C_s(x) = C_0$.

These two statements of the limits for initial and final solidification clearly agree with the lever-rule predictions for *equilibrium* freezing based on the phase diagram. Curiously, however, one would rather expect to recover the lever rule from a macrosegregation model when the physical solidification time becomes extremely long, or, equivalently, when the system's size is extremely small. In either event, true equilibrium, as a limit, should be approached only as $Fo_s \rightarrow \infty$, rather than the current finding based on the Brody–Flemings model, which indicates, *incorrectly*, that the 'lever rule' would occur when $Fo_s \rightarrow 1/2$! This obvious discrepancy occurs precisely for the same reasons that the solute mass balance underlying Eq. (5.19) provides an increasingly inadequate approximation to solute conservation as the solutal Fourier number becomes larger, and approaches $1/2$. Then, the mass balance formulated as Eq. (5.19) seriously underestimates the time needed for the system to reach full equilibrium. In this instance, the model fails by falsely predicting that global thermodynamic equilibrium occurs when $Fo_s = 1/2$, instead of providing the correct limit, which is reached when $Fo_s \rightarrow \infty$.

Numerical treatments of solute redistribution during solidification now correctly incorporate the diffusion behavior at both small and large Fourier numbers. Widely used numerical models were first developed by Clyne and Kurz [14], and by Onaka [15], and later by Voller [16, 17]. These numerical schemes account properly for back diffusion through the solid. Aspects of these diffusion models are currently incorporated into computer-based segregation codes. Foundries can employ proprietary computer codes that combine segregation models with accessible thermodynamic and diffusion data bases to provide them extremely accurate solute activities and mobility data for multi-component commercial casting alloys. These computer codes remain under development to assist the quality casting industry that must control solute segregation in its products.

5.6 Binary Alloy Segregation Curves

5.6.1 Gulliver–Scheil Segregation: Solubility Limits

Figures 5.4 and 5.5, discussed in Section 5.3, show pairs of plots calculated using the Gulliver–Scheil model for macrosegregation in binary alloys having various equilibrium segregation coefficients, considering both $k_0 < 1$ and $k_0 > 1$. Implicitly, the solutal Fourier number for these macrosegregation plots was zero, so discussion of solid-state diffusion was precluded. Note that the macrosegregation curves for $k_0 < 1$ all exhibit the expected rising concentration levels as the extent of freezing increased, measured by an increasing volume fraction, f_s. Quite often the solubility limit for the solidifying phase will be exceeded before reaching the end of solidification, where $f_s = 1$. Exceeding solubility limits as a result of chemical segregation can cause the unexpected onset of some secondary solidification reactions, such as the appearance of a eutectic or peritectic microstructural constituent, where none should appear according to the phase diagram. As macrosegregation theory shows, however, secondary solidification reactions should occur in solidifying almost all commercial casting alloys. These reactions greatly affect the cast microstructure and alter many of its mechanical properties through the appearance of nonequilibrium phases. Even cast dilute alloys can yield microstructures that exhibit second phases, precipitates, and 'unexpected' microconstituents that would not be predicted on the basis of the phase diagram. Also, especially in the case of more concentrated alloys, where second phases would normally be expected in the microstructure, the occurrence of non-equilibrium chemical macrosegregation alters the amount and form of these second phases. Thus, macrosegregation has a truly profound effect on the microstructural outcome of solidification processing, and, consequently, strongly influences and alters the properties and behavior of cast materials.

5.6.2 Influence of the Fourier Number

Figures 5.7 and 5.8 display macrosegregation curves plotted for various values of the solutal Fourier number, Fo_s. The left-hand ordinates of these figures show the concentration ratios for the melt, whereas the right-hand ordinates show the concentration ratios expected for the solid. Figure 5.7 contains typical macrosegregation curves for alloys in which $k_0 < 1$, whereas Fig. 5.8 contains macrosegregation curves for alloys for the less common case of $k_0 > 1$. Fourier numbers greater than $1/2$ are not plotted, because the segregation curves for that value already correspond to what should occur under global equilibrium as predicted by the lever law. (See again the discussion provided in Section 5.5.) This error is the direct result of overestimating the amount of solute transport through the solid.

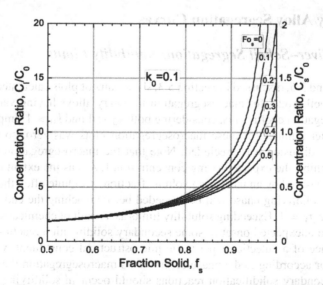

Fig. 5.7 Brody–Fleming macrosegregation curves, based on Eqs. (5.32) and (5.33) for binary alloys with $k_0 = 0.1$. Each *curve*, which rises with increasing fraction solid, is parameterized by its solutal Fourier number, plotted over the range $0 \leq Fo_s \leq 0.5$. The segregation in the melt appears as the ratio C_ℓ/C_0 on the left-hand ordinate, whereas the segregation in the solid appears as C_s/C_0 on the right-hand ordinate

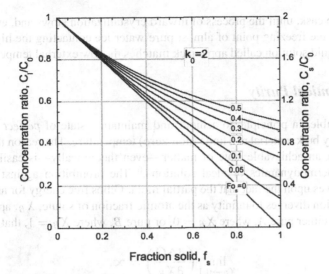

Fig. 5.8 Brody–Fleming macrosegregation curves, based on Eqs. (5.32) and (5.33) for binary alloys with $k_0 = 2$. Each *curve*, which falls with increasing fraction solid, is parameterized by its solutal Fourier number, plotted over the range $0 \leq Fo_s \leq 0.5$. The segregation in the melt appears as the ratio C_ℓ / C_0 on the left-hand ordinate, whereas the segregation in the solid appears as C_s / C_0 on the right-hand ordinate

5.7 Purification via Freezing

5.7.1 Fractional Crystallization

Fractional crystallization is a *physical*, i.e., non-chemical, purification process used to separate a dissolved substance from its solvent. It also can be thought of as a process to purify the major (solvent) phase by rejection of impurities (solutes) at a freezing interface. Fractional crystallization, usually involves random nucleation and growth of crystals, and has been employed the world over, from ancient to modern times, by salt workers to improve the quality and purity of salts extracted from brines produced both by deep mining and from the sea. The 'art' of fractional crystallization of diverse chemical substances was developed over centuries by alchemists, and their often secretive methods have evolved since medieval times to what have become the standard purification methods of modern inorganic chemistry and chemical engineering.

Another, perhaps more tasty, example of fractional crystallization is the interesting process for producing apple jack 'brandy'. Apple jack is a non-distilled, but highly alcoholic drink, produced by partially freezing a cask of fermented apple cider at extremely low wintertime temperatures. The water ice that crystallizes adjacent to the walls of the cask has low k_0 values (ca. 10^{-4}) for the remaining fruit sugars, acids, esters, and, especially, for the ethyl alcohol. The thickening ice layer concentrates these desirable 'impurities' into a smaller and smaller volume near the center of the cask, until the process of inward crystallization slows and, eventually, stops, when the freezing point of almost pure water ice contacting the highly concentrated liquid solution called apple jack matches the low external temperature.

5.7.2 Chemical Purity

It is impossible, in principle, to achieve and maintain a state of *perfect* chemical purity for any bulk material at a finite (non-zero) temperature. The reason that 100% purity is not an achievable state of matter—even theoretically—is easily proved using the thermodynamics of ideal solutions.[6] The prohibition against absolute purity devolves upon the fact that the partial molar Gibbs free energy for any binary $A - B$ solution diverges to infinity as the atomic fraction of solute, X_B, approaches the limits of either pure A, where $X_B = 0$, or pure B, where $X_B = 1$, that is

$$\lim_{X_B \to 0,1} \left(\frac{\partial \Delta G}{\partial X_B} \right) = \infty. \tag{5.38}$$

Equation 5.38 predicts that as states of greater and greater purity are achieved, it becomes (logarithmically) more 'costly', from an energetic standpoint, to increase the purity even further. As a corollary, the thermodynamic tendency for impurities to re-enter a material from the environment and lower its purity becomes increasingly likely as a material approaches perfect purity. This tendency to prevent the attainment of a state of perfect purity results directly from the increasing entropy of mixing per atom of impurity added. As the temperature, T, of the material increases, the free energy decrease available for adding an available impurity atom also increases as $-\Delta S \times T$, where the entropy change per impurity atom, or molecule of solute, for a binary solution is given by

$$\Delta S = -k_B \ln \left[(X_B)(1 - X_B) \right], \tag{5.39}$$

where k_B is Boltzmann's constant. One notes that the entropy change for impurity addition, Eq. (5.39), is the source of the logarithmic divergence mentioned above that appears in the partial molar Gibbs free energy as $X_B \to 0$.

5.7.3 Why Pure Materials?

Purity as a chemical property is arbitrarily specified by either the concentration of the impurity, C_B, or that of the corresponding solvent, C_A. In so-called 'high-purity'

---

[6] Modern exceptions to this thermodynamic 'Reinheitsgebot' (purity law) would seem to be laser cooling of atoms and Bose–Einstein condensation, in which clustering of identical 'slowed' atoms, or formation of a strange entangled quantum state both occur within a few nanokelvins of $T = 0$ K. Such clusters and quantum condensates may, in fact, be 100% pure. They exist, however, only where the product of their temperature and mixing entropy of an impurity atom is virtually zero.

materials the level of purity achieved is often specified by the molar concentration of the solvent species, e.g., $X_A = 0.9999999$, or, equivalently, 5–9s pure, based on the impurity content expressed in molar percentage, 99.99999%, or, equivalently, 10 ppm. The '5' appearing in 5–9s here refers to the number of nines to the right of the decimal point of the percent impurity concentration.

In a number of applications where one is interested in achieving small controlled solute additions, such as specified dopant-atom levels in semiconductors and ceramics, one usually stipulates the impurity level as the density of impurity atoms per unit volume. Thus, a semiconductor might require 10^{16} dopant atoms per cm^3 to form electronic p-n junctions in a microcircuit. That's roughly one dopant atom per million lattice atoms. To be meaningful, of course, this semiconductor itself must have initially contained far fewer electronically active impurity atoms than the required dopant-atom concentration itself. Thus, an initial purity level much better than 1 ppm is needed to form p-n junctions. Note, the relatively pure 5–9s material discussed above would actually be insufficiently pure to allow meaningful dopant atom action at 10^{16} cm^{-3}. This was the situation encountered about 60 years ago when the concept of a transistor, or solid-state semiconductor switch, was developed at Bell Telephone Laboratories. So-called 'chemically pure' germanium and silicon that were available at that time for building such devices were of inadequate purity to allow proper doping of donor or acceptor atoms to levels at which anticipated p-n junctions should form. Purer semiconductor crystals had to be developed.

5.7.4 Cyclic Unidirectional Solidification

The Gulliver–Scheil macrosegregation equations for the solid phase predict that directional freezing of a binary alloy leads to a redistribution of solute, which, if $k_0 < 1$, makes a portion of the initially solidified material purer than the material that freezes later in the process. The solidified fraction, $f_s^\star$, at which the local concentration in the freezing solid just exceeds the average starting composition is easily found by setting the LHS of Eq. (5.13) equal to unity, and then solving for the fraction solidified. One finds, after several steps of algebra, that the solidified fraction partially purified by continuous unidirectional freezing is given by the expression,

$$f_s^{\star(1)} = 1 - k_0^{\frac{1}{k_0 - 1}}, \qquad (5.40)$$

where $f_s^{\star(1)}$ denotes the linear fraction along the initially solidified column- or bar-shaped melt, beyond which all the solute removed from this 'purified' fraction is located.

Figure 5.5 demonstrates two related aspects of unidirectional freezing when used for the purposes of optimally purifying a material. As first suggested by Davies [18], the lower is the value of k_0, (1) the more material that can be solidified with solute concentrations remaining below that of the starting composition, C_0, and (2) the lower will be the *average* value of impurities in that re-frozen material. For the additional removal of residual impurities unidirectional freezing of the partially purified melt could be repeated again, and again, so long as the purified portion is separated cyclically after each freezing stage by being cut away from the impure remainder, then re-melted and mixed to a state of homogeneity prior to its next freezing stage.

As Eq. (5.40) indicates, the purified fraction depends only on one solidification parameter, namely, k_0.[7] Consequently, the purified fraction, $f_s^{\star(n)}$, which results from the nth stage of melting, homogenizing, and unidirectionally re-freezing the melt purified from the prior $(n-1)$th stage, merely repeats the Gulliver–Scheil solute redistribution. The purified fraction remaining after n stages of partial freezing is

$$f_s^{\star(n)} = \left(1 - k_0^{\frac{1}{k_0-1}}\right)^n .\tag{5.41}$$

Figure 5.9 provides a semi-logarithmic plot of Eq. (5.41) for ten stages of unidirectional freezing. It shows that as long as k_0 is small (< 0.01) one loses only a few additional percent of the starting material for each stage by discarding the solid fraction beyond $f_s^{\star(n)}$. For impurities with $k_0 > 0.1$, too much material would be sacrificed to make additional purification practical.

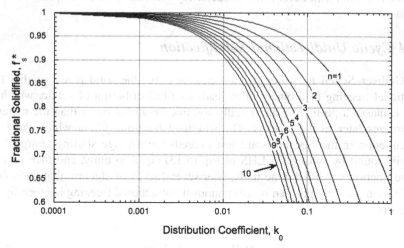

Fig. 5.9 Fraction of purified crystal produced after the nth stage of cyclic unidirectional freezing of a binary alloy with a distribution coefficient k_0. Unless $k_0 \ll 0.1$ too much material would be sacrificed after a few freeze cycles for only modest gains in average purity. See also Fig. 5.10

---

[7] When fractional crystallization is applied to relatively impure materials, the distribution coefficient is influenced by the stability of the solid–liquid interface. Strictly speaking, an effective distribution coefficient, k, should be substituted for k_0, as the latter requires a stable, planar interface. Discussions of various phenomena influencing k follow in subsequent chapters.

The average impurity concentration of the first solid frozen, $\langle C_s^{(1)} \rangle$, relative to the uniform impurity level of the starting melt, C_0, may be found by integrating the Gulliver–Scheil solute distribution, Eq. (5.13). This integral is taken from the starting location, $f_s = 0$, to the first impurity cut-off, $f_s^{\star(1)}$, where freezing any of the remaining melt fraction would add increments of solid containing more than the starting concentration of impurity, C_0. The scaled average impurity content for the first freezing pass is

$$\frac{\langle C_s^{(1)} \rangle}{C_0} = \frac{k_0}{f_s^{\star(1)}} \int_0^{f_s^{\star(1)}} (1 - f_s)^{k_0-1} df_s. \tag{5.42}$$

Carrying out the indicated integration yields,

$$\frac{\langle C_s^{(1)} \rangle}{C_0} = \frac{k_0}{f_s^{\star(1)}} \left(1 - \left(1 - f_s^{\star(1)} \right)^{k_0} \right). \tag{5.43}$$

Substituting Eq. (5.40) for the first cut-off fraction, $f_s^{\star(1)}$, appearing on the RHS of Eq. (5.43) shows that the average impurity content relative to the starting impurity content, for the first unidirectional freeze—up to the cut-off fraction—is also only a function of k_0, namely,

$$\frac{\langle C_s^{(1)} \rangle}{C_0} = \frac{1 - k_0^{\frac{k_0}{k_0-1}}}{1 - k_0^{\frac{1}{k_0-1}}}. \tag{5.44}$$

Repeating the process of melting the purified portion, now with the reduced relative impurity content given by Eq. (5.44), homogenizing it, and again unidirectionally freezing the second cut-off fraction yields an identical improvement of the relative impurity content. So, after the second unidirectional freeze one finds,

$$\frac{\langle C_s^{(2)} \rangle}{\langle C_s^{(1)} \rangle} = \frac{1 - k_0^{\frac{k_0}{k_0-1}}}{1 - k_0^{\frac{1}{k_0-1}}}. \tag{5.45}$$

Substitution of $\langle C_s^{(1)} \rangle$, using Eq. (5.44), into the LHS of Eq. (5.45) shows that the reduction of impurities from the second stage, relative to the material's initial impurity content, C_0, is

$$\frac{\langle C_s^{(2)} \rangle}{C_0} = \left(\frac{1 - k_0^{\frac{k_0}{k_0-1}}}{1 - k_0^{\frac{1}{k_0-1}}} \right)^2. \tag{5.46}$$

As mentioned above, providing k_0 is sufficiently small so that the loss of material is not too severe per stage of unidirectional freezing, one concludes from the progression of Eq. (5.44), (5.45) and (5.46) that the nth sequential stage of this process yields an average impurity concentration, relative to the starting level, that is given by

$$\frac{\langle C_s^{(n)} \rangle}{C_0} = \prod_{i=1}^{n} \frac{\langle C_s^{(i)} \rangle}{\langle C_s^{(i-1)} \rangle} = \left(\frac{1 - k_0^{\frac{k_0}{k_0-1}}}{1 - k_0^{\frac{1}{k_0-1}}} \right)^n . \tag{5.47}$$

Equation (5.47) lends itself to being plotted on double logarithmic coordinates as used in Fig. 5.10, and shows that sequential unidirectional freezing is capable of purifying materials efficiently, provided they contain impurities with only small k_0 values.

The maximum allowable rate of freezing is another important parameter, which is limited by the starting impurity concentration and, unfortunately, also depends inversely on the value of k_0. Freezing rate is neither included nor within the scope

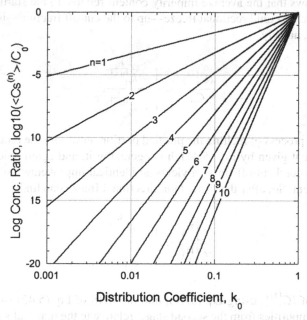

Fig. 5.10 Purification by n sequential unidirectional freezing (Eq. 5.47). The average relative impurity level for the nth cycle, $\langle C_s^{(n)} \rangle / C_0$, is plotted logarithmically against the value of the impurity's distribution coefficient, k_0. As these plots indicate, this purification method is extremely effective for removing impurities that have their $k_0 < 0.1$. Impurities having larger values of k_0, however, require many more cycles to achieve comparably large impurity reductions and, consequently, one would recover much smaller fractions of the solidified sample, making this purification process impractical for such impurities. See again Fig. 5.9

of the present analysis. Thus, both low k_0 values and elevated initial impurity levels combine to limit severely the freezing speed. These effects are related to the stability of the solid–liquid interface—a topic of importance in single crystal growth, and which will be developed in Chapters 9, 10, and 11. As shown later, instability of the traveling interface allows lateral rejection of the solute that greatly reduces the overall efficiency of impurity transport and removal.

5.8 Zone Refining

The fundamental difficulties encountered in early attempts to dope germanium (melting point $T_m = 938$ C) with electronically active impurities to produce p–n junctions for transistor switches were finally surmounted in 1951 by W.G. Pfann [19–26], the engineer who is credited with inventing zone refining. Later H.C. Theurer extended Pfann's method to silicon—a more technically desirable semiconductor with the much higher melting point of 1414 C—using a clever adaptation called float-zone (FZ) crystal growth [27], to be described briefly later. The main advantage of zone refining over, say, repeated fractional crystallization, is that zone refining is more suited to a spectrum of impurities with differing k-values. Today, however, zone refining as a purification process is used primarily to produce purified metals and semiconductor compounds, such as GaAs single crystals for large-area device substrates. Zone refining of Si is still applied commercially in preparing some special grades of ultra-pure Si single crystals required in electrical devices such as high-power rectifiers. Also, because ultra-pure silicon has excellent transparency to infra-red (IR) radiation, it has gained additional importance in fabricating optical components, such as IR filters, gratings, and lenses for terahertz applications.

The contemporary process for producing conventional electronic-grade single crystals of Si no longer uses zone refining. Electronic-grade Si crystals are currently grown with diameters exceeding 300 mm and weigh hundreds of kilograms. These huge, almost dislocation-free single crystals, are produced in commodity tonnages in 'silicon foundries' using Czochralski (CZ) crystal growth.[8]

5.8.1 Single Zone Pass: Pfann's Equation

The zone refining processes is shown schematically in Fig. 5.11. Here $\hat{x}$ is the running variable that locates the advancing solid–liquid interface, with $\hat{x} = 0$ denoting the left-hand side of the charge of material where freezing begins in the first zone. The physical length of the charge to be zone refined is arbitrarily set to L, and $z < L$ is the length of the molten zone. It proves convenient to introduce

---

[8] Czochralski and other standard methodologies for preparing single crystals and directionally solidifying castings are discussed briefly in Appendix D.

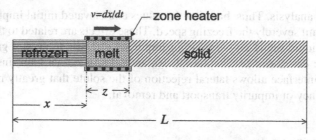

Fig. 5.11 Schematic representation of batch zone refining

dimensionless variables based on the overall physical length of the material charge, L, viz., $X = \hat{x}/L \leq 1$, and $Z = z/L < 1$. In a typical batch zone refiner, however, the zone length is small relative to the length of the charge, $z/L \approx 0.05 - 0.1$.

The initial stage of zone refining may be easily analyzed if five simplifying assumptions hold true:

1. Rejection of impurities is *unidirectional*, implying both a bar of uniform area, a, and a planar, stable[9] solid–liquid interface moving steadily in the $+x$ direction ($v = d\hat{x}/dt$=const.).
2. Transport of the impurity through the solid is negligible ($D_s = 0$).
3. The mass densities of the phases are equal ($\rho_s = \rho_\ell$).
4. The effective distribution coefficient, k, remains constant and is assumed to be less than unity throughout the process.[10]
5. The molten zone length, z, never varies, except at the termination of each full pass where the forward (melting) interface reaches the end of the material charge, i.e., where $\hat{x} > L - z$.

The impurity mass balance within the traveling molten zone contains two contributions:

(1) the time rate, $\dot{m}_1$, at which impurities are released behind the freezing solid–liquid interface and the solid re-freezes with them in solution,

$$\dot{m}_1 = -akC_\ell \left(\frac{d\hat{x}}{dt} \right),$$ (5.48)

(2) the time rate, $\dot{m}_2$, at which impurities remaining in the charge enter the molten zone at the melting solid–liquid interface,

---

[9] Stability of the solid–liquid interface is difficult to maintain when the starting charge is impure. Using an effective distribution coefficient, k, rather than k_0, compensates for non-planarity of the interface as well as for other extrinsic effects, such as melt convection, which also change k.

[10] Impurities with $k > 1$, although less common than those with $k < 1$, can also be removed effectively via zone refining. Impurities in cases where $k > 1$ would be concentrated in the first portions of the material to freeze, and would be reduced toward the end of the material.

$$\dot{m}_2 = aC_0 \left(\frac{d\hat{x}}{dt} \right). \tag{5.49}$$

The sum of the rates given in Eqs. (5.48) and (5.49) is the net rate of change of solute mass within the molten zone, namely,

$$aC_0 \left(\frac{d\hat{x}}{dt} \right) - akC_\ell \left(\frac{d\hat{x}}{dt} \right) = az \left(\frac{dC_\ell}{dt} \right). \tag{5.50}$$

If one cancels common terms on both sides of Eq. (5.50) and divides through by the charge length, L, to non-dimensionalize both the running variable $\hat{x}$, and the (constant) zone length, z, and then separates terms that are dependent on time, an ODE is produced relating the rate of change of impurity concentration in the molten zone, $d(C_\ell/C_0)dt$, with dimensionless zone speed, $dX/dt = (1/L)d\hat{x}/dt$, namely,

$$\frac{1}{Z} \left(\frac{dX}{dt} \right) = \frac{1}{1 - k(C_\ell/C_0)} \left(\frac{d(C_\ell/C_0)}{dt} \right). \tag{5.51}$$

Removal of time as an explicit independent variable from Eq. (5.51) allows application of initial conditions for the zone position and its starting concentration, which are $X = 0$ and $(C_\ell/C_0) = 1$, respectively. One may now forward integrate Eq. (5.51) to an arbitrary location in the charge, X, where the relative concentration increases to C_ℓ/C_0 as,

$$\frac{1}{Z} \int_0^X dX' = \int_1^{\frac{C_\ell}{C_0}} \frac{(dC_\ell/C_0)'}{1 - k(C_\ell/C_0)'}. \tag{5.52}$$

The integral on the LHS of Eq. (5.52) is X/Z, whereas the integral on the RHS is $\log_e \left[(1-k)/\left(1 - k\frac{C_\ell}{C_0} \right) \right]^{\frac{1}{k}}$. After taking antilogarithms and performing several algebraic rearrangements to solve for $\frac{C_\ell}{C_0}$, one obtains Pfann's expression for the impurity level in the molten zone,

$$\frac{C_\ell}{C_0} = \frac{1}{k} \left(1 - (1 - k)e^{-k\frac{X}{Z}} \right). \tag{5.53}$$

By virtue of local thermodynamic equilibrium at the freezing interface, $\hat{C}_s(X) = k\hat{C}_\ell(X)$. If solid-state diffusion is ignored, thus preventing any changes in the solid composition after solidification, one may multiply both sides of Eq. (5.53) by k and recover the corresponding distribution of impurities left in the frozen solid after the first zone refining pass,

$$\frac{C_s}{C_0} = 1 - (1 - k)e^{-k\frac{X}{Z}} \quad (X < 1 - Z). \tag{5.54}$$

Equation (5.54) properly describes the impurity distribution along the *initially* zone refined ingot, excepting the last zone length to freeze at the end of the material, where, instead, the Gulliver–Scheil distribution develops because the last molten zone is forced to steadily diminish in length as total freezing is accomplished. Thus, the Gulliver–Scheil impurity distribution initiates at the point where the forward (melting) interface within the last zone length reaches the position $X = 1 - Z$. Application of Eq. (5.13) shows that

$$\frac{C_s}{C_{init.}} = k(1 - f_s)^{(k-1)}, \tag{5.55}$$

where $C_{init.}$ denotes the uniform impurity concentration developed in the molten zone at the termination of the first pass, where $X = 1 - Z$. In the present context, the fraction solidified, f_s, that appears in Eq. (5.55), may also be expressed in terms of the scaled zone length, Z, and the (freezing) interface position, X, as

$$f_s = 1 + \frac{X - 1}{Z}, \quad [(1 - Z) \leq X \leq 1]. \tag{5.56}$$

The value of $C_{init.}$ at the beginning of normal freezing of the last zone length is given by Eq. (5.53) evaluated at the position $X = 1 - Z$. See again Fig. 5.11. One finds by applying Eq. (5.54) that the molten zone, as it begins termination of the first passage, contains the initial, uniform impurity concentration given by,

$$C_{init.} = \frac{C_0}{k}\left(1 - (1 - k)e^{-k\frac{(1-Z)}{Z}}\right). \tag{5.57}$$

The distribution of the impurities developed within the last zone length to freeze at the end of the material, thereby completing the first zone pass, is

$$\frac{C_s}{C_0} = \left(1 - (1 - k)e^{-k\frac{(1-Z)}{Z}}\right)\left(\frac{1 - X}{Z}\right)^{k-1}, \quad [(1 - Z) \leq X < 1]. \tag{5.58}$$

Figure 5.12 presents plots for several impurities with different values of k that combine Eqs. (5.54) and (5.58), which reveal the impurity distribution developed throughout a zone-refined charge of material after just a *single* completed zone pass. The zone refiner producing the theoretical distribution depicted in Fig. 5.12 operated with a material charge that was 10 zones in length.

The clear advantage of zone purification over unidirectional solidification is that only a portion of the last zone becomes more impure than the starting material, for all $k < 1$. Also, only a minor amount of material is held in the molten state at any instance, reducing the tendency for further pick-up of impurities or chemical reactions with, or contamination from, the container. Extraordinarily efficient purification can be achieved with the employment of narrower zones and longer charges of starting material.

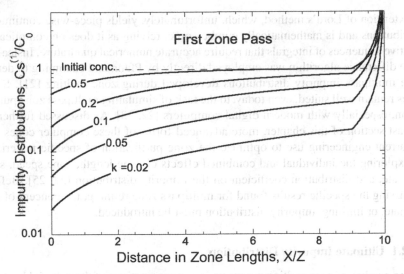

Fig. 5.12 Complete impurity distributions—plotted log-scale—after a single zone pass for different k values. The material charge is 10 zones in length, i.e., $Z = z/L = 0.1$, and $X/Z = x/z \leq 10$. Note how the impurities segregate in the last zone to freeze and suddenly increase in concentration

5.8.2 Multipass Zone Refining

Multipass zone refining is required to ultra-purify materials, and this process presents a more difficult challenge to analyze mathematically. Lord [28] solved the macrosegregation equations exactly for the special case of multipass zone melting of a material charge of semi-infinite length. His analysis also estimates the impurity distribution accurately in charges of finite length—at least for the first few zone lengths—provided that (1) they are far away from the last zone, and (2) only a few zone passes have been performed. Lord's treatment for semi-infinite charges avoids the complication attending the unidirectional freezing of the last zone for a charge of finite length. The impurity distribution within the last zone (of the first pass) was already considered in developing Eqs. (5.57) and (5.58). The complication encountered in multipass purification of real material charges (having finite lengths) is that each subsequently completed zone pass allows some of the impurities originally concentrated within the last zone to dissolve back into additionally remote zones as more zone passes occur. After a sufficient number of additional zone passes, a portion of the impurities from the last zone travel all the way back to the starting point, and thereby affect the impurity distribution throughout all the material. This complication precludes making acceptable estimates of the so-called 'ultimate impurity distribution', which is achieved in practice after passage of a large number (10–100) of molten zones. The ultimate distribution provides an important design criterion for optimizing the purification of materials, and will be treated in the next section.

An exact analysis of multipass zone purification in materials of finite length was eventually published by Braun and Marshall [20]. Their analysis is based on an extension of Lord's method, which, unfortunately, yields piece-wise continuous distributions and is mathematically cumbersome, relying as it does on complicated iterative sequences of integrals that require accurate numerical quadrature. Instead, a finite difference algorithm was employed directly by Pfann and Hamming to determine multipass impurity distributions developed during zone melting [21]. Such codes remain well suited, even today, to the task of simulating multipass zone purification, especially with modern digital computers [22, 23]. As discussed further in the last section of this chapter, more advanced forms of these computer codes are in current engineering use to optimize the zone purification of specific materials, by exploring the individual and combined effects of zone length, zone speed, stirring rate, and distribution coefficient on the impurity distribution [24, 25]. Before discussing the specific results found for multipass zone refining, the concept of the ultimate, or limiting, impurity distribution must be introduced.

5.8.2.1 Ultimate Impurity Distribution

As the number of zone melting passes, n, increases, portions of the material become purer. Depending on the overall length of the material charge, L, the length of the molten zone, z, and the value of k, there comes a point of diminishing returns in achieving further purification, and eventually an impurity distribution evolves that represents the maximum attainable purification. This distribution is the ultimate, or steady-state, impurity distribution, denoted as $C_s^\infty(x)$.

If a zone of width z, and uniform cross section travels through the charge, the impurity concentration freezing out at any point $x = \hat{x}$ at steady-state is $C_s^\infty(\hat{x})$. The steady-state average composition within the traveling molten zone, with its trailing (freezing) interface at $x = \hat{x}$, and leading (melting) interface at $\hat{x} + z$, is $\langle C_\ell^\infty(\hat{x}) \rangle$. Solute mass conservation and local equilibrium, moreover, require that the average steady-state composition of the melt must remain equal to $C_s^\infty(\hat{x})/k$.

Now, the average zone composition between $\hat{x}$ and $\hat{x} + z$ is defined by the definite integral,

$$\langle C_\ell^\infty(\hat{x}) \rangle = \frac{1}{z} \int_{\hat{x}}^{\hat{x}+z} C_s^\infty(x) dx, \qquad (5.59)$$

so the ultimate impurity distribution developed within any solidified zone is specified by the integral equation,

$$C_s^\infty(\hat{x}) = \frac{k}{z} \int_{\hat{x}}^{\hat{x}+z} C_s^\infty(x) dx. \qquad (5.60)$$

The solution to Eq. (5.60) is itself a functional, which is proportional to itself at all points x. A trial solution satisfying this equation is that of a general exponential function in x, namely,

$$C_s^\infty(\hat{x}) = \alpha e^{\eta(k)\hat{x}}, \qquad (5.61)$$

where the unknown coefficients needed to define the ultimate impurity distribution are α and the exponent, $\eta(k)$. Substitution of the RHS of the trial function, Eq. (5.61), into the integral equation, Eq. (5.60), gives

$$\alpha\, e^{\eta(k)\hat{x}} = \frac{k}{z} \int_{\hat{x}}^{\hat{x}+z} \alpha\, e^{\eta(k)x}\, dx.$$

(5.62)

Carrying out the integration indicated on the RHS of Eq. (5.62) shows that

$$\alpha\, e^{\eta(k)\hat{x}} = \frac{\alpha}{z}\, \frac{k}{\eta(k)}\left(e^{\eta(k)z} - 1\right) e^{\eta(k)\hat{x}},$$

(5.63)

which may be solved explicitly for the distribution coefficient, k,

$$k = \frac{\eta(k)z}{e^{\eta(k)z} - 1}.$$

(5.64)

Equation (5.64) provides the required relationship between the impurity distribution coefficient and the unknown exponent $\eta(k)$. If values of $\eta(k)$ are obtained by solving Eq. (5.64) for given values of k, or if Fig. 5.13 is used as a look-up 'table', then the minimum number of zone stages, $n^{\star}$, required to reach the ultimate distribution may be found after the charge length, L, is selected ($L > 1$). As a simple

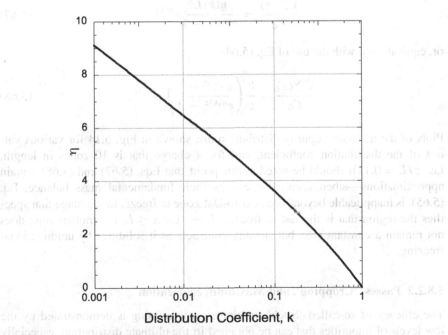

Distribution Coefficient, k

Fig. 5.13 Plot of Eq. (5.64) relating the exponent $\eta(k)$ to the effective distibution coefficient, k, for determining ultimate impurity distributions

inverse solution to Eq. (5.64) for $\eta(k)$, does not exist, a numerical solution, or just the graphical inverse, may be used to find η-values for any specific values of k. Figure 5.13 shows this relationship over a range of k values for a zone of arbitrary unit length ($z = 1$).

The coefficient α for the solution to the ultimate distribution may now be found by applying conservation of impurity mass over the entire material charge, assumed to be zone refined to its ultimate impurity distribution, $C_s^\infty(\hat{x}) = \alpha \exp[\eta(k)\hat{x}]$. Conservation of impurity mass for the entire charge is expressed by another definite concentration-distance integral extending over the entire length, L, of the material, namely,

$$C_0 = \frac{1}{L} \int_0^L \alpha\, e^{\eta(k)x}\, dx. \tag{5.65}$$

Evaluating the integral in Eq. (5.65) and then solving for α shows that

$$\alpha = C_0 \frac{\eta(k)L}{e^{\eta(k)L} - 1}. \tag{5.66}$$

Substituting $\eta(k)$ and the expression for α given by Eq. (5.66) into the general exponential solution for the ultimate impurity distribution yields the desired result

$$\frac{C_s^\infty(x)}{C_0} = \frac{\eta(k)L}{e^{\eta(k)L} - 1} e^{\eta(k)x}, \tag{5.67}$$

or, equivalently, with the use of Eq. (5.64),

$$\frac{C_s^\infty(x)}{C_0} = \frac{k}{z} \left(\frac{e^{\eta(k)z} - 1}{e^{\eta(k)L} - 1} \right) e^{\eta(k)x}. \tag{5.68}$$

Plots of the ultimate impurity distribution are shown in Fig. 5.14 for various values of the distribution coefficient, and for a charge that is 10 zones in length, i.e., $z/L = 0.1$. It should be noted at this point that Eqs. (5.67) and (5.68) remain approximations—albeit accurate ones—as their fundamental mass balance, Eq. (5.65), is inapplicable beyond the next-to-last zone to freeze. In the range that specifies the region that is the last to freeze, $(L - z) < x \leq L$, the molten zone does not remain a constant size, but steadily contracts as it solidifies by unidirectional freezing.

5.8.2.2 Passes, 'Cropping', and Maximum Separation

The efficacy of so-called 'batch' multipass zone refining is demonstrated by the low levels of impurities that can be obtained in the ultimate distribution, especially for impurities with $k_0 < 0.5$. In addition, multipass zone purification results in just moderate material 'wastage' after achieving the steady-state, ultimate impurity dis-

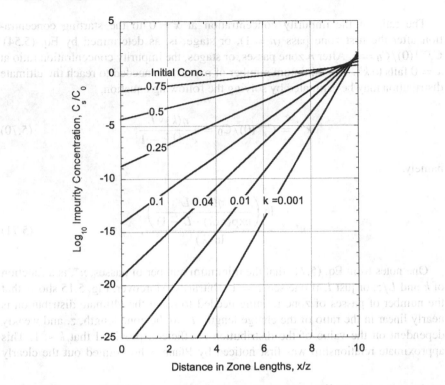

Fig. 5.14 Plot of Eq. (5.68) showing ultimate impurity distributions, C_s^∞/C_0, for various values of the distribution coefficient. The zones here may be considered to be of unit length ($z = 1$), and the total material charge is of length, $L = 10$, i.e., the charge is 10 zones in total length. Impressive impurity reductions can be obtained for low values of k_0, which would exceed the ability to quantify them by available chemical analytical methods

tribution in the charge. Wastage here is measured in terms of how much material must be removed, or 'cropped' from the ends of the charge, so as not to exceed the starting impurity level, C_0. Figure 5.16 contains a plot of the 'cross-over', or cropping points, $X^* = x^*/L$, which provide an estimate of the fraction of the starting charge made purer by multipass zone refining than its initial impurity level. As this plot shows, for $k < 1$, and irrespective of its k-value, at least half the charge receives some benefit, however minimal, from multipass zone melting.

In addition, there exists a maximum separation that is theoretically possible to achieve in any multipass batch zone refining process after establishing the ultimate distribution. The maximum separation may be defined, for $k < 1$, as the ratio of the impurity level achieved at $x = 0$, to the starting impurity level in the charge, i.e., $C_s^\infty(0)/C_0$. This concentration ratio is obtained from the ultimate distribution function, Eq. (5.67), by setting $x = 0$, and choosing a zone length, $z = 1$, for convenience, so that

$$C_s^\infty(0)/C_0 = \frac{\eta(k) \cdot L}{e^{\eta(k) \cdot L} - 1}. \tag{5.69}$$

The ratio of the impurity concentration at $x = 0$ to the starting concentration after the first zone pass ($n = 1$), or stage, is, as determined by Eq. (5.54), $C_s^{n=1}(0)/C_0 = k$. After n zone passes, or stages, the impurity concentration ratio at $x = 0$ falls to k^n. The minimum number of passes, n^*, needed to reach the ultimate distribution may be estimated by solving the following equation,

$$k^{n^*} = C_s^{\infty}(0)/C_0 = \frac{\eta(k) \cdot L}{e^{\eta(k) \cdot L} - 1},$$ (5.70)

namely,

$$n^* = \frac{\ln \left(\dfrac{\eta(k) \cdot L}{\exp(\eta(k) \cdot L - 1)} \right)}{\ln (k)}.$$ (5.71)

One notes from Eq. (5.71) that the minimum number of passes, n^*, is a function of k and L/z, or just L if one sets $z = 1$ arbitrarily. Moreover, Fig. 5.15 shows that the number of passes of zone refining needed to reach the ultimate distribution is nearly linear in the ratio of the charge length, L, to the zone length, z, and weakly dependent on the value of the distribution coefficient, provided that $k < 1$. This approximate relationship was first noticed by Pfann, who pointed out the clearly

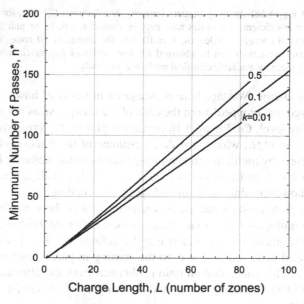

Fig. 5.15 Plot of Eq. (5.71) showing the dependence of the minimum number of zone melting passes, n^*, obtained from Eq. (5.71), which are required to achieve the ultimate impurity distribution in a charge of length L containing molten zones of length $z = 1$. The nearly linear relationship shown here remains valid for $k < 1$ only

analogous behavior between between the number of passes in zone refining needed to reach steady state, and the number of stages in (zero reflux) distillation.

5.8.2.3 Zone Leveling

In principle, the purified fraction of the material can be 'zone leveled' to produce a uniform crystal of intermediate purity, or even re-melted completely to form a new starting charge to begin a second cycle of multipass zone refining. Pfann's book gives numerous practical hints for improving the purity of crystals using multipass and multicycle zone melting [21]. The somewhat arduous combined processes of multicycle, multipass, zone refining, however, do provide the preferred method for preparing the purest crystalline solids. A few ultra-pure materials have been chosen by standards laboratories to establish so-called secondary reference temperature standards. These secondary standards provide remarkably reproducible melting points (± 0.002 K) that are useful for calibrating and interpolating thermometric equipment on the International Practical Temperature Scale [29].

The actual limits of purification achievable using zone melting depend on several additional factors, including zone size, speed of zone travel, efficiency of mixing within the molten zone, and, of course, any re-introduction of impurities from the container walls and atmosphere. All these important, practical, mostly extrinsic factors influence the ease of efficient purification by zone melting. They are all discussed in detail in Pfann's excellent monograph on this subject [21] (Fig. 5.16).

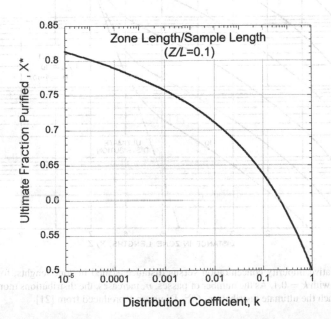

Fig. 5.16 Cross-over point for multipass zone refining giving the fraction of the starting charge receiving at least some purification after achieving the ultimate impurity distribution. Calculations are for a charge that is 10 zone lengths long with a starting impurity concentration of C_0. Even for k-values that are close to unity, more than half the material receives at least some purification

5.8.2.4 Multipass Refining Results

The relationship between multipass zone refining and the ultimate impurity distribution can be immediately appreciated by seeing the computations published by Pfann and Hamming for $k = 0.1$, and reproduced here in Fig. 5.17.

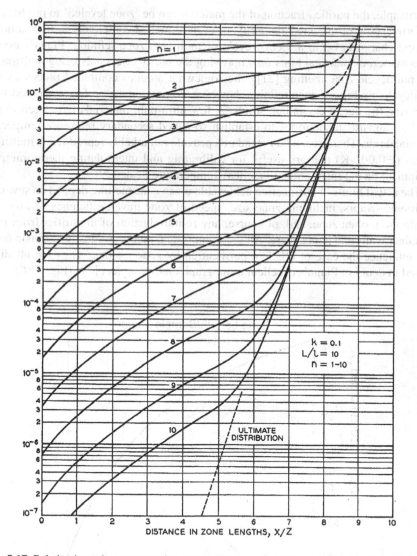

Fig. 5.17 Relative impurity concentration versus distance, scaled in zone lenghts, for multipass zone melting, with $k = 0.1$. As the number of passes, n, increases, the distributions more and more closely approach the ultimate distribution, C_s^∞. Diagram reproduced from [21]

Multipass results calculated for $k = 1.5$ are shown in Fig. 5.18. One sees that the impurity distribution for the nth zone passage becomes a closer and closer approximation to the ultimate distribution, as described approximately by Eqs. (5.67) and

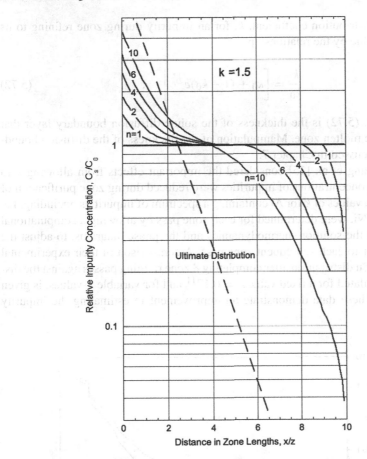

Fig. 5.18 Relative impurity concentration versus distance, scaled in zone lenghts, for multipass zone melting, with $k = 1.5$. As the number of passes, n, increase, the distributions straighten and rotate toward the ultimate distribution, C_s^∞. Data reproduced from [21]

(5.68). As is also noted in Fig. 5.17, with just $n = 10$ sequential zone passes, almost half of the resultant crystal has developed what appears to be a close approximation to the theoretical ultimate distribution. With a sufficiently large number for n, say $n^\star$, which depends on k, the multipass impurity distributions follow closely the ultimate distribution across the material. Estimates of $n^\star$ may be obtained in a straightforward manner.

Recent experiments reported by Ghosh et al. [24] on purifying by zone refining the group-III element of III–V compound semiconductors, such as GaAs, demonstrates convincingly that the number of passes needed to approach the ultimate distribution rises steeply as $k \to 1$, independently of whether k is larger or smaller than unity. In addition, these investigators explored the influences of zone speed on the ultimate distribution for impurities with a fixed k_0-value. They found an improvement of about 7-orders of magnitude in the steady-state impurity level by lowering the speed of advance, v, of the zones from 6 cm/h to 2 cm/h. Ghosh et al. related the effective distribution coefficient, k, for an impurity during zone refining to its equilibrium value by the relationship

$$\frac{k}{k_0} = \left[k_0 + (1 - k_0)e^{\frac{v\delta}{D_\ell}} \right]^{-1},$$ (5.72)

where δ in Eq. (5.72) is the thickness of the solute diffusion boundary layer that develops in the molten zone. Manipulation of the thickness of the diffusion boundary layer is discussed in Chapter 7.

Lastly, Cheung et al. [25] considered the important effects from allowing k to change as the concentrations of impurities were reduced during zone purification of aluminum. The values of k for Al containing a spectrum of impurities, including, Fe, Cu, Zn, P, and Ni, were determined for each zone pass by allowing a computational loop, based on the solution thermodynamics and the phase diagrams, to adjust the value of k prior to each subsequent zone pass. A comparison of their experimental results for the Cu distribution after completing 8 zone refining passes against the distributions calculated for a fixed value $k = 0.17$[11] and for variable k-values, is given in Fig. 5.19. These data demonstrate an improvement in estimating the impurity

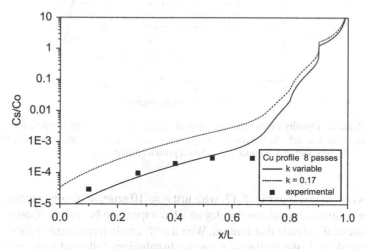

Fig. 5.19 Experimental measurement of the Cu impurity distribution after multipass zone refining of Al. Two distributions are calculated after 8 zone passes: the *dashed curve* is based on a fixed value for $k = 0.17$ for Cu in Al, whereas the *solid curve* is based on variable k-values adjusted after each zone pass. The variable k-values were computed by adding a thermodynamic computational loop to the multipass zone refining software. Figure reproduced from [25]

---

[11] The value of k_0 for Al-Cu binary alloys actually falls from about 0.14, at moderate Cu concentrations, to approximately 0.11 for nearly pure Al. The value of 0.17, chosen here for k by Cheung et al., may reflect its *effective* value appropriate to his zone purification experiments.

concentration after multipass zone refining by using properly adjusted values of k for each zone pass.

5.9 Summary

Macrosegregation is a ubiquitous solidification phenomenon that is caused by different equilibrium solubilities of a second component, or impurity, in a crystal and its melt at the same temperature and pressure. The differing solubility in solid and liquid is required to make the chemical potentials and activities of the solute *equal* in both phases. The equilibrium solubility ratio is expressed by the phase diagram's distribution coefficient, k_0. Most binary alloys have $k_0 < 1$, but many have $k_0 > 1$. None may have $k_0 = 1$.

The consequent rejection or additional incorporation of solute at a moving crystal-melt interface results in diffusion that promotes long-range chemical redistribution, or macrosegregation. Given that the time available for solidification in most realistic processes is insufficient for establishing complete equilibrium, the process obeys 'local equilibrium'. Local equilibrium at an atomically thin region surrounding the solid–liquid interface and mass balances applied over the well-mixed melt combine to provide the basis of Gulliver–Scheil segregation theory. Gulliver–Scheil theory predicts the basic macrosegregated state of directionally frozen alloys. It assumes for simplicity that diffusion in the solid is negligible. Macrosegregation theories were considerably enhanced by Brody and Flemings to incorporate solid-state diffusion, at least approximately. Later numerical refinements by Clyne and Kurz and by others now provide accurate robust segregation models.

Finally, purification achieved by directional crystallization is treated as an extension of controlled macrosegregation, especially for solutes with small k_0 values. Several processes that use sequential crystallization, including multiple unidirectional solidification and multi-pass zone refining, are discussed in detail. Their importance lies both in their practical production of ultra-pure substances of many kinds, including metals, semiconductors, and organic compounds, and their subtle employment of macrosegregation effects.

References

1. H.W.B. Roozeboom, *Z. Phys. Chem.*, **30** (1899) 385.
2. J.N. Lalena and D.A. Cleary, *Phase Equilibria, Phase Diagrams, and Phase Modeling*, Wiley, New York, NY (2005) 335.
3. R.T. DeHoff, *Thermodynamics in Materials Science*, Chap. 3, McGraw-Hill, Inc. New York, NY, 1993.
4. *Diffusion in Materials: Part 1*, H. Mehrer et al., Eds., Scitec Publications Ltd., Zürich-Ütikon, 1997, p. 185.
5. *Diffusion in Materials: Part 2*, H. Mehrer et al., Eds., Scitec Publications Ltd., Zürich-Ütikon, 1997, p. 1265.
6. E. Scheil, *Z. Metall.*, **34** (1942) 70.

7. M.E. Glicksman and R.N. Hills, Philos. Mag. A, **81** (2001) 153.
8. G.H. Gulliver, J. Inst. Metals, **9** (1913) 120.
9. E. Scheuer, Z. Metall., **23** (1931) 237.
10. G.H. Gulliver, Metallic Alloys: Their Structure and Constitution, 4th Ed., Charles Griffin & Company Ltd., London, 1921, p. 397.
11. M.E. Glicksman, Diffusion in Solids: Field Theory, Solid-State from an and Applications, *Wiley Interscience Publishers*, New York, NY, 2000.
12. H.D. Brody and M.C. Flemings, Trans. Metall. Soc. AIME, **236** (1966) 615.
13. M.C. Flemings, Solidification Processing, *McGraw-Hill*, New York, NY, 1974.
14. T.W. Clyne and W. Kurz, Metall. Trans. A, **12A**, (1981) 965.
15. I. Onaka, Trans. ISIJ, **26** (1986) 1045.
16. V.R. Voller, Modeling of Casting, Welding, and Advanced Solidification Processes VIII, *B.G. Thomas and C. Beckermann, Eds.*, The Minerals Metals & Materials Society, Warrendale, PA, 1986, p. 265.
17. V.R. Voller and S. Sundarraj, Int. J. Heat Mass Transf., **38** (1995) 1009.
18. L.W. Davies, Trans. AIME, **215** (1959) 672.
19. W.G. Pfann, Trans. AIME, **194** (1952) 747–753.
20. I. Braun and S. Marshall, Br. J. Appl. Phys., **8** (1957), 157.
21. W.G. Pfann, *Zone Refining*, 2nd Ed., Wiley Series on the Science and Technology of Materials, Wiley, New York, NY (1958, 1966) p. 38.
22. L. Burris, Jr., C.H. Stockman and I.G. Dillon, Trans. AIME, **203** (1955) 1017.
23. J.A. Spim, Jr., M.J.S. Bernadou and A. Garcia, J. Alloy Compd., **298** (2000) 299.
24. K. Ghosh, V.N. Mani and S. Dhar, J. Cryst. Growth, **312** (2009) 1521.
25. N. Cheung, R. Bertazzoli and A. Garcia, J. Cryst. Growth, **310** (2008) 1274.
26. W.G. Pfann, US Patents 2,739,088 and 2,739,045 (1956).
27. H.C. Theurer, US Patent 3,060,123 (1962).
28. N.W. Lord, Trans. AIME, **197** (1953) 1531.
29. M.E. Glicksman et al., J. Cryst. Growth, **89** (1988) 101.

Chapter 6
Plane-Front Solidification

6.1 Introduction

In Chapter 5 the importance of formulating solute mass balances during solidification was considered from the perspective of deriving the laws of chemical macrosegregation. Specifically, the Gulliver-Scheil equations were derived from an interfacial mass balance that ignored the influences of solute transport within the solid phase. In addition, it was also found convenient to impose the condition of 'perfectly efficient' convective mixing in the melt, an assumption that precludes the need to consider solute transport through the melt via atomic diffusion. Efficient mixing homogenizes the melt throughout the solidification process, keeping the residual melt concentration everywhere the same at each instant of time.

Next, the Brody-Flemings segregation formulas were derived by introducing an explicit form for the solidification time scale. Brody-Flemings analysis includes the influence of 'back-diffusion' of solute through the developing hot solid phase. The contribution to the solute mass balance from solid-state diffusion was formulated by using Fick's 1st law to approximate the solute flux diffusing backwards from the solid–liquid interface through the already-formed hot solid. That analysis has analogies to the earlier use in Chapter 4 of the quasi-static approximation needed to simplify heat transfer problems in castings solidifying at low Stefan numbers.

This chapter advances considerations of macrosegregation during solidification by introducing the time-dependent diffusion equation, Fick's 2nd law, for plane-front freezing. Solutions to the diffusion equation reveal transient and steady-state distributions of solute that develop both in the melt and in the solid. By contrast with the assumption made in earlier chapters that freezing melts are so well stirred that they remain homogeneous—with a uniform temperature and solute concentration everywhere—the fluid dynamic state of melts during plane-front solidification will instead be assumed as quiescent. Now lacking convective transport, solutes rejected from the interface mix into the melt *only* through the mechanism of liquid-state diffusion.

M.E. Glicksman, *Principles of Solidification*, DOI 10.1007/978-1-4419-7344-3_6,   145
© Springer Science+Business Media, LLC 2011

6.1.1 Steady-State Macrosegregation

We begin the study of unidirectional, 'plane-front' solidification of alloys, in which
the solute distributions during freezing also satisfy the diffusion equation. The diffusion equation itself is the analogous equation governing the transport of solute as
is the energy equation, Eq. (4.38), for predicting time-dependent temperature fields
during heat conduction. The mathematical analogy posed between solute diffusion
in either the melt or solid, and heat conduction is, in fact, perfect, at least from
the field-theoretic point of view. Moreover, the energy equation and the diffusion
equation are fully equivalent PDEs. The time-dependent diffusion equation in one
dimension [1], written in a single spatial coordinate, x, fixed to phase i is

$$\frac{\partial^2 C_i}{\partial x^2} = \frac{1}{D_i} \frac{\partial C_i}{\partial t} \quad (i = s, \ell), \tag{6.1}$$

where the fields C_ℓ and C_s are functions of x and t.

6.1.2 Steady-State Plane-Front Freezing

Equation(6.1) may be used to solve the problem of steady-state macrosegregation
developed in a quiescent binary alloy melt undergoing plane-front freezing at a constant speed, v. Instead of the condition of a 'well-mixed' melt, one assumes that
the melt remains motionless during solidification, so that solute transport proceeds
by the mechanism of atomic or molecular diffusion through the liquid phase ahead
of the moving solid–liquid interface. Even the slow advective motions of the melt,
unavoidably induced by the small mass density change that occurs upon melting or
freezing, will be ignored. The source of the solute atoms diffusing through the melt
is the redistribution of the second component occurring at the moving solid–liquid
interface, as required by the thermodynamic condition of local equilibrium. In the
less common situation in alloys where the distribution coefficient $k_0 > 1$, solute
would instead be removed from the melt at the solid–liquid interface and concentrated in the solid phase. Solute diffusion in the melt accompanying such 'negative'
or inverse macrosegregation depletes the concentration of solute from the adjacent
molten material. For the purposes of simplicity we shall assume that $D_s \ll D_\ell$,
signifying that little of the transport of solute away from the solid–liquid interface
occurs through the process of back-diffusion through the adjacent hot solid. Clearly,
a back-diffusion term, similar to that suggested in Chapter 5.4 in formulating the
Brody–Flemings macrosegregation equation, could be added in this example to
account for the effects from solid-state diffusion. Given, however, that for most
systems $D_s/D_\ell \approx 10^{-3}$, one accepts that diffusion in the melt is the overwhelm-
ingly important solutal transport process compared to back-diffusion, which shall
be ignored for the time being.

6.1.2.1 Co-moving Coordinates

It proves useful to introduce a second spatial coordinate, X, that travels at the speed of the interface, v, in the direction $+x$. This provides a co-moving coordinate system that may be imagined as affixed to the solid–liquid interface, at $x = \hat{x}(t)$. The corresponding origin of the co-moving coordinate always remains collocated at $X = 0 = x - \hat{x}(t)$. See Fig. 6.1 for details.

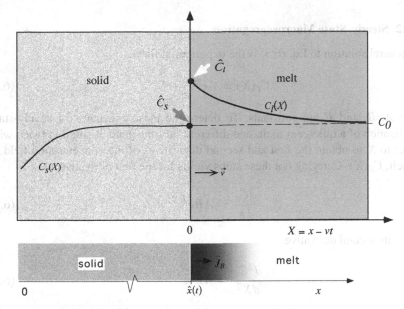

Fig. 6.1 *Upper:* Steady-state motion to the right of a planar solid–liquid interface during freezing of an alloy melt in form of a long bar of initial composition C_0. The solid–liquid interface travels at a speed v and is located at the position $\hat{x}(t)$, in the fixed coordinate, x, and remains positioned at the origin of the co-moving coordinate, $X = 0$. The development of a steady-state solute distribution in the solid and melt is indicated. *Lower:* Schematic of the process of solute rejection into the melt ($k_0 < 1$), with the solute flux vector, J_B located at the interface

The new coordinate system, X, is defined by the Galilean transformation,

$$X = x - vt. \tag{6.2}$$

This coordinate transformation confers the advantage that the time dependence expressed in Eq. (6.1) may be 'hidden', provided that the solid–liquid interface itself travels at the same constant speed when the system eventually achieves steady-state. Thus, for the restricted case of *steady-state* solidification, applying the coordinate transformation given by Eq. (6.2) to the spatial variable, x, in Eq. (6.1), transforms the diffusion equation to an ordinary second-order differential equation. The diffusion equation for the melt, Eq. (6.1), now expressed in the co-moving coordinate, X, becomes

$$\frac{d^2 C_\ell}{dX^2} + \frac{v}{D_\ell}\frac{dC_\ell}{dX} = 0 \quad (X \ge 0). \tag{6.3}$$

The application of the Galilean transformation, allowed here by the steady-state assumption, conveniently removes the explicit time variable from the diffusion equation. Equation (6.1), now reduced to an ODE, Eq. (6.3), allows immediate access to desired steady-state macrosegregation solutions.

6.1.2.2 Steady-State Macrosegregation

The general solution to Eq. (6.3) is the exponential form,

$$C_\ell(X) = Ae^{BX} + E, \tag{6.4}$$

where A, B, and E are constants. To determine these constants for steady-state solidification of a quiescent melt, one differentiates the general solution twice with respect to X to obtain the first and second derivatives of the concentration field in the melt, $C_\ell(X)$. Carrying out these steps yields for the first derivative,

$$\frac{dC_\ell}{dX} = ABe^{BX}, \tag{6.5}$$

and for its second derivative,

$$\frac{d^2 C_\ell}{dX^2} = AB^2 e^{BX}. \tag{6.6}$$

Substituting Eqs. (6.5) and (6.6) back into the general solution, Eq. (6.4), shows that for $X \ge 0$, i.e., in the melt phase only,

$$AB^2 e^{BX} + \frac{v}{D_\ell} ABe^{BX} = 0, \quad (X \ge 0). \tag{6.7}$$

Aside from the trivial solution $A = B = 0$, inspection of Eq. (6.7) suggests that the choice of the constant $B = -v/D_\ell$ will satisfy the general solution, Eq. (6.4), and also reduces it to the explicit exponential form,

$$C_\ell(X) = Ae^{-\frac{v}{D_\ell}X} + E. \tag{6.8}$$

As shown in Fig. 6.1, the 'far-field' boundary condition in the melt may be applied to Eq. (6.8). That is, assuming that the solidifying system is infinitely long, the solute concentration far away from the solid–liquid interface, that is as $X \to \infty$, approaches the melt's initial value, C_0. This far-field concentration boundary condition shows that

$$C_\ell(\infty) \to C_0 = A \cdot 0 + E, \tag{6.9}$$

and, thus, one concludes that $E = C_0$. Substituting this result into Eq. (6.8) gives

$$C_\ell(X) = Ae^{-\frac{v}{D_\ell}X} + C_0. \tag{6.10}$$

The steady-state diffusion solution, Eq. (6.10), is now determined to within one unknown coefficient, A. The constant A may be found by applying the natural boundary condition on the freezing melt. Namely, at steady-state the solute mass entering the melt at the solid-liquid interface, $X = 0$, must exactly equal the solute mass being rejected from the advancing solid. For the sake of specificity, one chooses that the solidifying system exhibits the usual phase diagram characterisitic where $k_0 < 1$. Note, however, that this natural boundary condition also holds true for inverse macrosegregation in alloys where $k_0 > 1$, and, therefore, as solidification progresses, solute would be steadily 'withdrawn' from the melt near the interface and incorporated within the solid phase to increase its concentration.

The flux of solute (designated component B) entering (or leaving) the melt at the solid–liquid interface may be determined by applying Fick's 1st law at the location $X = 0$, namely,

$$J_B = -D_\ell \left(\frac{dC_\ell}{dX}\right)_{X=0}. \tag{6.11}$$

The concentration gradient on the right-hand side of Eq. (6.11) may be found by differentiating Eq. (6.10), which when evaluated at the position $X = 0$ yields the value for the solute flux in the melt just head of the solid–liquid interface, namely $J_B = Av$. This flux, as mentioned above, must be equal to the amount of solute rejected per unit time, per unit area, from the advancing solid–liquid interface. Thus, one may set the rate of rejection of solute from the interface per unit unit area as equal to the diffusion flux of B atoms, J_B, so that

$$v(\hat{C}_\ell - \hat{C}_s) = Av. \tag{6.12}$$

This equality establishes the value of the unknown coefficient A as

$$A = \hat{C}_\ell - \hat{C}_s = \hat{C}_s \left(\frac{1}{k_0} - 1\right). \tag{6.13}$$

The assumptions underlying this steady-state diffusion analysis of macrosegregation require that when steady-state conditions are finally achieved,[1] the crystalline solid freezing from the melt must also have established a steady solute concentration exactly equal to that of the *initial* melt concentration, C_0. This is tantamount to

---

[1] Strictly speaking, the mathematical macrosegregation analysis shows that steady-state solidification take an 'infinite' amount of time and distance to be achieved. As will be demonstrated through various aspects of this analysis, practical limits can be established where the departure from the true steady-state may be ignored.

demanding that under steady-state conditions each unit mass of melt consumed supplies the same amount of solute contained within each mass unit of solid produced. By ignoring the small density difference between the phases, the equality of solute also holds for each unit volume of solid frozen. Thus, one finds that the system at steady-state requires $\hat{C}_s = C_0$, which leads to the result that

$$A = C_0 \left(\frac{1 - k_0}{k_0} \right). \tag{6.14}$$

Inserting the value for A given by Eq. (6.14) into Eq. (6.10) yields the solution, $C_\ell(X)_{ss}$, to the *steady-state* solute concentration fields in the melt, and in the solid, respectively. Specifically, as expressed in their co-moving coordinates, X, one finds that in the melt,

$$C_\ell(X)_{ss} = C_0 \left[1 + \frac{1 - k_0}{k_0} e^{-\frac{v}{D_\ell} X} \right] \ (X \geq 0), \tag{6.15}$$

whereas in the solid,

$$C_s(X)_{ss} = C_0 \ (X \leq 0). \tag{6.16}$$

One should be able to match qualitatively the mathematical results, Eqs. (6.15) and (6.16), to the sketch of these fields portrayed in Fig. 6.1. Equation (6.15) may also be compared to the results given by Chalmers on steady-state macrosegregation [2]. A form of the scaled steady-state solute excess from macrosegregation over the mean composition of the melt is plotted in Fig. 6.2. The ordinate in Fig. 6.2 is the (dimensionless) deviation, or excess, of the melt composition relative to its initial value at a given (dimensionless) distance, $\zeta = X v / D_\ell$, ahead of the interface. The dimensionless distance is measured in intervals of the 'characteristic diffusion length', D_ℓ / v. Once fully formed, the compositional disturbance plotted here travels along with the solid–liquid interface as a positive or negative 'spike' of concentration. This diffusion structure in the melt is also referred to as the solute boundary layer. As noted in Fig. 6.2, the concentration disturbance associated with the boundary layer falls off quickly—e.g., within about 3-5 dimensionless distance units—beyond which the melt remains at its initial composition, C_0, and is essentially unaffected by the diffusion-limited macrosegregation process occurring during freezing. For every additional unit of the characteristic diffusion distance, D_ℓ / v, or in dimensionless terms the exponential ('e-folding') distance, the compositional disturbance in the melt at that distance from the interface decreases by an additional factor of $1/e$. As is also suggested in Fig. 6.2, at three characteristic distances the concentration of the melt is elevated by only about 5% relative to its initial value. One dimensionless diffusion distance, $\zeta = 1$, for most molten alloys freezing at the typical rates encountered in commercial casting practice, is less than about 0.01 mm. Even for relatively slow solidification speeds, more typical of the freezing rates applied when growing large single crystals, the characteristic diffusion distance seldom exceeds about 1 mm.

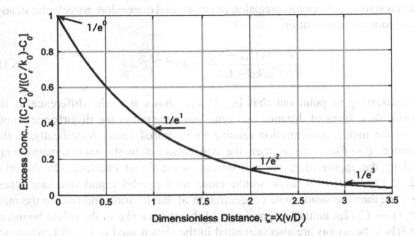

Fig. 6.2 Plot of the dimensionless 'excess' concentration of solute in the melt, $(C_\ell(X) - C_0)/[(C_0/k_0) - C_0]$ ahead of a solid–liquid interface during steady-state plane-front solidification. Subsequent exponential (e-folding) distances are marked at integer values of the dimensionless distance, $\zeta = Xv/D_\ell$, where the excess melt concentration at the interface ($X = 0$) decays back towards its initial concentration, C_0, by factors of e^{-n}, where $n = 0, 1, 2, 3 \cdots$

As a practical issue, the longer the length of the solute boundary layer, the more sensitive it behaves to any hydrodynamic interactions with the melt. Any mechanical or thermal disturbances of the melt that perturb it from quiescence will also alter the form of the boundary layer concentration spike, and, ultimately, affect the concentration of the crystal freezing out. Avoiding boundary layer fluctuations becomes a crucial issue in the production of high-quality single crystals for electronic or optical applications, where maintaining strict homogeneity of the product is essential. Methods of applying controls on the boundary layer thickness during crystal growth will be discussed in the next chapter.

6.1.2.3 Time-Dependent Forms

If the inverse of the Galilean transformation, Eq. (6.2), is applied to the steady-state macrosegregation distribution, Eq. (6.15), the explicit time dependence removed by using co-moving coordinates, X, may be recaptured. The concentration field in the melt expressed in the stationary coordinate, x, becomes a moving solute field, $C_\ell(x, t)$, namely,

$$C_\ell(x, t)_{ss} = C_0 \left[1 + \frac{1 - k_0}{k_0} e^{-\frac{v}{D_\ell}x} e^{\frac{v^2}{D_\ell}t} \right] \quad (x \geq \hat{x}). \tag{6.17}$$

It is now useful to introduce dimensionless parameters to represent time (Fourier number, $Fo = tv^2/D_\ell$) and distance ($\zeta = xv/D_\ell$), and then rearrange Eq. (6.17) so that the LHS becomes the scaled excess in the melt composition from its initial value, C_0. This procedure yields a particularly compact form of the time-dependent diffusion solution of macrosegregation in one spatial dimension: namely the steady-state exponential distribution,

$$\frac{C_\ell(x,t)_{ss} - C_0}{C_0/k_0 - C_0} = e^{-(\zeta - Fo)}. \tag{6.18}$$

It is interesting to point out that Eq. (6.18) shows it is the difference in the *dimensionless* forms of distance and time that determines the disturbance amplitude to the melt's concentration relative to its initial value. Specifically, if this difference, $\zeta - Fo$, is large, then the concentration in the melt remains unaffected by the approaching, but still distant, solid–liquid interface. On the other hand, if $(\zeta - Fo) \to 0$, as it would close to the solid–liquid interface where $x \to \hat{x}(t)$, then the steady-state concentration at that location and time in the melt, $C_\ell(x,t) \to C_0/k_0$, indicating proximity of the sharp spike in the solute boundary layer. These behaviors are also suggested in the sketch used in Fig. 6.1, where the lower portion of the diagram is a cartoon of the solute boundary layer near $x = \hat{x}(t)$, as well as the flux of solute rejected from the moving interface.

6.1.3 Solute Boundary Layers

It is instructive to re-write the steady-state concentration field in the melt, Eq. (6.15), in the co-moving coordinate, X, as

$$C_\ell(X)_{ss} = C_0 + \left(\frac{C_0}{k_0} - C_0\right) e^{-\frac{v}{D_\ell}X} \quad (X \geq 0). \tag{6.19}$$

Inspection of Eq. (6.19) shows that the composition profile developed at steady-state is an exponential distribution of amplitude $C_0/k_0 - C_0$ at $X = 0$, which is added to the starting melt composition, C_0. The coefficient of the second term on the RHS of Eq. (6.19), $\Delta C_0 \equiv C_0/k_0 - C_0$, represents the constant (tie line) difference in the phase compositions maintained across the solid–liquid interface. The exponential concentration field, will be approximated in the form of a 'solute boundary layer', using the foregoing equilibrium difference in composition.

Instead of dealing with an exponential distribution of solute, spread, at least in principle, over an infinite distance, one may usefully approximate the diffusion field by an equivalent *finite* boundary layer of thickness, δ_ℓ. Boundary layers are constructs used in many engineering applications as controlled approximations that capture special features of exact, but perhaps awkward, mathematical descriptions. They are used prevalently in heat and mass transfer, hydrodynamics, aerodynamics and lubrication theory. In macrosegregation, a useful boundary layer simply replaces the exact steady-state exponential distribution. In this particular instance, moreover, the interfacial solute boundary layer during steady-state solidification is configured to contain the identical total mass of solute as does the mathematically exact, infinitely decaying exponential field. A useful, but not unique, definition of such a 'mass equivalent' boundary layer is replacing the exponential solute distribution by the simple triangular structure sketched in Fig. 6.3.

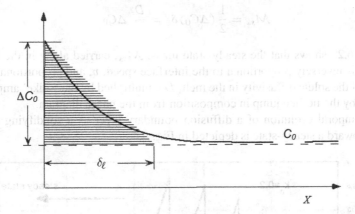

Fig. 6.3 A solute boundary layer of thickness, δ_ℓ. The triangular mass-equivalent boundary layer represents several major features of the exact exponential solute distribution, C_ℓ, that develops, in principle, over infinite time and length scales during steady-state plane-front solidification. Note that this triangular boundary layer maintains the correct tie-line jump in solute concentration, ΔC_0, at the solid–liquid interface, and linearly decreases the solute concentration in the melt from its local equilibrium concentration at the solid–liquid interface, $\hat{C}_\ell$, to the 'far-field' melt concentration, C_0

Essentially, the approximation chosen here to represent the solute field is a constant concentration gradient throughout the finite-thickness boundary layer. Mathematically, this particular boundary layer representation of the solute distribution is defined such that the integral of the solute mass under the entire exponential concentration distribution equals the integral of the mass under the triangular approximation. These integrals are respectively,

$$\Delta C_0 \frac{D_\ell}{v} \int_0^\infty \frac{v}{D_\ell} e^{-\frac{vX}{D_\ell}} dX = \int_0^{\delta_\ell} \Delta C_0 \left(1 - \frac{X}{\delta_\ell}\right) dX. \quad (6.20)$$

The infinite integral on the LHS of Eq. (6.20) represents the total 'excess' amount of solute (above the initial level C_0) that develops at steady-state. The value of this integral, is unity. Substituting 1 for the integral appearing on the LHS of Eq. (6.20), evaluating the integral on the RHS, and canceling common factors from both sides, yields the relationship between the boundary layer thickness, interface speed, and diffusivity,

$$\frac{D_\ell}{v} = \frac{\delta_\ell}{2}. \quad (6.21)$$

Equation (6.21) shows that the thickness, δ_ℓ, for the mass-equivalent triangular boundary layer is just twice the e-folding distance of the exact exponential solute distribution, namely, $\delta_\ell = 2D_\ell/v$. The total steady-state mass of solute contained within the triangular boundary layer is therefore,

$$\mathcal{M}_{ss} = \frac{1}{2}\left(\Delta C_0\right)\delta_\ell = \frac{D_\ell}{v}\Delta C_0. \tag{6.22}$$

Equation 6.22 shows that the steady-state mass, $\mathcal{M}_{ss}$, carried along in the boundary layer is inversely proportional to the interface speed, v, and proportional to the product of the solute diffusivity in the melt, D_ℓ, multiplied by the 'spike' amplitude, ΔC_0, set by the tie-line jump in composition from the phase diagram.

The temporal evolution of a diffusion boundary layer as a solidifying system evolves toward a steady-state is depicted in Fig. 6.4.

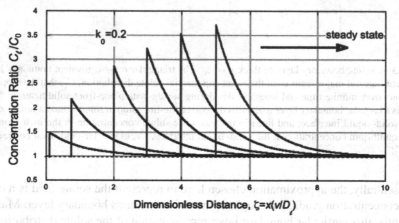

Fig. 6.4 Transient development of the solute diffusion profile for an alloy with a distribution coefficient $k_0 = 0.2$. At the farthest (dimensionless) distance from the interface calculated, viz., $\zeta = 5$, the concentration ratio at the spike has reached a value of approximately 3.7. The concentration ratio at the trailing edge of the solute distribution must build up toward a value of $C_\ell(\hat{x})/C_0 = 1/k_0 = 5$ before the steady-state is reached. The smaller the value of k_0, however, the slower the boundary layer develops, and the longer the transient period lasts. Low k_0 values pose practical difficulties for efficiently producing homogeneous single crystals by directional freezing, because of the long transients involved in approaching steady state

6.2 Transient Macrosegregation

The diffusion boundary layer structure at steady-state provides a deep clue as to how a large uniform melt of initial composition, C_0, begins to freeze in a unidirectional manner and gradually build up and approach the steady-state that has a boundary-layer structure described in Section 6.1.3. A steady state, however, requires the development of the solute boundary layer, implying the prior occurrence of an initial transient solidification period, and ultimately, the occurrence of a final transient, when the solidification process eventually consumes the last amount of melt. We shall explore the nature of these freezing transients in the following sections.

6.2.1 Initial Transients in Infinite Systems

The concentration gradient in the liquid phase at the solid–liquid interface may be found by differentiating Eq. (6.19) with respect to X, and evaluating the gradient at $X = 0$. That procedure gives

$$\left(\frac{dC_\ell}{dX}\right)_{X=0} = -\frac{\Delta C_0}{D_\ell/v}. \tag{6.23}$$

The boundary layer solute mass and the gradient expressed in Eq. (6.23) may each be generalized for any arbitrary concentration difference, $\Delta C_0 = \hat{C}_\ell - C_0$, that develops over time. Thus, the interfacial concentration gradient developed at any time during the formation period of the solute boundary layer is

$$\left(\frac{dC_\ell}{dX}\right)_{X=0} = -\frac{\hat{C}_\ell(t) - C_0}{D_\ell/v}, \tag{6.24}$$

where now the explicit time-dependences of the interfacial concentrations, $C_s(t)$ and $\hat{C}_\ell(t)$, are now noted from Eq. (6.24) and beyond. The boundary layer's solute mass generalizes as the time-dependent function,

$$\mathcal{M}(t) = \frac{D_\ell}{v}\left(\hat{C}_\ell(t) - C_0\right). \tag{6.25}$$

It also appears that a solute boundary layer builds up gradually, precisely because the rejection of solute atoms from the interface occurs initially at a greater rate than the flux of atoms diffusing away from the interface through the increasingly steeper concentration gradient being established in the adjacent melt. The solute flux rejected (or absorbed) at the solid–liquid interface during solidification at a constant speed, v, is

$$J_B^{seg} = v\left(\hat{C}_\ell(t) - \hat{C}_s(t)\right), \tag{6.26}$$

whereas the diffusion flux transported through the melt, found here by substituting the expression for the interfacial gradient, Eq. (6.24), into Fick's 1st law, is

$$J_B^{diff} = -D_\ell\left(\frac{dC_\ell}{dX}\right)_{X=0} = v\left(\hat{C}_\ell(t) - C_0\right). \tag{6.27}$$

The difference between these 'unbalanced' solute fluxes accumulates as either an excess or a deficit in the solute mass of the developing boundary layer, namely

$$J_B^{seg} - J_B^{diff} = v\left(C_0 - \hat{C}_s(t)\right) = v\left(C_0 - k_0\hat{C}_\ell(t)\right),\qquad(6.28)$$

where in the second equation on the RHS of Eq. (6.27) local thermodynamic equilibrium is applied that allows substitution of the now time-dependent tie-line relationship, $\hat{C}_s(t) = k_0\hat{C}_\ell(t)$.

The flux difference given by Eq. (6.28) steadily accumulates as a change in the excess, or deficit, solute mass within the developing boundary layer, $\mathcal{M}(t)$. Differentiation of Eq. (6.25) with respect to time yields the rate of change of the boundary layer solute mass. Taking the time derivative of the total boundary layer mass given by Eq. (6.25), and equating the result with the flux difference shown in Eq. (6.28), yields the transient mass balance relationship within the boundary layer,

$$\frac{d\mathcal{M}(t)}{dt} = \frac{D_\ell}{v}\frac{d\hat{C}_\ell}{dt} = \left(C_0 - k_0\hat{C}_\ell\right)\frac{d\hat{x}(t)}{dt},\qquad(6.29)$$

where the solidification speed, v, appearing in the denominator of the RHS of the first equality in Eqs. (6.29) is now written as the derivative of the interface position, $\hat{x}(t)$.

The variables in Eq. (6.29) may now be separated, and the resultant ODE solved by integration. The initial condition, $\hat{x}(t) = 0$ at $t = 0$, for the melt's concentration is that $C_\ell(0, 0) = C_0$, so the following definite integrals must be equal,

$$\int_{C_0}^{C_\ell(\hat{x})} \frac{d\hat{C}_\ell}{C_0 - k_0\hat{C}_\ell} = \frac{v}{D_\ell}\int_0^{\hat{x}(t)} d\hat{x}.\qquad(6.30)$$

Carrying out the integrations indicated in Eq. (6.30) yields, after a few steps of algebra, the approximate (i.e., boundary layer) solution to this problem, first derived by Tiller et al. [3] for the initial macrosegregation transient in the melt, namely,

$$C_\ell(\hat{x}(t)) = \frac{C_0}{k_0}\left(1 - (1 - k_0)\, e^{-\frac{k_0 v}{D_\ell}\hat{x}(t)}\right),\qquad(6.31)$$

and in the solid phase,[2]

$$C_s(\hat{x}(t)) = C_0\left(1 - (1 - k_0)\, e^{-\frac{k_0 v}{D_\ell}\hat{x}(t)}\right).\qquad(6.32)$$

---

[2] It is worth noting that Tiller's initial transient solution for solute macrosegregation is identical in structure to Pfann's solute distribution derived specifically for single-pass zone refining. Cf. Eq. (5.54).

This pair of transient macrosegregation expressions, Eqs. (6.31) and (6.32), provides approximations to the time-dependent solute concentrations at the moving solid–liquid interface. The results depend on two linked system parameters: (1) the equilibrium distribution coefficient, k_0, and (2) the dimensionless solidification distance, $\hat{x}(t)(k_0 v/D_\ell) = k_0 \zeta$, which appears in the exponential terms for the transient. Given that the interface velocity was assumed to remain steady during freezing, physical freezing distance, $\hat{x}$, and time, t, are simply related as $\hat{x} = t \times v$. A similarity equivalence thus exists between the dimensionless interface position appearing in the exponentials of Eqs. (6.31) and (6.32) and dimensionless freezing time, $\tau = t(k_0 v^2/D_\ell) = k_0 Fo$. In the case of the initial transient, the Fourier number, Fo, is multiplied by the distribution coefficient, k_0, which effectively 'stretches' or 'compresses' both the physical time scale, t, for evolving the transient, as well as the physical distance, x, needed to develop steady-state conditions. This can be problematic in the case of alloys freezing with small k_0 values. Indeed, bulk crystal growth processes applied to systems with very low k_0s often fail to produce acceptably uniform crystals by unidirectional freezing, as the initial macrosegregation transient is just too long for practical operations. Too much material would be solidified with a varying solute content, and therefore wasted, before an acceptably homogeneous crystal would be produced. In critical applications where uniform levels of solute doping are required, for example, semiconductor and opto-electronic crystals, preparation techniques other than directional freezing must be used.

Nonetheless, the similarity relation may be used to show the nature of transient macrosegregation distributions. Figure 6.5 contains plots for different values of k_0 that predict the transient distribution of macrosegregation that develops through a unidirectionally solidified bar of solid, or single crystal. These curves clearly show that small k_0 values cause the initial transient to stretch out, and take longer to achieve steady-state freezing. Quite often the initial portion of a unidirectionally solidified crystal must be discarded because of compositional non-uniformity caused by the initial transient. Techniques for controlling the initial transient during solidification are fundamental to the 'art' and technology of crystal growth. A widely employed technique for controlling transients in the solute concentration during crystal growth will be discussed in detail in Chapter 7.

The exact mathematical treatment of the initial transient segregation problem may be found by solving the time-dependent diffusion equation, Eq. (6.1), subject to appropriate initial and boundary conditions in both the solid and the melt. This approach does indeed provide exact results in the form of a more mathematically elaborate solution to the initial transient problem. By ignoring solid-state diffusion, Smith, Tiller, and Rutter obtained the macrosegregation solution satisfying Fick's 2nd law in the melt, and all the boundary and initial conditions already discussed here. Their result, published in 1955, is included here for completeness [4].

$$\frac{C_s(\zeta)}{C_0} = \frac{1}{2} + \frac{1}{2}\mathrm{erf}\sqrt{\frac{\zeta}{2}} + \frac{2k_0 - 1}{2}\mathrm{erfc}\left((2k_0 - 1)\sqrt{\frac{\zeta}{2}}\right). \qquad (6.33)$$

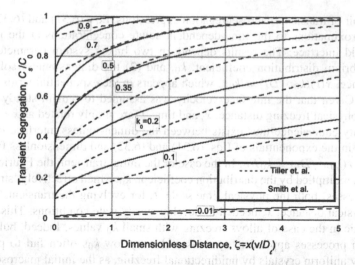

Dimensionless Distance, ζ=x(v/D)

Fig. 6.5 Plots of Tiller's initial transient solution, Eqs. (6.32), shown as *dashed curves*, and Smith et al.'s exact solution, Eq. (6.34), shown as *solid curves*. These transient macrosegregation curves are expressed as the concentration ratio C_s/C_0, versus dimensionless solidification distance, ζ. These are the predictions for initial transients affecting macrosegregation during plane-front solidification. *Curves* are plotted for various values of the equilibrium distribution coefficient, $k_0 < 1$. Note the increasing amount of freezing distance, ζ, required to achieve steady-state solidification as k_0 decreases. Both sets of solutions converge at small and large values of ζ. Extremely small k_0 values make impractical the preparation of homogeneous single crystals by directional freezing methods

The Smith et al. diffusion solution to the initial transient problem was derived using Laplace transform methods, which need not be reproduced here. In their original paper their solution is shown, but not derived. The interested reader can consult standard references on Laplace transforms, or follow derivations provided in diffusion textbooks [1]. Suffice it to say that the exact solution to the initial transient problem predicts a modestly different result from that based on the simpler mass-balance solution published by Tiller et al. [5], and derived in detail in this section. The exact diffusion solution and Tiller et al.'s mass-balance approximation must, of course, agree at extremely small and extremely large values of ζ, as the boundary conditions they satisfy become identical. As Fig. 6.5 confirms, however, the mass-balance solution always underestimates the time needed to reach steady-state freezing, because quasi-steady estimates are made that only satisfy Fick's 1st law, not the fully time-dependent diffusion equation.

6.2.2 Transients in Finite Systems

Each of the initial transient formulas, Eqs. (6.31), (6.32) and (6.33) were derived for solidifying systems of *semi-infinite* length. All real systems, of course, always terminate at some finite length, L. As suggested in the initial transient formulas, a new dimensionless parameter appears, called the macrosegregation Péclet number, which is defined as

$$Pe \equiv \frac{L}{D_\ell/v} = \frac{\text{physical system length}}{\text{characteristic diffusion length}}. \tag{6.34}$$

The macrosegregation Péclet number is the ratio of the size of the solidifying system to its melt's characteristic diffusion length. One may re-write Eqs. (6.31) and (6.32) for solidifying systems of *finite* size by inserting the definition of the macrosegregation Péclet number provided in Eq. (6.34), and using f_s as the fraction solidified of the finite melt's length. For the initial transient in the melt, one obtains,

$$\frac{C_\ell(\hat{x})}{C_0} = \frac{1}{k_0} + \left(1 - \frac{1}{k_0}\right) e^{-k_0 Pe \cdot f_s}, \tag{6.35}$$

and for the initial transient in the solid phase

$$\frac{C_s(\hat{x})}{C_0} = 1 - (1 - k_0) e^{-k_0 Pe \cdot f_s}, \tag{6.36}$$

One may connect these equations for the initial transients with the Gulliver-Scheil macrosegregation equations derived previously in Chapter 5. Specifically, if the Péclet number is relatively small, typically where $Pe \leq 1$, then the initial transient curve will approximate the predictions of Gulliver-Scheil. On the other hand, if the macrosegregation Péclet number is large, say $Pe > 10^3$, then the initial transient in a finite system will be virtually identical to predictions made for the initial transient in an infinite system, viz., Eqs. (6.31), (6.32) and (6.33). Finally, if the Péclet number falls within the range $1 < Pe < 10^3$, then numerical estimates are best used. Cases of crystal growth for which intermediate values of the macrosegregation Péclet number are encountered may be conveniently checked by referring to the publication by Verhoeven et al. [6], who have computed typical behaviors expected during the initial transients for many commonly encountered situations in crystal growth.

6.2.3 Final Segregation Transient

As solidification proceeds in a system of finite length, eventually the solid–liquid interface position will approach the system's far boundary, so that the solid–liquid interface $\hat{x} \rightarrow L$. Sometime prior to the interface actually reaching the very end of the melt, the diffusion boundary layer (See discussion in Section 6.1.3.) traveling ahead of it 'senses' that the system is of finite extent; the boundary layer then begins to interact more and more strongly with the approaching external environment that exists beyond the end of the molten phase. That environment in fact might be a crucible wall, a layer of some fluxing agent, or an adherent melt-gas film; but in any event, the on-going diffusion and solidification processes in the melt are increasingly affected and eventually interrupted.

Specifically, solute atoms reaching the end of the melt are either 'reflected' back into the melt, or undergo chemical reaction, such as oxidation. The solute actually turned back by the system's boundary causes the concentration within the boundary layer to rise rapidly, creating a second transient event called the *terminal transient*.

As a practical matter, however, terminal transients occurring in macrosegregation are not usually considered important in most industrial crystal growth and solidification processes. The reason that terminal transients are usually ignored in crystal growth is simply that other rather prosaic factors arise that degrade the quality obtained near the end of a crystal growth run. For example, as crystal growth processes consume the last vestiges of the melt phase, there is a rapid loss of thermal, shape, and dimensional control, making the final portion of many melt-grown crystals useless, except for being remelted again and then added to start another crystal growth run. In shaped castings, the final transients that occur near the periphery of the mold are of such a complicated nature that they are normally ignored. Terminal transients, however, can be responsible for severe 'inverse segregation', a wide variety of surface imperfections on castings, and even troublesome alteration of the near-surface mechanical properties and corrosion resistance.

For the sake of completeness, however, we include the terminal transient solution to the macrosegregation process during unidirectional freezing. The terminal transient event for unidirectional solidification was also solved by Smith et al. [4]. Their one-dimensional diffusion solution for the solute concentration in the solid toward the last portion of a unidirectional crystal growth process was shown to be described by the sum of infinite products,

$$\frac{C_s(f_s)}{C_0} = 1 + \sum_{n=1}^{\infty} (2n+1) \frac{\prod (n-k_0)}{\prod (n+k_0)} e^{-n(n-1)Pe(1-f_s)}. \qquad (6.37)$$

6.3 Numerical Studies in Finite Systems

The initial and terminal transients, Eqs. (6.33) and (6.37), taken together comprise a 'complete solution' to chemical macrosegregation in solidifying long bars. As mentioned earlier, however, if the system's physical length is not too large, as compared to the solute diffusion length in the melt, D_ℓ/v, thereby allowing the macrosegregation Péclet number to be less than ca. 10^3, then Eqs. (6.33) and (6.37) become inaccurate. A more discerning criterion to establish the range of validity of the macrosegregation theory by Smith et al. in real, finite-sized systems is that

$$Pe > \left| 1 + \frac{1}{k_0(1-k_0)} \right| \cdot \beta, \qquad (6.38)$$

where β is some numerical factor. Verhoeven et al. [6] carefully determined values for β using numerical methods, and established that for small values of the distribution coefficient, $k_0 \approx 0.01$, $\beta \approx 7$, whereas for larger values, such as $k_0 \approx 0.5$, they found that $\beta \approx 4$. Clearly, one does not need extremely accurate values of the parameter β to allow at least engineering estimates. Such estimates are based on the criterion given by Eq. (6.38) for the minimum values of a casting's Péclet number, below which the analytical macrosegregation equations developed here become inaccurate in finite-sized castings.

6.4 Summary

This chapter extends the discussion of macrosegregation by bringing into play the important phenomenon of solute diffusion and redistribution through the remaining melt to be solidified. These considerations of solute diffusion replace the initial assumption in segregation theory that the residual melt fraction remains 'perfectly mixed' throughout the entire freezing process—as was used throughout Chapter 5—by the contrasting assumption that the residual melt fraction remains 'perfectly stationary', i.e., in a quiescent, hydrostatic state.

Mathematical solutions were then derived for both steady-state and transient macrosegregation. For analytic simplicity and ease of application the steady-state diffusion-segregation equations were derived first for semi-infinite solidifying systems, where convenient exponential and error-function solutions are easily found and applied. Transient segregation solutions were then reviewed and divided into an initial transient—already encountered in the process of zone refining in the previous chapter—and a terminal, or final transient. The former is of major importance in achieving homogeneous solute distributions in unidirectionally solidified single crystals, whereas the latter has relatively minor technical significance in either the solidification of castings or in most crystal growth processes.

Introduction of the macrosegregation Péclet number allows quantitative discrimination between 'semi-infinite' segregation behavior, and segregation found in finite systems. Specifically, a value of about $Pe \geq 10^3$ was shown to provide the criterion for the residual melt acting as if it were of 'semi-infinite' length. Finally, the topic of macrosegregation in systems with finite length was briefly discussed, where use of either finite difference or finite element numerical solutions are advised for obtaining accurate estimates of the solute distributions.

References

1. M.E. Glicksman, *Diffusion in Solids: Field Theory, Solid-State from an and Applications*, Wiley Interscience Publishers, New York, NY, 2000.
2. B. Chalmers, *Principles of Solidification*, **133**, Wiley, New York, NY, 1964.
3. W.A. Tiller, K.A. Jackson, J.W. Rutter and B. Chalmers, *Acta Metall.*, **1** (1953) 428.
4. V.G. Smith, W.A. Tiller and J.W. Rutter, *Can. J. Phys.*, **33** (1955) 723.
5. W. Tiller, *Liquid Metals and Solidification*, ASM, Cleveland, OH, 1958, p. 276.
6. J.D. Verhoeven, W.N. Gill, J.A. Puszynski and R.M. Gindé, *J. Cryst. Growth*, **89** (1988) 189.

Chapter 7
Composition Control

7.1 Convection in Freezing Melts

Solute transport ahead of a unidirectionally freezing planar solid–liquid interface actually behaves in a manner intermediate to the two limiting situations already discussed. First, the melt was assumed in Chapter 5 to be well-mixed into a state of perfect homogeneity; next, in Chapter 6, solute transport was limited to liquid-state diffusion, and convective mixing was assumed to be totally supressed. Macrosegregation processes in solidifying binary melts are explored again in this chapter, but now subject to *both* controlled amounts of convective mixing plus solute diffusion acting within some related distance in the melt adjacent to the solid–liquid interface. A rudimentary, but extremely useful convection model was first conceptualized at Bell Telephone Laboratories for the preparation of transistor-grade semiconductor crystals with controlled doping levels. This model will be developed in detail, as it helps identify some of the basic fluid mechanical interactions that occur between a crystal-melt interface and its nearby convecting melt.

7.1.1 Mixing Limits

7.1.1.1 Perfect Mixing

The term 'perfect' mixing implies laminar and turbulent mixing occurring on hydrodynamic length scales that reach sufficiently small sizes compared to the characteristic diffusion length, D_ℓ/v. The diffusion length, a system parameter, was defined and discussed in Section 6.1.2. Although diffusion is the process primarily responsible for most mixing at atomic scales, diffusion lengths are macroscopic, and, typically, are of the order of millimeters. Thus, the eddy structures in a strongly stirred melt can easily provide efficient mixing of melt components on scales that are comparable to, or shorter than, this diffusion length. Stirring can therefore provide convenient composition control during crystal growth, and allow microstructure modifications, such as grain refinement and texture changes during casting.

A turbulent, well-mixed melt is incapable of supporting the macroscopic concentration gradients normally required for net solute transport via diffusion. The theory

M.E. Glicksman, *Principles of Solidification*, DOI 10.1007/978-1-4419-7344-3_7,

of turbulent mixing [1, 2], however, continues to pose some conceptual and computational difficulties for engineers, especially for determining the significant length
scales for mixing in a turbulent fluid velocity field acting within an enclosed space
such as a crucible or mold. Specifically, the difficulties encountered here involve
the fact that turbulent mixing occurs simultaneously, and sometimes chaotically,
over many length scales [3], down to the smallest eddies existing in the melt at the
lower limit of what is called the 'Kolmogorov turbulence scale' [1]. This scale of
turbulence is set by such factors as the fluid velocity, its viscosity, and the associated
spectrum of turbulent eddy energies. Turbulent flow does indeed efficiently enhance
solute transport throughout the melt, and mixes the solute with the solvent atoms
down to the shortest length scales allowed by the smallest eddies generated in the
flow. Transport mediated by atomic diffusion ultimately completes the process of
solute mixing by carrying it even further down to the melt's molecular limit. (See
discussions in Chapter 1 on the short-range molecular and atomic arrangements
found in melts.)

The presence of a solid–liquid interface, however, restricts the onset of melt
turbulence in its immediate proximity by imposing the well known 'no-slip' fluid
dynamic boundary condition at $x = \hat{x}$, where the momentum of the adjacent fluid,
and its kinetic energy, must fall to zero. The fluid molecules adjacent to the interface
are literally 'stuck' in place by short range interaction forces. The flow streaming
past the interface gradually increases its velocity with distance and approaches the
free-stream velocity. The imposition of no-slip as a fluid dynamic condition generates a zone adjacent to the crystal-melt interface called the 'momentum boundary
layer'. Fully developed turbulence is possible at distances beyond the momentum
boundary layer. Readers interested in further fluid mechanical details of melt boundary layers at solid–liquid interfaces and their influences on heat and mass transfer
should consult [3–5].

7.1.1.2 Quiescent Melts

A truly quiescent, or stationary, melt can support, at least principle, macroscopic
concentration gradients over unlimited distances. These concentration gradients
stimulate diffusive transfer of solute, as predicted by Fick's 1st law. In fact, just
as occurs in solids, truly static fluids also require atomic diffusion in order to transport solutes over macroscopic distances. The process of diffusive transport in melts,
however, occurs through *microscopic* atomic and molecular motions of solute atoms
within the relatively mobile liquid structure. As explained above, even in the presence of convective mixing, solute diffusion is still needed ultimately to complete the
intimate molecular mixing required to satisfy thermodynamic equilibrium. Fluids,
even at nominal 'rest' in normal terrestrial environments, are always subject to some
level of subtle convective motion. Exceptions to this statement require either rare, or
specialized, environments that offer 'microgravity' conditions [6–8]—where gravity
is nearly absent—as is found in outer space, far away from planetary-sized objects,
in low-Earth orbit, or under terrestrial free-fall in vacuum. Gravitationally-induced
mixing in such gently flowing fluids is called 'natural' convection [4]. Natural

convection, therefore, may be expected whenever a fluid supports concentration gradients and/or thermal gradients that cause thermo-solutal density gradients, which are only partially aligned with the local gravity vector.

The first theoretical model published that discussed extrinsic control of solute transport during crystal growth through control of both diffusive and convective mixing in the melt was published by Burton, Prim, and Slichter [9]. Their model, now referred to as BPS theory, captures with astounding simplicity several of the major features to be considered when the melt ahead of the solid–liquid interface is stirred either by buoyancy forces, such as gravity acting on density gradients, by mechanical means, such as rotors, pumps, and propellers, or by inducing relative motions, such as rotation or translation between the solid and its melt. Their model owes its remarkable simplicity, and limitations, to ignoring some of the formal requirements of fluid mechanics, and thus avoiding the attendant mathematical complexities associated with properly specifying the physics of vector fields in fluids. Models much more accurate than that of Burton, Prim, and Slichter have been published in considerable numbers since the development of BPS theory [5, 10]. Most of these convective macrosegregation models require numerical simulations usually limited to specific crystal growth configurations and casting processes. Numerical simulation of fluid interactions with attendant solute diffusion is called 'convecto-diffusion', and is now in common use by commercial crystal growers and foundries. These simulators are often proprietary, and use sophisticated solvers and computer sub-routines designed explicitly for efficiently and simultaneously solving the equations of incompressible fluid flow, heat transfer, and diffusion. However, these more elaborate and physically accurate models require many more system parameters than does BPS theory. Thus, for purely practical reasons, convection models for crystal growth and casting are usually applied to a restricted range of materials and processing parameters. Crystal growth and solidification processes, as already pointed out, operate over a much broader parameter 'universe', the basic rules for which are perhaps viewed and understood more easily through a simpler, albeit less accurate, solidification model.

As a consequence of its simplicity, BPS theory represents the transport of solute through partially stirred melts only rather crudely. Nevertheless, the BPS model remains in use by practical crystal growers, at least as a starting point to control their processes. In point of fact, crystal growers require estimates of how the controlled stirring of the melt influences the composition of the freezing solid. Also, as will be shown later in this book, partial stirring of the melt also influences the morphological stability of an advancing crystal-melt interface, and thereby directly affects the success of many crystal growth processes. Again, the reader is cautioned that BPS theory does not provide a hydrodynamically-faithful model of how the melt actually flows adjacent to a growing interface; the rigorous methods of fluid mechanics must be used to answer such questions.

Over the past three decades, the advent of powerful digital computational techniques and the development of sensitive, non-invasive, experimental diagnostics for fluid dynamical measurements allow a much more accurate portrayal of the flow states that occur during crystal growth. Such methods now provide engineers with great insight as to how to control solidification and crystal growth processes. These methods are currently incorporated for real-time process control in the semiconductor industry, where single crystals with tailored solute and impurity distributions are in tonnage demand. The reader, therefore, should consider the subsequent discussion of BPS theory as representing an important early step in what has become a half-century of significant engineering developments in crystal growth technology.

7.1.2 The BPS Model

The BPS model included for the first time a plane-front solidification model that allows convection in the melt. Macroscopic melt motions, as suggested in Fig. 7.1, must be supported by the continual input of energy to overcome dissipation of the fluid's kinetic energy to heat by the viscous forces acting within the melt. BPS theory also retains the already discussed concept of an interfacial region, through which solute transport to or from the advancing solid–liquid interface must occur initially by diffusion. Solute transport in the melt adjacent to the interface occurs initially by molecular diffusion down concentration gradients. The melt must remain nearly at rest over some small distance from the moving interface to support the presence of these gradients, as suggested in Fig. 7.2. Finally, at some greater distance away from the interface, the BPS model postulates that solute transport becomes augmented through efficient convective mixing, from which the melt remains essentially uniform in solute concentration.

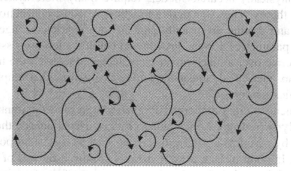

Fig. 7.1 Schematic of turbulent flows in a melt. Vortical fluid motions, indicated by *arrows*, are chaotic, spanning a large range of length scales from centimeters to fractions of millimeter. Kinetic energy is constantly dissipated during turbulent convection by internal viscous forces, so that energy must be continually fed into the melt, either mechanically, as kinetic energy, or through the action of gravitational (buoyancy) forces, which provide a reservoir of potential energy

The BPS approach uses a fictive 'diffusion boundary layer' of thickness δ, modeled by them as a quasi-stationary zone of fluid adjacent to the solid–liquid interface. Choosing a coordinate X, co-moving with the interface at $X = 0$, a quasi-stationary zone, $X \leq \delta$, is postulated, within which solute transport is limited to pure diffusion. Next, it is postulated that mixing in the melt suddenly resumes at distances beyond

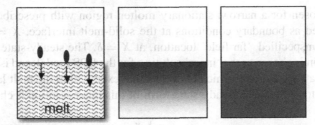

Fig. 7.2 Schematic of solute diffusion occurring in a quasi-stationary region of melt. Solute molecules, or atoms, added in the *left-hand panel* move microscopically in random directions caused by local thermodynamic fluctuations. As suggested in the *middle* and *right-hand panels*, if given sufficient time, the motions of myriad solute atoms will eventually spread them uniformly throughout the melt. Diffusion transport, in contrast with convection transport, is a spontaneous kinetic process that draws upon the chemical free energy available from the concentration gradients initially in the melt. Diffusion transport ceases only when the gradients (and their associated chemical free energy) disappear

the quasi-stationary zone. This 'outer' mixing is assumed to be 'perfect', in the sense that the melt is fully turbulent and flowing so vigorously on sufficiently short length scales that its time- and space-averaged solute composition may be approximated by the condition that $C_\ell(x, t) = $ constant.

Elementary considerations of fluid mechanics and heat transfer [3, 11], however, clearly show that a quasi-stationary zone is incorrect in principle. Specifically, their simultaneous application at the plane $X = \delta$ of two incompatible flow conditions persisting in the melt violates the principle of conservation of momentum, and also destroys the required continuity of the melt's velocity fields. To reiterate, the two flow conditions joined at $X = \delta$ are: (1) *no-mixing* (that is, transport by liquid-state diffusion for $X < \delta$), and (2) *perfect* (convective) *mixing* for the region $X > \delta$. Though certainly fictitious, the notion of a 'quasi-stationary' BPS melt zone may be viewed as a first—although technically incorrect—attempt to employ the necessary no-slip hydrodynamic boundary condition at a moving solid–liquid interface. Nevertheless, some important, qualitatively correct results derive from the BPS model, making its analysis worthwhile as an introduction to the complicated subject area of convecto-diffusive macrosegregation.

7.1.3 Solution to the BPS Transport Equations

7.1.3.1 Inner Solution

The mathematical solution to the diffusion equation within a stagnant boundary layer adjacent to the crystal-melt interface, expressed in a coordinate system, X, co-moving with that interface, may be approximated as an exponential solute distribution satisfying the steady-state diffusion equation. This solution was derived in detail in Chapter 6 as the steady-state form of the macrosegregation solution in a melt of semi-infinite extent. In BPS theory, however, the exponential diffusion solution is chosen for a narrow, stationary molten region with prescribed compositions imposed as boundary conditions at the solid-melt interface, $X = 0$, and at some, as-yet unspecified, 'far field' location, at $X = \delta$. The steady-state diffusion-limited solution is chosen as the 'inner' solution for the BPS model, and is presumed to be appropriate to hydrodynamic conditions that exist within the melt layer, especially if close to $X = 0$. The steady-state form of this inner solution is chosen as,

$$C_\ell(X) = A e^{-\frac{v}{D_\ell} X} + E. \tag{7.1}$$

The diffusive solute flux evaluated at the solid–liquid interface, $X = 0$, in the quasi-stationary zone, which is transported either to ($k_0 < 1$), or from ($k_0 > 1$) the bulk melt, is given by Ficks 1st law, applied at the plane $X = 0$, specifically,

$$J_B = -D_\ell \frac{\partial C_\ell}{\partial X} = D_\ell A \left(\frac{v}{D_\ell} \right) e^0, \tag{7.2}$$

therefore

$$J_B = A v. \tag{7.3}$$

The configuration of the inner diffusion-limited quasi-stationary layer and the adjoining outer well-mixed convecting melt are sketched in Fig. 7.3. Solute mass conservation at the advancing interface requires that the mass of B atoms released at, or absorbed by, a unit area of the interface is balanced by the diffusive solute flux flowing through the melt, given by Eq. (7.2). The moving interface is the source, or sink, of B atoms, because a tie-line jump in concentrations, ΔC_0, is continuously maintained across the crystal-melt interface, as prescribed by the phase diagram. The local mass balance at the solid–liquid interface—i.e., the Stefan solute mass balance—is expressed as,

$$A v = v \Delta C_0 = v \left(\hat{C}_\ell - \hat{C}_s \right). \tag{7.4}$$

The sign of the mass of solute rejected (or absorbed) on the RHS of Eq. (7.4), and the corresponding sign of the diffusion flux in Eq. (7.2) are consistent for either the case of positive macrosegregation, with $k_0 < 1$, or negative macrosegregation, with $k_0 > 1$. Equation (7.4) shows that the value of the unknown constant is given by $A = \hat{C}_\ell - \hat{C}_s$, which equals the concentration 'jump' set by the alloy's tie-line at the temperature of solidification. Substituting this value of A back into the diffusion solution, Eq. (7.1), yields

$$C_\ell(X) = \left(\hat{C}_\ell - \hat{C}_s \right) e^{-\frac{v}{D_\ell} X} + E, \quad (X < \delta). \tag{7.5}$$

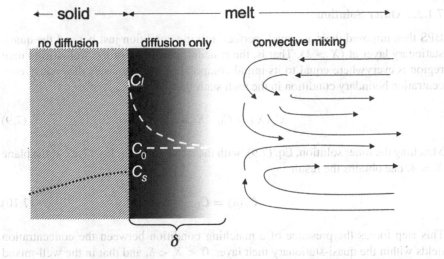

Fig. 7.3 Configuration of the quasi-stationary BPS melt zone, of thickness, δ, supporting the 'inner' diffusion-limited solution, and adjacent well-mixed 'outer' melt that has a uniform concentration of the solute, C_0. Local equilibrium prevails at the moving crystal-melt interface, where the tie-line jump, or discontinuity in concentrations, $\Delta C_0 = \hat{C}_\ell - \hat{C}_s$ is maintained in accord with the phase diagram's distribution coefficient, $k_0 = \hat{C}_s/\hat{C}_\ell$. Solute concentration in the crystal can be conveniently arranged by controlling the strength of the outer mixing

The remaining unknown constant, E, can now be determined from the boundary condition established at the solid–liquid interface located at $X = 0$. The phase diagram tie-line specifies at each temperature the jump in composition between solid and liquid where they are in contact, and at local equilibrium. Specifically, at the solid–liquid interface, $C_s(0) = \hat{C}_s$, and $C_\ell(0) = \hat{C}_\ell$. Evaluating Eq. (7.5) at the plane $X = 0$ yields

$$C_\ell(0) = \hat{C}_\ell = \hat{C}_\ell - \hat{C}_s + E, \qquad (7.6)$$

and therefore the constant E may be fixed as

$$E = \hat{C}_s. \qquad (7.7)$$

Burton, Prim, and Slichter found that by applying local equilibrium and the interfacial solute flux balance, and holding to the approximation of the assumed quasi-stationary fluid zone, the diffusion profile given by the *inner* transport solution, becomes

$$C_\ell(X) = \left(\hat{C}_\ell - \hat{C}_s\right) e^{-\frac{v}{D_\ell}X} + \hat{C}_s. \qquad (7.8)$$

7.1.3.2 Outer Solution

BPS then imposed their assumed 'perfect' mixing condition just beyond the quasi-stationary layer at $(X > \delta)$. That is, the concentration field of the well-stirred melt region is everywhere equal to its initial composition C_0. This sets the 'outer' concentration boundary condition in the melt such that

$$C_\ell(X) = C_0 \quad (X > \delta). \tag{7.9}$$

Matching the inner solution, Eq. (7.9), with the outer solution, Eq. (7.9), at the plane $X = \delta$, one obtains the result

$$C_\ell(\delta) = C_0. \tag{7.10}$$

This step forces the presence of a matching condition between the concentration fields within the quasi-stationary melt layer, $0 < X < \delta$, and that in the well-mixed region, $X > \delta$. Imposing the matching condition given by Eq. (7.10) on the inner diffusion solution, Eq. (7.8), yields the central result of BPS theory:

$$C_\ell(\delta) = C_0 = \left(\hat{C}_\ell - \hat{C}_s\right) e^{-\frac{v}{D_\ell}\delta} + \hat{C}_s, \quad (X = \delta). \tag{7.11}$$

Equation (7.11) actually provides the definition of the zone width, δ, of the quasi-stationary BPS boundary layer.

By non-dimensionalizing Eq. (7.11) as

$$e^{\frac{v}{D_\ell}\delta} = \frac{\hat{C}_\ell - \hat{C}_s}{C_0 - \hat{C}_s}, \tag{7.12}$$

one can easily show by taking logarithms of Eq. (7.12) that the quasi-stationary melt layer, consistent with the matched inner and outer transport solutions, has a thickness

$$\delta \equiv \frac{D_\ell}{v}\left[\ln\frac{\hat{C}_\ell - \hat{C}_s}{C_0 - \hat{C}_s}\right]. \tag{7.13}$$

It is interesting to note that the definition of δ provided in Eq. (7.13) just involves the usual diffusion length scale in the melt, D_ℓ/v, multiplied by the logarithmic ratio of the tie-line jump, ΔC_0, divided by the composition difference between the solid actually freezing out of the quasi-stationary melt layer and that for solid at the steady-state composition, C_0.

A cursory study of the behavior of the RHS of Eq. (7.13) shows that in order for the solid to freeze out with the steady-state macrosegregation conditions, the denominator of Eq. (7.13) would vanish, $(C_0 - \hat{C}_s) \to 0$, and the quasi-stationary layer would extend 'indefinitely', i.e., $\delta \to \infty$. In reality, the solute boundary layer would over take and limit the true extent of the BPS quasi-stationary layer. Instead, as shown by Eq. (6.21), $\delta \to 2D_\ell/v$.

Thus, BPS theory suggests that to crystallize a melt with a steady-state composition equal to C_0, the melt must be free of significant convection for distances from the solid–liquid interface that exceed $2D_\ell/v$. This would indeed be a difficult condition to satisfy, except perhaps in microgravity, or by greatly restricting the effective length of the adjacent melt. To achieve the other extreme of macrosegregation, with solidification occurring with 'perfect' stirring, where the solid composition $C_s \to k_0C_0$, would require that the turbulent melt region be reduced to a thin region near the solid–liquid interface, squeezing the quasi-stationary layer so that $\delta \to 0$. This limit is also elusive, given that melts have finite, i.e., non-zero viscosities. (See again Section 1.2.1 for a discussion of melt viscosity.) Thus, one finds that in order to approach the limiting condition $\delta \to 0$, an extremely large shear rate would be required, which, of course, might not be not possible under practical crystal growth conditions. Clearly, the extreme condition of reducing the momentum boundary layer to a negligible thickness would seldom be accomplished, precluding practical achievement in which $\hat{C}_\ell \to C_0$, and $(C_0 - \hat{C}_s) \to C_0(1 - k_0)$.

Obtaining solute levels in the solid that lie between the two hydrodynamic limits discussed above, viz., C_0 and k_0C_0, is possible by merely using some active stirring control. Melt convection is often conveniently controlled by setting either the rotation rates of the crystal during Czochralski crystal growth, or, during Bridgman crystal growth, by inserting baffles or restrictor plates immersed in the melt close to the solid–liquid interface. Some of the basic configurations of bulk crystal growth are described briefly in Appendix D.

7.2 Effective Distribution Coefficients

7.2.1 Definition and Role of Effective Distribution Coefficients

The BPS result presented in Eq. (7.12) can be fully non-dimensionalized by dividing its numerator and denominator by the overall composition of system, C_0. This second normalization gives the result

$$e^{\frac{v}{D_\ell}\delta} = \frac{\frac{\hat{C}_\ell}{C_0} - \frac{\hat{C}_s}{C_0}}{1 - \frac{\hat{C}_s}{C_0}}. \tag{7.14}$$

An effective distribution coefficient, specifically termed the BPS segregation coefficient, k_{BPS}, may now be defined as the independent parameter appearing on the RHS of Eq. (7.16),

$$k_{BPS} \equiv \frac{\hat{C}_s}{C_0}. \tag{7.15}$$

Engineers refer to k_{BPS} is an 'effective' distribution coefficient, because it relates the composition of solid freezing unidirectionally from a stirred melt of overall composition, C_0. Although the BPS segregation coefficient is an analogue to the equilibrium distribution coefficient, k_0, k_{BPS}, unlike k_0, is *not* fundamentally based. Specifically, k_0 relates the composition of solid, $\hat{C}_s$, in equilibrium with a melt of composition $\hat{C}_\ell$, whereas the BPS segregation coefficient depends both on the phase diagram and on complex hydrodynamic conditions that are established through the melt. The BPS segregation coefficient does, however, provide a useful operating parameter for crystal growth and zone refining, insofar as it can be used to correlate the composition of a crystal being pulled from a stirred melt of known composition.

A remarkable example of applying these ideas is the engineering behind an important solidification technology, namely, the growth of large (now exceeding 300 mm dia.) silicon crystals. Meter-long Si crystals are now grown routinely with controlled amounts of oxygen—an impurity picked up unavoidably by molten Si during crystal growth in silica (SiO_2) crucibles. The correct oxygen impurities levels can be maintained within limits during the entire growth process by stirring the molten Si through imposed crystal rotation or oscillation. Hundreds of kilograms of molten Si can be solidified into a nearly perfect single-crystal semiconductor with tightly controlled electronic dopant and impurity levels—a truly impressive technical feat accomplished by implementing this elegant control method.

One may combine Eq. (7.14) and Eq. (7.15), the definitions now developed for the BPS quasi-stationary zone thickness, δ, and the effective distribution coefficient, k_{BPS}, to yield

$$\frac{\hat{C}_\ell}{C_0} = (1 - k_{BPS})e^{\frac{v}{D_\ell}\delta} + k_{BPS}. \tag{7.16}$$

Multiplying both sides of Eq. (7.16) by C_0 and then dividing through by $\hat{C}_s$, yields

$$\frac{\hat{C}_\ell}{\hat{C}_s} = \frac{C_0}{\hat{C}_s}\left[(1 - k_{BPS})e^{\frac{v}{D_\ell}\delta} + k_{BPS}\right], \tag{7.17}$$

which upon simplifying by factoring k_{BPS} gives the result:

$$k_0 = \left(\frac{1 - k_{BPS}}{k_{BPS}}e^{\frac{v}{D_\ell}\delta} + 1\right)^{-1}. \tag{7.18}$$

Equation (7.18) is an interesting result, inasmuch as it relates the equilibrium distribution coefficient, k_0, directly to the value of k_{BPS}. Equivalently, Eq. (7.18) may be thought of as providing an explanation of the influence of melt convection on the variation of the effective distribution coefficient. If Eq. (7.18) is inverted, it provides the relationship between k_{BPS} and the key material and process control parameters: k_0 and δ,

$$k_{BPS} = \frac{k_0}{k_0 + (1 - k_0)e^{-\frac{v}{D_\ell}\delta}}. \tag{7.19}$$

One notes that Eq. (7.19) suggests that the macrosegregation of a particular solute in the solid phase indeed depends strongly on the phase diagram through the value of k_0, but also depends upon extrinsic hydrodynamic conditions captured approximately in the BPS quasi-stationary layer thickness.

7.2.2 Results from BPS Theory

Figure 7.4 for $k_0 < 1$, and Fig. 7.5, for $k_0 > 1$ are plots based on Eq. (7.19). The non-dimensionless width of the quasi-stationary zone, $(v/D_\ell)\delta$, is chosen as the abscissae in these figures. Each curve shown indicates how k_{BPS} varies for a specific value of k_0, which is selected as the plot parameter. Under intense stirring, where $\delta \to 0$, one should note that $k_{BPS} \to k_0$. By contrast, as stirring is stopped, and the melt become quiescent, $k_{BPS} \to 1$, and steady-state macrosegregation is restored, i.e., $\hat{C}_s \to C_0$. The independent variable, δ, can be controlled by selecting the vigor and mixing efficiency by which the melt is stirred. A comparison of Figs. 7.4 and 7.5 shows that the behavior for alloys with $k_0 < 1$ is qualitatively different from that obtained with $k_0 > 1$. Specifically, one sees that for alloys with k_0 values below unity, a value of $k_{BPS} \to 1$ would require a quasi-stationary zone width, δ, that varies with k_0. By contrast, for alloys with $k_0 > 1$, to achieve a k_{BPS} near unity requires just a fixed width of the quasi-stationary zone equal to approximately two to three characteristic diffusion distances $(3D_\ell/v)$.

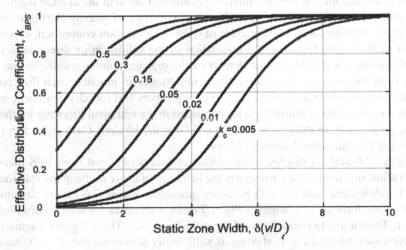

Fig. 7.4 Plots of Eq. (9.21) giving the effective distribution coefficient, k_{BPS}, as a function of the quasi-stationary zone width, δ, for alloys with different $k_0 < 1$

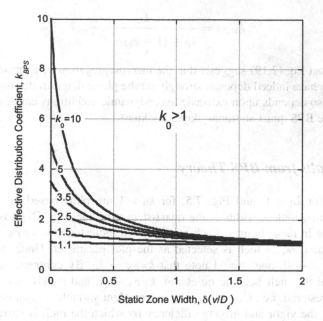

Fig. 7.5 Plots of Eq. (7.19) giving the effective distribution coefficient, k_{BPS}, as a function of the quasi-stationary zone width, δ, for alloys with different $k_0 > 1$

The stirring may be accomplished through purely mechanical means by using an externally driven impeller, or by rotating the crystal itself relative to the melt. Magneto-hydrodynamic boundary layer control at the crystal-melt interface is also possible in liquid conductors, such as molten metals and semiconductors. Through the application of a rotating magnetic field, electrically conductive melts may be pumped, and the mixing near the interface controlled through the applied magnetic field strength and its rotation rate. The mere presence of gravity during crystal growth will usually cause some amount of natural, or buoyant convection, depending on such detailed factors as the orientation of the solidification direction to the Earth's gravitational field, and the density of the rejected solutes relative to that of the solvent. Because of their wide-ranging importance in practice, melt-flow control methods are studied extensively by crystal growers and foundry engineers, and reported in hundreds of journal articles published in the technical literature for these fields, particularly in the *Journal of Crystal Growth*, *Modern Casting*, and in the *Journal of Crystal Growth Design* [12–14].

Figures 7.6 and 7.7 display some additional results obtained from BPS theory, which allow one to observe particularly the behavior of k_{BPS} as the quasi-stationary zone width widens, and the melt becomes quiescent. Figure 7.6 covers situations where alloys have $k_0 < 1$, whereas Fig. 7.7 deals with cases for which alloys have $k_0 > 1$. Shown are the variations of the ratio k_{BPS}/k_0. These figures emphasize that for cases where $k_0 < 1$, stirring, if sufficiently strong so that $\delta \to 0$, would drive the values of the distribution coefficient ratios toward unity, and, as expected $k_{BPS} \to k_0$. In the opposite limit of a nearly quiescent melt with a large value of δ, the curves in Fig. 7.6 approach their limiting values of $1/k_0$, so that one sees clearly that $k_{BPS} \to 1$.

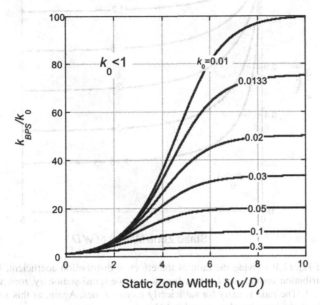

Fig. 7.6 Plot of Eq. (7.19) giving the ratio of the effective distribution coefficient, k_{BPS}, to the equilibrium distribution coefficient, k_0, as a function of the quasi-stationary zone width, δ, for alloys with $k_0 < 1$. Note how the ratio increases toward the value $1/k_0$ as the zone widens

In the less common case of alloys for which $k_0 > 1$, the ratio of $k_{BPS}/k_0 \to 1$ as the stirring becomes strong. The form of the BPS curves plotted in Fig. 7.6 also suggests that for alloys with values of k_0 closer to unity it will be more difficult to control the composition of the solid by stirring control alone, as these k_{BPS} values change very little over a large range of δ. In normal crystal growth practice, however, stirring and other types of flow control are often employed as effective crystal growth controls that conveniently vary the value of k_{BPS} and, in turn, set some desired composition of the product.

7.3 Summary

The control of macrosegregation and solute or doping distributions during bulk crystal growth is now a standard practice in commercial production of semiconductors. The initial discussions of macrosegregation in well-mixed melts found in Chapter 5, followed by solute distributions from diffusion transport in static melts found in Chapter 6, were, in a sense, joined by the concept of convection control in the melt through forced stirring. This method, conceived by Burton, Prim, and Schlichter, is known as BPS control theory. The hydrodynamic state of the melt was shown by those authors to influence the width of the solute diffusion boundary layer, δ, adja-

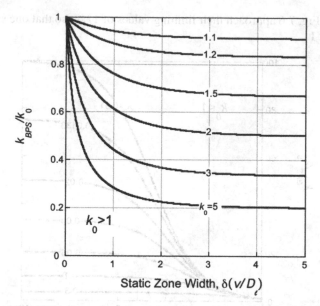

Fig. 7.7 Plot of Eq. (7.19) giving the ratio of the effective distribution coefficient, k_{BPS}, to the equilibrium distribution coefficient, k_0, as a function of the quasi-stationary zone width, δ, for alloys with $k_0 > 1$. The ratio is unity for sufficiently narrow zones. Again, as this zone becomes sufficiently wide, the ratio approaches the value $1/k_0$.

cent to the solid–liquid interface. The magnitude of δ in turn modifies the effective distribution coefficient called the BPS distribution coefficient, $k_{BPS} = C_s/C_0$.

This effective distribution coefficient, k_{BPS}, is of practical interest to the crystal grower, as it relates the composition of the crystal, C_s, to the melt composition, C_0 in which it is growing. Under intense stirring, where $\delta \rightarrow 0$, $k_{BPS} \rightarrow k_0$. By contrast, as stirring is halted, and melt flow slows, $k_{BPS} \rightarrow 1$, and steady-state macrosegregation is restored, i.e., $\hat{C}_s \rightarrow C_0$. Crystal growth in stationary melts maintained under microgravity conditions, where buoyancy forces are negligible, were investigated in the 1990s using space flight experiments. These studies proved the difficulty of achieving strict diffusion-limited crystal growth under terrestrial conditions where Earth's gravity usually stimulates natural convection during solidification. Consequently, melt-flow control remains a constant engineering challenge to crystal growers attempting to produce homogeneous bulk crystals from the melt.

References

1. A.N. Kolmogorov, *Izv. Akad. Nauk SSSR Ser. Fiz.*, **VI**, No. 1–2, (1942) 56.
2. L. Prandtl, *Über ein neues Formelsystem für die ausgebildete Turbulenz*, Nachrichten Akad. Wiss. Göttingen, 1945.
3. A. Bejan, *Convection Heat Transfer*, Wiley Interscience Publication, Wiley, New York, NY, 1984, p. 271.

4. S. Kou, *Transport Phenomena and Materials Processing*, Chap. 3, Wiley, New York, NY, 1996.
5. D.R. Poirier and G.H. Geiger, *Transport Phenomena in Materials Processing*, Chap. 8, TMS, Minerals, Metals Materials Society, Warrendale, PA, 1994.
6. M.E. Glicksman and S-C. Huang, *Adv. Space Res.*, **1** (1981) 25.
7. M.E. Glicksman, S.R. Coriell and G.B. McFadden, *Ann. Rev. Fluid Mech.*, **18** (1986) 307.
8. M.E. Glicksman, 'Solidification research in microgravity', *ASM Handbook* **15**, *Casting*, ASM International, Metals Park, OH, 2009.
9. J.A. Burton, R.C. Prim and W.P. Slichter, *J. Chem. Phys.*, **70** (1953) 1987.
10. R. Tönhardt, *Convective Effects on Dendritic Solidification*, Ph.D. dissertation, Royal Institution of Technology, Stockholm, 1998.
11. F.M. White, *Fluid Mechanics*, 2nd Ed., McGraw-Hill Book Company, New York, NY, 1986.
12. *Journal of Crystal Growth*, North-Holland Publishing, Elsevier, Amsterdam, 1967–2010.
13. *Modern Casting*, American Foundry Society, Schaumburg, IL, 1938–2010.
14. *Journal of Crystal Growth Design*, The American Chemical Society, Philadelphia, PA , 2001–2010.

Part III
Solid–Liquid Interfaces: Capillarity, Stability, Nucleation

Chapter 8
Crystal-Melt Interfaces

8.1 Capillarity

The technical term, 'capillarity', applies broadly to a large class of interfacial phenomena that deal with a range of thermodynamic influences associated with internal and external surfaces. Geometric aspects of surfaces and interfaces also connote associated metric quantities, including volumes, areas, lengths, and curvatures, all of which determine the relative importance of capillary phenomena in determining the behavior of a thermodynamic system.

The influence at different scales of the geometry of crystal-melt interfaces on solidification and crystal growth processes are certainly not exceptions. In fact, the thermodynamic behavior of solid–liquid interfaces, in general, tends to be simpler than those exhibited by solid–solid interfaces, which are capable of supporting more complicated interfacial stresses and strains, and complex defects. Our selective interest in discussing the capillarity (surface science) of crystal-melt interfaces at this juncture remains primarily to explore a few of its many roles in determining the equilibrium and kinetic behavior of solid–liquid interfaces during solidification and crystal growth. These include three subsequent chapters devoted to such basic topics as the influence of geometry, energy, and speed on the stability of solid–liquid interfaces during solidification. Also, as will be shown subsequently, capillarity induces a range of other interesting and important phenomena during, and even after, solidification, including subtle shifts in the equilibrium temperature and solubility at curved crystal-melt interfaces, and associated microstructural changes such as phase coarsening and texture modification.

8.1.1 Background

Capillary phenomena arise during solidification because one or more of the following factors are operative during freezing of a melt to produce a crystal:

1. *Evolution of surfaces and interfaces.* The properties of a bulk phase are altered near its surfaces. Surfaces and internal interfaces are regions where a phase is brought into contact with another, where a phase simply ends, as it does at

M.E. Glicksman, *Principles of Solidification*, DOI 10.1007/978-1-4419-7344-3_8,
© Springer Science+Business Media, LLC 2011

its *external* bounding surfaces, or where a crystalline phase exhibits a sudden alteration in the orientation of its lattice vectors, namely at various *internal* boundaries. All real (finite) material bodies have external surfaces that enclose them, and, in addition, might also contain internal interfaces, e.g., sub-grain, grain, twin, interphase, and anti-phase boundaries. All of these surfaces and interfaces exhibit associated capillarity effects because they introduce surface tension, which is a small[1] excess free energy per unit of surface or interfacial area that alters the material's behavior in substantial ways.

2. *Surface-to-volume ratio.* When a phase has a small surface-to-volume ratio, S/V, the effects on its thermodynamic properties from imbalanced interfacial or surface forces also remain small. The validity of this statement applied to any material body having some linear scale, $\mathcal{L}$, depends on the facts that the number of atoms or molecules at the surface or interfaces increases in proportion to $\mathcal{L}^2$, whereas the number of atoms in the bulk phase increases as $\mathcal{L}^3$. The surface-to-volume ratio of a phase particle provides an inverse metric measure, viz., $S/V \approx \mathcal{L}^{-1}$, and thus capillarity effects become increasingly significant as S/V rises, or, equivalently, as $\mathcal{L}$ decreases.

3. *Length scales in solidification.* The length scales of many macroscopic interfacial features produced during solidification develop from interactions occurring among solute and thermal transport fields and the small-scale geometry of the interface. As is well-known from examples such as solid-state precipitates, eutectoid microstructures, and polycrystals, solidified materials may also contain one or more phase distributions exhibiting relatively small length scales, or, equivalently, contain microconstitutents with large S/V ratios. Other common examples of interfacial features encountered in solidification processes include, for example, crystalline nuclei, interfacial cells, dendrites, eutectics, grain boundary grooves and ridges, gas channels and pores. Thus, there are a wide distribution of length scales, ranging from microscopic (near molecular) to macroscopic (above the wavelengths of visible light) that ultimately develop over the course of many solidification processes. The multiplicity of such length scales in solidification microstructures is caused fundamentally by complex interactions occurring among short-range interatomic forces acting at solid–liquid interfaces, with relatively long-range transport fields arising from solute diffusion and heat transfer that usually accompany solidification and crystal growth.

4. *Mesoscale structures.* Capillarity effects achieve their main importance in solidification processes because solidified alloys exhibit microstructures that contain 'mesoscopic' (mid-scale) features that evolve during the motion of *unstable* solid–liquid interfaces. More specifically, solidification processes often produce complicated microstructures with minimum lineal dimensions as small as 0.01

---

[1] Surface and interfacial energies, expressed on a per atom basis, are often not more than about 10^{-3} times the bulk free energy per atom developed by supercooling a melt ca. 1 K, or slightly supersaturating a crystalline phase. Although seemingly inconsequential, surface energies remain important in many crystal growth and solidification processes.

μ m (10 nm) to as large as several hundred micrometers (≈ 1 mm). Therefore, the S/V-values encountered in ordinary solidification structures tends to include the range between $10^2 - 10^6$ cm^{-1}, although some exceptions are known that do lie outside this (reciprocal) size range. Mesoscopic length scales combined with the high temperatures and rapid diffusion rates through freezing alloy melts often lead to significant capillary-mediated interactions, and, as a consequence, a remarkable array of interesting microstructures.

Figure 8.1 provides a good example of curved, interconnected solid–liquid interfaces in a partially solidified Al-15wt%Cu alloy casting. Voorhees (P.W. Voorhees, Private Communication, 2001) and his co-workers imaged serial-sections of this casting, which was held for different periods of time in the solid-plus-liquid region of the phase diagram, at about 5 K above the Al-Cu eutectic temperature. They produced striking three-dimensional reconstructions of the serial sections [1]. These reconstructed microstructures consisted initially of dendrites of the primary Al-Cu solid solution, surrounded by the Cu-rich interdendritic melt. The microstructures were subsequently analyzed quantitatively to extract their three-dimensional length-scales distributions at different holding times. Figure 8.1 shows two of their reconstructions, one of which was held in the solid-plus-liquid state for a relatively short time of 600 s, the other for an extended period, 6×10^4 s. It is easily appreciated from this study that capillarity acting along the curved solid–liquid interfaces dramatically affects the evolving cast structure.

Fig. 8.1 Three-dimensional reconstructions of serial-sectioned specimens ($1 \times 1 \times 3$ mm) of a partially-solidified Al-15wt%Cu casting alloy. The upper reconstruction is that for a specimen annealed for 600 s at 553 C in the solid-plus-liquid region, and the lower reconstruction is that for a specimen similarly held at that temperature for 6×10^4 s. As is evident from the interconnected appearance of these cast structures, capillarity causes significant changes in the distribution of the interfacial length scales. These distributions reflect capillarity-induced coarsening of the microstructures. Reconstructions courtesy of P.W. Voorhees (Private Communication, 2001)

8.2 Planar and Curved Interfaces

Constitutional supercooling, the most basic version of crystal growth stability theory, will be developed in the next chapter to provide an inital—albeit rudimentary— estimate of the stability of an advancing *planar* interface. As such, deviations from planarity, caused either by intrinsic fluctuations or extrinsic disturbances, need not be included to develop the constitutional supercooling model. Indeed, one need not include any aspects of interfacial curvature, provided that the question of stability is considered solely on the basis of constitutional supercooling. To proceed further, however, and to develop a more comprehensive understanding of interface stability and cast microstructure evolution in general, one must also consider a few basics of both the geometrical and kinetic attributes of curved and moving interfaces.

The presence of bounding (curved) surfaces on materials modifies their thermodynamic and the kinetic behaviors, as already mentioned in general terms in Section 8.1. In order to extend and deepen that discussion, and allow quantitative interpretation of observed solidification structures, one should first consider a few geometrical principles based on the mathematics of curved surfaces. In particular, the subject of 'interface kinematics' is introduced, allowing some deformation modes associated with melting or freezing processes to be quantified through application of the rules of differential geometry. By familiarizing the reader with these kinematic rules, deformations of a solid–liquid interface in response to thermal and solutal fields may be analyzed. This somewhat specialized approach provides a quantitative basis to understand the origin of important solidification structures, prompted by broader interests to quantify the description of all three-dimensional microstructures. This intriguing topic in materials science has expanded greatly in recent years because of impressive strides made in computer simulation of many types of solid–liquid and solid-state microstructures, and our increasing need to verify such computations through observation and measurements.

8.2.1 Surface Patches

Any small smooth region of a surface containing an arbitrary point located in three-dimensional space may be represented geometrically by a Monge coordinate patch. This surface patch is so named after the French geometer Gaspard Monge, who is considered as the 'father' of differential geometry [2]. A Monge patch is a local representation of a geometric surface, or interface, scaled small enough to capture *uniform* geometrical properties about some point (x_1, x_2, x_3), as portrayed in Fig. 8.2. The specification of such a coordinate patch includes the orientation in space of the unit surface normal, $\mathbf{N}$, with respect to these Cartesian coordinates. Each patch also contains a pair of orthogonal *surface* coordinates, u, v, that have an arbitrary rotational degree of freedom about the surface normal. Two orthogonal tangent vectors to the surface normal, $\mathbf{T}^{(u)}$ and $\mathbf{T}^{(v)}$, may be chosen arbitrarily to designate the orientation on the surface for the two surface coordinate axes.

A sharp solid–liquid interface, as will be shown, can be modeled geometrically as an assembly of small surface patches. These patches permit characterization of the curvature at every point on the interface, by specifying a set of unique geometric properties to be outlined in the following section [3].

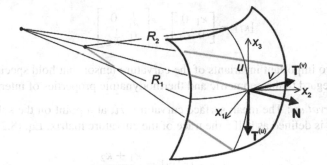

Fig. 8.2 Coordinate (Monge) patch defining the geometry of a small piece of surface. The outward pointing normal vector **N** at $(0, 0, 0)$ defines the orientation of the patch, and $\mathbf{T}^{(u)}$ and $\mathbf{T}^{(v)}$ are orthogonal surface tangent vectors to **N**, that set the orientation about the normal for the surface coordinates (u, v). As indicated, the oriented triplet of vectors defines two *principal* radii of curvature, R_1 and R_2, on the patch. Their associated curvatures, $\kappa_1 = R_1^{-1}$ and $\kappa_2 = R_2^{-1}$, are elements that diagonalize the curvature matrix

8.2.2 Curvatures

Curvature, κ, is a geometric property of a plane curve that measures the instantaneous change in the orientation angle, $\Delta\theta$, of the local normal vector with respect to displacement along the curve's arc length, ΔS; that is, $d\theta/dS$. Two such curvatures, κ_u and κ_v may be defined by the orthogonal surface traces along the surface coordinates, u, v. The so-called *principal planes* at the surface point are found by choosing intersection traces on the surface that yield the maximum and minimum radii of curvature, R_1 and R_2. As shown below, the associated curvatures, $\kappa_1 = R_1^{-1}$ and $\kappa_2 = R_2^{-1}$, become the elements that diagonalize the curvature matrix of the surface at that point.

The Monge patch itself must be chosen on a scale sufficiently small to allow these radii and their associated curvatures to be considered as constants over the area of the patch. A real solid–liquid interface, therefore, may be analyzed geometrically by considering it to consist of a collection of patches, each, strictly speaking, defining the local surface geometry only at one point on the interface [2].

The two curvatures on the patch, each defined along a surface coordinate, will be designated as a matrix element,

$$\kappa_i = \frac{\partial \theta_i}{\partial S_i} \quad (i = 1, 2). \tag{8.1}$$

The curvatures defined in Eq. (8.1) are scalar measures that define the elements of a 2×2 matrix called the curvature tensor, shown next in its principal orientation on the surface, where it is diagonalized to expose the two principal curvatures, κ_1, and κ_2, as

$$[\kappa_i] = \begin{bmatrix} \kappa_1 & 0 \\ 0 & \kappa_2 \end{bmatrix} = \begin{bmatrix} \frac{1}{R_1} & 0 \\ 0 & \frac{1}{R_2} \end{bmatrix}. \tag{8.2}$$

There are two important invariants of the curvature tensor that hold special significance with regard to the geometric and thermodynamic properties of interfaces.

1. *Mean curvature*: The mean surface curvature, $\mathcal{H}$, at a point on the solid–liquid interface, is defined as half [2] the trace of the curvature matrix, Eq. (8.2),

$$\mathcal{H} \equiv \frac{1}{2} tr\,[\kappa_i] = \frac{\kappa_1 + \kappa_2}{2}. \tag{8.3}$$

   Equivalently, the mean curvature can be expressed with the radii of curvature as

$$\mathcal{H} = \frac{1}{2}\left(\frac{1}{R_1} + \frac{1}{R_2}\right). \tag{8.4}$$

   The mean curvature has the special attribute that it relates the change in area, A, of a moving interface, to the volume, V, 'swept' out by a curved interface expanding normal to itself, much like blowing up a balloon. Specifically, the area and swept volume are related to the mean curvature through the variational parameter,

$$\mathcal{H} = \frac{1}{2}\left(\frac{\partial A}{\partial V}\right). \tag{8.5}$$

   The mean curvature is of direct significance to the thermodynamic behavior encountered when a curved interface advances or retreats during freezing or melting. Specifically, the mean curvature relates the local area and volume changes associated with phase change, and through these geometric quantities one may determine, as will be done shortly, the free energy changes attributed to interfacial motion and interfacial shape change.

2. *Gaussian curvature*: The Gaussian curvature, named after the German mathematician and geometer Karl Friedrich Gauss, is defined as the determinant of the curvature matrix,

$$\mathcal{K} \equiv \det[\kappa_i] = \kappa_1\kappa_2. \tag{8.6}$$

Note that the Gaussian curvature bears the unit of $[\text{length}]^{-2}$, which is different from the unit of the mean curvature $[\text{length}]^{-1}$. Clearly, the Gaussian and

---

[2] The scientific literature on capillary phenomena also employs interface curvatures termed 'mean curvature' and 'total curvature', but defined instead as the total trace of the curvature tensor. This difference from the definition given in the text—viz., a factor of 2—if not explicitly noted or specified, can be a source of confusion.

mean curvatures represent entirely different aspects of the interfacial geometry, despite their both being called surface 'curvatures'. The Gaussian curvature yields another differential coefficient of the coordinate patch. This geometrical coefficient corresponds to the rate of change of the spatial orientation (solid) angle, $\Omega = \theta_u \theta_v$, subtended by the coordinate patch with respect to its change in area,

$$\mathcal{K} \equiv \left(\frac{\partial \Omega}{\partial A}\right)_{\kappa_1, \kappa_2} = \frac{1}{R_1} \times \frac{1}{R_2}. \tag{8.7}$$

The Gaussian curvature exhibits the interesting property of remaining constant for simple bending of a surface, such as the smooth (unwrinkled) deformation of a flat sheet into a cylinder or cone. Note, it is the invariance of the Gaussian curvature under bending that is connected to the topological fact that one cannot smoothly bend a flat surface into, say, a sphere, (or vice versa) without stretching, tearing, or wrinkling the surface. By contrast, a cylindrical surface is easily bent into another cylindrical or conical surface of arbitrary radius, including a plane. The reason for this geometrical behavior is that a cylinder, a cone, and a plane *all* have the same Gaussian curvatures—viz., $\mathcal{K} = 0$. Another basic property of the Gaussian curvature (but not the mean curvature) is that any simply connected 3-dimensional object—one lacking 'handles' or 'holes'—has a total spherical image of 4π steradians. The spherical image of a body obeys a topological conservation law in three spatial dimensions called the Gauss–Bonnet theorem, namely,

$$\iint \mathcal{K} dA = 4\pi, \tag{8.8}$$

where the integration indicated in Eq. (8.8) is taken over the object's enclosing surface, including any corners, vertices, and edges. Note that Eq. (8.8) is equivalent to the statement that mapping all the unit normal vectors of an object's bounding surface fills the area (4π) of the unit sphere of orientations. The Gauss–Bonnet theorem is a fundamental topological property of Euclidean three-space, and applies to bodies of arbitrary external shape. Because the Gaussian curvature does not normally appear in surface thermodynamic formulas, it is often overlooked. It has, nevertheless, numerous applications in describing the geometrical and topological properties of surfaces and objects.

8.2.3 *Kinematics of Interfacial Deformation*

When curved interfaces advance, recede, or deform during freezing, the local principal curvatures change in complicated ways. The kinematic equations governing curvature changes for a smooth interface subject to continuous deformation was derived by Drew [4], who found that

$$\frac{d\mathcal{H}}{dt} = -\left(2\mathcal{H}^2 - \mathcal{K}\right)v - \frac{1}{2}\left(v_{uu} - v_{vv}\right), \tag{8.9}$$

and

$$\frac{d\mathcal{K}}{dt} = -2\mathcal{H}\mathcal{K}v - \mathcal{H}\left(v_{uu} + v_{vv}\right) + \sqrt{\mathcal{H}^2 - \mathcal{K}}\left(v_{uu} - v_{vv}\right). \tag{8.10}$$

The coefficients v_{uu} and v_{vv}, appearing on the right-hand side of Eqs. (8.9) and (8.10), are the second derivatives of the normal interface velocity, v, with respect to the principal surface coordinates. See again Fig. 8.2. Equations (8.9) and (8.10) are kinematic equations for the mean and Gaussian curvatures, respectively, that must be satisfied under general conditions of interface deformation. When a solid–liquid interface is *uniformly* displaced normal to itself, so that the normal velocity is everywhere uniform, the second derivatives of the normal speed vanish in Eqs. (8.9) and (8.10). The mean and Gaussian curvatures then satisfy a special kinematic relationship,

$$\frac{d\mathcal{H}}{d\mathcal{K}} = \frac{2\mathcal{H}^2 - \mathcal{K}}{2\mathcal{H}\mathcal{K}}. \tag{8.11}$$

Equation (8.11) is a non-linear first-order ODE relating the mean and Gaussian curvatures under *uniform normal* motion of an interface, e.g., the growth or melting of

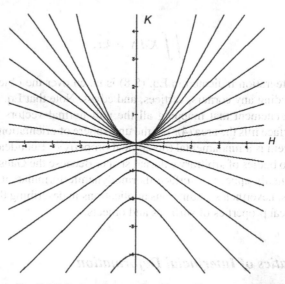

Fig. 8.3 Solutions to Eq. (8.11) for interfaces undergoing uniform normal motion. Spherical interfaces are located only along the upper-most parabola $K = H^2$. Cylindrical interfaces appear along the positive H-axis, where $K = 0$, and cylindrical tubes along the negative H-axis. Saddle-shaped interfaces follow trajectories in the negative half-plane, for which $K = 1/(R_1 R_2) < 0$, as the two radii of curvature have opposite signs

a spherical crystal. Figure 8.3 shows solutions for this ODE, which are 'trajectories' over the H-K plane for any curved interface undergoing uniform normal motion. In general, of course, interfacial motions can be more complicated than uniform normal motion.

Voorhees and co-workers developed experimental techniques to track the changing distribution of interface curvatures after various periods of partial solidification. They quantified the curvature distributions associated with the microstructural changes observable in Fig. 8.1. Their curvature data for these specimens, plotted as probability distributions on the H-K plane, are displayed in Fig. 8.4. The curvature distributions in Fig. 8.4 quantitatively capture geometrical details of the changes over time caused by partial solidification and coarsening of these Al-Cu microstructures.

8.2.3.1 Classical Capillarity

The classical approach of treating interfaces and determining their energetic behavior was developed by J. Willard Gibbs [5, 6]. Gibbs, the great American thermodynamicist of the mid-nineteenth century, pointed out that phases in heterogeneous thermodynamic equilibrium require a transition zone, which he analyzed as a 'sharp' interface that he termed a 'dividing surface' [8]. The Gibbsian dividing surface is a mathematical construct—not at all an attempt to describe physical interfaces composed of atoms or molecules. The dividing surface is imagined as located somewhere within a system's phase transition zone, which itself has a small physical thickness, λ. A dividing surface placed between two otherwise homogeneous phases is suggested in Fig. 8.5.

The dividing surface is located such that the excess, or deficit, of some particular property, ϕ, may be attributed entirely to that surface. In this manner, Gibbs's now classical treatment of capillarity effects and surface excesses of solutes is capable of accounting for the free energy, or adsorbed solute, associated with steep structural or chemical changes at an interface [9]. Different dividing surfaces can be defined depending on which excess quantity is of interest. The dividing surface defined by specifying a zero mass excess is one often used in interfacial thermodynamics.

If an interface is planar (zero curvature) or nearly so, as shown in Fig. 8.5, then the precise position of the dividing surface is unimportant, as a displacement or motion of a flat surface changes neither its area nor its surface energy. However, when an interface is curved, as sketched in Fig. 8.6, locating the precise location of its dividing surface can be non-trivial. The sensitivity to exactly where the dividing surface is placed arises because the area of a curved surface changes as the phase it encloses increases, or decreases, in volume. The volume 'swept' by an expanding curved interface depends on both its curvature and displacement. When curvatures are present, but remain moderate in magnitude, as in Fig. 8.6, their radii, R, remain large compared to the width of the interfacial transition zone, ($R \gg \lambda$). Under such conditions little uncertainty exists in choosing the proper location of the dividing surface, and negligible error would occur in determining where its excess energy or mass vanish. If, on the other hand, the interfacial curvature is large, or if the

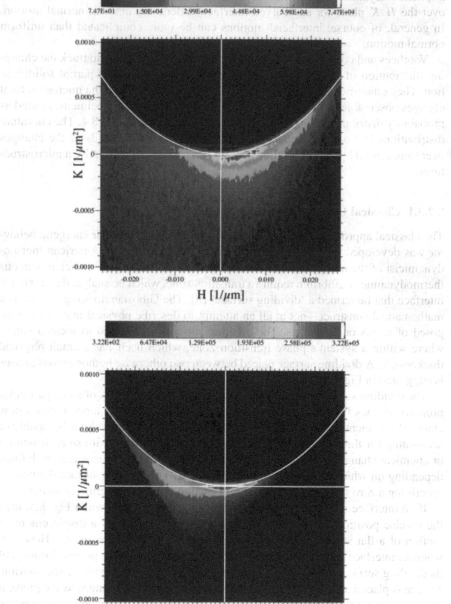

Fig. 8.4 Data showing the curvature distributions measured by Mendoza et al. [7] by serial sectioning the two partially solidified microstructures shown as three-dimensional reconstructions in Fig. 8.1. The probability density function $p(\mathcal{H}, \mathcal{K})$ is defined so that $p(\mathcal{H}, \mathcal{K})d\mathcal{H}d\mathcal{K}$ represents the probability that a randomly chosen interface point has a mean curvature between $\mathcal{H}$ and $\mathcal{H} + d\mathcal{H}$ and a Gaussian curvature between $\mathcal{K}$ and $\mathcal{K} + d\mathcal{K}$. The lower curvature distribution, measured

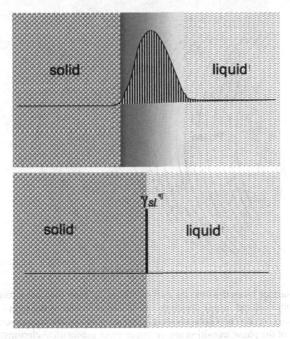

Fig. 8.5 *Upper:* Variation of the Gibbs energy density with distance between solid and liquid phases in equilibrium. The energy density is elevated through the narrow interphase transition, because the intervening material is neither equilibrium solid nor equilibrium liquid. *Lower:* Extrapolating any uniform property, in this case the Gibbs energy density of each phase, to a common plane of separation (the Gibbs dividing surface), produces a singularity in the excess energy density, as it 'collapses' onto the dividing surface. Conservation of that distributed free energy density yields the classical surface energy, γ_{sl}, which equals the integrated area under the continuous distribution shown in the upper sketch

interface transition zone itself were excessively wide or diffuse, $(R \approx \lambda)$ then the Gibbs dividing surface no longer provides a useful artifice to estimate excess interfacial quantities such as the surface free energy, $\gamma_{1,2}$, or solute adsorption. Computer simulations, using molecular dynamics, show that for many intermolecular potentials the Gibbs dividing surface, or other sharp interface approximations, work well down to surprisingly small interfacial radii, on the order of $R \approx 10$ nm. This length

---

Fig. 8.4 (continued) at a longer holding time, clearly shows an overall increase in the interfacial length scales, as the distribution coalesces closer to the origin, where $\mathcal{H} \to 0$, and the radii of curvature become larger relative to the upper distribution. Also, there is a noticeable drift over time from initially positive average mean curvatures (dendrites) in the upper distribution toward a preponderance of negative mean curvatures (enclosed melt pockets and channels) in the lower distribution. The detailed reasons for this drift, or overall 'flow', of the microstructure's curvature distribution through $\mathcal{H}$-$\mathcal{K}$ space during holding in the solid-plus-liquid state are not understood at present [10]

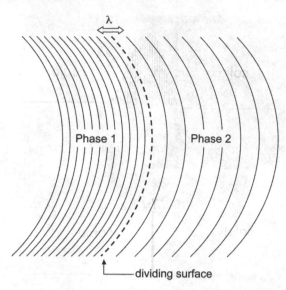

Fig. 8.6 Dividing surface located within the *curved* interfacial zone between a pair of phases. The transition zone width depicted in this sketch is still small compared to the radius of curvature of the interface, and so $\lambda \ll R$. The Gibbs dividing surface, or sharp-interface description, provides an adequate model of the phase transition region under these geometrical circumstances

scale already is well within the nanoscale region of most materials, where classical thermodynamics may be inapplicable for other reasons.

As portrayed in Fig. 8.6, crystal-melt boundaries, and in fact solid–liquid interfaces in general, tend to involve steep, rather narrow, transitions, insofar as their property and structural gradients occur over just a few (1–10) atomic spacings. So, if the radii of curvature of interfacial structures are larger than about 100 atom spacings (about 10 nm, or larger), one may safely use the classical approach. More specifically, interfacial structures that develop at solid–liquid interfaces during solidification usually have radii of curvature that are large compared to their transition zone thicknesses, λ. Consequently, such structures have properties that are usually independent of their radii. For example, the 'excess' interfacial energy, $\gamma_{s\ell} \times A$ just depends on the total area, A, of the crystal-melt interface, and *not* on the interface's radius of curvature. The portrayal of solid–liquid interfaces as sharp, as opposed to diffuse, allows simple analyses of dynamical interfacial stability to be developed for curved interfaces in subsequent chapters.

8.2.3.2 Mass Density Distributions

Differing mass density distributions at a solid-melt interface are suggested in Fig. 8.5. These distributions contrast a diffuse, continuous density variation through the interfacial transition zone, with that of a sharp, step-like change based on uniform, bulk densities. The density distribution shown in Fig. 8.5 locates the dividing surface of zero excess mass just slightly to the left of center of the transition zone. The precise position of the surface of zero excess mass in a system, however, depends on subtle details of the actual mass density distributions within the phases adjacent to the crystal-melt interface. The total number of atoms encountered per unit area of the solid–liquid interface, for a sample of the phases extending over a large distance $\Lambda \gg \lambda$ (where λ again denotes the thickness of the solid–liquid interface) from either side of the sharp interface location, $\hat{x}$, may be written as

$$n_{tot} = \int_{\hat{x}-\Lambda}^{\hat{x}} \left(\frac{dn_s}{dx} \right) dx + \int_{\hat{x}}^{\hat{x}+\Lambda} \left(\frac{dn_\ell}{dx} \right) dx, \tag{8.12}$$

If the *average* densities of the bulk solid and melt phases are designated as ρ_s and ρ_ℓ, respectively, then the mass excess, Γ, per unit area, may be calculated by comparing the mass density distributions on the left and right of the solid–liquid interfaces as suggested in Fig. 8.5. The differences of the total number of atoms, n_s and n_ℓ, respectively, in the solid and liquid phases adjacent to the interface, compared to those in the bulk phases far from the interface, is given by

$$\frac{\Lambda \rho_s N_A}{M} - \int_{\hat{x}-\Lambda}^{\hat{x}} \left(\frac{dn_s}{dx} \right) dx = \int_{\hat{x}}^{\hat{x}+\Lambda} \left(\frac{dn_\ell}{dx} \right) dx - \frac{\Lambda \rho_\ell N_A}{M}. \tag{8.13}$$

The symbol N_A appearing in Eq. (8.13) is Avogadro's number, and M is the atomic weight of the material. The excess interfacial mass, $\Gamma(\hat{x})$, can be forced to vanish by appropriately choosing the location, $\hat{x}$, of the solid–liquid interface. Here $\hat{x}$ denotes where the total mass indicated in the interfacial transition zone, Eq. (8.12), becomes identical to the masses of solid and liquid of uniform 'bulk' density extended up to the Gibbs dividing surface. This choice results in zero excess mass. Specifically, this position is found by combining Eqs. (8.12) and (8.13) and making the excess mass vanish.

$$\Gamma(\hat{x}) \equiv n_{tot} - \frac{\Lambda N_A}{M} (\rho_s + \rho_\ell) = 0. \tag{8.14}$$

8.2.4 Interfaces

8.2.4.1 Cahn–Hilliard Theory

Thermodynamic properties are often imagined to be uniform right up to the plane of contact between two homogeneous phases. The finite ranges of molecular interactions, however, preclude the appearance of a true discontinuity of the intensive or specific properties, including, for example, the densities of the energy, mass, entropy, etc. A thin transition zone or interfacial layer, perhaps only a few atoms or molecules wide, always exists.

Figure 8.7 represents the interfacial transition as a thin, slightly 'diffuse' region. In reality, the solid-to-liquid transition also exists over some suitably small length scale, such that certain quantities, e.g., the chemical potential of each species, will

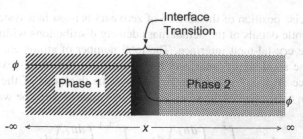

Fig. 8.7 Diffuse interfacial transition zone between a pair of phases. The transition for crystal-melt interfaces occurs typically over a few molecular distances. The variation of some property, ϕ, with distance is shown from one phase to the other. The property of interest is depicted as uniform throughout the homogeneous portions of both phases, but also exhibits steep gradients across a narrow interfacial transition. The transitional structure has spatially *inhomogeneous* properties and excess interfacial energy. The variation of the physical property, ϕ, is assumed in Cahn–Hilliard theory to be smooth and continuous, as indicated here

be uniform and equal throughout the homogeneous regions of each phase that are far from the transition zone. It has also been clearly recognized since the time of Gibbs [5] that the variation of properties across an interface is not arbitrary. The transition zone, at equilibrium, is configured such that the system's total free energy is minimized, subject to the constraint that the interface between the phases exists, and that the chemical potentials of each species are pairwise equal when the interface is at rest.

J. Cahn and J. Hilliard [11] showed generally that the free energy, G_{tot}, of an *inhomogeneous* binary system can be represented in one spatial dimension, x, as a Taylor series expansion in the local concentration, $C_B(x)$, using even powers of its spatial derivatives through the phase. A form of the Cahn–Hilliard expansion follows:

$$F_{tot} = \mathcal{A}\, n_v \int_{-\infty}^{\infty} \left[f(C_B) + K_1 \left(\frac{\partial C_B}{\partial x} \right)^2 + \ldots h.o.t. \right] dx. \tag{8.15}$$

Here $f(C_B)$ is the classical Helmholtz free energy per mole[3] of a *homogeneous* phase consisting of a binary solution of composition C_B. See Fig. 8.8 showing the Helmholtz potential difference $\Delta f(C)$ between the homogeneous solution and its weighted components A and B to form the solution. n_v is the number of moles of this mixture per unit volume; $\mathcal{A}$ is the cross sectional area of the interface formed between two phases, denoted generally as 1 and 2; K_1 is the energy coefficient of the squared gradient term, and sets the energy 'penalty' for having concentration

---

[3] See Appendix A on the definition of the Helmholz free energy or potential, $F(T, V, N_B)$, which is the Legendre transform of the internal energy, $U(S, V, N_B)$, where temperature replaces the entropy as the independent variable. This choice assumes that the system is maintained both at a constant temperature and constant volume by its surroundings.

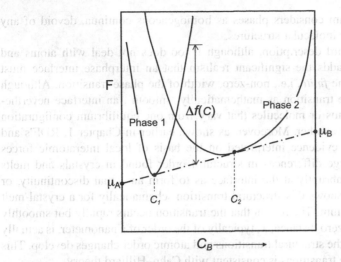

Fig. 8.8 Helmholtz free energy versus composition at a fixed temperature and volume. At any composition, C_B, the quantity $\Delta f(C)$ represents the difference between the free energy of the components, A and B, that comprise the phases, given by their weighted chemical potentials, μ_A and μ_B, and the free energy of the homogeneous solution phases, indicated by the *U-shaped curves*. The compositions C_1 and C_2 are the tie-line compositions expected for each phase at an interface formed between them at local equilibrum

gradients in the material; and *h.o.t.* stands for all 'higher-order terms in the Taylor series not included in this expansion.

In the subsequent discussion of this important representation of compositionally inhomogeneous materials, phase 1 is identified as a homogeneous crystalline solid for $x \ll 0$, and phase 2 is the corresponding homogeneous melt for $x \gg 0$. The questions to be considered here—the answers to which Cahn–Hilliard theory provides considerable insight—are, (1) What is the nature of an 'interphase' or diffuse interface transition zone, and (2) What dynamics occur during subsequent phase separation?

Solutions to Cahn–Hilliard theory, Eq. (8.15), and other variations of it developed over the past half century, collectively called phase-field theories, now allow phase separation, in general, and solidification kinetics, in particular, to be analyzed at a much higher level of detail and sophistication. Phase field simulations run on large parallel machines are capable of reproducing complex three-dimensional solidification structures.

Note, that the gradient term appearing in Eq. (8.15) becomes significant only where properties, in this instance the composition, C_B, and the phase structures are changing rapidly with distance. Such rapid changes occur also at a 'sharp' solid–liquid interface, where chemical segregation, treated under local equilibrium conditions already discussed in Chapters 5, 6, 7, and 8, required a discontinuity in composition. That compositional discontinuity was also predicted from the constant-pressure phase diagram based on length of the tie-lines. The reader is reminded that a phase diagram considers phases as homogeneous continua, devoid of any particular atomic or molecular structures.

The Cahn–Hilliard description, although it too does not deal with atoms and molecules per se, adds the significant realism that an interphase interface must take into account the *finite*, i.e., non-zero, width of the phase transition. Although they treat the phase transition as mathematically 'smooth', an interface neverthe-less consists of atoms or molecules that vary from one equilibrium configuration and composition to another. Moreover, as shown earlier in Chapter 1, RDF's and other experimental evidence interpreted on the basis of local interatomic forces suggest that the large differences in structural order found in crystals and melts do not develop so abruptly at the interface as to form an actual discontinuity, or 'jump'. Figure 8.9 shows this structural transition schematically for a crystal-melt interface at equilibrium. The reason that the transition occurs rapidly but smoothly is that a small, non-zero distance, λ, typically of the order of a nanometer, is actually needed over which the structural transitions and atomic order changes develop. This smoothness in these transitions is consistent with Cahn–Hilliard theory.

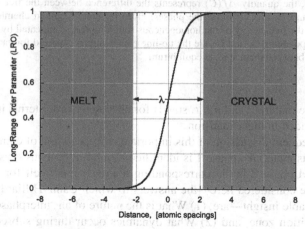

Fig. 8.9 Spatial variation of the long-range order parameter between a crystal and its equilibrium melt phase. The transition occurs over a microscopic distance, λ, that is typically several atomic spacings, or about a nanometer

8.2.5 *Interfacial Energy*

The excess free energy per unit area of the interface, $\gamma_{1,2}$, associated with the transi-tion region between phases 1 and 2 may be found, at least in principle, by subtracting the specific Helmholtz free energy per unit area for a uniform system consisting of homogeneous phases from the total Helmholtz free energy, F_{tot}, represented by Eq. (8.15), for an inhomogeneous system containing the transition region or diffuse interface. This small free energy difference may be shown to be

$$\gamma_{1,2} = \frac{F_{tot}}{A} - \int_{-\infty}^{\infty} \left[(1 - C_B)\mu_A^i + C_B\mu_B^i \right] dx \quad (i = 1, 2). \tag{8.16}$$

The coefficients, μ_A^i, and, μ_B^i, appearing in Eq. (8.16) are the standard chemical potentials (Joules/mole) for components A and B in either the liquid or solid ($i = 1, 2$, respectively). Note that the integral appearing in Eq. (8.16) defines quite generally the total free energy per unit area of an infinite system consisting of solid and liquid brought into contact at $x \approx 0$, and allowed to relax and reach thermodynamic equilibrium. Substituting for F_{tot} from the Cahn–Hilliard gradient expansion, Eq. (8.15), and ignoring higher-order terms contributing to the energy, yields

$$\gamma_{1,2} = n_v \int_{-\infty}^{\infty} \left[\Delta f_i(C_B(x)) + K_1 \left(\frac{\partial C_B}{\partial x} \right)^2 \right] dx \quad (i = 1, 2), \tag{8.17}$$

where the quantity $\Delta f_i(C_B(x))$, appearing in Eq. (8.17) and in Fig. 8.8, denotes the free energy difference between $f_i(C_B)$ for the ith homogeneous binary solution and the compositionally-weighted chemical potentials for its components.

$$\Delta f_i(C_B(x)) = f_i(C_B) - [(1 - C_B)\mu_A + C_B\mu_B] \quad (i = 1, 2). \tag{8.18}$$

The compositions of the homogeneous phases, 1 and 2, are C_1 and C_2, respectively, as indicated in Fig. 8.8. Here the usual condition for chemical equilibrium between uniform phases has been applied: that is, the partial molar free energies, or chemical potentials, of the components in each phase are made equal. For component B,

$$\left(\frac{\partial f_1}{\partial C_B} \right)_{T,V} = \left(\frac{\partial f_2}{\partial C_B} \right)_{T,V}, \tag{8.19}$$

and for component A,

$$\left(\frac{\partial f_1}{\partial C_A} \right)_{T,V} = \left(\frac{\partial f_2}{\partial C_A} \right)_{T,V}. \tag{8.20}$$

Equations (8.19) and (8.20) are the mathematical forms of the well-known common tangent graphical construction that individually equates the chemical potentials of B and A atoms for each phase. The bulk equilibrium compositions, viz., C_1 and C_2, provide the important property that $\Delta f_i(C_B) = 0$. If the gradient terms appearing in Eq. (8.15) were also zero, and the system lacks an inhomogeneous transition zone, or interface, then one would find accordingly that the Cahn–Hilliard theory predicts $\gamma_{1,2} = 0$. The interpretation is evident: if steep chemical gradients are not present in a solid solution, there would not be a free energy 'penalty' via the gradient terms in the Cahn–Hilliard free energy expansion.

As pointed out originally by Gibbs, a transition zone between contacting *dissimilar* phases will always be required where phases contact each other. Therefore, the energy and property gradients can not be negligible everywhere, especially within the transition zone that separates what would be otherwise 'bulk' homogeneous phases. Thus, compositional inhomogeneities invariably occur near interfaces, and these inhomogeneities are intimately connected with a positive value of the 'energy penalty' or interfacial excess free energy, $\gamma_{1,2}$.

The Cahn–Hilliard formula offers a useful representation of the total free energy of a phase containing interfaces and their associated local gradients. The application of this theory, however, requires that the free energy-composition curves be known, or modeled, and that some estimate be provided of the spatial variation of composition or structure through the phases. These quantities can, in principle, be calculated from specific thermodynamic solution models, or estimated from molecular dynamics via numerical methods. Usually, the phase free energies are not known accurately in advance. Modern phase-field computations, for example, provide a thermodynamically consistent method for estimating gradients based on specific mathematical forms for the free energy-concentration relationships. Phase-field computations and numerics based on molecular dynamics are now routinely implemented on high-speed computers, but their detailed exposition is beyond the scope of this book. The interested reader should consult the excellent monograph on phase-field theory and associated methods by Emmerich [12].

8.2.5.1 Computational Approaches

When the excess free energy changes sensitively with the position of the dividing surface, and the position of the dividing surface itself becomes uncertain, or, simply, when classical experiments based on the Gibbs dividing surface cannot be carried out for a variety of practical reasons, one is forced to implement more fundamental formulations that account for atomicity, such as discussed earlier in Section 1.4. To carry through the necessary detailed calculations of the solid-melt free energy using fundamental approaches, however, invariably requires detailed knowledge about the material's interatomic potential. Fortunately, alternative methods based on first principles or molecular dynamic computer simulations were recently introduced by J.J. Hoyt et al., [13, 14] for obtaining ab initio estimates of interphase excess free energies. The new technique employs a statistical mechanical analysis of fluctuations occurring on a simulated crystal-melt interface, and produced good results for a few transition metals (Fe and Ni) for which well-established embedded atom potentials exist, and for phases modeled with a 'hard-sphere' atomic potential, which may be used as an approximation for weakly attracting noble gas atoms such as He, Xe, and Kr. Figure 8.10 shows a 'snap shot' of a crystal-melt interface simulated with molecular dynamics in pure Ni at its melting point (1,450 C). Computational methods similar to those employed by Hoyt and Asta should prove useful for determining accurate values for $\gamma_{s\ell}$ for many materials. As $\gamma_{s\ell}$ itself is an anisotropic surface property that is difficult to measure accurately by direct experimental techniques, especially at high temperatures, this simulation methodology offers an alternative, providing that good interatomic potentials exist. The source of the difficulty in measuring $\gamma_{s\ell}$ is due simply to the fact that interfacial energies, per atom, are extremely

Fig. 8.10 Simulation of the crystal-melt interface in Ni. The ordered *light gray* 'atoms' form an FCC crystal in a [100] orientation on the plane of the page, and the more randomly distributed *dark gray* 'atoms' form the adjacent melt phase. The crystal-melt interface shows atomic-scale fluctuations in its location, based on the imposed simulated temperature gradient that places the average interface on a plane about midway through the simulated area. Adapted from [13]

small, having magnitudes only about 10^{-3} that of typical bulk thermodynamic properties such as the enthalpy of fusion, and the anisotropies found in $\gamma_{s\ell}$ are themselves but a few percent of the mean value.

8.3 Gibbs–Thomson Effect

8.3.1 Equilibrium at Curved Interfaces

The free energy changes associated with the motion (normal displacement) of a patch of *curved* interface involve both area and volume changes of the phases. Consider a crystal-melt interface described by the usual geometrical rules for a smooth mathematical surface derived in Section 8.2.3. The rate of free energy change for a moving, curved,[4] solid-melt interface may be written as three distinguishable energetic contributions:

$$\frac{dG}{dt} = \left(\frac{\partial G_s}{\partial V_s}\right)\frac{dV_s}{dt} + \left(\frac{\partial G_\ell}{\partial V_\ell}\right)\frac{dV_\ell}{dt} + \left(\frac{\partial G_s}{\partial A_{s\ell}}\right)\frac{dA_{s\ell}}{dt}, \tag{8.21}$$

where, as indicated in Eq. (8.21), the total time-rate of change of the free energy ($\dot{G}$) may be split into volume-related terms for each phase, and an area-dependent term assigned arbitrarily to the solid. $\dot{G}$ depends on *both* area and volume changes that are inherently associated with the deformation and motion of the curved surface representing the physical crystal-melt interface.

The volumetric and areal free energy coefficients appearing on the right-hand side of Eq. (8.21) are, respectively, for a one-component (unary) system solidifying or melting at constant melt pressure,

---

[4] When an energetically isotropic, *planar* interface is displaced normal to itself, additional freezing or melting occurs. The energy changes associated with this displacement include only the swept volume contributions, as the interfacial area, and its energy, remain constant. In the case of a curved interface, however, *both* the area and volume change with displacement, in accord with the kinematics required by the mean curvature, $H = dA/dV$, so additional energy changes occur.

$$\frac{\partial G_i}{\partial V_i} = -\frac{S_i}{\Omega_i}(T - T_m) \quad (i = s, \ell), \tag{8.22}$$

and

$$\frac{\partial G_s}{\partial A_s} = \gamma_{s\ell}. \tag{8.23}$$

Equation (8.22) gives the free energy changes in each phase per unit volume of solid and liquid undergoing transformation at a curved interface at constant melt pressure. The temperature, T, may be below or above its equilibrium melting point, T_m depending on the curvature. Note that this coefficient equation is consistent with the fact that entropies are positive, so the free energy changes would be negative only if T increases relative to T_m, and they would be positive only if T decreases relative to T_m.

Equation (8.23) is the definition of the interfacial energy, $\gamma_{s\ell}$, needed to form a unit area of solid–liquid interface. By convention, interfacial energies are always positive, and must increase with increasing interfacial area. These differential expressions may be considered accurate, assuming: (1) that the solid and liquid phases are incompressible; (2) that the curvature of the solid is not too large[5]; and (3) that the rates of freezing or melting are not so large that local interfacial equilibrium fails.

Substituting the thermodynamic definitions for the volumetric and areal free energy coefficients, Eqs. (8.22) and (8.23), into the time rate of free energy change associated with the motion of the interface, Eq. (8.21), yields

$$\frac{dG}{dt} = 0 = \left(\frac{-S_s}{\Omega_s}\right)(T_{eq} - T_m)\frac{dV_s}{dt} + \left(\frac{-S_\ell}{\Omega_\ell}\right)(T_{eq} - T_m)\frac{dV_\ell}{dt} + \gamma_{s\ell}\frac{dA_{s\ell}}{dt}. \tag{8.24}$$

Note that on the RHS of Eq. (8.24) the symbol T_{eq} has replaced the arbitrary melt temperature, T_m, appearing in Eq. (8.22). This substitution stresses that only the *equilibrium* temperature is relevant when setting the rate of total free energy change for a moving curved interface equal to zero. Moreover, for any arbitrary rate of freezing or melting, $\dot{V}_s$, mass conservation requires that the number of atoms transferred to, or from, one phase must be equal and opposite to the number transferred from, or to, the other. Mass conservation is expressed as

$$\frac{1}{\Omega_s}\frac{dV_s}{dt} = -\frac{1}{\Omega_\ell}\frac{dV_\ell}{dt}. \tag{8.25}$$

Substituting the LHS of Eq. (8.25) into the RHS of Eq. (8.24) and factoring out the time rate of volume change for the solid from each term leaves

---

[5] Standard capillarity equations, such as the Gibbs–Thomson relationship, have been shown to be accurate provided $\mathcal{H} < 2 \times 10^5$ cm^{-1}.

$$\frac{dG}{dt} = 0 = \left[\left(\frac{-S_s}{\Omega_s} \right) (T_{eq} - T_m) + \left(\frac{S_\ell}{\Omega_s} \right) (T_{eq} - T_m) + \gamma_{s\ell} \frac{dA_{s\ell}}{dV_s} \right] \dot{V}_s. \quad (8.26)$$

Insofar as the rate of freezing or melting, $\dot{V}_s$, is arbitrary, the condition of local interfacial equilibrium, Eq. (8.26), will be satisfied, if and only if the quantities within the square brackets on the right-hand side of Eq. (8.26) vanish. Thus, for local equilibrium to hold during freezing or melting at a curved crystal-melt interface

$$\frac{\Delta S_f}{\Omega_s} (T_{eq} - T_m) + \gamma_{s\ell} \left(\frac{dA_{s\ell}}{dV_s} \right) = 0, \quad (8.27)$$

where the quantity $\Delta S_f = S_\ell - S_s$ in the first term on the LHS of Eq. (8.27) is the molar entropy of fusion. The differential coefficient that appears in the second term, $dA_{s\ell}/dV_s$, was already shown in Section 8.2.2 to be just twice the mean interface curvature, $\mathcal{H}$, as defined for a smooth surface in Eq. (8.5). Substituting Eq. (8.5) into Eq. (8.27) shows that

$$\frac{\Delta S_f}{\Omega_s} (T_{eq} - T_m) + 2\gamma_{s\ell}\mathcal{H} = 0. \quad (8.28)$$

Solving Eq. (8.28) for the equilibrium interface temperature as a function of the local interfacial mean curvature, yields an important equation in classical surface thermodynamics called the Gibbs–Thomson relationship [9]:

$$T_{eq} = T_m - \frac{2\gamma_{s\ell}\Omega_s}{\Delta S_f}\mathcal{H}. \quad (8.29)$$

Equation (8.29) shows that the shift of the local equilibrium temperature, T_{eq}, from the ordinary melting temperature, T_m, at a curved solid–liquid interface is proportional to the mean curvature, $\mathcal{H}$. As the mean curvature itself is a signed quantity, the interfacial equilibrium temperature is depressed at locally convex interfaces, and raised at locally concave ones.

The Gibbs–Thomson relationship is often applied to capillarity problems in solidification using the principal radii at a point on the solid–liquid interface. Substituting (half) the sum of the principal curvatures for the mean curvature, using Eq. (8.4), yields an equivalent form of the Gibbs–Thomson relationship expressed in terms of the principal radii of curvature,

$$T_{eq} = T_m - \frac{\gamma_{s\ell}\Omega_s}{\Delta S_f} \left(\frac{1}{R_1} + \frac{1}{R_2} \right). \quad (8.30)$$

8.3.2 Chemical Potentials at Curved Interfaces

The change in equilibrium temperature, $T_{eq}(\mathcal{H})$, at curved solid–liquid interfaces in pure substances, relative to that at a planar interface, suggests that the chemical potential will be altered as well. The chemical potential, $\mu(\mathcal{H})$—which is the partial molar free energy of the single component comprising a pure phase with a curved interface—shifts relative to its reference chemical potential, μ_p, at a planar interface separating pure solid from pure liquid. Although the chemical potential of the solid and melt must match at each point along a curved interface to satisfy thermodynamic equilibrium, a complication occurs in that phases meeting at curved interfaces are *not* at the same pressure. This aspect of two-phase equilibria at curved interfaces will be discussed later for alloys in Section 8.3.4.

The relative free energies of phases at planar and curved solid–liquid interfaces are compared in Fig. 8.11 for the condition of constant pressure in the liquid. At a planar interface, the normal melting point, T_m, satisfies the standard condition of thermodynamic equilibrium, i.e., one lacking capillary shifts. When the interface is curved, the equilibrium temperature shifts to $T_{eq}(\mathcal{H})$, as specified by Eq. (8.29), and the molar free energies of the solid and liquid each change by an amount ΔG. The chemical potential shift at the curved solid–liquid interface may be calculated easily as

$$\mu_i(\mathcal{H}) = \mu_p + \Delta G \quad (i = s, \ell),$$ (8.31)

where the free energy change at constant melt pressure caused by a small interface temperature change is

$$\Delta G = \Delta S_f \left(T_m - T_{eq}\right), \quad P_{melt} = \text{const.}$$ (8.32)

Substituting Eqs. (8.32) and (8.29) into Eq. (8.31) yields the result that

$$\mu_i(\mathcal{H}) = \mu_p + 2\gamma_{s\ell}\Omega_s\mathcal{H} \quad (i = s, \ell).$$ (8.33)

Equation (8.33) shows that the chemical potential of both phases shifts equally, either up or down, in proportion to the mean interfacial curvature, $\mathcal{H}$. The mean curvature is a signed quantity, so that a positive curvature corresponds to an interface that is convex toward the melt, like the bump sketched in Fig. 10.1 in Chapter 9. By contrast, when a concavity, such as a bowl-shaped depression, develops on a solid–liquid interface the mean curvature of the interface becomes negative. See again the interfaces in Fig. 8.1. The equilibrium temperature for concavities *rises* relative to the ordinary melting point at a planar interface, and the chemical potential of both phases *falls* relative to its value at a planar interface.[6] Thus, at the bulk melt-

---

[6] Kulkarni and DeHoff provide a thorough discussion of both pressure changes and chemical potential shifts at curved interfaces in their important paper on this subject [15]. Those authors

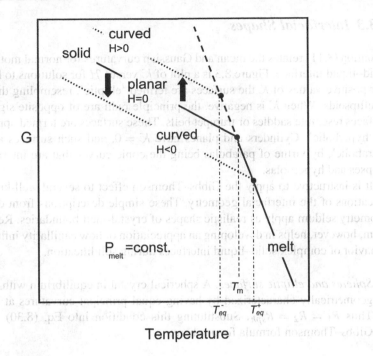

Fig. 8.11 Free energy lines for a *curved solid* and its surrounding melt versus temperature. The Gibbs–Thomson influence of interfacial curvature is shown to decrease the melting temperature for positive mean curvature, $\mathcal{H} > 0$, and to raise the melting temperature for negative mean curvature, $\mathcal{H} < 0$. The melting temperature and energy shifts are small, typically involving changes of less than 10^{-2} K from the 'bulk' melting point, T_m. As shown, T_m is the normal melting temperature, which applies to a *planar* interface, $\mathcal{H} = 0$, at a pressure, P_{melt}

ing temperature, the tendency on complex interfaces with features having positive and negative mean curvatures would be that depressions tend to solidify and fill in, whereas bulges tend to melt and shrink away. Indeed, morphological 'smoothing' induced by thermal interactions are observed commonly on annealed interfaces with initially complicated topographies.

In Section 8.3.4 capillarity effects in alloys will be discussed, and chemical potential changes induced by curvature will be shown to influence the chemical solubility as well as the equilibrium temperature. The reader should note that the influences of capillarity on equilibrium properties are not incorporated directly on phase diagrams. Despite the fact that the effects arising from capillarity may be considered as being 'minor' relative to bulk thermodynamic properties, the impact of capillarity on solidification processes and cast microstructures are actually profound and of wide importance.

---

point out inconsistencies from classical capillarity developed for liquid- or solid-gas interfaces that have been promulgated through papers and reviews on capillarity-induced shifts between condensed phases such as solid–liquid systems.

8.3.3 Interfacial Shapes

Equation (8.11) relates the mean and Gaussian curvatures for normal motions of the solid–liquid interface. Figure 8.3 is a plot of $\mathcal{K}$ versus $\mathcal{H}$ for solutions to Eq. (8.11). For positive values of $\mathcal{K}$ the surfaces are termed 'elliptic', resembling the surfaces of ellipsoids. When $\mathcal{K}$ is negative, the principle radii are of opposite sign, and the surfaces resemble saddles or trumpet bells. These surfaces are termed appropriately as 'hyperbolic'. Cylinders and planes have $\mathcal{K} = 0$, and such surfaces are termed 'parabolic', by virtue of parabolas being the conic curves that are intermediate to ellipses and hyperbolas.

It is instructive to apply the Gibbs–Thomson effect to several well-known classifications of the interfacial geometry. These simple descriptions from differential geometry seldom apply as realistic shapes of crystal-melt boundaries. Recognizing them, however, helps in developing an appreciation of how capillarity influences the behavior of complex solid–liquid interfaces during solidification.

1. *Spheres and elliptic surfaces*: A spherical crystal in equilibrium with its melt is geometrically characterized by having equal principal curvatures at all points. Thus $R_1 = R_2 = R_{sph}$. Substituting this condition into Eq. (8.30) yields the Gibbs–Thomson formula for a sphere:

$$\hat{T}_{eq} = T_m - \frac{\gamma_{s\ell}\Omega_s}{\Delta S_f}\frac{2}{R_{sph}}. \tag{8.34}$$

   Besides its use for spherical precipitates, Eq. (8.34) also applies to special, isolated points on more complicated interfaces. For example, the poles of oblate and prolate spheroidal surfaces are considered 'umbilicals', at which points the principal radii of curvature are equal. Other points on the surfaces of spheroids have differing principal radii, and are termed elliptic surfaces, so one must use Eq. (8.30) in describing their behavior.

2. *Cylinders and parabolic surfaces*: Cylindrical interfaces are generated by sweeping a straight line generatrix in the third dimension about a closed curve. The right-circular cylinder is perhaps most familiar, with the generatrix being normal to the tangent of a circle. The principal radii consist of the radius of the circle, $R_1 = R_{cyl}$, and the radius of the generatrix, $R_2 = \infty$. Geometers classify surfaces with one infinite radius of curvature as 'parabolics'. Substituting the value of these radii into Eq. (8.30) yields the Gibbs–Thomson relationship for a long, right-circular cylindrical crystal:

$$\hat{T}_{eq} = T_m - \frac{\gamma_{s\ell}\Omega_s}{\Delta S_f}\frac{1}{R_{cyl}}. \tag{8.35}$$

Equation (8.35) finds application in describing capillary effects at the interfaces of needle-like, or long 'stringy' particles. Such interfaces occur commonly in the directional solidification of eutectics and monotectics, and in the melting of heavily cold-worked materials that contain filamentary second-phase particles.

3. *Minimals and hyperbolic surfaces*: Another interesting class of surfaces are 'minimals', which minimize their surface areas subject to certain constraints—as do spheres—but, additionally, they exhibit *zero* mean curvature at all points—whereas spheres do not. Minimals accomplish this geometrical feat by having $1/R_1 = -1/R_2$, so that the mean curvature is zero, and the Gaussian curvature is negative. The curious feature of a curved interface, or a portion of an interface configured as a minimal surface, is that it acts thermodynamically as though it were planar, as $\mathcal{H} = 0$. With reference again to Fig. 8.3, minimal surfaces are all located along the negative $\mathcal{K}$-axis. Geometrically complex solid–liquid interfaces, as occur in dendritic 'mushy zones', are not minimals, but contain localized areas, called 'necks' and 'unduloids' that are hyperbolic surfaces, some of which approximate the behavior of minimal surfaces. Such regions tend to resist further melting, relative to nearby convex portions of the solid–liquid interface that have lower equilibrium temperatures, and become persistent features in the microstructure of castings.

4. *General solid–liquid interfaces*: When viewed overall, solid–liquid interfaces can exhibit almost every imaginable form of surface geometry [16, 17]. Figure 8.12 is a schematic illustration of a small portion of a branched dendritic interface. The interface can continually alter its geometrical character, both in space and in time, as freezing or melting occur. The changes shown in the two probability distributions in Fig. 8.4 capture the evolution of the statistical geometry of a complex solidification interface. Quantifying these results is a task currently beyond our grasp, or within the predictive capabilities of any solidification model. These data may be understood only as a statistical summary of the interface's geometric tendencies, as interpreted through the kinematics of coordinate patches, as described in Section 8.2.3. Advanced numerical techniques, particularly those based on phase-field and level-set methods, can simulate the time evolution of equivalently complex solid–liquid interfaces with surprising fidelity. The description of those modeling methods and their application to solidification microstructures, however, also lie beyond the intent of this chapter. The interested reader should see [12].

8.3.4 Kelvin's Equation

Closely associated analogies to the Gibbs–Thomson effect, which relates interface curvature to the equilibrium temperature, are the influences of curvature on a material's equilibrium vapor pressure and component chemical potentials. The thermodynamic basis for predicting these capillary-based relationships was also discovered by William Thomson, Lord Kelvin, in 1871, [18]. Kelvin's equation, as it is often called, provides the fundamental basis for explaining the subtle influences of curvature on a condensed phase and its vapor. The fundamental activity relationships so

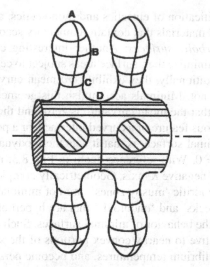

Fig. 8.12 Sketch of a small branched dendritic region exhibiting a variety of local geometrical characteristics. Limited areas on this complex interface may be described approximately by uniform (Monge) coordinate patches: Location A is an umbilical region, which acts like a portion of a sphere, with equal principal radii of curvature. Location B is a general convex region, termed geometrically 'elliptical'. The principal radii at B have the same sign, but are unequal. Location C is a concave interfacial region, termed geometrically 'hyperboloidal'. The principal radii at C have opposite signs. Hyperbolic interfaces act similarly to minimal surfaces if the two principal radii of curvature have nearly equal magnitudes. Location D is a nearly cylindrical, or geometrically 'parabolic' region. Here, one of the principal radii is much larger than the other. Sketch adopted from [16]

derived by Lord Kelvin then allow predictions on shifts in the equilibrium solubility of a solute.

8.3.4.1 Pure Substances

Figure 8.13 shows two reservoirs; one containing bulk material on the left, and the other containing finely divided droplets of the same material on the right. The vapor spaces above the bulk and finely-divided condensed phase show the equilibrium pressures and temperature. A series of *isothermal* steps are now taken to determine the change in chemical potential upon transferring material from bulk to droplet form.[7]

---

[7] The exploration of curvature on vapor pressure and chemical potential to be demonstrated here uses isothermal steps applied to a pure condensed phase and its vapor. Previously, the Gibbs-Thomson equation was derived for two condensed phases (solid plus liquid) using isobaric freezing or melting processes to determine the curvature-induced shift in the melting temperature. The conditions imposed on the intensive variables and the nature of the phases are both critically important to these derivations and their correct thermodynamic interpretation [15].

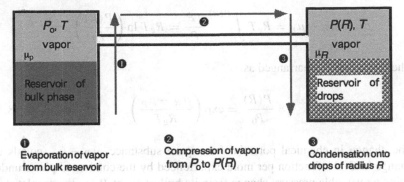

Fig. 8.13 Thermodynamic cycle to explore Kelvin's equation, by relating the change of vapor pressure, or chemical potential, with curvature. The left reservoir contains the 'bulk' pure phase (lacking curvature or impurities) at its equilibrium chemical potential, $\mu_p(T, P_0)$. Here μ_p represents the reference chemical potential of the single component at temperature T and pressure P_0. The right reservoir contains a large quantity of fine droplets of the same pure substance with radius R, and chemical potential μ_R. Three steps, described briefly in this figure, allow mass to be *reversibly* transferred from the 'bulk' reservoir to the 'droplet' reservoir, to account for the total work required

1. Reversible evaporation of a unit mass (1 mole) under equilibrium temperature, T, and pressure, P_0, causes no change in the left-hand reservoir's Gibbs function, or chemical potential, so $\Delta\mu_1 = 0$.

2. The mole of vapor produced in step 1 must now be compressed isothermally, from $P_0 \rightarrow P(R)$ to allow slow, frictionless transfer of this molar mass to the right-hand reservoir.[8] The change in the Gibbs function, or chemical potential for isothermal reversible compression is

$$\Delta\mu_2 = \int_{P_0}^{P(R)} V_{vap} dP.$$

3. The vapor (vap), now compressed to $P(R)$, and transferred to the right-hand reservoir may be reversible condensed into a molar mass of fine droplets, again without any changes in the system's Gibbs function. Thus, $\Delta\mu_3 = 0$.

4. Summing the energy and work changes to convert one mole of bulk condensed phase via its vapor phase to one mole of droplets with radii equal to R gives,

$$\Delta\mu_1 + \Delta\mu_2 + \Delta\mu_3 = 0 + \int_{P_0}^{P(R)} V_{vap} dP + 0.$$

If the vapor is ideal, then $V_{vap} = R_g T/P$, where R_g is the universal gas constant, which if substituted into the integral in the preceding step yields

---

[8] The conceptual devices (frictionless pump and valves) for accomplishing this isothermal *reversible* compression and transfer are not indicated in Fig. 8.13.

$$\mu_R - \mu_p = R_g T \int_{P_0}^{P(R)} \frac{dP}{P} = R_g T \ln\left(\frac{P(R)}{P_0}\right).$$

The result may be rearranged as

$$\frac{P(R)}{P_0} = \exp\left(\frac{\mu_R - \mu_p}{R_g T}\right). \tag{8.36}$$

5. The change in chemical potential for a pure substance, $\mu_R - \mu_p$, equals the change in Gibbs function per mole experienced by the condensed phase undergoing a reversible pressure change from its bulk state, at P_0, to the droplet state, at P_{drop}, where P_{drop} is the *internal* pressure found within a drop of radius R. Thus,

$$\mu_R - \mu_p = \int_{P_0}^{P_{drop}} \Omega_{drop} dP = \Omega_{drop}\left(P_{drop} - P_0\right), \tag{8.37}$$

where the molar volume for the condensed phase, Ω_{drop}, is essentially constant over the small pressure changes considered here. For exactness, the condensed phase may be assumed to be incompressible, so there is no need to draw any distinction between the molar volumes of the droplet, Ω_{drop}, and that of the bulk phase, Ω.

The free-body diagram in Fig. 8.14 may be used to calculate the internal pressure of a droplet, P_{drop}. Here all the forces acting on the droplet are exposed, including: (1) the external pressure, $P(R)$, pushing inward on a droplet's surface; (2) the surface tension, γ, pulling tangentially at every point over its surface; and (3) the internal pressure pushing outward everywhere within the droplet. For any meridional cross-section, such as shown in Fig. 8.14, the summation of these net forces must balance.

The sum of forces acting on the free body is

$$\pi R^2 P_{drop} - 2\pi R \gamma - \pi R^2 P(R) = 0, \tag{8.38}$$

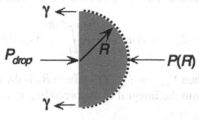

Fig. 8.14 Free-body diagram of a spherical droplet with radius R, subject to its surface tension, γ, acting on the curved interface between the condensed and vapor phases. The internal pressure is P_{drop}, whereas the pressure outside the droplet is its vapor pressure $P(R)$

and one finds in solving for P_{drop} from Eq. (8.38) that the interior pressure in a droplet is the exterior pressure, $P(R)$, plus what is termed the Laplace–Kelvin pressure 'jump', $2\gamma/R$. Thus,

$$P_{drop} = P(R) + \frac{2\gamma}{R}. \tag{8.39}$$

Substituting this result into Eq. (8.37) gives

$$\mu_R - \mu_p = \Omega \left(P(R) - P_0 + \frac{2\gamma}{R} \right), \tag{8.40}$$

where recognizing that the pressure difference $P(R) - P_0 \ll 2\gamma/R$, allows this small pressure difference to be be neglected in the pressure sum on the RHS of Eq. (8.40). The chemical potential shift becomes well approximated as

$$\mu_R - \mu_p \approx \Omega \frac{2\gamma}{R}. \tag{8.41}$$

8.3.4.2 Alloys

Equation (8.41) may be generalized beyond the result derived for a pure substance to obtain the shift in chemical potential, $\Delta\mu_i$, of a dilute component, i, present in an alloy particle having mean curvature $\mathcal{H}$.

$$\Delta\mu_i = \mu_i(\mathcal{H}) - \mu_p^i = 2\Omega\gamma \cdot \mathcal{H}, \tag{8.42}$$

where μ_p^i is the reference chemical potential of the component in the bulk phase. Equation (8.42) indicates that the chemical potential of a component at a curved interface shifts proportionately with the particle's mean curvature.

Substituting that approximation back into the numerator of the exponential term in Eq. (8.36) yields the ratio of the vapor pressures in the droplet state to that in the bulk reference state, namely,

$$\frac{P(R)}{P_0} = \exp\left(\frac{2\gamma\Omega}{R_g T} \mathcal{H} \right). \tag{8.43}$$

This is a form of Kelvin's equation. The vapor pressure ratio of a component surrounding a droplet of the phase, compared to that in equilibrium with the bulk phase, also equals the ratio of the component's thermodynamic activities, $a_i(\mathcal{H})/a_i^0$. Thus, one may express Kelvin's equation for the ith component in the vapor as an identical activity ratio of the component in the curved phase particle to that in the bulk phase,

$$\frac{a_i(\mathcal{H})}{a_i^0} = \exp\left(\frac{2\gamma\Omega}{R_g T} \cdot \mathcal{H} \right). \tag{8.44}$$

Figure 8.15 shows a plot of Kelvin's equation, where the mean curvature, $\mathcal{H}$, is normalized by multiplying it by the material's characteristic capillary length, $2\gamma\Omega/R_gT$. For many materials, this capillary length is small, ca. 10^{-7} cm. Thus, realistic solidification interfaces have scaled mean curvatures that have typical magnitudes, $|H| \ll 0.1$.

As a consequence of the small scaled mean curvature, Kelvin's equation may be linearized by a one-term series expansion of the exponential term as

$$\frac{a_i(\mathcal{H})}{a_i^0} \approx 1 + \frac{2\gamma\Omega}{R_gT} \cdot \mathcal{H}. \tag{8.45}$$

The associated compositional shift in solubility, ΔC, for a dilute solution, induced by the mean curvature of a solid-melt interface may also be approximated as

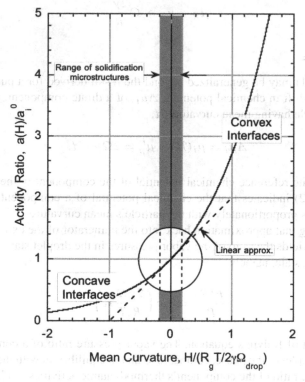

Fig. 8.15 Plot of Kelvin's equation showing the influence of (dimensionless) mean curvature on a component's thermodynamic activity. This plot reveals how convex interfaces ($\mathcal{H} > 0$), such as the surfaces of droplets, raise the activity, vapor pressure, and solubility of a component, i, whereas concave interfaces ($\mathcal{H} < 0$) depress the activity. Note that the curvature range of realistic solidification structures is extremely small, justifying the usual linear approximation for Kelvin's equation

$$\frac{\Delta C}{C} \approx \frac{a_i(\mathcal{H}) - a_i^0}{a_i^0} = \frac{2\gamma\Omega}{R_g T} \cdot \mathcal{H}. \tag{8.46}$$

8.4 Summary

This chapter opens with an elementary introduction to the basic differential geometry of compact surfaces. The kinematics of displacing a small patch of interface area, thereby causing freezing or melting, is described in terms of two invariants of the curvature tensor: its trace and determinant. The former equals twice the mean curvature, H, whereas the latter is the Gaussian curvature, K. These curvature quantities change in a prescribed manner as a surface evolves in accord with kinematic relationships relating surface area and 'swept' volume derived for normal displacements. An example developed by Voorhees et al. stresses the statistical nature of how solid–liquid interfaces in a casting alloy change over time. Then, surface thermodynamics, specifically, classical capillarity theory, allows introduction of the physical, rather than geometrical, aspects of solid–liquid interfaces, including their specific surface energy and the mass and order parameter distributions that occur at crystal-melt interfaces.

Next, the classical Gibbsian description of the interphase transition zone that employs a sharp 'dividing surface' is presented. That discussion is followed up by a more modern description of 'diffuse interfaces' described through Cahn–Hilliard theory. Some recent computational methods using molecular dynamics are mentioned, especially those by Hoyt and Asta that provide estimates of the solid–liquid interfacial energy and its anisotropy based on applying the fluctuation-dissipation theorem to the molecular configuration of a crystal-melt interface composed of pure Ni.

Returning to classical capillarity, the Gibbs–Thomson relationship is derived, connecting the equilibrium melting point of a material to its mean curvature, and the related Kelvin equation that gives the change in chemical potential of a component with curvature, and the corresponding shift in solubility. These basic interfacial relationships are used in subsequent chapters where nucleation, dendritic and eutectic solidification, and phase coarsening, will be discussed.

References

1. J. Alkemper and P.W. Voorhees, *J. Micros.*, **201** (2001) 388.
2. M.M. Lipschutz, *Differential Geometry*, McGrawHill Book Co., New York, NY, 1969.
3. D. Struik, *Lectures on Classical Differential Geometry*, Addison-Wesley, Reading, MA, 1950.
4. D.A. Drew, *SIAM J. Appl. Math*, **50** (1991) 649.
5. *The Scientific Papers of J. Willard Gibbs, Thermodynamics*, **I**, Dover Publications, Inc., New York, NY, 1961.
6. J.W. Gibbs, *Trans. Conn. Acad.*, **III** (1875–1976) 108.
7. R. Mendoza, J. Alkemper and P.W. Voorhees, *Metall. Mater. Trans.*, **34A** (2003) 481.
8. J.W. Gibbs, *Trans. Conn. Acad.*, **III** (1877–1978) 343.

9. R.T. DeHoff, *Thermodynamics in Materials Science*, Chap. 12, McGraw-Hill, Inc., New York, NY, 1993.
10. J. Alkemper, R. Mendoza and P.W. Voorhees, *Adv. Eng. Mater.*, **4** (2002) 694.
11. J. Cahn and J. Hilliard, *J. Chem. Phys.*, **28** (1958) 258.
12. H. Emmerich, *Phase Field Methods*, Springer, Berlin, 2003.
13. J.J. Hoyt, M. Asta and A. Karma, *Phys. Rev. Lett.*, **86** (2001) 5530.
14. J.J. Hoyt, M. Asta and A. Karma, *Mater. Sci. Eng. Rep.*, **41** (2003) 121.
15. N. Kulkarni and R.T. DeHoff, *Acta Mater.*, **45** (1997) 4963.
16. S.P. Marsh and M.E. Glicksman, *Mater. Sci. Eng.*, **A238** (1997) 140.
17. M.E. Glicksman and S.P. Marsh, Geometric Statistical Mechanics for Non-Spherical Bubbles and Droplets, *Drops and Bubbles*, T.G. Wang, Ed., American Institute of Physics, Conference Proceeding **197**, New York, NY, 1989, p. 216.
18. W.T. Thomson, *Philos. Mag.*, **42** (1871) 448.

Chapter 9
Constitutional Supercooling

9.1 Introductory Remarks

Models of solid–liquid interfaces were described up to this point as both planar in geometry and unchanging over time. Some of the critical factors were mentioned that were needed in continuous, stable, plane-front solidification of alloys. For example, the crystal and its molten phase were assumed to remain in contact across a featureless, mathematical plane, designated simply as $\hat{x}(t)$. Transport processes occurring throughout such solidifying systems were explicitly treated as 1-dimensional in their spatial dependence, with fluxes of heat and solute treated in one spatial direction. The solute redistribution and implied heat flow found under initially assumed conditions of unidirectional solidification were calculated by using matter and energy balances, respectively. Indeed, the balances, or conservation laws, so developed formed the foundations of several important solidification models used earlier in Chapters 4, 5, 6 and 7.

Local, unidirectional mass balances at the solid-melt interface were again incorporated in formulating both the steady-state and transient solute concentration fields, but now satisfying the time-dependent energy and diffusion equations—again, however, limited to just a single spatial variable. These time-dependent solidification problems included Neumann's moving boundary solution and the initial and final solute transients developed during macrosegregation. Thus, each of these models of crystal growth and solidification entailed solidification occurring under the highly restrictive assumption that the geometry of the solid–liquid interface remained *planar*, and was unchanging over time. Solidification interfaces that implicitly were considered to be 'stable' over time will be shown in this chapter that they need not be so. The onset of interfacial instability—i.e., the development of solidification structures with higher spatial dimensionality—will now be considered in order to formulate more realistic aspects of cast alloy structures and crystal growth defects.

The main issues to be addressed in this chapter are to develop: (1) an initial understanding of the kinetics of moving solid–liquid interfaces, along with (2) insights into the influences and interplay of a few materials parameters and process variables that either support or reduce interfacial stability. In order to accomplish these goals, two aspects of solidification, treated thus far as totally independent phenomena, will

M.E. Glicksman, *Principles of Solidification*, DOI 10.1007/978-1-4419-7344-3_9,

be linked. To be specific, thermal fields developed during freezing, considered under topics of heat transport, and concentration fields caused by macrosegregation in the melt near moving solid-melt interfaces, will now be considered *simultaneously*. Stated slightly differently, macrosegregation was treated as some vaguely 'isothermal' process, with the temperature fields at the solid–liquid interface determined solely by the interfacial composition. Now, by joining the influences of the temperature and concentration fields, more realistic—and far more interesting—aspects of alloy solidification may be exposed, analyzed, and discussed.

9.2 Background

The idea that a solid-melt interface has structure, and might not be best modeled as a 'featureless plane', was abundantly evident to most early foundry metallurgists and other scientific observers of both man-made and natural crystallization processes. These investigators observed and reported direct evidence of the occurrence of complex three-dimensional solidification structures, such as dendrites and euhedral crystals present in alloy castings, ice, and igneous rocks. It was, however, mainly through the painstaking and insightful experiments conducted over a half-century ago by Professor Bruce Chalmers [1] and his research students, to whom the field of solidification owes so much. The Chalmers school led in the systematic study of small-scale (sub-millimeter) solidification patterns that were discovered on the solid–liquid interface, catalogued, and analyzed. Chalmers and co-workers then proceeded to quantify their observations of solidification patterns and uncover correlations among the alloy characteristics and the freezing conditions that led to observably stable or unstable crystal-melt fronts. These quantitative correlations were eventually codified during the mid-1950s under a theory now known as 'constitutional supercooling' [2].

As shown in subsequent sections, constitutional supercooling was the forerunner of modern interfacial dynamics. Dynamical theories of the stability of solid–liquid interfaces, to be considered in more detail in the next two chapters, were begun by others about a decade after publication of Chalmers and his co-authors' own research on that subject. It is the author's opinion that the body of research culminating in the concept of constitutional supercooling was the key that unlocked many subtle and unsuspected aspects of the solidification process, and which ushered in the modern age of solidification science.

9.3 Decanting Studies

It was already confirmed by 1950 that nonplanar, three-dimensionally patterned interfaces develop during solidification, especially in impure substances and alloys. The critical experiments that helped expose and systematically record these interesting structures involved so-called 'decanting' studies of an actively advancing crystal-melt interface. Decanting a solid–liquid interface requires steadily direction-ally freezing a material, and then, after some solidification, rapidly accelerating the already-formed crystalline solid away from the remaining melt. The decanted, or exposed, interface that terminates the frozen solid phase could, after cooling, be viewed and measured under an optical microscope. As will be discussed next, a number of system parameters and experimental variables entered these decanting experiments, including the crystal structure of the host solvent metal, the solute element and its concentration, the interface speed, and the thermal gradient under which freezing occurred.

9.3.1 Interfacial Instabilities in Single Crystals

The original decanting studies were conducted on single crystals of metallic alloys by B. Chalmers and R.W. Cahn (unpublished, 1947). Their findings can be summa-rized in three key results, which proved that a solidifying interface could maintain its stability indefinitely, provided that:

1. The initial concentration of solute in the alloy remains sufficiently small.
2. The speed of freezing is sufficiently slow.
3. A sufficiently steep (positive) thermal gradient is provided at the solid–liquid interface.

However, if the alloy being solidified had too high a concentration of solute, or impurities, or if the rate of solidification was too fast relative to the magnitude of the applied thermal gradient, then the single crystal interface did not remain pla-nar. Chalmers et al. [3] reported that under such 'unstable' growth conditions they observed the development of distinctive interfacial patterns throughout the course of solidification. These interfacial patterns were described as periodic, square, or hexagonal arrays of 'pits', sequential parallel rows of corrugations or 'cells', and in yet more concentrated alloys, ultimately, elaborate small-scale structures developed, which appeared as complex tree-like crystals called dendrites.[1] A highly schematic sequence of sketches, depicting developing interfacial instabilities on decanted alloy crystals is shown as the upper series in Fig. 9.1. A corresponding sequence showing the behavior expected for *stable* crystal growth is given in the lower series included in Fig. 9.1.

---

[1] The arborescent form of solidification instabilities, called dendrites, occur ubiquitously in cast and welded structures. They profoundly influence the performance and properties of the solidified material; indeed, their appearance during crystal growth can cause difficulties in obtaining homo-geneous single crystals of doped semiconductors and oxides. The fundamentals of dendritic growth will be treated separately in a later chapter, because of its importance as the ultimate 'instability' and its wide-spread appearance in many solidification microstructures.

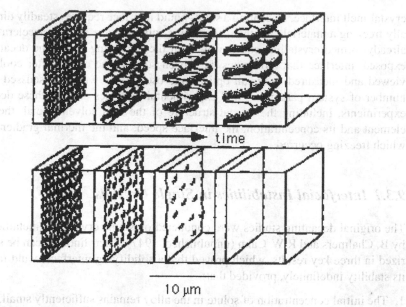

10 µm

Fig. 9.1 *Upper series:* Unstable, and *Lower series:* stable, evolution at an advancing solid-melt interface during freezing of an alloy. The *upper sequence* of sketches suggests that when a planar interface becomes morphologically unstable, small defects, or perturbations, amplify with time to form an array of new interfacial structures. The stable interface portrayed in the *lower sequence*, by contrast, returns to planarity by remelting any earlier formed protuberances. Adapted from [4]

9.3.2 *Interfacial Instabilities in Polycrystals*

It is worth pointing out that solid–liquid interfaces in polycrystalline materials (as contrasted with those in single crystals, discussed in Section 9.3.1) become unstable by passing through a different, slightly more complicated, sequence of morphological changes. When real-time *in situ* observing methods were eventually perfected in the mid-1960s—as opposed to just decanting the interface at a given instant— they allowed direct observations of the dynamic evolution of the advancing (poly-crystalline) solid–liquid interface. The morphological sequence of events following onset of instability of an initially planar interface was finally recorded in detail. As described briefly next, interfacial instabilities in polycrystals are influenced strongly by a variety of crystalline defects that are not present in more perfect single crystals.

One such study, by Schaefer and Glicksman [5], used *in situ* optical microscopy to record the temporal development of morphological instabilities in polycrystals. A transparent, body-centered cubic (BCC) compound, succinonitrile, $CN-(CH_2)_2-NC$, abreviated here as SCN, was the test substance. This organic compound (ca. 99% pure) was selected for investigation because its solidification behavior, at least concerning constitutional supercooling and instability formation, is essentially that seen in cubic metal crystals using decanting.

A metallic sample would, of course, be opaque. So, starting with a featureless, planar, and initially stationary polycrystalline interface in a transparent, slightly impure SCN 'alloy', Schaefer and Glicksman initiated solidification. They observed the following sequence of changes at the crystal-melt interface. The patterns are described below, for the sake of clarity, as a series of individual 'stages', each representing a distinctive departure from the interface's initial, almost featureless planar form. These stages actually developed progressively at different times and locations on the multi-grained SCN interface, all within a total area of about 1 cm^2. Some of the major stages in their formation included:

1. A few isolated pit-like depressions appear at what was initially thought to be random locations. Later, each of these isolated pits was found to be associated with the location of a triple-junction, which is the vertex formed by the intersection of three grain boundaries, each oriented normal to the solid–liquid interface. Their intersection with the solid-melt interface forms a 'triple-line' trace.[2] Where each triple line intersects the interface, a deep funnel-shaped pit forms.

2. The isolated pits then connect by three V-shaped grooves radiating away from each grain vertex. These grooves follow the network of the traces formed by the grain boundaries separating the individual crystallites comprising the polycrystalline solid. See Fig. 9.2.

3. As the interface continues to advance, it grows ahead more rapidly near the grooves mentioned in stage (2), forming pairs of raised mounds, or ridges, running parallel to each other on either side of the grain boundary grooves. For clarity, the reader should also refer to Fig. 9.3, which shows a schematic of the three-dimensional form of the solid–liquid interface at this stage in its progression toward instability.

4. A shallow valley then forms adjacent to each line of ridges, followed by a second pair of parallel ridges, much reduced in height than was the first pair that developed on the moving interfacial plane. See upper right micrograph in Fig. 9.2, and the third sketch from the left in Fig. 9.3.

5. At the grain vertices, where the ridges intersect with the triple-junction pits, three distinct 'pre-dendritic' disks form at the termination of each ridge. With progressive solidification, these disks extend the disturbance from the triple-junction along the three intersecting ridge lines, producing a distinctive periodic 'chain of hillocks'. See Fig. 9.2, upper right-hand micrograph, and the right-hand schematic in Fig. 9.3 .

6. The still featureless regions within individual grains, each now bordered by grain boundary grooves, steadily develop additional isolated craters at what again appear to be random locations. These newly cratered sites occur where

---

[2] A triple line actually consists of three, intersecting grain boundary grooves. The theory explaining why grooves develop on a polycrystalline solid–liquid interface is complicated—as it involves interactions among the interface shape, the grain boundary, and the applied thermal gradient—the combined effects from which are explained quantitatively in Appendix B.

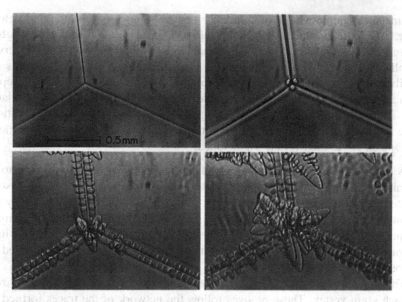

Fig. 9.2 Influence of grain boundary triple lines on the stability of an advancing solid-melt inter-face. Note that these are all high-angle grain boundaries by virtue of the 120° dihedral angles between them. *Upper left:* Formation of the initial triple-line structure, where three grain boundary grooves intersect. *Upper right:* three 'pre-dendritic' disks appear close to the disturbance provided by the triple junction. In addition, a pair of ridges have formed that parallel the original traces of the grain boundary grooves. *Lower left:* Periodic bumps, or 'hillocks', extend along the ridges. These hillocks represent yet a finer scale, set in a direction orthogonal to the ridges. *Lower right:* Pre-dendritic disks have grown into short dendrites that protrude into the melt, and the hillock structure is coarsening and becoming more chaotic

individual dislocation lines threading through the grains intersect with the inter-face. See time-sequence of micrographs in Fig. 9.4.

7. As solidification proceeds further, a circular ring, resembling a tiny 'volcano', gradually forms around each of these isolated craters. Eventually, weaker con-centric rings, or moats, form to produce distinctive duplex ring-structures.
8. Next, a complicated tracery of shallow grooves develops over the entire inter-face, demarking a myriad of low-angle 'mosaic' boundaries that intersect the solid–liquid interface. See micrographs in Fig. 9.5.
9. The multitude of intersecting grooves, ridges, and rings all interact and become so entangled that the solid–liquid interface takes on an overall roughened, or textured appearance.
10. In the final stage of polycrystalline interfacial instability, protuberances extend farther out into the melt from sites where the original grain vertices developed pre-dendritic disks and hillocks, as described in stage (5). These protuberances sprout periodic bulges that extend as side arms, and branches. They can even-tually form tertiary instabilities that develop additional higher-order branches. These tree-like, reticulated structures are dendrites, and because of their impor-tance in the microstructures of cast and welded materials they will be discussed

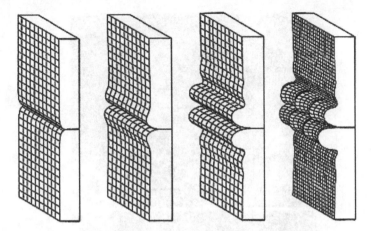

Fig. 9.3 Schematic of the local shape and dimensional changes occurring on an advancing crystal-melt interface in the vicinity of a grain boundary groove. *Left:* Initial grain boundary groove appears as a two-dimensional feature on the interface. *Second:* Ridges form and bulge out into the melt. *Third:* Ridges continue to amplify, producing adjacent valleys and weak parallel secondary ridges that propagate the instability outward over the interface from the grain boundary. *Right:* Hillocks then form as a periodic ripple along the ridges, breaking their symmetry by modulating the interface in the third dimension

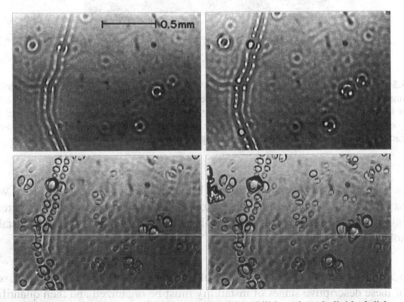

Fig. 9.4 Time sequence of the formation of localized instabilities where individual dislocations thread through the interface. *Upper pair:* Ring-like structures form on the interface in the form of volcano-like features, some of which eventually develop a double-ring, or moat structure. *Lower pair:* At a later time, the ring structures fill in and protrude farther into the melt as blob-like protuberances, some of which eventually evolve into dendrites. Note the general texture, or roughening, that develops on a scale of ca. 100 μm within the background of each grain

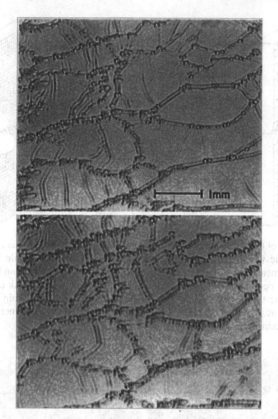

Fig. 9.5 Two sequential micrographs of the same area on a growing single crystal-melt interface, showing the development over time of a dense network of instabilities. These instabilities originate at initially weak disturbances that eventually destabilize the entire solid-melt interface. The tracery of tiny hillocks and ridges is caused by the initial weak disturbances of low-angle (mosaic) boundaries and individual dislocations intersecting the interface

in more quantitative detail in Chapter 13. Dendrites form in cast alloys, as well as in supercooled pure substances; they represent the end-stage of an increasingly complex evolution of interfacial instabilities evolving from an unstable solid-melt interface.

To advance beyond a mere taxonomy of unstable interfacial morphologies, several of these descriptive stages of instability must be organized and then quantified in terms of interfacial dynamics. These morphogenic stages may be reduced to a few mathematical models that greatly improve the reader's grasp of the basics of interface dynamics, and permit some (limited) quantitative predictions to be made that may be tested in controlled experiments. Such 'testable' models form the hallmark of modern solidification science, and specifically provided the foundations of modern morphological stability theory.

9.4 Constitutional Supercooling

9.4.1 Introduction

Assume, once again, for the sake of mathematical simplicity that solute diffusion does not occur in the hot solid, and that local equilibrium prevails at the moving solid–liquid interface. For the purposes of this discussion, furthermore, one may also choose alloys for which the equilibrium distribution coefficient, $k_0 < 1$, although the results sought will be completely general, and also include the less common situation where $k_0 > 1$.

Temperature fields, $T(x, t)$, during freezing will be understood as being present in the neighborhood of the solid–liquid interface, although the only aspect of these fields with which one need be concerned are the local temperature and thermal gradient at, or adjacent to, the interface. Also without loss of generality, the thermal conductivities of the solid and melt phases are taken as being equal. This assumption allows consideration of just a single value of the temperature gradient, G, applied at the interface. Without benefit of this simplification one would have to deal with different values of the thermal gradient acting on either side of the solid–liquid interface. In a later chapter, the effect of different solid and liquid thermal conductivities in real materials will be explored. More detailed considerations shall show that a suitably-weighted average of these thermal gradients also suffices to explain the effect of applied thermal gradients on the stability of solid–liquid interfaces.

9.4.2 The Interfacial Solute Mass Balance

First, and without direct concern of the thermal field, one formulates the *steady-state* solute mass balance as a Stefan balance: i.e., the flux of solute rejected from the solid–liquid interface, J_{rej}, equals the solutal flux, J_{diff}, transported away from the interface by diffusion through the melt. These fluxes are, respectively,

$$J_{rej} = \left(\hat{C}_\ell - \hat{C}_s \right) \cdot v, \qquad (9.1)$$

from interface mass conservation, and from Fick's law,

$$J_{diff} = -D_\ell \frac{\partial C_\ell}{\partial X}. \qquad (9.2)$$

Equating the fluxes given by Eqs. (9.1) and (9.2) provides the interfacial solute mass mass balance, namely,

$$-D_\ell \frac{dC_\ell}{dX} = \left(\hat{C}_\ell - \hat{C}_s \right) v. \qquad (9.3)$$

Equation (9.3) may be solved for the solute concentration gradient that develops in the melt adjacent to the interface, namely,

$$\left(\frac{\partial C_\ell}{dX}\right)_{\hat{X}} = \frac{v}{D_\ell}\hat{C}_s\left(1 - \frac{\hat{C}_\ell}{\hat{C}_s}\right). \qquad (9.4)$$

Local equilibrium at the solid–liquid interface allows use of the equilibrium distribution coefficient, $k_0 \equiv \hat{C}_s/\hat{C}_\ell$, which when substituted into Eq. (9.4) yields the expression for the interfacial solute concentration gradient, namely,

$$\left(\frac{\partial C_\ell}{dX}\right)_{\hat{X}} = \frac{v}{D_\ell}\left(1 - \frac{1}{k_0}\right)\hat{C}_s. \qquad (9.5)$$

It should be noted that this interfacial solute concentration gradient is identical to that found earlier for the steady-state solute boundary layer described by Eq. (6.23).

The partial phase diagram in Fig. 9.6 illustrates a tie line at the momentary interfacial temperature, $\hat{T}$, the corresponding interfacial phase composition, $\hat{C}_s$ and $\hat{C}_\ell$, and the melting temperature, T_m, for the pure component A at $C_B = 0$. All these quantities are related through the two constitutive equations for the solidus and the liquidus 'lines'. The linearized form of these equations is,

$$\hat{T} = T_m + m_i\hat{C}_i, \ (i = s, \ell), \qquad (9.6)$$

where the coefficients m_i in Eq. (9.6) represent the differential expressions

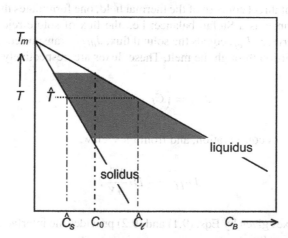

Fig. 9.6 Edge of a binary phase diagram, identifying the solidus and liquidus, and a typical tie-line representing the jump in phase compositions at the interface, $\hat{C}_\ell - \hat{C}_s$. The alloy undergoing solidification is of average composition C_0, and the tie line shows local solid-melt equilibrium at an intermediate temperature, $\hat{T}$, between the solidus and liquidus

$$m_i \equiv \frac{\partial \hat{T}}{\partial \hat{C}_i} \quad (i = s, \ell). \tag{9.7}$$

The quantity m_ℓ defined in Eq. (9.7) is the liquidus slope. In the case of non-dilute alloys that are away from the edges of the phase diagram, one may still apply Eq. (9.7) using the local liquidus slope, $m_\ell(C_B)$.[3] The quantity m_s in Eq. (9.7) is the corresponding solidus slope at any given interface temperature.

9.4.3 Constitutional Gradient

In 1953, Tiller et al. [3] proposed that by multiplying Eq. (9.5), giving the solute concentration gradient in the melt adjacent to the interface, by the liquidus slope, Eq. (9.7), the functional dependence is obtained between the interface temperature, the melt composition, distance, and time. This dependence should be viewed mathematically as $\hat{T}(C_\ell(\hat{X}(t)))$. Such a nested set of functional dependences provides a basis for what Tiller et al. defined as the *constitutional gradient*, $\mathcal{G}$.

$$\mathcal{G} \equiv \left(\frac{d\hat{T}}{dX} \right)_{\hat{X}} \tag{9.8}$$

Equation (9.8), however, contains a subtle mathematical difficulty, inasmuch as their definition of the constitutional gradient requires that a spatial derivative be taken of a interfacial quantity, despite the fact that this (hatted) temperature is strictly *local*. Thus, the mathematical operation shown in Eq. (9.8) appears logically suspect. This difficulty may be overlooked at present, given that it will be re-examined more thoroughly later and resolved. Alternatively, one may temporarily justify the seemingly peculiar mathematical definition assigned to $\mathcal{G}$ through an argument based on the chain differentiation of the LHS of Eq. (9.5):

$$\mathcal{G} \equiv \left(\frac{d\hat{T}}{dX} \right)_{\hat{X}} = \left(\frac{d\hat{T}}{d\hat{C}_\ell} \cdot \frac{dC_\ell}{dX} \right)_{\hat{X}}. \tag{9.9}$$

Equation (9.9) may be justified if $\hat{C}_\ell = C_\ell(\hat{X})$, which requires that the composition of the melt is *continuous* in the limit as $X \to \hat{X}$ from above. The validity of the concept of the constitutional gradient, $\mathcal{G}$, ultimately rests on the fact that the length scale over which a solid–liquid interface interacts with the adjacent melt is

---

[3] With multicomponent alloys, however, the solidus and liquidus are surfaces (or hyper-surfaces) on the phase diagram, instead of lines as in the present case of binary alloys. Multicomponent alloys exhibit more complicated dependences among temperature, pressure, and the phase compositions, making the analysis of constitutional supercooling for them more difficult. Again, in the interest of accessing easy-to-understand analytical results, quantitative discussion of constitutional supercooling will be limited to the case of binary alloys.

extremely small compared to the distances over which diffusion-limited macroseg-regation occurs. The actual 'sampling' of temperatures and concentrations adjacent to a solid–liquid interface occurs by fluctuations on molecular scales as small as 1 nm, and by diffusion on macroscopic scales of 1 mm.

Indeed, substituting Eqs. (9.5) and (9.7) into Eq. (9.9), along with $k_0 = \hat{C}_s/\hat{C}_\ell$, yields the operational definition for the constitutional gradient suggested by Tiller and Rutter [6],

$$\mathcal{G} \equiv v \frac{m_\ell}{D_\ell} \left(1 - \frac{1}{k_0} \right) \hat{C}_s. \tag{9.10}$$

The constitutional gradient describes a distribution of 'liquidus temperatures' just ahead of the solid–liquid interface. These liquidus temperatures are, of course, fic-tive, insofar as the liquidus relationship between concentration and temperature applies strictly *at* the solid–liquid interface, where the solid phase and its melt exchange atoms, equalize their chemical potentials for each component, and main-tain local thermodynamic equilibrium. The constitutional gradient, therefore, may be thought of as an estimate of future, or extrapolated, *liquidus* temperatures over a small region of the melt adjacent to the physical solid–liquid interface. The liquidus temperatures predicted with the constitutional gradient may be visualized as occur-ring as though the solid–liquid interface were, somehow, inserted at each location to create momentarily the required local equilibrium temperature and solute composi-tion as required by the phase diagram. The 'insertion' of the interface is physically accomplished by fluctuations that give rise to minute perturbations that extend and distort portions of the interface. The perturbed solid–liquid interface, in this manner, briefly 'probes' the thermodynamic conditions[4] in the adjacent melt.

9.4.4 Stable Interfaces

If the actual temperature gradient existing at the solid–liquid interface, $G(X) \equiv dT/dX$, is steeper than the constitutional gradient, $\mathcal{G}$, then stability is pre-dicted for the interface, insofar as the melt ahead of the interface has a lower melting temperature and a higher chemical potential than does the solid, and, therefore, is thermodynamically stable relative to the crystal. See Fig. 9.7. Here the liquid at *all* points ahead of the interface is stably maintained above its thermodynamic equilibrium temperature. Thus, the actual temperature field, $T(X)$, is everywhere hotter than the predicted distribution of 'future' liquidus temperatures, $T_\ell(X)$. The condition of constitutional stability shown in Fig. 9.7 is guaranteed by the simple condition that the actual interfacial thermal gradient, $G(X)$, is steeper than its con-stitutional gradient, $\mathcal{G}$, so the inequality for stability holds, namely, that $G(X) > \mathcal{G}$.

---

[4] It is actually testing the local Gibbs free energy and chemical potentials in a melt, with spatially varying solute concentration and temperature, against that of the contacting solid.

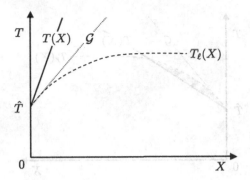

Fig. 9.7 Stable configuration of the applied temperature field, $T(X)$, relative to the constitutional gradient, $\mathcal{G}$, which is the *dotted line* showing the limiting slope of the extrapolated liquidus temperatures, $T_\ell(X)$ as $X \to 0$

Thus, to maintain a stable, planar solid–liquid interface the following inequality applies to the thermal gradient:

$$\frac{d}{dX}(T(X))_{\hat{X}} > \mathcal{G}. \tag{9.11}$$

9.4.5 Unstable Interfaces

If the temperature gradient applied to the interface is shallower than the constitutional gradient, $\mathcal{G}$, instability is predicted. See configuration of the actual temperature and the constitutional gradient shown in Fig. 9.8.

Instability is predicted to occur because the melt just ahead of the interface remains below its equilibrium liquidus, $T_\ell(x)$, and, therefore, is slightly supercooled or supersaturated. Perturbations of the interface, due perhaps to any of a variety of fluctuations occurring in the temperature, composition, or pressure, can cause the formation of a small forward-pointing projection of solid. Such a fluctuation can induce the supercooled melt to undergo further solidification at that point. Additional growth of such a protuberance, as suggested in Fig. 9.9, causes further intrusion of the solid into the supercooled liquid, causing yet more growth, and so forth.

Thus, a solid–liquid interface will be unstable and, eventually, will depart from planarity if the following inequality applies to the thermal gradient:

$$\frac{d}{dx}T(\hat{x}) < \mathcal{G}. \tag{9.12}$$

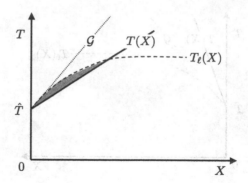

Fig. 9.8 Unstable configuration of the temperature field, $T(X)$, relative to the constitutional gradient, $\mathcal{G}$. The *dashed curve* shows the distribution of extrapolated liquidus temperatures, $T_\ell(x)$, ahead of the advancing interface; the slope of the *dotted line* as $X \to 0$ is the constitutional gradient, $\mathcal{G}$. The *solid line* represents the actual temperature field in the melt, $T(X)$, the slope of which is the thermal gradient, $G(X)$. Here the thermal gradient is shown lower than the constitutional gradient, $\mathcal{G}$, causing the *lens-shaped region* below $T(X)$ to be 'constitutionally' supercooled immediately ahead of the advancing interface

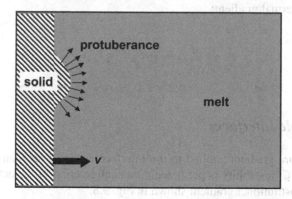

Fig. 9.9 A disturbance on an unstable solid–liquid interface projects a small distance into supercooled melt. The metastable liquid is supersaturated relative to the solid, so the protuberance can grow farther, releasing latent heat and rejecting solute (*small arrows*) into the surrounding melt. An interface subject to such a disturbance will evolve into more complex structures over time. See again Section 9.3.2 for a description of the sequence of interfacial morphologies that are observed to develop in constitutionally unstable alloys

9.4.6 Marginal Stability

Figure 9.8 shows that a zone of 'constitutionally supercooled' liquid exists adjacent to the solid–liquid interface whenever the actual temperature gradient is less than the constitutional gradient. A bump, or protuberance, if developed anywhere on the interface, might grow into the supercooled liquid and become even larger. A wide spectrum of perturbations is possible on real crystal-melt interfaces. A solid–liquid interface is continually 'testing' its stability against this spectrum of disturbances.

Nothing extraordinary is required to induce interfacial instability—just a miniscule localized perturbation—provided that constitutionally supercooled melt exists
ahead of the interface.

A critical condition may be defined among the solidification variables that is
termed 'marginal stability'. An interface is considered marginally stable when the
zone of constitutionally supercooled liquid just vanishes, that is, when the thermal
gradient, G, applied to the interface exactly equals the constitutional gradient, $\mathcal{G}$.
Such an interface is conditionally, or marginally, stable, as any weakening of the
interfacial temperature gradient, or steepening of the constitutional gradient, such as
caused by a slight acceleration of the interface, could cause the onset of instability.

The criterion for marginal stability is found by substituting Eq. (9.10) into the
right-hand side of the equality $G^\star = \mathcal{G}$. Here $G^\star$ is the critical minimum thermal
gradient, below which constitutional supercooling and interfacial instability occur.
Given the alloy composition, solute diffusivity, the alloy system's properties and the
solidification speed, v, the value of $G^\star$ may be determined as,

$$G^\star = v\frac{m_\ell}{D_\ell}\left(\frac{k_0 - 1}{k_0}\right)\hat{C}_s. \tag{9.13}$$

The minimum thermal gradient needed to insure interfacial stability, as indicated
in Eq. (9.13), is proportional to the interface speed and its concentration, and is
inversely proportional to the solute's diffusivity in the melt. The minimum gradient,
$G^\star$, can be non-dimensionalized by dividing both sides of Eq. (9.13) by the quantity $(m_\ell C_0 v)/D_\ell$, and then recognizing that the quantity $\hat{C}_s/C_0 = k_{BPS}$. See again
Chapter 7 on composition control.[5] The variables in Eq. (9.13) can be rearranged in
a convenient dimensionless form to estimate the critical, minimum, dimensionless
thermal gradient, $\Gamma^\star$, to achieve stable, plane-front solidification behavior, specifically,

$$\frac{-G^\star}{m_\ell C_0 v/D_\ell} \equiv \Gamma^\star = \left(\frac{1-k_0}{k_0}\right)k_{BPS}. \tag{9.14}$$

If the RHS of Eq. (7.19), Chapter 7, is substituted for k_{BPS} in Eq. (9.14), an interesting relationship is produced that connects the minimum thermal gradient for stable
growth with the value of the equilibrium distribution coefficient, k_0, and the BPS
melt zone width, δ, namely,

$$\Gamma^\star = \left[\frac{k_0}{1 - k_0} + e^{-\frac{v\delta}{D_\ell}}\right]^{-1}. \tag{9.15}$$

---

[5] The reader is reminded that Chapter 7 shows that the effective distribution coefficient, k_{BPS},
depends on the state of convection in the melt, which can be varied by mechanical or electromagnetic stirring to decrease the width of the 'static' melt zone. The allowed range of k_{BPS} is
$k_0 \le k_{BPS} \le 1$, and its precise value depends on the intensity of convection achieved in the melt.

The variation of Γ^* is plotted versus the normalized zone width, $v\delta/D_\ell$, for several values of k_0 in Fig. 9.10. The minimum thermal gradient for stable crystal growth responds sensitively to the value of k_0, and its required value for stable crystal growth can be lowered considerably just by stirring the melt and narrowing the BPS zone width. Figure 9.10 clearly adds justification for using active stirring control to stabilize a moving solid–liquid interface, which is enhanced by reducing the width of the solute boundary layer. Convection control, such as mechanical or magnetic stirring of the melt, can achieve practical stable interface behavior during crystal growth in systems that might otherwise resist the preparation of high-quality homogeneous single crystals. Also, as a general engineering consideration, the lower the thermal gradient and the higher the growth speed, the less costly will be the processing of a melt into a single crystal by unidirectional freezing.

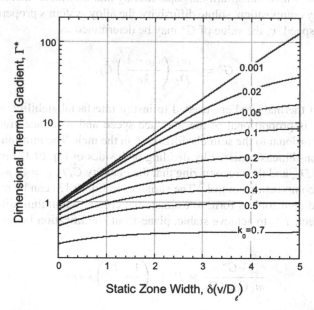

Fig. 9.10 Semi-logarithmic plot of Eq. (9.15) showing the dimensionless minimum thermal gradient for stable solidification versus the normalized BPS static melt zone width, $\delta(v/D_\ell)$. The magnitude of Γ^* is efficiently reduced by stirring the melt to decrease the static melt zone width. The minimum thermal gradient is remarkably sensitive to the system's k_0 value

Achieving steep enough temperature gradients during unidirectional crystal growth from the melt is, of course, limited by practical thermal engineering of furnaces. For example, the maximum temperature gradient is limited by the available heater power, efficiency of the furnace insulation, thermal conductivity of the melt, and by other materials limitations imposed by the selection of the heating element, crucible, or crystal growth ampoule. It becomes clear, therefore, that there always exist stringent limits imposed on the maximum allowable interface speed and solute concentration to achieve stable conditions for single-crystal growth.

Lastly, alloys with small values of k_0 solidify with relatively wide diffusion boundary layers, and, thus, pose additional difficulties to be solidified in a stable mode. See again Fig. 9.10. The exception to this statement might be achieved by using directional crystal growth at low speeds with extremely dilute alloys.

9.4.7 Bulk Crystal Growth: Limiting Forms for Stability

Equation (9.13) is perhaps best applied with the speed of the solid–liquid interface, v, moved to the left-hand side of the equation. This particular form of the marginal criterion for constitutional supercooling gathers all the solidification control variables on the LHS, whereas all the materials parameters and the system's solute concentration appear together on the RHS. Thus, in papers on bulk crystal growth, Eq. (9.13) regularly appears written in the form,

$$\frac{G^*}{v} = \frac{m_\ell}{D_\ell}\left(\frac{k_0 - 1}{k_0}\right)\hat{C}_s. \tag{9.16}$$

Equation (9.16) expresses the marginal criterion for constitutional supercooling in terms of the interfacial composition, $\hat{C}_s$, in the solid. If one assumes that various extreme states of mixing exists in the melt, as was discussed in the development of BPS composition control theory, in Chapter 7, then values for Γ^* in Eq. (9.15) can be solved under these limits:

1. *Quiescent melt with diffusion transport only.* When the melt is motionless, i.e., all convective motions cease, then the solutal boundary layer developed from macrosegregation becomes relatively wide, so that $\delta \to 2D_\ell/v$, $k_{BPS} = 1$, and $\Gamma^* \to (1 - k_0)/k_0$. The transport of solute through the melt can occur only by diffusion. Under these conditions, the solute concentration in the freezing solid, $\hat{C}_s$, approaches C_0, which is the steady-state solute distribution depicted earlier in Fig. 6.1. Equation (9.16) may be expressed in terms of the steady-state (overall) concentration of the alloy, C_0, as

$$\frac{G^*}{v} = \frac{m_\ell}{D_\ell}\left(\frac{k_0 - 1}{k_0}\right)C_0. \tag{9.17}$$

   Equation (9.17) is the limiting form of the marginal constitutional supercooling criterion derived originally by Tiller, Jackson, Rutter, and Chalmers [3].

2. *Efficiently stirred melt, convective transport.* If the melt is stirred vigorously and efficiently, then macrosegregation is confined to a relatively thin boundary layer, $\delta \approx 0$, and, $k_{BPS} = k_0$, and $\Gamma^* \to (1 - k_0)$. The solute concentration in the solid at the interface, $\hat{C}_s$, under this limiting set of crystal growth conditions approaches $C_0 k_0$. The marginal stability criterion for constitutional supercooling, Eq. (9.16), takes the limiting form,

$$\frac{G^\star}{v} = \frac{m_\ell}{D_\ell} (k_0 - 1) C_0. \tag{9.18}$$

3. *Stirring control and interface stability.* A comparison of Eq. (9.17) with Eq. (9.18) shows that they merely differ by a factor of k_0. Assuming that $k_0 < 1$, however, it becomes clear that the value of $G^\star/v$ needed to achieve marginal stability during crystal growth in the limit of solute transport by diffusion alone is usually larger than when convectively-assisted transport is present. See again Fig. 9.10. In short, stirring can improve interfacial stability. Indeed, crystal growers have known through experience that rotating a crystal being drawn from the melt improves the chances of avoiding interfacial instabilities or 'breakdown' of the growth front. A planar interface reduces lateral solute segregation, internal stresses and crystalline defects, and also avoids the unwanted formation of cells and dendrites. Cellular and dendritic structures are undesirable in single crystal growth because they induce complex patterns of chemical *microsegregation*—an important subject which will be covered in Chapters 13 and 14. Microsegregation not only causes chemical inhomogeneities that can degrade the properties of a crystal, but is a major source of microstraining that causes shape distortion, cracking, and dislocations to form.

Thus, the influence of stirring on interface stability may be succinctly summarized as,

$$\left(\frac{G^\star}{v}\right)_{\delta \to \infty} > \left(\frac{G^\star}{v}\right)_{\delta} > \left(\frac{G^\star}{v}\right)_{\delta = 0}. \tag{9.19}$$

9.5 Verification of Constitutional Supercooling

It is accepted that constitutional supercooling provides a useful concept for quantifying the criterion of morphological stability. Marginal stability, as defined from the theory of constitutional supercooling, has achieved its popular utility with engineers and crystal growth technologists because of its ease of application and conceptual simplicity. In fact, experimental correlations amassed over the years convincingly demonstrate the soundness of this simple criterion for achieving stable crystal growth. As shown later in Chapters 10 and 11, more sophisticated stability analyses reveal that constitutional supercooling theory provides the correct mathematical criterion for judging the stability of a solid–liquid interface in the limit of disturbances acting over long length scales. It is now well established, however, that most solid–liquid interfaces become relatively more resistant to disturbances, by requiring smaller $G^\star/v$ values, as their disturbances act over smaller and smaller length scales. Constitutional supercooling theory itself, unfortunately, provides neither guidance nor any deeper insights on the critically important issues of the length and time scales for the development of interfacial disturbances.

9.5.1 Experiments

Both decanting experiments and *in situ* interface observations, detailed earlier in Section 9.3, led early researchers on solidification toward the realization that the solid–liquid interface could become unstable and develop an assortment of complex two- and three-dimensional solidification structures. Similar experiments were eventually used to test and quantify the criterion of marginal stability based on constitutional supercooling theory. Specifically, a series of dilute Sn-Pb alloys were examined by Chalmers and co-workers by unidirectionally solidifying them at various rates under different temperature gradients. The frozen alloys were decanted after partial solidification and then studied microscopically to determine the morphological conditions (planar or patterned) at the solid–liquid interfaces during freezing. The morphological observations obtained from these experiments were categorized by the investigators as either 'stable' or 'unstable'.

9.5.1.1 Stability Maps

Stability maps are two-dimensional cross plots, based on Eq. (9.17), for selected solidification variables: for example, by choosing the thermal gradient, G, as the ordinate and the solidification velocity, v, as the abscissa, and then fixing the alloy composition, (C_0, and k_0), a map may be produced predicting the combinations of all these process variables over which different basic solidification microstructures are expected. Figure 9.11 shows such a stability map for achieving a wide range of cast microstructures.

A quantitative test of the predictions from Eq. (9.13) was developed by plotting the ratios of the thermal gradient measured in the melt to the interfacial growth speed, G_ℓ/v, against the composition of the alloy, C_0. A type of morphological stability 'map' could be developed by noting which points plotted on the plane C_0 versus G_ℓ/v corresponded to either stable or unstable interfaces. Figure 9.12 shows experimental results obtained by Chalmers [1].

The dashed borderline, separating stable from unstable conditions was located theoretically by inserting the material properties, alloy concentrations, and thermochemical data from the Sn-Pb phase diagram into Eq. (9.13). The dashed line represents the condition of marginal stability, and successfully separates stable solidification conditions from unstable ones. As shown, this early test of constitutional supercooling satisfactorily supports the theory.

9.5.1.2 Critical Freezing Speed

An alternative approach to testing constitutional supercooling theory and the criterion of marginal stability is to rearrange Eq. (9.13) in the following form:

$$m_\ell \left(\frac{k_0 - 1}{k_0} \right) \hat{C}_s = D_\ell G v^{-1}. \tag{9.20}$$

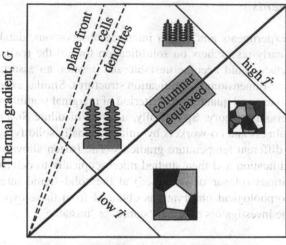

Solidification front velocity, v

Fig. 9.11 Stability-microstructure map based on constitutional supercooling. The regions of this map are approximately divided into areas of 'high' and 'low' cooling rates, $\dot{T}$, and 'columnar' and 'equiaxed' structures. The sector labeled 'plane front' is a region of interface stability. 'Cells' occupy only a very narrow sector of solidification conditions, followed at lower gradients and higher solidification speeds by the onset of dendritic growth. At yet higher casting speeds and still lower thermal gradients, equiaxed grain formation is predicted. Adapted from [7]

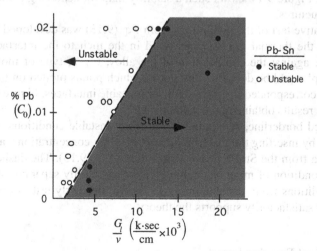

Fig. 9.12 Morphological stability map for dilute Sn-Pb alloys testing predictions based on the theory of constitutional supercooling. The data points designate experimentally verified stable (*solid symbols*) and unstable (*open symbols*) solidification conditions. The criteria for stability or instability were judged on the basis of whether or not the interface remains featureless (planar) during solidification. The *dashed line* denotes marginally stable solidification conditions and is based on the theoretical prediction separating unstable interface behavior from stable behavior

Taking logarithms of both sides of Eq. (9.20) yields an expression that proves useful for experimental testing, namely,

$$\ln\left[m_\ell\left(\frac{k_0 - 1}{k_0}\right)\hat{C}_s\right] = \ln(D_\ell G_\ell) - \ln(v). \qquad (9.21)$$

If the quantity in square brackets on the LHS of Eq. (9.21)—basically the alloy concentration, C_0—is plotted against the solidification speed, v, on double logarithmic coordinates, then the marginal stability criterion separating stable and unstable interface behavior appears as a straight line with a slope of minus one. Figure 9.13 displays such data acquired for dilute Pb-Sn alloys growing close to the condition of marginal stability. The agreement is excellent between these data and the theoretically predicted separatrix, which appears as the dashed line with a slope of -1. A systematic departure from the theoretical separatrix occurs particularly as the rates of solidification rise above about 10 μm/s. The additional conclusion that may be drawn from these data, at least tentatively, is that the basic theory of marginal stability provides the most accurate predictions at slow freezing rates.[6]

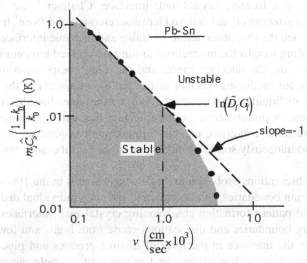

Fig. 9.13 Stability map for dilute Pb-Sn alloys based on the analysis provided by Eq. (9.21). The data designate experimentally verified marginally stable solidification conditions. The *dashed line* denotes marginally stable solidification conditions and has the theoretically predicted slope of -1. The experimental data, especially at slower speeds, agree well with this form of constitutional supercooling theory

---

[6] At higher growth rates the value for the distribution coefficient may change due to a variety of extrinsic effects, such as melt convection and lateral segregation. Even intrinsic interface kinetic phenomena at higher growth speeds can change the k-value for an impurity or solute element. In Chapter 17, rapid solidification processes that operate far from equilibrium will be considered in detail. Under rapid solidification conditions, not surprisingly, constitutional supercooling theory is incapable of predicting the observed stability behavior.

Constitutional supercooling still provides the best starting point for estimating the required process parameters needed to insure interfacial stability, especially for conventional crystal growth processes where growth rates tend to be slow to moderate. In fact, as more sophisticated theories of interface stability are developed subsequently in Chapter 10 [8, 9], they will serve mainly: (1) to broaden our understanding of additional factors that influence interfacial dynamics and evolutionary stability, and (2) to allow the application of a more methodical approach for analyzing rapid solidification phenomena, where local equilibrium at the interface no longer provides a useful kinetic approximation for the behavior of the solid–liquid interface.

9.6 Summary

Constitutional supercooling was the paradigm-altering phenomenon introduced in the 1950s by B. Chalmers to explain the formation, or non-formation, of complex microstructures at a freezing crystal-melt interface. Chalmers' use of decanting studies, where a growing crystal was suddenly accelerated and 'torn' from its adjacent melt, provided the first observations of stable and unstable interfaces. Constitutional supercooling weighs the interactions among the applied temperature gradient during solidification, the interface speed, and the alloy composition to arrive at a simple criterion for predicting whether or not the melt adjacent to the interface is above or below its liquidus. In the former condition the super-liquidus melt remains stable to a bump, or protuberance, on the interface that extends momentarily into the melt. In the latter condition, the sub-liquidus melt is 'constitutionally supercooled' and spontaneously solidifies on any interface protuberance making it even larger.

Decanting observations evolved into in situ experiments in the 1960s, where the influences of grain boundaries, sub-boundaries, and even individual dislocations on the stability and pattern formation at advancing crystal-melt interfaces were studied. Intersecting boundaries and dislocations cause both high- and low-amplitude disturbances on the interface in the form of distinct grooves and pits. These persistent local disturbances then trigger the formation of a whole range of interfacial forms, including ridges, hillocks, rings, blobs, 'volcanos', and pre-dendritic disks.

Constitutional supercooling, despite its simplicity, allows estimates to be made regarding establishing processing conditions—i.e. choosing speed, thermal gradient, and composition—that are consistent with stable growth of alloy crystals from their melt. As will be demonstrated in the next chapter, entitled 'Linear Morphological Stability', the theory of constitutional supercooling actually provides the exact answer to the question of stability to long-wavelength disturbances on the solid–liquid interface. In this sense, Chalmers' early concept, constitutional supercooling, was subsumed by a later more comprehensive theory.

References

1. B. Chalmers, *Trans. AIME*, **200** (1956) 519.
2. J.W. Rutter and B. Chalmers, *Can. J. Phys.*, **31** (1953).
3. W.A. Tiller, K.A. Jackson, J.W. Rutter and B. Chalmers, *Acta. Metall.*, **1** (1953) 428.
4. W. Kurz and W. Fisher, *Fundamentals of Solidification*, Trans Tech Publications, Aedermannsdorf, 1989.
5. R.J. Schaefer and M.E. Glicksman, *Metall. Trans. A*, **A1** (1970) 1973.
6. W.A. Tiller and J.W. Rutter, *Can. J. Phys.*, **34**, (1956) 96.
7. M. Rettenmayr and H.E. Exner, 'Directional Solidification of Crystals', in *Encyclopedia of Materials: Science and Technology* , Elsevier Science Ltd., Amsterdam, ISBN: 0-08-0431526, 2000.
8. W.W. Mullins and R.F. Sekerka, *J. Appl. Phys.*, **34** (1963) 323.
9. W.W. Mullins and R.F. Sekerka, *J. Appl. Phys.*, **35** (1964) 444.

Chapter 10
Linear Morphological Stability

In Chapter 8 the foundational subject of interface capillarity was developed to explain why curved crystal-melt boundaries exhibit modified thermodynamic properties compared to their planar counterparts, especially regarding their equilibrium temperature and solubility. Having discussed the thermodynamic aspects of crystal-melt interfaces, we return again to a study of their kinetic behavior, and concentrate more on the dynamical aspects of morphological stability during freezing.

10.1 Introduction

Criteria for interfacial stability were developed in Chapter 9. The focus of that discussion was that the stability, or instability, of an initially planar interface arises as a predictable *quantitative* outcome from considerations of whether or not constitutional supercooling develops at the interface. In brief, the stability of such an interface depended simply on whether or not the liquidus temperatures extrapolated ahead of the moving interface were below (stable) or above (unstable) the actual temperatures established at points in the melt.

Constitutional supercooling theory, however, provides but a limited interpretation of what is more broadly termed *morphological stability*, the latter also requiring consistent solute diffusion as well as heat transfer fields throughout the solid and the melt phases. The theory of constitutional supercooling also lacks several significant physico-chemical effects to be introduced through considerations of capillary phenomena on curved crystal-melt interfaces. Consequently, constitutional supercooling theory remains incapable of describing the full range of interfacial dynamics, nor does it include important material parameters that bear upon the general question of interfacial stability.

Dynamical theories of linear morphological stability were developed in the mid-1960's by Mullins and Sekerka [1, 2], and independently by V.V. Voronkov [3]. Their theories were elaborated by many contributors over the next 25 years, and now incorporate aspects of interfacial stability in a panoply of diverse solidification and crystal growth phenomena. Morphological stability now includes the roles played by anisotropy of the crystal-melt interfacial energy, departures from local equilibrium

M.E. Glicksman, *Principles of Solidification*, DOI 10.1007/978-1-4419-7344-3_10,

induced by rapid solidification, plus a host of extrinsic effects that are imposed by melt hydrodynamics, external electric and magnetic fields, gravity, etc. A detailed monograph on the subject of interface stability by S.H. Davis [4] is suggested to the reader interested in a comprehensive review of mathematical and experimental details for linear and non-linear stability theories of solid–liquid interfaces.

10.2 Perturbation Theory

The general approach used in *linear* stability theories is to postulate the ubiquitous presence on simply shaped interfaces of small-amplitude fluctuations, or perturbations. These weak perturbations are described mathematically using modal analysis, by 'testing' the stability of the system against the shape distortions added by the 'modes'. Each modal distortion of the interface adds some independent geometrical departure from an assumed initial 'base state'—generally chosen in the form of a simple interface shape, such as a plane, cylinder, or sphere. To develop the theory, mathematically convenient modal forms for these disturbance are employed.

Long before the problem of morphological stability of a crystal-melt interface was considered, the identical mathematical approach, viz., linear perturbation theory using modal analysis, was successfully applied to other physical problems involving dynamical stabily: e.g., electrical, acoustic, and mechanical oscillators, elastic beam and column theory, and fluid flow. Indeed, in the spirit of linear stability, an interface is deemed 'linearly stable' if and only if small amplitude disturbances disappear at *all* possible length scales.

We choose the interface to be growing at some steady rate, v_0, on which is imposed at time $t = 0$ a single, one-dimensional, spatially periodic ripple. The interface ripple has a small amplitude, δ_0, and an arbitrary wavenumber, $k = 2\pi/\lambda$. Here, λ is the wavelength of the ripple disturbance. See Fig. 10.1 for a sketch of such a perturbation mode distorting an otherwise infinite, planar, crystal-melt interface. The subsequent philosophy, applied at least within the current linear mathematical framework, is to determine whether or not the low-amplitude ripple will grow over time in amplitude, $\delta(t)$, and 'destabilize' the interface, or, alternatively, decay over time, and leave the initial planar interface unchanged and, therefore, morphologically 'stable'.

The disappearance of the disturbances occurs under specified kinetic conditions: including a fixed steady-state interface speed, solute concentration, and applied thermal gradient. Achieving stability, or instability, under these conditions rests on the outcome from testing, in a mathematical sense, whether the interface relaxes back to its unperturbed form or 'base state', or deviates still further from that initial form, usually at an exponentially increasing rate. The base state for the interface must be chosen with considerable care, as it represents the geometric form of the interface that must being tested mathematically. More specifically, the interface is considered to be linearly unstable if any of the wavenumbers over the entire modal spectrum of perturbations (i.e., weak sinusoidal ripples of *all* wavenumbers) undergo

M.E. Glicksman, Principles of ... Solidification, DOI 10.1007/978-...   © Springer Science+Business Media, LLC 2011

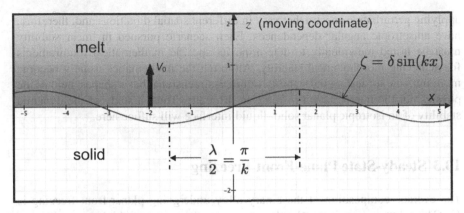

Fig. 10.1 Sinusoidal perturbation of a planar crystal-melt interface growing steadily at speed v_0 in the $+z$-direction. The interface is of infinite extent, and lies parallel to stationary x-y planes normal to the page. Co-moving coordinate, Z, travels with the base state of the interface, which always occupies the plane $Z = 0$. The interface disturbance shown in this figure is chosen in the form of a low-amplitude sine wave, $\zeta(x) = \delta \sin(kx)$, extended parallel to the interface in the $\pm x$-directions. The disturbance, or perturbation, represented by this sine wave has a wavelength, λ, and a wavenumber, k, where $\lambda = 2\pi/k$

amplification. On the other hand, an interface is termed *marginally stable* if the perturbation amplitude, $\delta(t)$, is found to remain constant in time, i.e., the amplitude of the applied disturbance neither grows nor shrinks relative to the moving interface. Marginal states, besides separating conditions leading to either linear instability or stability, also will play a role later in the formation of more complex, evolved solidification forms, including rolls, cells, and even dendrites. These microstructurally important outcomes of instabilities triggered at an advancing solid–liquid interfaces will be treated in full detail later in Chapters 13 and 14.

10.2.1 Stability of Planar Interfaces

For the sake of simplicity and clarity, we consider an initially planar solid–liquid interface advancing at a steady-state speed as the unperturbed form, or base state, $Z = 0$. The steady advance of the interface implies that a steady-state macrosegregation field is developed through the melt.

The temperature field at and around the moving interface is also considered to be steady, an assumption termed the 'frozen temperature-field' approximation [5]. In essence the frozen temperature approximation adds the assumption that the temperature field acting across the interface is steady and fully relaxed, despite the motion of the interface and its liberation of latent heat. These assumptions combined with a perturbation spectrum—chosen in the form of low-amplitude sine waves—reduces the mathematical stability analysis to that of a 2-dimensional thermo-chemical field-analysis problem. Other more geometrically complicated cases may be treated similarly, such as analyzing the stability of cylindrical or spherical interfaces, or applying perturbation spectra that vary in different spatial directions and, therefore, have anisotropic angular dependences. Each scenario pursued in linear stability must be tested individually to determine its specific mathematical requirements for achieving morphological stability. Although the mathematics usually become more intricate for testing the stability of three-dimensional basis shapes, neither the physics nor the interpretation increase in difficulty. Thus, a study of the dynamic stability of an isotropic planar solid–liquid interface will suffice here.

10.3 Steady-State Plane-Front Freezing

The average composition of the binary alloy undergoing planar-front freezing is C_0. The solute composition, C_0, also represents the uniform initial composition of the melt far from the steadily advancing interface. (See again Chapter 6, where the steady-state diffusion solutions for macrosegregation in one spatial dimension are developed.) Figure 10.1 shows a section of the Cartesian coordinate frame that is anchored to, and co-moving with, a weakly perturbed, sinusoidally shaped crystal-melt interface with the profile $\zeta(x)$. This interface is advancing into the melt at a steady base speed, v_0, measured relative to any fixed Cartesian plane, $z = const$.

The length scale λ denotes the wavelength of the sinusoidal disturbance along the x-direction of the interface, whereas δ denotes its amplitude as measured relative to the co-moving base plane, $Z = 0$. The wavenumber of the disturbance, k_x, is defined in the usual way from wave physics, namely, $k_x \equiv 2\pi/\lambda$, where k_x is the number of waves per unit length along the x-axis. Note, that the wavelength, λ, and its reciprocal measure, the wavenumber, $k_x = 2\pi/\lambda$, set the transverse spatial scale of the perturbation wave distorting the interface, whereas the perturbation ampli-tude, δ, provides the longitudinal spatial scale normal to that interface. If additional perturbations were needed, say along the y-direction (into the plane of the page), then sine waves specified by their wavenumber, k_y, could be added. For the sake of simplicity one may assume that the interface is isotropic, and dynamical stability against disturbances in the x-direction is identical to that for any other direction on the interface. Thus, a sinusoidal disturbance with its wavenumber k_x, varied between 0 and ∞, suffices to 'test' the system, as the x-direction itself is chosen arbitrarily on the interface plane. This method will expose wavenumber regions which are stable and unstable.

10.3.1 Boundary Conditions

The following boundary conditions are imposed on the sinusoidally perturbed solid–liquid interface, regarding its thermal and solutal environment during steady-state solidification:

1. *Far-field concentration*: The concentration field in the melt, far from the solid–liquid interface, must approach the initial, uniform concentration of the alloy,

$$C_\ell(Z \to \infty) = C_0. \tag{10.1}$$

2. *Far-field temperature in the melt*: The temperature field in the melt rises to some fixed temperature, T_∞, far above the solid–liquid interface, $T_\infty > T_m$,

$$T_\ell(Z \to \infty) = T_\infty. \tag{10.2}$$

3. *Far-field temperature in the solid*: The temperature in the solid falls off indefinitely in the solid below the solid–liquid interface. This far-field boundary condition, although in some sense unrealistic, does not affect the answer about whether or not interfacial stability occurs,

$$T_s(Z \to \infty) \to -\infty. \tag{10.3}$$

10.3.2 Transport Field Equations

The following two-dimensional field equations are to be solved in the melt and solid:

1. *Steady-state temperature field in the melt*:

$$\frac{\partial^2 T_\ell}{\partial x^2} + \frac{\partial^2 T_\ell}{\partial Z^2} + \frac{v_0}{\alpha_\ell} \frac{\partial T_\ell}{\partial Z} = 0, \ (Z > \zeta(x)). \tag{10.4}$$

2. *Steady-state temperature field in the solid*:

$$\frac{\partial^2 T_s}{\partial x^2} + \frac{\partial^2 T_s}{\partial Z^2} + \frac{v_0}{\alpha_s} \frac{\partial T_s}{\partial Z} = 0, \ (Z < \zeta(x)). \tag{10.5}$$

3. *Steady-state concentration field in the melt*:

$$\nabla^2 C_\ell + \frac{v_0}{D_\ell} \frac{\partial C_\ell}{\partial Z} = 0, \ (Z > \zeta(x)). \tag{10.6}$$

4. *Diffusion transport of solute in the solid is ignored, $D_s \approx 0$*:

$$C_s = \hat{C}_s(Z), \ (Z < \zeta). \tag{10.7}$$

10.3.3 Perturbation Analysis: The Concept

Regarding solutions to these differential equations for the transport fields given in the last section, one may pose the following key questions: Do the ripples and disturbances that one could imagine as slightly perturbing a moving planar solid–liquid interface all eventually disappear? Would some, or perhaps even a single 'danger-ous' mode, be able to persist and grow, and thereby destabilize the interface? The answers to those questions, as will now be obtained through formal perturbation analysis, provide access to the next level of understanding morphological stability, and also introduce the reader to some important additional issues for discussion and study.

Seeking these solutions within the framework of linear analysis allows their mathematical construction as a sum of two types of terms: the first arises from the choice of the basis state—i.e., the planar, or 'unperturbed' interface, whereas the second arises from the choice of the mathematical form of the disturbance modes, chosen here, arbitrarily, as sinusoidal ripples[1] of arbitrary wavenumber, k_x. Thus, conceptually, one imagines weak interfacial ripples of *all* wavenumbers, $0 < k_x < \infty$, appearing from time-to-time on the moving interface. The modal decomposition of random disturbances (fluctuations) into sinusoidal waves is, of course a mathematical construct based on standard Fourier analysis. The justifi-cation for using Fourier sine modes to represent the set of interface disturbances appearing on a solid–liquid interface, although technical in its full mathematical detail, reduces mainly to the straightforward concept that a complete, independent set of these eigenfunctions may be summed, or superposed, to represent *any* dis-turbance. Such analyses were employed successfully for more than a century in various engineering applications including heat conduction, diffusion, mechanical vibrations, and signal analysis.

10.3.4 The Transport Solutions

The two thermal fields in the solid and liquid, respectively, and one solutal field in the liquid are each described by combining the basis (unperturbed) and the perturbed solutions to Eqs. (10.3) through (10.6).

1. *Solute diffusion in the melt*:

$$C_\ell(x, Z) = C_\ell(Z) + \tilde{C}_\ell(x, Z) \cdot \delta \quad (Z \geq \zeta). \tag{10.8}$$

2. *Temperature field in the melt*:

$$T_\ell(x, Z) = T_\ell(Z) + \tilde{T}_\ell(x, Z) \cdot \delta \quad (Z \geq \zeta). \tag{10.9}$$

3. *Temperature field in the solid*:

---

[1] Sine wave ripples were specified in Fig. 10.1 arbitrarily by choosing the origin, $x = 0$, in such a manner that the phase of the interfacial disturbance wave corresponds to a sine function. If the ori-gin of the x-coordinate were shifted to the right or left by $\lambda/2$, then the perturbation would appear as a cosine function. Judicious selection of the form of the perturbation mode and its coordinate allows convenient disposal of this arbitrary phase shift.

$$T_s(x, Z) = T_s(Z) + \tilde{T}_s(x, Z) \cdot \delta \quad (Z \leq \zeta). \tag{10.10}$$

The functions $C_\ell(Z)$, $T_\ell(Z)$, and $T_s(Z)$, appearing as the first terms on the RHS of Eqs. (10.8), (10.9), and (10.10) are the solutal and thermal fields, respectively, for the planar, or basis form of the interface. These are the fields that develop ahead of, and behind, the unperturbed interface, so they apply where ζ, and δ are zero. These unperturbed solutions for the solute distribution and the temperature, as noted, depend *only* on the distance, Z, from the interface plane $Z = 0$, and do not depend on the lateral position, x. The x-coordinate, of course, should not affect the planar basis solution.

By contrast, the second terms, $\tilde{C}_\ell(x, Z)$, $\tilde{T}_\ell(x, Z)$, and $\tilde{T}_s(x, Z)$, each represent the corresponding perturbed transport fields associated with the ripples. The non-planar interface, $\zeta(x, Z)$, naturally, depends on both spatial coordinates. In linear analysis, the unperturbed solutions[2] for the macrosegregation and temperature fields may be added together with their corresponding perturbed solutions to form the total, or superposed, solutions. Again, in the spirit of linear analysis, one assumes that each perturbed field response is proportionate in strength to the disturbance amplitude, δ. Moreover, $\delta(t)$ itself is assumed to be initially extremely small, as compared, say, to a sinusoidal disturbance of unit amplitude. Thus, the interface perturbations have amplitudes $\delta(0) \ll 1$. Any terms appearing in the subsequent analysis that are of higher order, say proportional to δ^2, or δ^4, may be safely ignored, as long as the linear terms contributed by the ripple are still considered to be small. Clearly, as soon as a perturbation grows for a while, non-linear contributions would appear, and the linear approximation would fail. Non-linear stability theories needed to treat such problems tend to become technical, from a mathematical standpoint, and consequently they remain today an active research field in applied mathematics [6].

10.3.4.1 Concentration Field in the Melt

The solution to Eq. (10.5), valid to linear order, for the solute field developed at steady-state in the melt ahead of the perturbed solid-melt interface is

$$C_\ell(x, Z) = C_0 \left[1 + \frac{1 - k_0}{k_0} \right] e^{-\frac{v_0}{D_\ell} Z} + \delta \left[b + C_0 \frac{1 - k_0}{k_0} \frac{v_0}{D_\ell} \right] \sin(k_x x) e^{-k_C Z}, \tag{10.11}$$

where the term k_C appearing in the last exponential term on the RHS of Eq. (10.11) is a function of the wavenumber, k_x, and defined as

---

[2] In the physics literature on electron wave propagation in solids the unperturbed wave solutions are called 'plane waves', which remains apt terminology for the temperature and solute diffusion fields associated with the planar interface.

$$k_C \equiv \frac{v_0}{2D_\ell} + \sqrt{\left(\frac{v_0}{2D_\ell}\right)^2 + k_x^2}. \tag{10.12}$$

The constant b appearing in the second term on the RHS of Eq. (10.11) is unknown, and k_0 is the alloy's equilibrium distribution coefficient. Note that the total solutal field is comprised of the two terms we expect:

1. The so-called 'plane wave' term is a function only of Z. This term represents the one-dimensional macrosegregation field present ahead of a steadily solidifying planar interface, so that the terms in Eq. (10.11) are identical to those in the steady-state plane-front diffusion solution developed earlier in macrosegregation theory. Please see again Chapter 6 for the details of this 'unperturbed' basis solution for the solute field.
2. The 'perturbation' term depends on both the Z- and x-coordinates. This term is a periodic sine-wave solute distribution in x, which is exponentially damped, or reduced, in the $+Z$ direction. In fact, careful scrutiny of this term shows that the perturbation decays faster with distance from the interface than does the unperturbed plane-wave macrosegregation profile. Moreover, the higher is the wavenumber of the periodic disturbance, k_x, the more rapidly does this perturbed component of the total concentration field diminish with increasing distance from the solid–liquid interface.

10.3.4.2 Thermal Field in the Melt

The solution to Eq. (10.4) for the thermal field developed at steady-state ahead of the perturbed solid-melt interface is similarly found to be

$$T_\ell(x, Z) = \hat{T} + G_\ell \frac{\alpha_\ell}{v_0} \left(1 - e^{-\frac{v_0}{\alpha_\ell}Z}\right) + \delta\,(a - G_\ell) \sin(k_x x)e^{-k_{T_\ell}Z}, \quad (\zeta > 0), \tag{10.13}$$

where the function k_{T_ℓ} appearing in the last exponential on the right-hand side of Eq. (10.13) is defined as

$$k_{T_\ell} = \frac{v_0}{2\alpha_\ell} + \sqrt{\left(\frac{v_0}{2\alpha_\ell}\right)^2 + k_x^2}. \tag{10.14}$$

In Eq. (10.13), the term a remains as an unknown constant of the solution; $\hat{T}$ is the equilibrium temperature of a planar solid-melt interface with solid phase composition C_0; and α_ℓ is the thermal diffusivity of the melt. The term G_ℓ denotes the thermal gradient in the melt just ahead of the interface $Z = 0$. Note that Eq. (10.13) has a mathematical structure similar to Eq. (10.11), although it describes the perturbed temperature field in the melt $(Z > \zeta)$, rather than the solute concentration.

10.3.4.3 Thermal Field in the Solid

The solution to Eq. (10.5) for the thermal field developed at steady-state in the solid phase is, once more, found as a sine-exponential solution added to the plane-wave segregation solution, namely,

$$T_s(x, Z) = T_m + G_s \frac{\alpha_s}{v_0} \left(1 - e^{-\frac{v_0}{\alpha_s} Z} \right) + \delta (a - G_s) \sin(k_x x) e^{-k_{T_s} Z}, \quad (Z \leq \zeta).$$

$$(10.15)$$

In Eq. (10.15), α_s is the thermal diffusivity of the solid, and the term G_s denotes the thermal gradient in the solid behind the solid–liquid interface. The function k_{T_s} appearing in the second term on the RHS of Eq. (10.15) is defined as

$$k_{T_s} = \frac{v_0}{2\alpha_s} - \sqrt{\left(\frac{v_0}{2\alpha_s} \right)^2 + k_x^2}. \qquad (10.16)$$

The solution describing the temperature field in the solid, Eq. (10.15), has the peculiar feature that at large negative values of Z—i.e., far behind the solid–liquid interface—the temperature becomes unbounded and negative! This unsettling aspect of the solid's temperature field can, fortunately, be disregarded. Inasmuch, as the temperature field in the solid phase remains of interest only near the solid-melt interface, its mathematical behavior at large negative Z-values is irrelevant to the present enquiry concerning the interface's stability.

In a real crystal growth system, of course, the temperature could neither rise nor fall exponentially for unlimited distances. The solid's temperature in real solidification and crystal growth processes usually just falls off toward the ambient temperature, whereas the melt temperature usually rises to some realistic limit set by conditions in the crystal growth furnace. The exponential solutions encountered here for these temperature fields should be accepted as accurate mathematical approximations to the behavior of the real fields only in the vicinity of the interface.

The functions representing the characteristic reciprocal distances for the exponential decay of disturbances in the transport solutions—viz., k_C, k_{T_ℓ}, and k_{T_s}—all exhibit similar behavior with respect to the interface speed, v_0, the wavenumber of the perturbation, k_x, and to their relevant diffusivities. Specifically, one finds in the high wavenumber limit, at sufficiently low interface velocities, v_0, Eqs. (10.12), (10.14), and (10.16) all reduce as,

$$k_C = k_{T_\ell} = -k_{T_s} \to k_x.$$

By contrast, in the so-called 'long-wave' or large-scale limit, where $k_x \to 0$, each of these functions reduce to the following limits:

- $k_C \to v_0/D_\ell$,
- $k_{T_\ell} \to v_0/\alpha_\ell$,
- $k_{T_s} \to 0$.

10.3.5 Local Interfacial Equilibrium

Local equilibrium requires, even at a perturbed—and, therefore, curved—interface, $\zeta(x)$, that the temperatures of the solid and liquid phases must be equal, so that $\hat{T}_s = \hat{T}_\ell = \hat{T}$. The temperature, $\hat{T}$, for the perturbed, and therefore curved, interface can be obtained by introducing the Gibbs–Thomson equilibrium condition developed in Chapter 8. The interface temperature may be set equal to the liquidus (or the solidus) temperature of the alloy, corrected for the influence of curvature, caused in this instance by the presence of the perturbation on the interface. One may arbitrarily select the liquidus temperature, expressed by the first two terms appearing in Eq. (10.17), and add the small capillary correction term derived earlier for the Gibbs–Thomson effect as Eq. (8.29).

$$\hat{T} = T_m + m_\ell \hat{C}_\ell - \frac{2\gamma_{s\ell}\Omega_s}{\Delta S_f}\mathcal{H}. \tag{10.17}$$

The presence of the perturbation causes the interfacial temperature to vary just slightly[3] as the mean curvature, $\mathcal{H}$, changes along the interface from point to point. In the case of a small-amplitude, two-dimensional sinusoidal perturbation along the x-direction, each mode imposes a low-amplitude 'cylindrical' shape undulation with a wavelength given by

$$\lambda = \frac{2\pi}{k_x}. \tag{10.18}$$

This type of perturbation may be described more appropriately as a 'ruled' surface, which has zero Gaussian curvature, $\mathcal{K} = \kappa_1\kappa_2$, because the smaller principal curvature equals zero. That is, the principal radius of curvature, $R(y)$, in the plane normal to the page of Fig. 10.1 is infinite at all positions, whereas the other radius, $R(x)$, in the plane parallel to the page varies with distance x along the trace of the interface, $\zeta(x)$. Specifically, the mean curvature, $\mathcal{H}$, for such a sinusoidally ruled surface is

$$\mathcal{H} = \frac{1}{2}\left(\frac{1}{R(x)} + \frac{1}{\infty}\right). \tag{10.19}$$

Substituting Eq. (10.19) into Eq. (10.17) yields an expression for the interfacial temperature function of the perturbed interface,

$$\hat{T} = T_m + m_\ell \hat{C}_\ell - \frac{\gamma_{s\ell}\Omega_s}{\Delta S_f}\frac{1}{R(x)}. \tag{10.20}$$

---

[3] Interface temperature variations induced by capillarity are generally less than about 10^{-2} K.

10.3.6 Linearization of the Curvature

The mean curvature along a planar interface perturbed by a single sinusoidal mode, $\zeta(x) = \delta \sin(k_x x)$, can be found by applying the standard differential calculus relationship for the reciprocal radius of curvature, $R(x)^{-1}$, of a smooth curve, $\zeta(x)$. The reciprocal radius of curvature, is the in-plane curvature, given in the x-Z coordinates shown in Fig. 10.1 as

$$\frac{1}{R(x)} = \frac{-\zeta(x)_{xx}}{\left[1 + (\zeta(x)_x)^2\right]^{\frac{3}{2}}}. \tag{10.21}$$

The single and double subscript notation used in the curvature formulas indicates levels of partial differentiation with respect to x. If the perturbation amplitude, δ, is assumed to be a small quantity compared to unity, then it also holds that the slope of the interface, $\zeta(x)_x = d\zeta/dx$, is also everywhere a small number, $\epsilon(x)$. Thus, the rippled interface has at *all* points a small slope, $\zeta(x)_x \ll 1$, that supports the following linearization of the interfacial curvature curvature given previously by Eq. (10.21):

$$\frac{1}{R(x)} = \frac{-\zeta(x)_{xx}}{\left[1 + (\zeta(x)_x)^2\right]^{\frac{3}{2}}} \approx -\zeta(x)_{xx}. \tag{10.22}$$

Substituting the sinusoidal form of the interface, $\zeta(x)$, into Eq. (10.22), then performing the differentiations indicated in Eq. (10.22), and reducing them to linear order, yields the mean curvature of the solid–liquid interface as

$$\frac{1}{R(x)} = \mathcal{H} = \delta k_x^2 \sin(k_x x) = k_x^2 \zeta(x). \tag{10.23}$$

Equation (10.23) shows, quite generally, that the curvatures of interfacial disturbances, such as sinusoidal ripples, increase quadratically with their wavenumber.

10.3.7 Interfacial Flux Balances

The fluxes of both heat and solute must be conserved through appropriate Stefan balances taken at the perturbed solid–liquid interface, $Z = \zeta(x)$.

The thermal flux balance at the interface is given by

$$\frac{\Delta H_f}{\Omega_s} \mathbf{n} \cdot \mathbf{v}(x) - k_s \frac{\partial T_s}{\partial Z} + k_\ell \frac{\partial T_\ell}{\partial Z} = 0 \quad [Z = \zeta(x)], \tag{10.24}$$

where the gradients in the solid and melt, appearing in Eq. (10.24), are evaluated from their respective thermal fields in each of the phases surrounding the solid–

liquid interface along the trace $Z = \zeta(x)$. The quantity $\mathbf{n} \cdot \mathbf{v}(x)$ is the local *normal* speed of the interface, which in general varies from point-to-point along the sine-wave.

The solutal flux balance at the moving interface may be written in similar form as,

$$\left(\hat{C}_s - \hat{C}_\ell\right) \mathbf{n} \cdot \mathbf{v}(x) - D_\ell \frac{\partial C_\ell}{\partial Z} = 0 \quad [Z = \zeta(x)]. \tag{10.25}$$

The solute balance formulated in Eq. (10.25) contains terms for macrosegregation occurring at the solid–liquid interface that do not include any correction for back-diffusion in the solid.

10.3.8 Solutions to the Field Equations

The general solutions Eqs. (10.11), (10.13) and (10.15) may be combined with the local equilibrium interface conditions and the Stefan mass and energy balances to determine the two unknown coefficients a and b. If, for convenience, the thermal conductivities of the solid and liquid are assumed to be equal, one finds, after a few straightforward steps of algebra, that

$$a = m_\ell b - \frac{T_m \gamma_{s\ell} \Omega_s}{\Delta H_f} k_x^2 = m_\ell b - T_m \Gamma k_x^2, \tag{10.26}$$

where the symbol $\Gamma \equiv \gamma_{s\ell} \Omega_s / \Delta H_f$ is the material's characteristic capillary length scale—typically about 1 nm for most materials, so that $\Gamma \approx 10^{-7}$ cm.

A lengthy expression for the constant b is, similarly, found to be

$$b = \frac{2\mathcal{G}' T_m \Gamma k_x^3 + \mathcal{G}' k_x (G_s + G_\ell)}{2 k_x m_\ell G_c + (G_s - G_\ell)\left[k_C - \frac{v_0}{D_\ell}(1 - k_0)\right]} +$$
$$\frac{\mathcal{G}'\left(k_C - \frac{v_0}{D_\ell}\right)(G_S - G_\ell)}{2 k_x m_\ell G_c + (G_s - G_\ell)\left[k_C - \frac{v_0}{D_\ell}(1 - k_0)\right]}. \tag{10.27}$$

The quantity $\mathcal{G}'$ appearing in Eq. (10.27) is the solute concentration gradient in the melt ahead of the interface, evaluated on the unperturbed base plane $Z = 0$. Note that $\mathcal{G}'$ was previously defined in Chapter 9, in Eq. (9.17), as the 'constitutional' solute gradient, namely,

$$\mathcal{G}' \equiv \frac{G^\star}{m_\ell} = -\frac{v_0}{D_\ell}\left(\frac{1 - k_0}{k_0}\right) C_0. \tag{10.28}$$

The differences in the thermal gradients in the solid and melt, respectively, *averaged* over the crystal-melt interface, $G_s - G_\ell$, may be determined by formulating a heat balance over one wavelength of the sine-wave ripple, λ. This global energy balance appears as the following expression that integrates the periodic latent heat effects along the x-direction on the interface. Recalling the simplifying assumption made that $K_s = K_\ell = K$, one finds that the difference in the average thermal gradients acting the solid and liquid leads to the simple expression,

$$G_s - G_\ell = \frac{1}{\lambda} \int_0^\lambda \frac{\Delta H_f}{\Omega_s k_x} (\mathbf{n} \cdot \mathbf{v}) dx = \frac{\Delta H_f}{\Omega_s k_x} v_0. \tag{10.29}$$

Equation (10.29) essentially relates the mean interfacial speed, v_0, in the $+Z$-direction, to the steady-state difference in the averaged thermal gradients. The explicit difference of the averaged thermal gradients surrounding the solid–liquid interface may be obtained by averaging the difference of the two gradient fields over one wavelength, λ. This connects the interface speed, v_0, to the average differences of the thermal gradients in the solid and liquid,

$$G_s - G_\ell = \frac{\Delta H_f}{\Omega_s k_x} v_0 = \frac{1}{\lambda} \int_0^\lambda \left[\frac{\partial T_s(x, z)}{\partial z} - \frac{\partial T_\ell(x, z)}{\partial z} \right] dx. \tag{10.30}$$

10.4 Stability Criteria

The *local* interface speed is defined as the mean speed, v_0, plus the time variation imposed by the evolving interface shape, so

$$v(x) = v_0 + \dot{\delta} \sin(k_x x), \tag{10.31}$$

where the symbol $\dot{\delta}$ denotes total time differentiation of the perturbation amplitude. Clearly, the behavior of the interface, regarding its stability, depends solely on the sign of its time derivative, $\dot{\delta}$.

$$\dot{\delta} \equiv \frac{d}{dt} \delta(t). \tag{10.32}$$

If the solutions for the temperature fields, Eqs. (10.13) and (10.15), are differentiated to obtain the interfacial thermal gradients, which are, in turn, substituted back into the kinematic equation for the interface, Eq. (10.32), along with the value for $\dot{\delta}$ found by differentiating the interface shape, $\zeta(x)$, one obtains after several algebraic steps

$$v(x) = \frac{\Omega_s K}{\Delta H_f} (G_s - G_\ell) + \frac{\Omega_s K}{\Delta H_f} k_x [2a - (G_s + G_\ell)] \delta(t) \sin k_x x. \tag{10.33}$$

Substituting Eq. (10.29) into Eq. (10.33) simplifies that expression for $v(x)$ as

$$v(x) = v_0 + \frac{\Omega_s K}{\Delta H_f} k_x \left[2a - (G_s + G_\ell)\right] \delta(t) \sin k_x x. \tag{10.34}$$

Substituting Eq. (10.31) into the LHS of Eq. (10.34), and canceling terms, shows that the time-rate of change of the amplitude of the sinusoidal disturbance is given by

$$\dot{\delta} = \delta(t) \left(\frac{\Omega_s K}{\Delta H_f}\right) k_x \left[a - \frac{1}{2}(G_s + G_\ell)\right]. \tag{10.35}$$

10.4.1 Amplitude Evolution

The evolution of a perturbation mode may be written in product function form by separating the time and wavenumber dependences, namely,

$$\dot{\delta} = \delta(t) \cdot \mathcal{F}(k_x), \tag{10.36}$$

where the wavenumber and the material constants all reside in the function $\mathcal{F}(k_x)$, and the perturbation amplitude, $\delta(t)$ expresses the time dependence. It must be noted that $\mathcal{F}(k_x)$ can be a complex function, containing both real, $\Re e[\mathcal{F}(k_x)]$, and imaginary, $\Im m[\mathcal{F}(k_x)]$ parts. The central result of linear perturbation analysis as described by Eq. (10.36), is that the time rate of change of the disturbance amplitude takes the simple *linear* form $\delta(t) \cdot \mathcal{F}(k_x)$. This result implies an equivalent simple dynamical expression for the rate of growth (or decay) of a single-mode sinusoidal ripple of wavenumber, k_x. Equation (10.36) may be written as the ODE for the evolution of the perturbation,

$$\frac{\dot{\delta}}{\delta(t)} = \mathcal{F}(k_x), \tag{10.37}$$

and then integrated through time as

$$\int_{\delta_0}^{\delta} \frac{d\delta}{\delta} = \int_0^t \mathcal{F}(k_x) dt. \tag{10.38}$$

Carrying out the integrations indicated in Eq. (10.38) yields a exponential time expression for the amplitude evolution of a sinusoidal ripple with wavenumber, k_x, namely,

$$\delta = \delta_0 e^{\mathcal{F}(k_x)t} = \delta_0 \left(e^{\Re e[\mathcal{F}(k_x)]t} \cdot e^{i \Im m[\mathcal{F}(k_x)]t}\right), \tag{10.39}$$

where in Eq. (10.39) the real and imaginary parts of $\mathcal{F}(k_x)$ are separated into individual exponentials: the first giving the tendency for the perturbation to either grow or decay, depending on the sign of the $\Re e[\mathcal{F}(k_x)]$, and the second allowing for oscillations of the disturbance over time. The important short-term behaviors predicted from Eq. (10.39) for an interface perturbed by a single sine-mode disturbance may be deduced as follows:

- *Decay of the perturbation amplitude*: Stability occurs if $\Re e[\mathcal{F}(k_x)] < 0$.

- *Growth of the perturbation amplitude*: Instability occurs if $\Re e[\mathcal{F}(k_x)] > 0$.

- *Oscillation of the perturbation*: Running waves develop if $\Im m[\mathcal{F}(k_x)] \neq 0$ and $\Re e[\mathcal{F}(k_x)] = 0$.

- *Time-independent (neutral state) behavior of the perturbation*: 'Marginal' stability occurs when $\Re e[\mathcal{F}(k_x)] = \Im m[\mathcal{F}(k_x)] = 0$.

Substituting the values for the constants a and b in the transport solutions given in Eqs. (10.26) and (10.27) into the expression for the amplitude function, $\mathcal{F}(k_x)$, gives the result:

$$\frac{\dot{\delta}}{\delta} = \mathcal{F}(k_x) = \frac{\mathcal{N}(k_x)}{\mathcal{D}(k_x)}, \tag{10.40}$$

where the numerator of Eq. (10.40), $\mathcal{N}(k_x)$, is equal to

$$\mathcal{N}(k_x) = -T_m \Gamma k_x^2 - \frac{G_s + G_\ell}{2} - m_l C_0 \frac{v_0}{D_\ell} \frac{1 - k_0}{k_0} \left(\frac{k_C - \frac{v_0}{D_\ell}}{k_C - \frac{v_0}{D_\ell}(1 - k_0)} \right), \tag{10.41}$$

and the denominator of Eq. (10.40), $\mathcal{D}(k_x)$, is given by

$$\mathcal{D}(k_x) = \frac{\Delta H_f}{2 K \Omega_s k_x} - \frac{m_\ell C_0 \frac{v_0}{D_\ell} \frac{1 - k_0}{k_0}}{k_C - \frac{v_0}{D_\ell}}. \tag{10.42}$$

10.4.2 Criteria for Interfacial Stability

Clearly, the gist of linear stability analysis is to determine the sign of $\mathcal{F}(k_x)$, specifically the sign of its real part. Both of the terms that appear in the denominator of Eq. (10.40), $\mathcal{D}(k_x)$, are *always* positive, as confirmed by inspection of each term in Eq. (10.42). Hence, it becomes clear that the sign of $\mathcal{F}(k)$ depends only on the sign of its numerator, $\mathcal{N}(k)$, given by Eq. (10.41). The numerator contains three terms, each arising from a different physical effect occurring at the solid–liquid interface. Specifically, the numerator consists of the following terms:

1. *Capillarity*: The first term on the RHS of Eq. (10.41), representing capillarity, i.e., curvature of the rippled interface, is *always* negative, and increases quadratically with increasing wavenumber. Thus, for any sinusoidal ripple, $-T_m \Gamma k_x^2 < 0$. The negative contribution of the capillary term contributes to

interfacial stability, especially at large wavenumbers, or small length scales of the perturbation.

2. *Thermal gradient*: As long as the melt remains hotter than the solid, the sign of the second term on the RHS of Eq. (10.41), which is associated with the average thermal gradient at the solid-interface, is also *always* negative. Consequently, $-(G_s + G_\ell)/2 < 0$. This term is proportional to the magnitude of the thermal gradient, and also tends to stabilize the interface independent of the disturbance's wavenumber or length scale.

3. *Solute diffusion*: Solute rejection, or incorporation, at the interface depends on whether or not the distribution coefficient $k_0 < 1$, or $k_0 > 1$. Irrespective of whether the macrosegregation results in solute rejection from, or inclusion into, the solid, this diffusion term is *always* positive, and destabilizing, so

$$-m_l C_0 \frac{v_0}{D_\ell} \frac{1-k_0}{k_0} \left(\frac{k_C - \frac{v_0}{D\ell}}{k_C - \frac{v_0}{D_\ell}(1-k_0)} \right) > 0.$$

Although the detailed dependence of the magnitude of the diffusion term on its various factors seems complicated, it increases linearly with alloy concentration. Thus, as a rough estimate, concentrated alloys tend to freeze with unstable solid–liquid interfaces more than do dilute alloys or nominally pure materials. The reader may recall that this is the same conclusion reached with the theory of constitutional supercooling.

The algebraic total of all three terms in appearing in the numerator of $\mathcal{F}(k_x)$ is, of course, what determines its algebraic sign, and, therefore, adjudicates the dynamical stability of the solid–liquid interface.

Dynamical stability of a moving solid–liquid interface is clearly a multifaceted issue when capillarity is considered. Constitutional supercooling, developed earlier in Chapter 9, will be shown to be recoverable from dynamical stability theory in the limit of $k_x \to 0$. Thus, for disturbances acting over long length scales, capillarity effects weaken, and conclusions based on the far simpler criterion of constitutional supercooling suffice. Readers interested in a more detailed article on the influence of wavenumber on the interface instability patterns should see [7].

10.4.3 Marginal Stability

As briefly mentioned in Section 10.4.1, an interface will exhibit neutral, or 'marginal' stability provided that the real part of $\mathcal{F}(k_x) \leq 0$ for *all* disturbances. That is, one must test the sign of the numerator, $\mathcal{N}(k_x)$, over the entire range of wavenumbers $0 \leq k_x \leq \infty$.

Marginal stability, the intermediate condition between exponential growth and exponential decay, conveniently separates stable from unstable interfacial states. The marginally stable state itself is defined by the criterion that three distinct physical effects acting at the interface cancel: (1) The capillary term, $-T_m \Gamma k_x^2$, which is always negative and stabilizing; (2) The average applied thermal gradient, $-(G_s + G_\ell)/2$, which is always negative and stabilizing; and (3) The solute diffusion term, which is always positive, and destabilizing. Thus, marginal stability requires that

$$\mathcal{F}(k_x) = 0 = -T_m \Gamma k_x^2 - \frac{G_s + G_\ell}{2} - m_l C_0 \frac{v_0}{D_\ell} \frac{1 - k_0}{k_0} \left(\frac{k_C - \frac{v_0}{D\ell}}{k_C - \frac{v_0}{D_\ell}(1 - k_0)} \right).$$

(10.43)

The critical condition that $\mathcal{F}(k_x) = 0$ prevents any perturbation present on the interface from either growing or decaying. This delicate requirement will be used later in Chapter 13 when discussing the growth and associated length scales of microstructural features such as cells and dendrites. Equation (10.43) provides a good example of a kinetically-based mathematical selection criterion for initially estimating the length scales of features in cast microstructures. Such selection criteria consist of the fundamental materials and process parameters that interact during solidification and allow either a featureless stable interface to grow, or the evolution of an unstable, complex freezing pattern, such as were uncovered originally by Chalmers' decanting experiments [8], and discussed previously in Chapter 9.

Despite the careful mathematical analysis employed here, and the sophistication of its ingredients, linear morphological theory only discriminates between stability and instability. Its predictive limitations lie in part in the small slope, small amplitude, small curvature approximations for the interfacial disturbances. Consequently, linear theory is capable of predicting neither the shapes nor the definitive length scales of any unstable interfacial features such as shown in Fig. 9.2. Non-linear analyses of stability, as will be surveyed briefly in the next chapter, along with companion numerical studies reviewed in detail by Davis [9], are needed to obtain more definitive information on the evolution and behavior of 'mature' interfacial instabilities.

If Eq. (10.43) is taken to the limit of long wavelength disturbances (i.e., ripples with small wavenumbers), and the interface is assumed to travel at slow speeds (so that latent heat production is negligible compared to the heat flux conducted through the solid–liquid interface) then $G_s \approx G_\ell \equiv G$, and Eq. (10.43) reduces to

$$\lim_{k_x \to 0} F(k_x) = 0 = -G - m_l C_0 \frac{v_0}{D_\ell} \frac{1 - k_0}{k_0} \times \left(\frac{k_C - \frac{v_0}{D\ell}}{k_C - \frac{v_0}{D_\ell}(1 - k_0)} \right)$$

(10.44)

By using the definition of k_C from Eq. (10.12), one can show that at slow freezing speeds the diffusion length ahead of the solid–liquid interface is typically much larger than the wavelength of the perturbation. Thus, one usually finds that $D_\ell/v_0 \gg \lambda$. The combination of large diffusion lengths relative to the wavelength of the disturbances further implies that $k_C \gg v_0/D_\ell$. Under this approximation, the term off-set in parentheses in the marginal stability criterion, Eq. (10.44), becomes ≈ 1. The condition of marginal stability reduces identically to the form derived from constitutional supercooling theory in Chapter 9, namely,

$$\mathcal{F}(k_x) = 0 = -G - m_\ell C_0 \frac{v_0}{D_\ell}\left(\frac{1-k_0}{k_0}\right). \tag{10.45}$$

10.5 Low Wavenumber Limit

Now recalling the definition of the critical constitutional gradient, $G^\star$, (See Eq. (9.17)). one sees that this quantity is equal, but opposite, to the second term on the right-hand side of Eq. (10.45). Moreover, the critical constitutional gradient is just $G^\star = -m_\ell \mathcal{G}'$, where $\mathcal{G}'$ is the constitutional solute gradient that reappears in the more general dynamical equations, Eqs. (10.27) and (10.28).

It is interesting to note that dynamical stability, derived here by using the more rigorous mathematics of linear stability, reduces in the limit of long-wavelength disturbances to the simpler condition found through the principles of constitutional supercooling. This is indeed a gratifying result, inasmuch as it teaches that constitutional supercooling is the correct limiting form of the more general, and extensible, treatment afforded by linear stability analysis. Thus, when curvature effects are relatively unimportant, and freezing speeds are sufficiently slow, the stability criteria based on constitutional supercooling may be used to assess interfacial stability during crystal growth as a reliable and controlled approximation.

10.6 General Stability Analysis

Solidification and crystal growth are often carried out under more general conditions where the speed of freezing might neither be slow, nor capillarity effects be neglected. Under these broader circumstances, the full expression of dynamical stability, given in Eq. (10.43), must be considered. Once again, only the numerator, $\mathcal{N}(k_x)$, in Eq. (10.40) need be considered to determine the sign of $\mathcal{F}(k_x)$, because its denominator, $\mathcal{D}(k_x)$, remains positive under any and all solidification circumstances.

Equation (10.43) may be written more compactly as

$$\mathcal{F}(k_x) = 0 = -T_m \Gamma k_x^2 - \bar{G} - m_\ell(g(k_x) + 1)\mathcal{G}', \tag{10.46}$$

where the *average* thermal gradient applied at the solid–liquid interface is defined as $\bar{G} \equiv (G_s + G_\ell)/2$, and where the function, $g(k_x)$, is introduced to capture all the wavenumber dependence of the diffusion term, namely,

$$g(k_x) \equiv \frac{k_C - \frac{v_0}{D_\ell}}{k_C - \frac{v_0}{D_\ell}(1 - k_0)} - 1. \tag{10.47}$$

The function $g(k_x)$ is plotted in Fig. 10.2 as a function of the wavenumber of the perturbation. Here, for illustrative purposes, the distribution coefficient $k_0 = 0.1$ and the diffusion length is $v_0/D_\ell = 102$ cm^{-1}. Note that for these choices $k_x \ll v_0/2D_\ell$ (in this case, $k_x \leq 10$ cm^{-1}, so $\lambda \geq 6$ mm) then $g(k_x)$ remains near its lower limit, viz., $g(k_x) \to -1$, and does not change significantly for lower wavenumbers (longer wavelengths). By contrast, if $k_x \gg v_0/2D_\ell$, (in this case $k_x \geq 102$ cm^{-1}, so $\lambda \leq 60$ μm) then $g(k_x)$ approaches its upper limit, viz., $g(k_x) \to 0$, but changes only slightly for larger wavenumbers and shorter wavelengths. The $g(k_x)$ function, although always less than or equal to zero, reflects the destabilizing influences of solute diffusion as it swings between -1 and 0 with increasing wavenumber (decreasing wavelength) of the perturbing ripple.

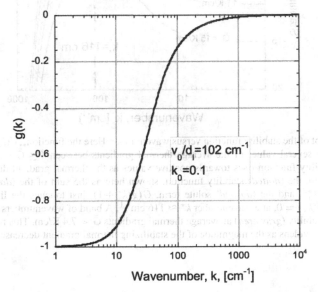

Fig. 10.2 Plot of $g(k_x)$ versus wavenumber. This function is always negative, with its value between the limits -1 and 0. At both large and small wavenumbers, $g(k_x)$ becomes insensitive to the wavenumber

To observe how each of these interfacial contributions affects the stability criterion, one selects the following typical set of additional solidification parameters: $m_\ell = -10$ K/wt%, $C_0 = 0.001$ wt%, and $T_m \Gamma = 5 \times 10^{-5}$ K/cm. A series of positive average thermal gradients is chosen for the purposes of calculating the stability functions: specifically, $G = 4, 6, 7.4, 11,$ and 16 K/cm. Figure 10.3 also shows how all the terms in Eq. (10.46) contribute to the overall behavior of the function $\mathcal{N}(k_x^*)$, Eq. (10.46), and, thereby, determine the sign of the interfacial stability function, $\mathcal{F}(k_x)$. Note how the interface is stabilized against *all* perturbations for thermal gradients greater than about 7.4 K/cm, because at higher thermal gradients $\mathcal{N}(k_x)$ remains negative for all perturbation wavenumbers. For slightly smaller thermal gradients than 7.4 K/cm, however, a small band of wavenumbers appears near k_x^* within the unstable (gray) region, where $\mathcal{N}(k_x) > 0$. The band of unstable wavenumbers spreads out even further as the applied thermal gradient decreases, further destabilizing the interface to perturbations.

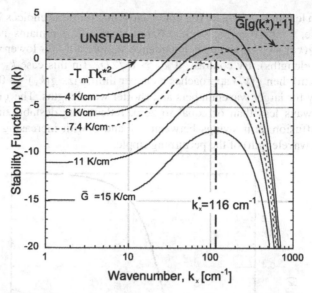

Fig. 10.3 Plot of the stability function versus wavenumber. Here the function, $\mathcal{N}(k_x)$, Eq. (10.46), is plotted for several values of the average thermal gradients between $4 \le \bar{G} \le 15$ K/cm. Note how the stability function rises towards positive values as the thermal gradient decreases. When $\bar{G} \approx 7.4$ K/cm, the (*dotted*) stability function, shown here as the sum of the (*dashed*) capillary term, $-T_m \Gamma k_x^{*2}$, and the (*dashed*) solute term, $\bar{G}(g(k^* + 1)$, just touches the line of marginal instability, $\mathcal{N}(k_x) = 0$, at a wavenumber $k^* \approx 116$ cm^{-1}. A band of wavenumbers spread into the region of instability (*gray area*) at average thermal gradients $\bar{G} < 7.4$ K/cm. This band of unstable wavenumbers widens as the magnitude of the stabilizing thermal gradient decreases further

10.6.1 Marginal Wavenumber

Inspection of Fig. 10.3 shows that as the thermal gradient decreases, keeping fixed values for the interface velocity and solute concentration, the $\mathcal{N}(k_x)$ function rises, eventually becoming zero. The tangency condition occurring at $\mathcal{N}(k_x) = 0$ denotes the 'margin of stability'. Finally, the discriminator, $\mathcal{N}(k_x)$, then becomes positive, and by doing so, crosses the stability boundary. For the case presented in Fig. 10.3, perturbations with a wavenumber of $k_x^* \approx 116$ cm^{-1} become marginally stable at an average thermal gradient of about 7.4 K/cm. These interfacial ripples have a critical wavelength of $\lambda^* \approx 90$ μm, or about 0.1 mm. Such disturbances would in principle persist, rather than decay back to a plane or grow exponentially into cells.

When $\mathcal{N}(k_x) = 0$, the marginal wavenumber, k_x^*, can be expressed implicitly through the following relationship,

$$k_x^* = \frac{2\pi}{\lambda^*} = \sqrt{\frac{-m_\ell \mathcal{G}[g(k_x^*) + 1] - \bar{G}}{T_m \Gamma}}. \tag{10.48}$$

Equation (10.48) can, of course, be solved numerically. However, this expression can be simplified under certain conditions of solidification, and such simplifications will be employed later in Chapter 13 when the behavior of dendritic structures is studied.

10.7 Experimental Validation

Obtaining quantitative measurements of the initially unstable modes evolving on an interface under conditions close to marginal instability has proven challenging. Generally, the experimental problems encountered in observing interfacial instabilities are that unstable interfacial behavior is associated with, and usually initiated near, persistent, locally strong interfacial disturbances caused by grain boundary grooves,[4] contact menisci, and dislocations. See also, for example, the variety of in situ photomicrographs of crystal-melt instabilities shown in Chapter 9. In addition, measurements of the marginal wavenumber are made difficult by the rapid (exponential) growth of unstable interfacial disturbances, which can cause initially unresolved interface perturbations to pass quickly through the range of amplitudes where linear stability actually holds. The ability to correlate the onset of such miniscule instabilities under well-controlled solidification conditions, as mentioned earlier, is limited to circumstances where the amplitude, $\delta(t) \ll \lambda$. Consequently, there is usually little observation time available for collecting valid data to compare with theory once an instability is triggered.

Nevertheless, several investigators have found some quantitative agreement with theory. For example, checks on the critical wavenumber performed in slightly impure CBr$_4$ by de Cheveigné et al. [10] showed agreement with linear theory within a factor of 2. More recent, carefully designed experiments were conducted by Losert et al. [11], who also found what they considered as 'good agreement' with a non-linear version of stability theory by Warren and Langer [12, 13]. Losert et al. used as their test alloy, succinonitrile (SCN), which crystallizes as a BCC crystal, containing traces of coumarin 152 (CM152), a fluorescent dye that allowed direct optical measurement of the solute diffusion fields. Figure 10.4 a-d, shows a series of micrographs taken of this alloy during steady directional freezing at increasing rates in a thin solidification cell. One observes the progressive development of low-amplitude ripples, which are clearly not pure sine waves, but contain higher harmonics determining their shape. Careful scrutiny of Fig.10.4b shows that the 'valleys' are curved more sharply than are the peaks. Moreover, the wave length of the ripple varies slightly across the field of view, with the central region displaying a smaller wavelength than the outer regions of this micrograph. At higher growth speeds, Fig. 10.4c the ripples have amplified into deep finger-like cells. Figure 10.4d, observed at the highest growth speed, shows the interface fully evolved into aligned dendrites. The cell-to-dendrite transition is of considerable interest in

---

[4] See Part V, Appendix B for additional details.

interpreting cast microstructures of dilute alloys, but its explanation remains enigmatic. That topic will be appear again in Chapter 13.

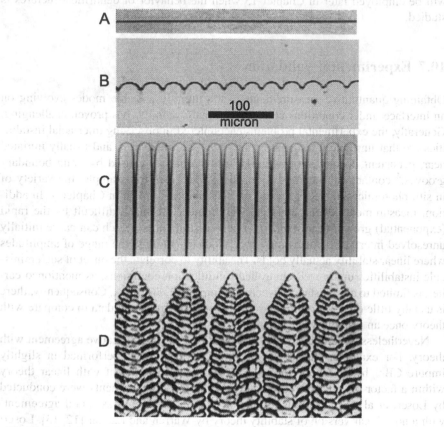

Fig. 10.4 Steady-state solidification fronts of dilute SCN-CM152 alloy advancing at increasing growth rates from a→d. **a** Stable plane-front growth; **b** Periodic ripple, with higher harmonic content; **c** Deep cells; **d** Aligned dendrites. Adapted from [11]

10.8 Absolute Stability

An unexpected, and interesting transition is predicted from linear theory to occur in the morphological stability of solid–liquid interfaces. This transition occurs only at high interfacial speeds—well beyond those used in Fig. 10.4. Theory suggests that the wavenumbers for marginal stability, k^*, will be driven toward increasingly large values when the growth speed achieves a sufficiently high value. At higher wavenumbers the wavelength and ripple radii decrease, and, eventually, stabilizing effects caused by capillarity become sufficiently large (recall that stabilization against a perturbation increases quadratically with wavenumber as k_x^2,

Eq. (10.46)) that the interface—provided that it is driven fast enough—re-stabilizes! This remarkable re-stabilization at high interfacial speeds is termed 'absolute stability'. Absolute stability was a surprise when it was first uncovered in early analyses of linear morphological stability, as the phenomenon had neither been reported experimentally, nor had it been predicted from constitutional supercooling theory. It is a capillarity-induced feature of dynamic interface stability.

Constitutional supercooling, a model of interface stability lacking capillarity effects, as mentioned, is incapable of predicting this interesting, somewhat counterintuitive effect at high freezing speeds. Figure 10.5 shows in detail the behavior of the $\mathcal{N}(k_x)$ function, plotted here for the same parameters as were used in Fig. 10.3, except for the interface speeds, v_0, which are increased in Fig. 10.5 by factors of several hundred—reaching the range of several millimeter per second . Such high solidification speeds are perhaps encountered in the early stages of casting alloys in a cooled mold, and in welding, but they would seldom, if ever, be used for purposes of bulk crystal growth. The average thermal gradient, $\langle G \rangle$, chosen in plotting Fig. 10.5 was held constant at 15 K/cm for each curve. Note that the solid–liquid interface in this case achieves absolute stability for solidification speeds above ≈ 0.07 cm/s. The marginal wavenumber at absolute stability for these higher growth rates is $k_x^\star \approx 1,550$ cm^{-1}, which corresponds to interfacial ripples with a wavelength of only $\lambda^\star \approx 40$ μm. Note that the marginal wavenumber, $k_x^\star$, calculated at absolute stability is almost 10 times larger than it was in the same alloy described earlier

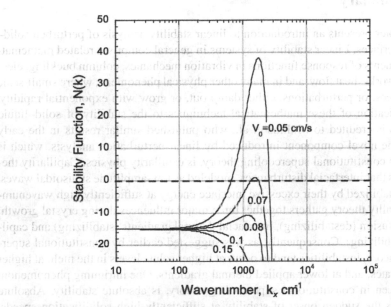

Fig. 10.5 Behavior of $\mathcal{N}(k_x)$ near the transition to absolute stability. In this example, when the interface speed, v_0, *exceeds* approximately 0.07 cm/s, all wavenumbers are stabilized because $\mathcal{N}(k_x) < 0$, and the interface remains planar. Absolute stability is often difficult to achieve in practice, because of the relatively high freezing speeds required

in Section 10.6.1, where for morphological instability to occur at relatively low solidification speeds, $k_x^\star \approx 116\,\text{cm}^{-1}$ and $\lambda^\star \approx 0.5$ mm.

Solidification of supercooled melts conveniently provides sufficiently high solidification rates, but such freezing processes lead only to morphologically unstable interfaces, because the average thermal gradient, $\langle G \rangle$, is *negative* whenever solidification occurs in melt that is supercooled. Consequently, achieving freezing rates of even a few millimeters per second with a *positive* thermal gradient proves technically challenging [14].

Modern processing equipment is required in order to induce absolute stability in an alloy. However, when absolute stability is achieved, then solute rejection at the interface leads to the maintenance during freezing of *planar* interfaces experiencing uniform, directional macrosegregation. *Microsegregation*, with its attendant solidification defects and localized chemical inhomogeneities is effectively eliminated under conditions of absolute stability. Absolute stability has, in fact, now been successfully and routinely achieved in cases of rapid surface melting using high-powered optical lasers and focused electron beams to sweep over and melt thin superficial layers on materials (usually metals) that have high enough thermal conductivity to permit sufficiently rapid re-solidification [15].

10.9 Summary

This chapter presents an introduction to linear stability analysis of perturbed solid–liquid interfaces. Linear stability of systems in general comprises related mathematical treatments of response functions in vibration mechanics, column buckling, electrical networks, heat flow, and in many other physical phenomena, where small-scale disturbances, or perturbations, either damp out, or grow with exponential rapidity. The application of these mathematical techniques to the stability of solid–liquid interfaces is credited to Mullins et al., who published similar results in the early 1960s. The novel component introduced by linear perturbation analysis, which is lacking in constitutional supercooling theory, is capillarity physics. Capillarity theory shows that interfacial disturbances, modeled as low-amplitude sinusoidal waves, become stabilized by their excessive interface energy at sufficiently high wavenumbers. Stability theory gathers together three major influences during crystal growth: solute diffusion (destabilizing), interfacial thermal gradients (stabilizing) and capillarity (stabilizing). Consequently, as was suggested earlier by constitutional supercooling theory, instabilities tend to occur at higher solute levels in the melt, at higher freezing rates, and at lower applied thermal gradients. One intriguing phenomenon, not foreseen in constitutional supercooling theory, is absolute stability. Absolute stability is the sudden onset of stability at sufficiently high solidification speeds, provided that the average gradient remains positive. Achieving absolute stability in practice is difficult but possible, using directed energy sources such as lasers and electron beams on metallic surfaces.

Tests of linear stability theory show that all these qualitative aspects are borne out. A fundamental difficulty in quantitatively checking specific predictions derived from linear morphological theory is that interfacial amplitudes *must* remain small, despite the onset of exponential amplification in time. These factors conspire to reduce the available time-scale over which linearity is retained well enough to make a meaningful experimental test. For these reasons, non-linear theories of interface stability have been proposed that do provide testable predictions. Some numerical results obtained for these non-linear approaches are surveyed in the following chapter.

References

1. W.W. Mullins and R.F. Sekerka, *J. Appl. Phys.*, **34** (1963) 323.
2. W.W. Mullins and R.F. Sekerka, *J. Appl. Phys.*, **35** (1964) 444.
3. V.V. Voronkov, *Sov. Phys. Solid State*, **6** (1964) 2278.
4. S.H. Davis, *Theory of Solidification*, Cambridge Monographs on Mechanics, Cambridge University Press, Cambridge, UK, 2001.
5. J.S. Langer, *Rev. Mod. Phys.*, **52** (1980) 1.
6. G.J. Merchant, R.J. Braun, K. Brattkus and S.H. Davis, *SIAM J. Appl. Math.*, **52** (1992) 1279.
7. P. Metzener, 'Long-wave and short-wave oscillatory patterns in rapid directional solidification', *Interactive Dynamics of Convection and Solidification*, P. Ehrhard, D.S. Riley and P.H. Steen, Eds., Kluwer Academic Publishers, Dordrecht, 2001, p. 13.
8. B. Chalmers, *Principles of Solidification*, Chap. 8, 253, Wiley, New York, NY, 1964.
9. D. Riley and S.H. Davis, *IMAJ. Appl. Math.*, **45** (1990) 267.
10. S. de Cheveigné, C. Guthmann and M.-M Lebrun, *J. Cryst. Growth*, **73** (1985) 242.
11. W. Losert, B.Q. Shi and H.Z. Cummins, *Proc. Natl. Acad. Sci. USA*, **95** (1998) 431.
12. J.A. Warren and J.S. Langer, *Phys. Rev. A*, **42** (1990) 3518.
13. J.A. Warren and J.S. Langer, *Phys. Rev. E*, **47** (1993) 2702.
14. D.A. Huntley and S.H. Davis, *Acta Metall. Mater.*, **41** (1993) 2025.
15. M. Gremaud, M. Carrard and W. Kurz, *Acta Metall. Mater.*, **38** (1991) 2587.

Chapter 11
Non-linear Stability Models

11.1 Limitations of Linear Stability

The linear theory of morphological stability, as shown in Chapter 10, predicts neither the explicit form nor the evolution dynamics of an unstable solid–liquid interface. Instead, linear theory predicts only the system's *initial* unstable behavior—viz., evolution through exponential growth or decay—of a single-mode interface perturbation in the form of a sine-wave ripple. The marginal stability criterion associates the most 'dangerous' wavenumber, k^*, with its subsequent time dependence, $\delta_0 \exp(\mathcal{F}(k))$, which depends solely on the sign of the stability function, $\mathcal{F}(k)$. In sum, any *linearized* theory of morphological stability will prove to be incapable of predicting the kinetic details or the morphological specifics of an evolving interfacial disturbance. More specifically, linear analysis cannot predict the form, amplitude, or long-term time dependence of an initially unstable interface.

One must instead extend the treatment of dynamical stability to encompass non-linear effects to learn more about these interesting and important details of alloy solidification, where the amplitude and slopes of the solid-melt interface do not necessarily remain small. Studies by Warren and Langer helped establish key aspects of the physics of non-linear interface stability [1, 2]. Nonlinear analyses of interface instability were reviewed extensively in a highly mathematical monograph by Davis et al. [3]. These analyses include detailed predictions of stability and pattern formation, and require up to third-order terms appearing in the expansion of the system's amplitude equation. Numerical studies based on such analyses are best used to answer questions related to specific technical issues of non-linear morphological evolution and dynamical stability that extend well beyond the scope of this textbook.

A few non-linear two-dimensional calculations will now be discussed that address the evolution of higher-amplitude interfacial disturbances, which often appear when solidification occurs under conditions beyond the margin of stability. As shall be shown, the resulting interfacial instabilities are predicted to develop into forms covering the taxonomy of solidification structures that closely approximate the formation of the cellular grooves and all the patterns already described in Chapter 9. Those interfacial patterns were observed 'in vivo' as evolving unstable interfacial structures in dilute transparent organic alloys [4], as well as 'post mortem' through decanting studies in metals [5, 6].

M.E. Glicksman, *Principles of Solidification*, DOI 10.1007/978-1-4419-7344-3_11,

11.2 Interfacial Patterns Just Beyond Marginal Stability

McFadden and Coriell employed advanced numerical methods to calculate the morphological shapes of two-dimensional sinusoidal disturbances, as the stability function, $\mathcal{F}(k)$, increased slightly beyond the margin of stability, where $\mathcal{F}(k^\star) \geq 0$. In fact, such calculations can be run long enough to evolve the interface shape and even ascertain the approximate, steady-state, nonplanar forms into which these unstable interfaces evolve. Steady-state unstable interfacial forms *cannot* be calculated from linear perturbation theory, which, per force, is strictly limited to extremely low-amplitude ripples exhibiting small physical slopes. In typical numerical simulations of interfacial evolution, where many independent parameters might be operative during freezing, computed unstable shapes, per force, become limited to just a specific alloy system at a chosen solute composition, C_0. McFadden and Coriell studied binary Al-Ag alloys freezing at rates and compositions just slightly above their marginally stable limit. Their calculations have also been extended to 3-dimensions by Boisvert [7].

Figure 11.1 shows the computed profiles of a highly dilute aluminum-silver binary alloy interface freezing at steady-state. The system is solidifying slightly beyond the limit of marginal stability, where $\mathcal{F}(k) = +0.5\%$. The specific parameters of the alloy interface are: $k_0 = 0.4$, $\lambda = 6 \times 10^{-3}$cm, $\langle G \rangle = 100$ K/cm. Note the relatively steep, high-amplitude interface profile, $\zeta(X)$, and the corresponding (normalized) solute segregation profile in the solid, $C_s(X)/C_0$. The interface shape and solute profile still tend to 'mirror' each other's form—that is, the peaks and valleys of the segregation profile closely match the valleys and peaks of the interface profile to within some vertical scale, or magnification factor. The segregation profile computed shows approximately a twofold variation in the solute concentration from that at the cell tips to that at the groove centers. The average macrosegregation levels predicted from Gulliver–Scheil theory, based on a total steady-state mass balance, are also included on Fig. 11.1. The Gulliver–Scheil predictions indicate the steady-state average concentration ratio expected across the entire planar interface.

11.3 Patterns Further from the Margin of Stability

The cell walls shown in Fig. 11.2 have already developed extremely steep sides under freezing conditions in this alloy that are substantially beyond the limit of marginal stability. Note that in Fig. 11.2 the predicted interface shape, $\zeta(x)$, no longer appears to be sinusoidal in shape. Steep walls surround narrow, relatively deep, 'trailing cells', permit *lateral* segregation to occur, which locally traps high concentrations of solute. In actual solidification processes, such regions are subject to the formation of non-equilibrium phases.

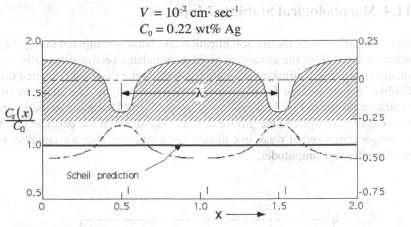

Fig. 11.1 Two-dimensional steady-state interfacial forms on the solid-melt interface of the dilute alloy, Al-0.22 wt%Ag, growing at a speed of 100 μm/s, which is slightly (+0.5%) above the limit for marginal stability, where $\mathcal{F}(k^*) = 0$. The Gulliver-Scheil average macrosegregation level is shown as the *solid line* at $C_s(X)/C_0 = 1$, whereas the steady-state composition oscillates about the Gulliver-Scheil average. The steep-walled grooves that develop allow lateral solute segregation, and account for the peak-and-valley segregation pattern. If the pattern observed experimentally in Fig. 10.4b is compared to the computed (*shaded*) shape shown here, a pleasing similarity may be noted in their higher-harmonic content. Both patterns are for steady-state unstable interfaces just beyond their respective margins of stability

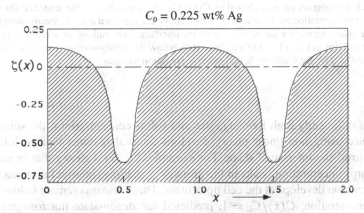

Fig. 11.2 Steady-state interface profile of the solid–liquid interface computed for Al-0.255wt% Ag alloy, 13% beyond the limit of marginal stability. The cell tips in the pattern, $\zeta(X)$, are more round compared to the morphology in Fig. 11.1, whereas the grooves are narrower and more steeply sloped. The general appearance of this unstable interface no longer resembles the initial low-amplitude sine-wave ripple. The opportunity for lateral segregation to occur in the deep trailing grooves greatly increases the solute concentration there, eventually allowing precipitates and other non-equilibrium constituents, such as eutectic, to form upon further cooling

11.4 Morphological Stability Maps

The segregation curve for this solidification condition is computed in Fig. 11.3. The solute rejected from the advancing solid accumulates between the cells, reaching steady-state at concentration values almost 40% higher than that predicted using the Gulliver–Scheil average for plane-front macrosegregation. The shape of the concentration profile in Fig. 11.3 exhibits almost a fivefold variation in the solute level between the cell tips and the groove centers. The shape of the periodic interface is no longer comprised of weak sine modes, but contains higher wavenumber Fourier modes with large amplitudes.

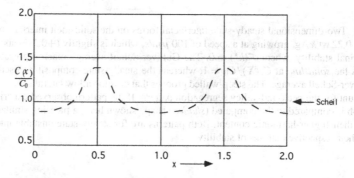

Fig. 11.3 Segregation profile, plotted as $C(x)/C_0$, corresponding to the interface shown in Fig. 11.2. The concentration peaks now extend 40% above the expected average steady-state macrosegregation value, and align with the grooves on the interface. The 'valleys' of segregation curve align with the cell tips in Fig. 11.2, and are about 10% below the steady-state average macrosegregation level, predicted by the Gulliver–Scheil macrosegregation analysis

At a sufficiently high freezing rates and solute concentrations, the solute profile computed using non-linear theory develops unusual features that no longer simply 'mirror' the interfacial shape. For example, Fig. 11.4 shows that in addition to the sharp concentration peaks in the grooves, a shallow (relative) maxima in solute concentration develops at the cell tip centers. These maxima remain below the average concentration, $C(x)/C_0 = 1$, predicted for steady-state macrosegregation by Gulliver–Scheil theory. The segregation profile clearly has evolved into a relatively complex form. Here the interfacial parameters are its wavelength, $\lambda = 6 \times 10^{-4}$ cm, and its growth speed, $v_0 = 1$ cm/s.

A linear 'stability map' may be computed for a range of dilute binary Al-Ag alloy, as is shown in Fig. 11.5. This stability map consists of the alloy's 'marginal curve', defined by the locus of pairs of values of the alloy concentration, C_0, and the solidification speed, v_0, that yield the marginal stability condition $\mathcal{F}(k) = 0$, for which $k = k^*$.

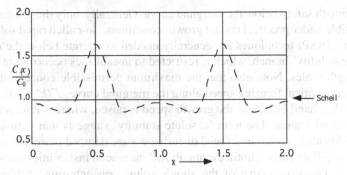

Fig. 11.4 Solute profile corresponding to freezing the alloy Al-0.22wt% Ag with $\mathcal{N}(k)$ about 15% above the marginal stability limit. The segregation pattern computed here shows sharp concentration peaks at the groove locations flanking the sides of the cell tips. These peaks rise more than 60% above the mean value for plane-front macrosegregation. A 20% reduction in solute concentration occurs in the valleys. A smaller local concentration maximum forms at the cell tip, the value for which lies below the Gulliver–Scheil steady-state average of $C(x)/C_0 = 1$

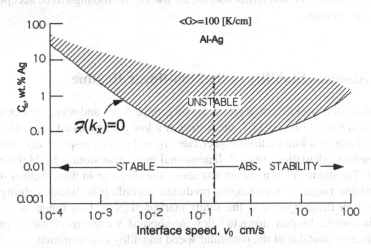

Fig. 11.5 Linear morphological stability map showing the critical alloy concentrations in dilute binary Al-Ag alloys versus a 6-decade logarithmic scale of solidification growth speeds, v_0. The *bowel-shaped* stability curve is the locus of marginally stable states. The *left-hand branch* of the marginal curve approaches the predictions of constitutional supercooling theory as the growth speed, $v_0 \to 0$; the *right-hand branch* is associated with the onset of 'absolute stability' that occurs at relatively high interface speeds, which in this example becomes possible when $v_0 > 0.2$ cm/s

Stability maps are used by some commercial crystal growers to insure the avoidance of morphological instability and the possible occurrence of *microsegregation*,[1] which can render useless single crystals grown for electronic and optical applications, where chemical homogeneity is critically important. One must conduct such

---

[1] The important topic of microsegregation will be treated later in Chapter 14.

crystal growth safely below the marginal curve. Generally, only the left-hand branch is accessible under practical crystal growth conditions. So-called rapid solidification processing (RSP) techniques are generally needed to operate below the right-hand 'absolute stability' branch, which is restricted to instabilities occurring at extremely short length scales. Note also that the maximum permissible concentration of the solute at any critical freezing speed along the marginal curve, $\mathcal{F}(k^\star) = 0$, *decreases* along the left-hand branch as the growth speed is raised, whereas it *increases* along the right-hand branch. The term 'absolute stability' suggests that with sufficiently high solidification speed, interfacial disturbances are damped more and more effectively by capillary forces, until, eventually, *all* interfacial instabilities are suppressed *absolutely*, i.e., irrespective of the alloy's solute concentration. Achieving sufficiently high interface speeds to induce the onset of absolute stability is, however, technically difficult in concentrated alloys. Although chemical segregation could in principle be avoided under absolute stability conditions, the attendant crystalline defect levels in rapidly grown crystals, caused by dislocations, stacking faults, twin formation, mosaicity, and lattice strains, are usually far too high to be acceptable for many applications.

11.5 Absolute Stability in the Non-linear Regime

The influence of solidification speed, v_0, on groove depth and wavenumber are summarized in Figs. 11.6 and 11.7, the former at a low growth speed ($v_0 = 0.01$ cm/s) and the latter at a hundredfold higher rate ($v_0 = 1$ cm/s), respectively. These data are based on calculations from 2-dimensional numerical studies by McFadden and Coriell. The shaded region shown just above the abscissæ in these figures indicates the unstable range of wavelengths predicted according to linear stability theory (denoted on these figures as the limits marked 'L.S.'). This range of instability extends from the longest, down to below the shortest, wavelengths that are predicted to be *linearly* unstable at the particular speed and alloy concentration.

The steady-state depth of the interfacial cells is not a simple function of λ, but tends to exhibit a complicated variation, especially at low freezing speeds. In addition, as indicated in Fig. 11.6, even the predicted range of unstable wavelengths computed for the non-linear model does not agree fully with that calculated from linear theory. Linear theory predicts stability for some small wavelengths (below about $\lambda = 50$ μm), whereas the non-linear calculations predict that an unstable cellular structure will develop. A so-called 'overstable' region, appearing as an isolated point in Fig. 11.6, is actually an oscillatory (time periodic) instability of the interface occurring slightly below the lowest linearly-predicted unstable wavelength. Here, however, a large amplitude perturbation is required initially to excite the oscillatory, overstable mode. The non-linear spectrum becomes quite complicated compared to that predicted using linear theory.

At a rapid rate of freezing, as portrayed in Fig. 11.7, the steady-state cell depth rises smoothly from either stability limit toward a single maximum near $\lambda = 5$ μm.

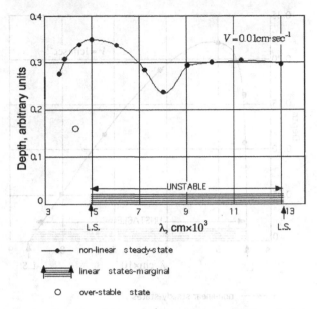

Fig. 11.6 Low-speed regime ($v_s = 0.01$ cm·s^{-1}) in non-linear stability calculations for an Al-Ag alloy. Linear stability (L.S.) theory fails to predict the correct range of disturbance wavelengths that occur in this alloy. Finite depths occur at all wavelengths because the instabilities here form sub-critically, i.e., below the marginal stability point predicted from linear theory

The wavelength limits in this case, however, are accurately predicted from linear theory, although the cell depths themselves must be determined using numerical computations based on non-linear stability theory.

11.6 Plan-Forms of Instabilities

Numerical computations of 3-dimensional nonlinear morphological stability, as just discussed, permit quantitative estimates to be made of the steady-state forms of both the concentration distribution, $C(x, y)$ and the associated 3-dimensional interface shape, $H(x, y)$. The computation of nonlinear states is practical, provided that the conditions of solidification are chosen not too far away from marginal stability.

Figure 11.8 (top sketch) shows the projection of the steady-state solute field developed just ahead of a solid–liquid interface on the plane $Z = 0$. The calculations are for an Al-Ag alloy growing under the same conditions as described in Fig. 11.1. The lower sketch in Fig. 11.8 shows the computed 3-dimensional form of the steady-state interfacial instability, which, in this particular case, is comprised of a periodic arrangement of pit-like dimples, arranged in an hexagonal pattern. Interfacial structures similar to those portrayed in Fig. 11.9 has been observed in rapidly

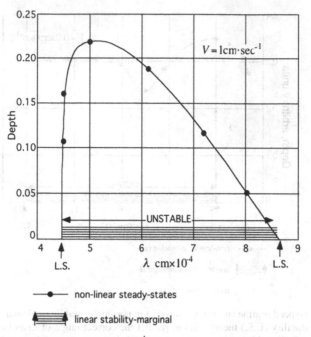

Fig. 11.7 High-speed regime ($v_s = 1.0$ cm·s^{-1}) for non-linear stability calculations for an Al-Ag alloy. In this case, linear theory accurately predicts the range of possible disturbance wavelengths. Note that as predicted from linear stability (L.S.) theory the groove depths from non-linear theory also fall to zero at both the short-wavelength limit and at the long-wavelength limit

decanted, dilute metallic alloys growing just above the margin of stability [5]. Figure 11.9 shows several other concentration distributions and computed planforms for instabilities developing on the solid–liquid interface. On the left are shown a series of 'roll' patterns, whereas on the right are so-called 'muffin-pan' instabilities. Similar periodic patterns have been observed experimentally at fluid-gas interfaces subjected to hydrodynamic convection.

The results shown in Figs. 11.8 and 11.9 are among the best examples of computations of non-linear morphological evolution. Such computations, unfortunately, require fast computers, specialized codes, and a considerable amount of experience and judgment regarding numerical convergence and error control. These types of computations are best left to experts. Nevertheless, the results of non-linear calculations are supported by direct observations of similar instabilities formed above and beyond the margin of stability. Although non-linear calculations must individually treat specific alloys growing under a set of prescribed solidification conditions, they do lend credence to the simpler linear morphological stability theory that has proven difficult to verify quantitatively with experiments attempting to measure the exponential onset and form of pure modes with the critical wavenumber, $k^\star$.

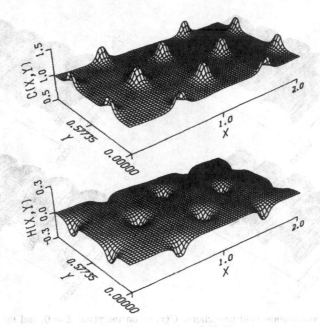

Fig. 11.8 *Top:* 2-dimensional concentration projection, $C(X, Y)$ at the plane $Z = 0$, on an unstable solid–liquid interface; *Bottom:* 3-dimensional form of a 'dimpled' hexagonal node structure, $H(X, Y)$, calculated from non-linear morphological stability theory. The concentration field clearly mirrors the interface shape. Compare with conditions listed in Fig. 11.1 (after McFadden et al. [7])

11.7 Summary

Non-linear stability theory allows detailed rendering and visualization of the three-dimensional plan forms of mildly unstable interfacial patterns, as well as estimates of the lateral solute segregation that is produced by their formation. Exposing these same features is precluded within the scope of linear theory, where one is limited to finding the margin of stability, and locating the most dangerous perturbation mode. Non-linear theories, by their mathematical nature, require numerical solution to elucidate specific stability results for a specific selection of the many solidification and system parameters. Such is the complexity of non-linear interface stability. A few such simulations were shown for Al-Ag binary alloys using results published by McFadden and Coriell, who computed interfacial forms reminiscent of the experimental observations reported many years earlier by Chalmers and others. These non-linear results underscore the impressive progress already made through advanced numerical and computational methods, and foretell that much greater progress will unfold in the years to come.

Non-linear stability maps gain importance in that modes of instability not revealed by linear theory become discoverable, including over-stable and oscillatory states. They also show consistency in the appropriate limits of solidification driving forces with standard linear stability theory, reinforcing confidence in our

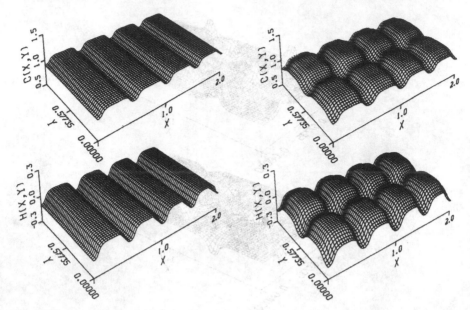

Fig. 11.9 Concentration field projections, $C(x, y)$ on the plane $Z = 0$, and the corresponding interface forms, $H(x, y)$, of unstable interfaces computed for Al-Ag alloys. *Left:* 2-dimensional 'roll' structure; *Right:* 'muffin-pan' structure with 3-dimensional modulations (after McFadden and Coriell [8])

current understanding of morphological stability. The plan forms and segregation patterns that are computed show sensitivity to the anisotropy of the interfacial energy selected for the crystal, and to the presence of any melt flow that can alter the diffusion fluxes. Also, numerical challenges are apt to be encountered in treating non-linear stability problems that are far from stability, so considerable experience is needed even where these computer codes have already been developed.

References

1. J.A. Warren and J.S. Langer, *Phys. Rev. A*, **42** (1990) 3518.
2. J.A. Warren and J.S. Langer, *Phys. Rev. E*, **47** (1993) 2702.
3. D. Riley and S.H. Davis, *IMAJ. Appl. Math.*, **45** (1990) 267 .
4. R.J. Schaefer and M.E. Glicksman, *Metall. Trans. A*, **A1** (1970) 1973.
5. B. Chalmers, *Trans. AIME*, **200** (1956) 519.
6. J.W. Rutter and B. Chalmers, *Can. J. Phys.*, **31** (1953).
7. G.B. McFadden, R.F. Boisvert and S.R. Coriell, *J. Cryst. Growth*, **84**, (1987) 371.
8. G.B. McFadden and S.R. Coriell, *Physica D*, **12** (1984) 253.

Chapter 12
Nucleation Catalysis

12.1 Introduction

Melting and solidification are reciprocal examples of first-order phase transformations, i.e., transformations exhibiting discontinuities in the first-order thermodynamic properties of the 'parent' and 'daughter' phases, such as the volume, entropy, and enthalpy. First-order phase changes can resist initiation—even from a highly supercooled or supersaturated metastable parent phase—as energetic barriers develop that inhibit the formation of viable embryos of the daughter phase, which once formed, can grow spontaneously. Stable nuclei require ordering of numerous atoms or molecules, composition change, creation of large interfacial area per unit volume, and elastic straining, all of which require the accumulation of 'excess' free energy. The additional free energy needed to accomplish nucleation is provided by thermodynamic fluctuations that momentarily allow a miniscule region of the metastable parent phase to depart significantly from its average order, composition, and free energy.

Several comprehensive reviews for phase nucleation phenomena, occurring both in condensed matter and gases, are available [1–3]. These reviews cover in depth what are now considered as the 'classical' theories of phase nucleation, developed originally by several German physical chemists in the 1920s and 1930s. Included among them are Volmer [4], Weber [5], Flood [6], Farkas [7], and Becker and Döring [8]. Those early contributions were followed by important contributions by Russian physicists, especially Zeldovich [9] and Frenkel [10] in the 1940s, and later by American scientists during the 1950s, including Turnbull [11, 12] and Fisher [13].

Discussed later in this chapter are athermal nuclei that have great importance in grain refinement processes that occur at small supercoolings, which is more typical of melt conditions encountered during ordinary casting. Athermal nuclei respond *deterministically* to the melt temperature, as contrasted with the *stochastic* behavior of nuclei formed by the classical fluctuation process. An up-to-date and comprehensive review of the field of nucleation catalysis in condensed systems that includes many aspects of crystallization and nucleation phenomena during solidification is the monograph published by Kelton and Greer [14].

M.E. Glicksman, *Principles of Solidification*, DOI 10.1007/978-1-4419-7344-3_12,
© Springer Science+Business Media, LLC 2011

12.2 Fluctuation-Dissipation

First-order transformations always require two well-distinguished steps to complete the phase transition process:

1. *Nucleation of the daughter phase*: Statistical mechanics teaches that thermodynamic fluctuations continually disturb and 'test' the microstates of any atomic or molecular system at any non-zero absolute temperature. It is the background spectrum of these fluctuations which ensures that a heterogeneous system achieves and maintains minimum free energy throughout each of its macroscopic phases. However, a thermodynamic fluctuation of sufficient strength, or amplitude, is needed within the metastable parent phase, α, to form initially what is termed the 'embryonic' volume of the more stable daughter phase, β.

   At any finite temperature, a spectrum of thermal fluctuations—most of which are weak, but a few are much stronger—continually appear throughout a phase. These fluctuations arise from, and then disappear back into the phase's Maxwell–Boltzmann molecular energy distribution. Many small fluctuations occur with high probability, and lead briefly to local free energy excursions and to phase 'embryos' that decay spontaneously. Weak fluctuations are merely 'turned back', and their slight excess energy rapidly dissipates within the system's internal energy distribution. Larger fluctuations, although relatively rare events, can raise a small region of a metastable phase to a sufficiently high energy to overcome the energetic barrier resisting the transition to an overall lower free energy phase state. In this uncertain, or so-called stochastic manner, a strong thermal fluctuation might overcome the energetic 'resistance' against lowering a metastable system's free energy, and spontaneously form a viable nucleus of the new lower-energy phase. Once formed, the nucleus of the daughter phase spontaneously increases its volume by consuming the parent phase, and leads the system away from metastability toward the state of thermodynamic equilibrium.

2. *Propagation of heterophase interfaces*: Formation of a daughter phase nucleus and then propagation of its heterophase boundary with the parent phase—such as a solid–liquid interface formed between the supercooled or supersaturated melt phase, α, and the conjugate crystalline phase, β—comprise the major sequential steps required in order to produce macroscopic amounts of the new phase. The basic thermodynamic and structural properties of an interface at equilibrium between a parent and its daughter phases were described in Chapter 8. Excess free energy accumulates where steep property gradients occur across the interphase region, which suggests that the local free energy change to form the embryonic daughter phase would be positive unless the parent melt phase was sufficiently supercooled so that the transformation to the crystalline daughter phase resulted in a sufficiently large compensatory negative free change.[1]

---

[1] Various models of interphase transition zones are described in Section 8.2.4, and the reader should also see the Gibbsian model of a sharp interface, portrayed in Fig. 8.5, and that for a diffuse interface in Fig. 8.7.

The interplay of positive energy changes to form the heterophase interface and allow local composition changes and elastic strains, and the negative free energy changes from transforming the embryonic volume of $\alpha \rightarrow \beta$ will be formulated in accord with classical nucleation theory. Once a stable embryonic fluctuation forms, the metastability of the parent phase assures continued propagation of the phase front, via crystal growth, which as discussed here allows the initiation of bulk solidification. In principle, the reverse sequence of superheating the solid and achieving nucleation of the melt phase occurs similarly. In practice, it is much easier to nucleate melt from the solid than accomplish the reverse.

12.3 Heterogeneous Equilibrium

Phases in heterogeneous equilibrium, connected by a first-order transition, are separated by a narrow interfacial region of atomic dimensions (*ca.* 0.1–1 nm), across which rapid changes occur in the properties. Figure 12.1 illustrates, for example, how the molar volume of the interphase region changes rapidly from point-to-point along a normal passing through the transition region between 'bulk' α and β phases.

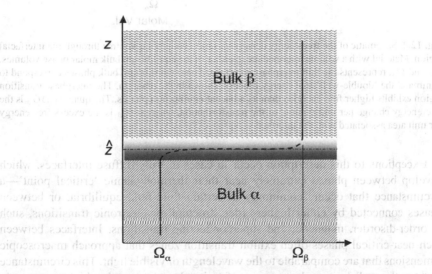

Fig. 12.1 Distance versus molar volume normal to an interface between bulk parent (α) and daughter (β) phases. The interfacial transition usually occurs over a few atomic distances centered around the plane $\hat{z}$. The location, $\hat{z}$, approximates the position of the Gibbs 'dividing surface'—a plane that defines the mathematical surface containing zero excess, or missing, molar volume. As interphase transitions in most condensed systems tend to be extremely narrow, they limit the additional free energy associated with matter having transitional properties intermediate to the two equilibrium phases, and 'squeeze' the associated excess free energy, $\gamma_{\alpha\beta}$ into a thin interfacial layer

The transition region defined in Chapter 8 includes the volume over which the free energy of the intervening material between bulk phases is constrained to higher energy states relative to the equilibrium phases, i.e., bulk α and bulk β, which exist beyond some small distance from the interphase region itself. Figure 12.2, shows that the transition from one equilibrium phase to the other is confined spatially to a narrow region called the interphase, or interface, which reduces the total volume of 'intermediate states', with their associated higher free energy densities.

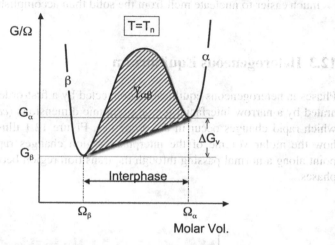

Fig. 12.2 Schematic of the free energy density variation with molar volume through the interfacial region. Material with a volume intermediate to either of the equilibrium bulk molar phase volumes, Ω_α and Ω_β, represents the thin interphase interface. These equilibrium bulk phases correspond to minima of the 'double-well' free energy function shown in the diagram. The interphase transition region exhibits higher free energies than its adjacent equilibrium phases. The quantity ΔG_v is the free energy change per unit volume of the transformation, whereas $\gamma_{\alpha\beta}$ is the excess free energy per unit area associated with the interface

Exceptions to this description occur in cases of truly diffuse interfaces, which develop between phases extremely near their thermodynamic 'critical point'—a circumstance that occurs commonly in certain fluid-fluid equilibria, or between phases connected by either higher-order structural or electronic transitions, such as order-disorder, magnetic, and superconducting transitions. Interfaces between such near-critical phases often exhibit transition zones that approach macroscopic dimensions that are comparable to the wavelength of visible light. This circumstance leads to the well-known phenomenon of 'critical opalescence', which occurs only when the temperature, pressure, and composition of the fluids are extremely close to their critical point. Nucleation can also occur in systems with diffuse interfaces, but other transformation pathways, such as spinodal decomposition via unstable diffusion waves, are also possible. Considering that crystal-melt systems never (to our knowledge) exhibit critical points, nucleation and growth are *always* the required processes for solidification and melting.

12.3.1 Free Energy Changes for Nucleation

12.3.1.1 Volumetric Free Energy

The molar free energy difference for a phase change, $\alpha \to \beta$, carried out at an arbitrary temperature and pressure favoring the transformation, is defined as $\Delta G_{\alpha\beta} \equiv G_\beta - G_\alpha$. It is convenient to define the volumetric free energy change of the transformation, ΔG_v, based on the system's average molar volume, $\Omega = (\Omega_\beta + \Omega_\alpha)/2$.

$$\Delta G_v = 2\left(\frac{G_\beta - G_\alpha}{\Omega_\beta + \Omega_\alpha}\right) = \frac{G_\beta - G_\alpha}{\Omega}. \tag{12.1}$$

In the case of melt-to-crystal, or crystal-to-melt phase changes, there is usually a few percent difference in the molar volumes of a crystal and its melt, so the average molar volume, Ω, differs little from the individual molar phase volumes.

For *reversible* melting or solidification occurring at the equilibrium melting temperature, T_m, $\Delta G_v = 0$. In the case of nucleating a *pure* crystalline phase, $\beta = s$, from its supercooled melt, $\alpha = \ell$, at a temperature $T_n \leq T_m$, the supercooling developed for nucleation may be defined as

$$\Delta T_n \equiv T_m - T_n \geq 0. \tag{12.2}$$

The volumetric free energy change favoring solidification associated with supercooling the melt to its nucleation temperature is easily found from the molar entropy of melting as

$$\Delta G_v = \int_{T_m}^{T_n} \frac{\Delta S_f}{\Omega} dT \approx -\frac{\Delta S_f}{\Omega} \Delta T_n. \tag{12.3}$$

The volumetric free energy change for formation of the crystalline solid is negative, provided that the melt phase is metastable, and, in this case supercooled or supersaturated. For moderate levels of supercooling, the magnitude of the (negative) volumetric free energy increases nearly linearly with the supercooling. Specifically, Eq. (12.3) remains accurate over the range of supercoolings for which the value of the molar entropy change, ΔS_f, remains reasonably constant. In the case of extremely large supercoolings, when $\Delta T_n > 100$ K, some correction may be warranted to account for the changing specific heat and entropy of the deeply supercooled melt. For the present purposes of describing nucleation phenomena in mildly supercooled melts, Eq. (12.3) suffices for estimating the bulk free energy change per unit volume.

12.3.1.2 Interfacial Free Energy

Creation of a nuclear volume of β phase requires the formation of a distinct α/β interface of area $\mathcal{A}$. The additional free energy, $\Delta G_{\mathcal{A}}$, required to form a solid–liquid interface that surrounds a nucleus of area $\mathcal{A}$ is

$$\Delta G_{\mathcal{A}} = \gamma_{s\ell}\, \mathcal{A}. \qquad (12.4)$$

The free energy change associated with increasing the area of *any* interface is positive,[2] and thus each element of area of the nuclear interface reversibly 'stores' a small excess free energy, $\gamma_{s\ell}$, as suggested in Fig. 12.2.

Equation (12.4) remains accurate as long as the nuclear volume is not too small, permitting the assumption that the nucleus consists of a large number of atoms or molecules. In fact, molecular dynamics computer studies [15] carried out on cylindrical clusters of noble-gas 'atoms' interacting through a Lennard-Jones 6–12 potential demonstrate that Eq. (12.4) remains accurate for nuclei containing as few as several hundred atoms. As a nuclear cluster gets smaller than roughly 10 molecules in diameter, its specific surface energy, $\gamma_{s\ell}$ begins to decrease, leading to error in Eq. (12.4).

The properties of nanoscale atomic clusters, $(2 \leq n \leq 200)$ were intensively studied experimentally over the past two decades. Curiously, it was found that the energies of small molecular or atomic clusters do not change smoothly with respect to the cluster's area or diameter. Some so-called 'magic number' metallic clusters (such as a clump of 55 atoms of Au) exhibited unusually strong surface catalytic properties [16]. Other 'magic number' clusters exhibit exceptionally low energies, high strengths, and superior stability [17], whereas other clusters exhibit diminished stability and, consequently, rarely occurred in these experiments.

Indeed, clusters that are extremely small may not even resemble their bulk crystalline phase structure. The reason is simply that in the case of crystalline phases some minimum number of unit cells is clearly needed to at least approximate the average coordination number of nearest and next-nearest neighbor atoms. In addition, a cluster consisting of too few unit cells of the crystal structure will have most of its atoms located at or near its surface, where the normal interatomic bonding pattern is disrupted, leading to atypical atomic arrangements within the bulk phase. Definitive rules do not exist at present to judge whether a small nuclear cluster is similar to, or dissimilar from, its associated bulk crystalline phase. When cluster diameters exceed about 100 nm, however, they are no longer considered as being nanoscopic, and it becomes reasonable to assume that such particles approximate closely their bulk phases in most respects.

12.4 Homogeneous Nucleation

Homogeneous nucleation in a metastable melt phase is induced by thermodynamic fluctuations occurring randomly throughout the melt, uninfluenced by the presence of any *extrinsic* surfaces, such as internal interfaces provided by dispersed colloidal

---

[2] A negative interfacial energy, if it ever occurred, would have the bizarre effect of potentially promoting endless spontaneous area increase—a phenomenon, fortunately, never encountered in phase transformations.

particles, or through contact with crucible or mold walls needed to support the melt. Homogeneous nucleation occurs strictly by thermodynamic fluctuations unaided by other effects.

12.4.1 Free Energy Barrier

The total free energy change needed to form a spherical nucleus of radius R consists of (1) volumetric, and (2) surface contributions, as dictated by the usual Euclidean measures of its volume and area, with each contribution multiplied by the appropriate free energy coefficient,

$$\Delta G_{tot} = \frac{4\pi}{3} R^3 \Delta G_v + 4\pi R^2 \gamma_{s\ell}. \tag{12.5}$$

Irrespective of the precise shape of the nuclear volume, the two energy terms appearing in Eq. (12.5) for a 'homogeneously' formed nucleus within the melt will always depend on the cube and square, respectively, of some convenient linear measure of its shape. Substituting Eq. (12.3) into Eq. (12.5) gives the free energy of formation of a spherical nucleus formed within a metastable melt supercooled by an amount, ΔT, below its equilibrium temperature,

$$\Delta G_{tot} = -\frac{4\pi}{3} \frac{\Delta S_f}{\Omega} \Delta T_n R^3 + 4\pi \gamma_{s\ell} R^2. \tag{12.6}$$

The free energy of formation of the nucleus, according to Eq. (12.6), is plotted in Fig. 12.3 (in units of electron-volts) for different levels of the supercooling.

The individual coefficients of the two terms (entropy per unit volume and interfacial energy) on the right-hand side of Eq. (12.6) were chosen as typical values for pure metals. Specifically, the volumetric energy coefficient was selected as -2.4 J/cm^3-K, or -1.5×10^{19} eV/cm^3-K, and the surface energy is 10^{-5} J/cm^2, or 6.3×10^{13} eV/cm^2. One sees that the formation free energy for a nucleus passes through a maximum—the energy barrier—at the 'critical' radius, $R^\star$. The maximum in the barrier height is defined formally by the necessary and sufficient (minimax) conditions that $dG_{tot}/dR = 0$, and $d^2 G_{tot}/dR^2 < 0$. The vanishing of the first derivative of the free energy, given by Eq. (12.6), occurs at the critical radius, $R = R^\star$, where

$$\frac{d\Delta G_{tot}}{dR} = 0 = -4\pi \frac{\Delta S_f}{\Omega} \Delta T_n R^{\star 2} + 8\pi \gamma_{s\ell} R^\star. \tag{12.7}$$

Solving Eq. (12.7) for the critical radius gives,

$$R^\star = \frac{2\gamma_{s\ell}\Omega}{\Delta S_f \Delta T_n}. \tag{12.8}$$

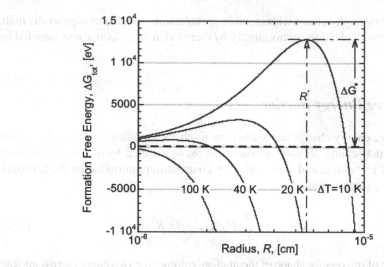

Fig. 12.3 Free energy, ΔG_{tot}, in electron-volts versus radius, R, for the homogeneous formation of a spherical nucleus of a crystalline phase from its supercooled melt. *Curves* representing Eq. (12.6) are plotted for several values of the melt's supercooling. Note that the free-energy barrier for nucleation, $\Delta G^\star$, rises rapidly as the supercooling decreases. The critical radius, $R^\star$, marking the top of the free energy barrier, depends inversely on the melt supercooling. The volumetric and surface free energy coefficients in Eq. (12.6) that were selected for these curves are typical of those for pure metals

The second condition needed to specify a relative maximum in the free energy change for forming a critical homogeneous nucleus is that the second derivative of the energy is negative. This requirement can be checked by twice differentiating Eq. (12.6), namely,

$$\frac{d^2 G_{tot}}{dR^2} = -8\pi \frac{\Delta S_f \Delta T_n}{\Omega} R + 8\pi \gamma_{s\ell}. \qquad (12.9)$$

Substituting the value of $R = R^\star$, Eq. (12.8), into Eq. (12.9) imposes the necessary condition that the first derivative vanishes, so one finds that

$$\frac{d^2 G_{tot}}{dR^2} = -8\pi \gamma_{s\ell} < 0, \quad (R = R^\star). \qquad (12.10)$$

The inequality shown on the RHS of Eq. (12.10) satisfies the requisite mini-max requirement for the occurrence of an energy *maximum* at the critical radius. The value of the maximum free energy of formation, $\Delta G^\star$, can now be determined by substituting Eq. (12.8) for the critical radius, $R^\star$, back into the general free energy expression, Eq. (12.6). One thereby obtains as the energetic barrier height,

$$\Delta G^\star = \frac{16\pi}{3} \left(\frac{\Omega}{\Delta S_f \Delta T_n} \right)^2 \gamma_{s\ell}^3. \qquad (12.11)$$

Insofar as $R^\star$ decreases inversely with the supercooling, the nuclear volume also decreases. Specifically, the number of atoms comprising the critically-sized homogeneous nucleus, $n^\star$, is found straightforwardly as

$$n^\star = \frac{32\pi}{3} N_A \Omega^2 \left(\frac{\gamma_{s\ell}}{\Delta S_f \Delta T_n}\right)^3, \qquad (12.12)$$

where N_A is Avogadro's number, and Ω is the atomic volume. Equation (12.12) shows that the number of atoms in the critical nucleus decreases rapidly as the inverse third power of the supercooling. Rewriting Eq. (12.12) in terms of the critical free energy barrier shows that

$$n^\star = -\frac{2\Delta G^\star}{\Omega \Delta G_v} = \frac{2\Delta G^\star}{(\mu_s - \mu_\ell)}. \qquad (12.13)$$

The first equality in Eq. (12.13) indicates that the number of atoms involved in a *critical* fluctuation, $n^\star$, is equal to the ratio of twice the barrier free energy to the free energy difference gained per atom between being in the supercooled melt and being in the solid nucleus.[3] Equivalently, this is the ratio of $2\Delta G^\star$ to the chemical potential difference between atoms in the supercooled melt and those in the crystalline phase.

12.4.1.1 Critical Heterophase Fluctuations

Nucleation, as treated classically, is a statistical phenomenon that depends on random occurrences of thermodynamic fluctuations. It becomes evident upon reflection of that fact that it is far more probable for a fluctuation in the melt to organize successfully (as a crystalline nucleus) a relatively small number of atoms at a large supercooling, than it is to organize a large number of atoms at a small supercooling. The quantitative formulation from statistical mechanics of the chance for a fluctuation successfully forming a homogeneous nucleus of critical size in a supercooled melt is provided by the Boltzmann probability, $p(n^\star)$, where[4]

$$p(n^\star) = e^{-\Delta G^\star / k_B T}. \qquad (12.14)$$

Substituting the expression for the maximum free energy change, Eq. (12.11), into the Boltzmann probability, Eq. (12.14), yields the probability of a critical fluctuation occurring in the melt,

---

[3] Again, the reader is cautioned that classical nucleation theory makes no distinction between the phase comprising the nuclear volume and the bulk crystalline solid. That approximation of classical nucleation theory, as discussed in Section 12.3.1, works reasonably well even into the nanoscopic regime of nuclear clusters.

[4] The simplified exponential form of the Maxwell-Boltzmann probability distribution used in Eq. (12.14) to calculate the probabilty of a successful fluctuation remains valid as long as $\Delta G^\star \gg k_B T \approx 0.09$ ev, at $T = 1,000$ K, which is true as shown by the examples applied above to Eq. (12.15).

$$p(n^*) = \exp\left[-\frac{16\pi}{3k_B T}\left(\frac{\Omega}{\Delta S_f \Delta T_n}\right)^2 \gamma_{s\ell}^3\right]. \tag{12.15}$$

Referring back to Fig. 12.3, one notes that the value of ΔG^* is approximately 3240 ev when the supercooling $\Delta T_n = 20$ K. The barrier free energy drops rapidly to about 130 ev as the supercooling increases to $\Delta T_n = 100$ K. Inserting these values of the free energy barrier into Eq. (12.14) yields the following probabilities at an arbitrary, but reasonable, melt temperature of $T = 1000$ K:

- *Melt supercooling of $\Delta T_n = 20$ K:*
  A critical homogeneous nucleus in this melt consists of about 4 million atoms, for which the Boltzmann probability is $p(n^*) \approx 10^{-4 \times 10^4}$. At the risk of understating this situation, this represents a miniscule probability that has forty thousand zeros after its decimal point! If one expects say 10^{37} fluctuations per second per cm^3 of supercooled melt, that is 3×10^{44} per year, or about 4×10^{54} over the estimated entire age of the universe. Multiplying even this cosmologically large number of attempts, by the Boltzmann probability, $p(n^*)$, still effectively yields *zero* as the chance of ever observing *homogeneous* nucleation in this moderately supercooled melt.
- *Melt supercooling of $\Delta T_n = 100$ K:*
  A critical homogeneous nucleus now requires about 31,000 atoms, for which the fluctuation probability is $p(n^*) \approx 10^{-150}$. This also represents a vanishingly small probability. With 10^{37} fluctuations occurring per second per cm^3 of supercooled melt, a homogeneous nucleation event probably would still not be observed.
- *Melt supercooling of $\Delta T_n = 200$ K:*
  A critical homogeneous nucleus now requires only about 3,900 atoms, for which the Boltzmann probability is $p(n^*) \approx 6 \times 10^{-37.5}$. Again, if one accepts the occurrence of about 10^{37} fluctuations per second per cm^3 of supercooled melt, a homogeneous nucleation event would be expected to occur at the rate of about $10^{-0.5} \approx 0.31$ [s $\cdot$ cm^3]$^{-1}$, or one nucleation event roughly every 3 s. A supercooling of 200 K is difficult, but not impossible, to obtain experimentally in an isolated ultra-clean melt. Thus, homogeneous nucleation is a phenomenon generally expected only in deeply supercooled melts.

It seems evident, therefore, that the mechanism of *homogeneous* nucleation would never operate during ordinary solidification practice, where the supercooling typically needed to observe the onset of copious nucleation in a melt is in the range of about 1–5 K. Thus, despite the extremely rapid decrease of n^* with increasing supercooling, homogeneous nucleation would seldom, if ever, occur in melts, excepting under the unusual circumstance of achieving extraordinarily large supercoolings, amounting to several hundreds of Kelvins. Essentially, one learns from this simple exercise in stochastics that at more moderate levels of supercooling far too many atoms, or molecules, must be organized as a crystalline nucleus than might reasonably be produced by a thermodynamic *heterophase* fluctuation. Clearly, a different nucleation mechanism is at work in practical casting situations.

As will be shown next, by contrast, fluctuations occurring in the melt that involve the presence of *extrinsic* substrates, such as container walls or dispersed particles, can drastically reduce the number of atoms required to form a critical nucleus, and increase the probability for nucleation. Thus, the presence of a suitable, i.e. catalytic, substrate provides a more efficient pathway to initiate the $\ell \to s$ transformation. Later, in Section 12.6.4, the important relationship between heterogeneous nucleation and *athermal* nucleation will be discussed, as even heterogeneous nucleation cannot account fully for the time-independent, copious nucleation induced progressively at rather small supercoolings when highly efficient nucleation catalysts for grain refinement are present.

12.5 Heterogeneous Nucleation

12.5.1 Background

All molten metals and alloys contain a distribution of colloidal matter of sufficiently small sizes (ca. diameters $\ll 0.1 \, \mu$m) that remain in indefinite suspension and dispersal by random thermal jostling of the surrounding melt molecules—a phenomenon called Brownian motion. Melt-borne colloidal particles are subject to chaotic pressure-momentum pulses delivered to their surfaces through random impacts by the surrounding melt molecules. Each cubic centimeter of a melt contains millions of such microscopic 'motes'. Some, as might be expected, are naturally occurring as native oxides, nitrides, silicides, carbides, etc., others may just be uncharacterized 'dirt and debris' introduced inadvertently from the surrounding crucible walls or from other extrinsic sources that contact the melt, such as fluxes, thermal probes, or even from the 'protective' gases often used to keep a melt from oxidizing or evaporating excessively.

A necessary condition for such motes to be effective nucleation catalysts, and act as substrates conducive towards the nucleation of the crystalline phase, is that they remain crystalline at the nucleation temperature. Thus, effective catalytic particles must have melting points that are higher than the melting point of the phase to be nucleated, and, preferably, their melting point should even be higher than the maximum temperature to which the melt is subjected prior to supercooling. The presence of such extrinsic nucleation substrates provides a myriad of heterogeneous catalytic sites on which the crystalline phase can nucleate with a reduced energy barrier compared to homogeneous nucleation within the melt.

The classical theory of heterogeneous nucleation in liquid-gas systems was also developed in the mid-1930s, by Becker and Döring [8], and then successfully checked against experiments for nucleation of liquid droplets from the supercooled vapor by Volmer and Flood [6]. For an exhaustive survey of early research on heterogeneous nucleation and its particular application to solid–liquid systems, the interested reader is encouraged to read the excellent review by Turnbull [18].

12.5.2 Spherical Cap Model

A basic model for describing heterogeneous nucleation was developed by Volmer and Flood, and called the spherical cap model. Those investigators considered a planar substrate surface within the supercooled melt (ℓ-phase) upon which fluctuations occur, forming, from time-to-time, 'embryos' of the crystalline s-phase. Figure 12.4 provides a sketch of their spherical cap model.

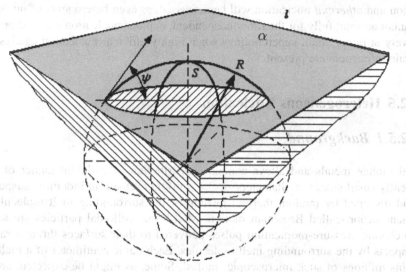

Fig. 12.4 *Cap-shaped* nucleus of crystalline solid s-phase of radius R residing on a foreign substrate, α, within a supercooled melt, ℓ. The oval area indicates the 'footprint' exerted by the cap on the substrate (*gray surface*) where the melt-substrate interface is replaced by the solid-substrate interface. The *dashed extension* of the spherical cap does not exist, and is shown only to indicate the volume of s that is 'avoided' by introducing the catalytic substrate, C. The equilibrium angle of repose of the solid phase on the substrate, when in intimate contact with the melt, ℓ, is the contact angle against the melt, ψ

The total Gibbs free energy change, ΔG_{tot}, needed to form such a spherical cap of s-phase on the planar substrate C, as shown in Fig. 12.4, consists of three terms:

$$\Delta G_{tot} = v_s \Delta G_v + \mathcal{A}_{s\ell}\gamma_{s\ell} + \mathcal{A}_{s\alpha}(\gamma_{s\alpha} - \gamma_{\ell\alpha}), \qquad (12.16)$$

where ΔG_v is the free energy change per unit volume, defined in Eq. (12.3), to form the cap-shaped nucleus of s-phase from the supercooled liquid; $\gamma_{s\ell}$ is the solid–liquid interfacial free energy; $\gamma_{s\alpha}$ is the interfacial energy between s-phase and the substrate; $\gamma_{\ell\alpha}$ is the interfacial free energy between the substrate and the liquid phase, ℓ.

The geometrical quantities characterizing the cap include v_s, $\mathcal{A}_{s\ell}$, and $\mathcal{A}_{s\alpha}$— all appearing as coefficients of the free energy of formation, Eq. (12.16). These quantities may be defined as follows for a spherical cap of radius R, and contact angle ψ:

The volume of this heterogenous nucleus is

$$v_s = \frac{\pi R^3}{3} \left(2 + \cos^3 \psi - 3 \cos \psi \right). \tag{12.17}$$

The area of the 'dome'-shaped solid–liquid interface is

$$A_{s\ell} = 2\pi R^2 \left(1 - \cos \psi \right). \tag{12.18}$$

The 'footprint' area of the nucleus on the planar substrate is

$$A_{s\alpha} = \pi R^2 \sin^2 \psi. \tag{12.19}$$

If Eqs. (12.17), (12.18), and (12.19), defining the nuclear cap's geometric quantities, are inserted back into the free energy expression, Eq. (12.16), one obtains the free energy change for heterogeneous nucleation, namely,

$$\Delta G_{het} = \frac{\pi R^3}{3} \left(2 + \cos^3 \psi - 3 \cos \psi \right) \Delta G_v$$
$$+ 2\pi R^2 \left(1 - \cos \psi \right) \gamma_{s\ell} + \pi (R^2) \sin^2 \psi \left(\gamma_{s\alpha} - \gamma_{\ell\alpha} \right). \tag{12.20}$$

If the RHS of Eq. (12.20) is grouped, and the equation is normalized by the free energy per unit volume of the supercooled melt, ΔG_v, one obtains the dimensionless free energy change, $\mathcal{G}(R, \psi)$,

$$\frac{\Delta G_{het}}{\Delta G_v} \equiv \mathcal{G}(R, \psi) = \frac{\pi R^3}{3} \left(2 + \cos^3 \psi - 3 \cos \psi \right)$$
$$+ \pi R^2 \left[\frac{2\gamma_{s\ell}}{\Delta G_v} \left(1 - \cos \psi \right) + \frac{\gamma_{s\alpha} - \gamma_{\ell\alpha}}{\Delta G_v} \sin^2 \psi \right]. \tag{12.21}$$

Equation (12.21) shows that $\mathcal{G}(R, \psi)$ is a function of two independent geometrical variables. The critical cap-shaped nucleus must therefore be found in terms of an extremum in the free energy surface at the critical coordinates R^*, ψ^*. Elementary calculus allows determination of the position of the free-energy extremum by setting the partial derivatives of $\mathcal{G}(R, \psi)$ simultaneously equal to zero, namely,

$$\frac{\partial \mathcal{G}(R, \psi)}{\partial R} = 0, \tag{12.22}$$

and

$$\frac{\partial \mathcal{G}(R, \psi)}{\partial \psi} = 0. \tag{12.23}$$

Evaluating the partial derivatives shown in Eqs. (12.22) and (12.23), and setting them equal to zero, yields, respectively,

$$\frac{\partial \mathcal{G}(R^\star, \psi)}{\partial R} = 0 = \pi R^{\star 2} \left(2 + \cos^3 \psi - 3 \cos \psi\right)$$
$$+ 2\pi R^\star \left[\frac{2\gamma_{s\ell}}{\Delta G_v} (1 - \cos \psi) + \frac{\gamma_{s\alpha} - \gamma_{\ell\alpha}}{\Delta G_v} \sin^2 \psi\right]. \tag{12.24}$$

and

$$\frac{\partial \mathcal{G}(R, \psi^\star)}{\partial \psi} = 0 = \pi R^3 \left(-\cos^2 \psi^\star \sin \psi^\star + \sin \psi^\star\right)$$
$$+ 2\pi R^2 \left[\frac{\gamma_{s\ell}}{\Delta G_v} (\sin \psi^\star) + \frac{\gamma_{s\alpha} - \gamma_{\ell\alpha}}{\Delta G_v} \sin \psi^\star \cos \psi^\star\right]. \tag{12.25}$$

The starred quantities appearing in Eqs. (12.24) and (12.25) represent the critical values of the radius, $R^\star$, and the equilibrium contact angle, $\psi^\star$, of an s-phase nucleus that simultaneously satisfies the free energy extremum specified by Eqs. (12.22) and Eqs. (12.23).

Solving for $R^\star$ in Eqs. (12.24), and R in Eqs. (12.25), gives the intermediate result that

$$R^\star = \frac{-4\gamma_{s\ell} + 2(1 - \cos \psi)(\gamma_{s\alpha} - \gamma_{\ell\alpha})}{(2 + \cos \psi)(1 - \cos \psi)\Delta G_v}, \tag{12.26}$$

and

$$R = \frac{-2\left[\gamma_{s\ell} + (\gamma_{s\alpha} - \gamma_{\ell\alpha})\cos \psi^\star\right]}{\sin^2 \psi^\star \Delta G_v}. \tag{12.27}$$

Next, one forces the simultaneous condition that $R = R^\star$ and $\psi = \psi^\star$, and sets equal the RHS of Eqs. (12.26) and (12.27), and applies the following trigonometric identity,

$$\sin^2 \psi^\star = \left(1 + \cos \psi^\star\right)\left(1 - \cos \psi^\star\right). \tag{12.28}$$

These steps yield the relationship among the critical contact angle, $\psi^\star$, and the three interface energy densities,

$$\frac{1 + \cos \psi^\star}{2 + \cos \psi^\star}\left[2\gamma_{s\ell} + \left(1 + \cos \psi^\star\right)(\gamma_{s\alpha} - \gamma_{\ell\alpha})\right] = \gamma_{s\ell} + \cos \psi^\star (\gamma_{s\alpha} - \gamma_{\ell\alpha}). \tag{12.29}$$

For convenience, define $\mathcal{M} \equiv 1 + \cos \psi^\star$, allowing Eq. (12.29) to be rewritten in the form

$$\frac{\mathcal{M}}{1+\mathcal{M}}\left[2\gamma_{s\ell}+\mathcal{M}\left(\gamma_{s\alpha}-\gamma_{\ell\alpha}\right)\right]=\gamma_{s\ell}+\left(\mathcal{M}-1\right)\left(\gamma_{s\alpha}-\gamma_{\ell\alpha}\right). \quad (12.30)$$

Solving Eq. (12.30) for $\mathcal{M}$ yields

$$\mathcal{M}=1-\frac{\gamma_{s\alpha}-\gamma_{\ell\alpha}}{\gamma_{s\ell}}. \quad (12.31)$$

Reinserting the definition of $\mathcal{M}$ into the LHS of Eq. (12.30) gives the important result that the extremum in the free energy is located at the equilibrium value of the contact angle, $\psi^\star$, the cosine of which is given by the surface energy ratio,

$$\cos\psi^\star=-\frac{\gamma_{s\alpha}-\gamma_{\ell\alpha}}{\gamma_{s\ell}}. \quad (12.32)$$

Equation (12.32) is merely a variant of Young's formula at the triple junction border circling the contact 'footprint' between the solid, liquid, and substrate. Young's formula gives the condition for local mechanical equilibrium between the nucleus, melt, and substrate. The contact angle may be found from that formula as,

$$\psi^\star=\cos^{-1}\left(\frac{\gamma_{\ell\alpha}-\gamma_{s\alpha}}{\gamma_{s\ell}}\right). \quad (12.33)$$

The balance of these interfacial tensions is depicted by the vectors shown in Fig. 12.5. Now, if Young's formula, Eq. (12.33), is inserted into Eq. (12.26), one obtains an expression for the nucleus's critical radius, $R^\star$. The following expression for the critical radius is found after several steps of algebra,

$$R^\star=-\frac{2\gamma_{s\ell}}{\Delta G_v}. \quad (12.34)$$

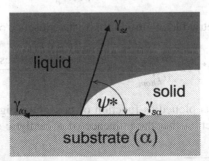

Fig. 12.5 Balance of interfacial tensions along the *triple line* of contact surrounding a *cap-shaped* nucleus, which is a portion of a sphere. This balance of surface tension vectors defines the equilibrium contact angle, $\psi^\star$ for heterogeneous nucleation

It is perhaps surprising to note that this expression proves that the critical radius of the cap-shaped nucleus is *independent* of the surface energy change over the 'footprint' area of the heterogeneous nucleus. That is, the substrate, per se, does not affect the critical radius of the s-nucleus, only its critical volume!

The critical radius and the equilibrium contact angle are, in fact, uncoupled, i.e., independent parameters defining the critical cap-shaped nucleus. The contact angle, $\psi^\star$, is also assumed to be independent of the supercooling,[5] whereas the critical radius, $R^\star$, depends on both the supercooling and the solid–liquid interface energy density.

The activation free energy, $\Delta G^\star_{het}$, to form a critical cap-shaped nucleus on the planar substrate is found by substituting back into Eq. (12.20) the expressions found for the critical radius, Eq. (12.26), and for the critical contact angle, Eq. (12.28). The free energy barrier required to form a heterogeneous nucleus is found, after several steps of algebra, to be

$$\Delta G^\star_{het} = \frac{16\pi \gamma_{s\ell}^3}{3\Delta G_v^2}\left[\frac{2 + \cos^3 \psi^\star - 3\cos \psi^\star}{4}\right]. \tag{12.35}$$

The structure of Eq. (12.35) shows that it may be separated into a term of geometric origin and a term of thermodynamic origin. The geometrical term becomes unity when the contact angle $\psi^\star = \pi$, and the nuclear cap becomes a full sphere. Inasmuch as $\psi^\star \to \pi$, the substrate and nucleus lose contact, which represents the limiting case of homogeneous nucleation, as already developed in Section 12.4.

Consequently, Eq. (12.35) for the barrier free energy resisting heterogeneous nucleation may be thought of as the product function,

$$\Delta G^\star_{het} = \Delta G^\star_{hom} \cdot f(\psi^\star), \tag{12.36}$$

where the thermodynamic factor is

$$\Delta G^\star_{hom} = \frac{16\pi \gamma_{s\ell}^3}{3\Delta G_v^2}, \tag{12.37}$$

and the purely geometric factor introduced by the presence of the planar substrate is

$$f(\psi^\star) \equiv \frac{\Delta G^\star_{het}}{\Delta G^\star_{hom}} = \left[\frac{2 + \cos^3 \psi^\star - 3\cos \psi^\star}{4}\right]. \tag{12.38}$$

Equation (12.38) is plotted as function of $\cos \psi^\star$ in Fig. 12.6. It is clear from the form of this function that the influence of a catalytic substrate in promoting

---

[5] Clearly, over a large range of supercooling the three interfacial energies, $\gamma_{s\ell}$, $\gamma_{\ell\alpha}$, and $\gamma_{s\alpha}$, shown as surface tension vectors in Fig. 12.5, can change individually, which would alter the equilibrium value of $\psi^\star$.

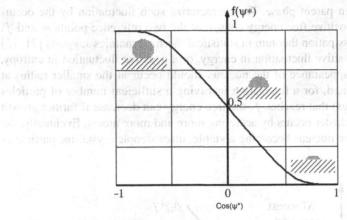

Fig. 12.6 Plot of the substrate's critical contact angle function, $f(\psi^\star)$ versus $\cos \psi^\star$. The inserted sketches show the influence of the contact angle on the nuclear shape and size at several points along the curve. As $\cos \psi^\star \to 1$ the number of atoms in the critical nucleus decreases rapidly to a monolayer wetting the substrate

nucleation is to stimulate the process simply by reducing the volume of the critical nucleus and, as a consequence, its number of atoms or molecules.

The presence of an effective catalytic substrate greatly reduces the number of atoms required to reach 'critical' size. Specifically, the smaller the contact angle, the more effective is the substrate as a catalyst. The contact angle, $\psi^\star$, depends sensitively on several factors, including: (1) the chemical bonding of the substrate itself (e.g., metallic, covalent, or ionic); (2) the orientation to the melt of its exposed crystallographic planes, affecting the interplanar spacings and bond density [19]; (3) the microtopography of the exposed crystal planes, including microsteps and ledges; (4) electrostatic effects including double-layers; and (5) perhaps most importantly, any chemisorption or reactions between melt atoms or molecules residing on the substrate surface [20].

12.5.3 Heterophase Fluctuations on a Substrate

The form of the free energy contributions in heterogeneous nucleation, and the total free energy change as expressed, say, in Eq. (12.20), is sketched in Fig. 12.7 as function of the radius, with the contact angle and supercooling both fixed. Note that ΔG_{tot} is the summation of two positive surface energy contributions arising from the cap-shaped solid–liquid interface and the circular contact area between the nucleus and the substrate, plus a negative contribution proportional to the volume of the nucleus and its (negative) free energy density ΔG_V. The free energy density of the nuclear volume is just minus the entropy of fusion, ΔS_f times the supercooling, ΔT. Consider a thermodynamic fluctuation in the local free energy due to the rearrangement of atoms to form a crystal-like embyronic cluster in the supercooled molten parent phase. We characterize such fluctuation by the occurrence of a local, positive free energy change at the two reference points m and f. The fluctuation-dissipation theorem in statistical thermodynamics suggests [21, 22] that to relieve a positive fluctuation in energy, or a negative fluctuation in entropy, remelting and disappearance of the nucleus should occur at the smaller radius at m. On the other hand, for a fluctuation involving a sufficient number of particles (atoms or molecules) that reaches f, the free energy can decrease if further growth of the embryonic cluster occurs by accreting more and more atoms. Eventually, the growing nanoscopic nucleus becoming a viable, macroscopic crystalline particle of the solid phase.

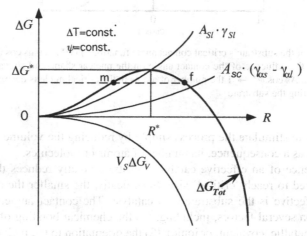

Fig. 12.7 Gibbs free energy change to form a *cap-shaped* nucleus, with contact angle ψ^*, at a fixed supercooling, $\Delta T_n > 0$. Point m designates the embryonic radius of an unsuccessful fluctuation that decays spontaneously; whereas point f indicates the radius and free energy change for a supercritical fluctuation capable of continued growth into the surrounding supercooled melt

12.6 Grain Refinement

Grain refinement is frequently required to improve the mechanical strength of cast alloys, and to provide enhanced ductility and good surface appearance. These issues are so important to commercial end users of non-ferrous alloys that metals processors are willing to provide effective grain-size control during solidification. Industrial control of cast-metal grain size, particularly in non-ferrous alloys of aluminum and titanium [23, 24], will employ modification of the melt chemistry, using additions of inoculant-forming components, or by direct injection of inoculant particles.

An approach used in the light metals industry to achieve grain-size control is to add a 'master alloy' to the main melt mass—i.e., to dissolve into the melt small quantities of a compatible alloy that is highly enriched in nucleants, or nucleant formers. The master alloy addition yields many billions of microscopic crystals of higher melting point crystalline compounds, some of which act as effective catalysts for the formation of primary-phase nuclei. In addition, such master alloys also adds trace solutes, such as Ti, to enhance the efficacy and activity of the inoculants [25]. It is generally believed that Ti solute atoms segregate ahead of the solid–liquid interface and thereby promote constitutional supercooling. The constitutionally supercooled melt favors subsequent nucleation around nearby TiB_2 particles, effectively refining the grains, and also limiting the continuous growth of columnar dendritic crystals.

The great advantage of using inoculants to control as-cast grain size is that their addition to large melt volumes being routinely cast is easy and safely performed in the cast shop. Moreover, the master alloy itself adds relatively little to the final cost of the grain-refined casting or ingot. The only significant disadvantages of employing inoculants for grain-size control are that without good inoculant dispersal, nonuniform grain size could result, and the addition of dispersed foreign particles, or introduction of master alloy components that react to form effective inoculant particles could have deleterious influences on the material's fatigue properties. The latter difficulty may be reduced to some extent by only using the minimum amount of master alloy, as larger additions do not reduce the cast grain size any further beyond some minimum level characteristic of the alloy and the casting temperature. A thorough, modern review by Perepezko [26] specifically covers nucleation phenomena in alloy melts, and presents grain refinement theory and practice for the light metals.

12.6.1 Inoculants

Desirable characteristics of effective nucleation catalysts include:

i. *Good wetting of the liquid phase:*
Partial wetting of the inoculant, C, by the melt, ℓ, implies some intimate physical contact caused by bonding affinity between them, or some chemical reactivity. This affinity can be expressed by the inequality, $\gamma_{v\alpha} > \gamma_{\ell\alpha} + \gamma_{\ell v}$. Here $\gamma_{v\alpha}$ is the interfacial energy between the vapor and the substrate; $\gamma_{\ell v}$ is the interfacial energy between the liquid and the vapor; and, again, $\gamma_{\ell\alpha}$ is the interfacial energy between the melt and the substrate. See Fig. 12.8 showing the partial wetting characteristics of a suitable substrate surface for nucleation when exposed to the melt and its vapor. If, however, the reactivity between the melt and inoculant is too high, then a deleterious phenomenon known as 'fading' may occur. Fading is an undesired effect that can occur during melt processing and casting, because dispersed catalytic particles, which differ in their mass density from the melt, often concentrate by gravitational sedimentation, or might gradually react with, or dissolve back into, the melt [27]. Such interactions of inoculants with the melt eventually cause inoculant particles to lose some of their catalytic activity for nucleating the solid, or cause inhomogeneous grain refinement throughout

the volume of a casting or ingot because of their uneven spatial distribution in the melt. Some fading even occurs for inoculants with high catalytic potency, but proper industrial practice, usually based on experience, such as reducing the holding time before pouring the melt, plus adjustments of the melt chemistry do minimize the loss of active inoculants.

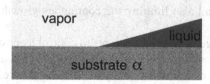

Fig. 12.8 Partial wetting of an inoculant by the melt. The contact configuration between the melt, ℓ, its vapor, and the substrate, α, is critical for achieving adequate catalytic stimulation of the nucleation process. If the contact angle were to be too large, limited wetting and reduced catalytic efficiency would ensue. If the contact angle were near zero—a sign of reactivity between the melt and substrate—reaction and dissolution might occur, limiting the active lifetime of the inoculant and allowing rapid fading to occur [27]

ii. *Small solid-substrate contact angle, ψ^**: Effective inoculation of the solid requires that $\gamma_{\ell\alpha} > \gamma_{s\alpha}$, so that the appearance of a solid nucleus beneficially *reduces* the total interfacial energy between the substrate and the nucleus. See again Fig. 12.4 for the geometric characteristics of a cap-shaped nucleus. A strong positive interaction between the wetted substrate and the heterophase fluctuation forming the solid nucleus assures a small solid-substrate contact angle, which effectively reduces the free energy barrier to nucleation.

iii. *Minimize $\gamma_{s\alpha}$*: A key factor that must be satisfied to reduce the interfacial energy, $\gamma_{s\alpha}$, between the solid and inoculant, and achieve good grain refinement efficiency is matching the chemical bonding in the solid to that of the substrate. These factors tend to augment the chemical affinity between the solid phase and the substrate. This is satisfied when the chemical bonding types in the substrate and solid are similar. Thus, for example, covalent, or ionic solids don't effectively nucleate molten metals. However, metallic additions are often much too reactive with alloy melts to provide effective nucleation—i.e., they fade much too rapidly to provide useful inoculants. Compounds such as titanium diboride, TiB_2 and titanium aluminide, $TiAl_3$, for example, prove highly effective for grain refining most aluminum alloys, but not all. The exact chemical make-up and processing condition of these compounds and the composition of their master alloys, such as Al-Ti-B for grain refining cast aluminum alloys, are usually kept proprietary by the companies that sell them to the light metals foundries that use them.

iv. *Lattice matching*: The lattice mismatch between the interfacial planes of the solid phase and those exposed on the substrate must be small. Unless the match between the interplanar spacings is within a small fraction of 1%, accommodation strains would develop between the solid nucleus and the substrate, thereby increasing the total free energy needed to overcome the nucleation barrier. Lattice mismatch, per se, between a substrate and the primary crystalline phase might prevent efficient catalysis from occurring, but mutual chemical affinity probably dominates as the major factor.

12.6.2 Substrate Area, Contact Angle, and Phase Spreading

After a fluctuation produces a heterophase nucleus, additional adjustments (relaxation) of the shape of the embryonic mass are possible. The fact that substrates offer only a distribution of limited interface areas—a topic investigated by Fletcher [29–31]—has consequences on the statistical considerations of their capture rate, as fluctuations occur throughout the melt. Moreover, the initial contact line established between the heterogeneously nucleated daughter phase and the available substrate can change in time, especially if a wetting layer—only one or two atoms thick—called the precursor film—spreads out across the available substrate area [32]. The phenomenon of spreading kinetics of a nuclear cap of Pb residing on a Cu substrate of defined crystallographic orientation was simulated by Webb et al. using molecular dynamics computations [28]. Some results of the computed shape changes for a constant-volume cap spreading via an expanding precursor film on a planar substrate are shown in Fig. 12.9.

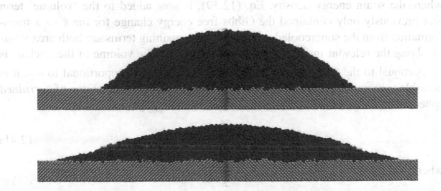

Fig. 12.9 Molecular dynamic simulation of a Pb nucleus, with fixed volume, spreading across a {100} plane of a Cu substrate at 700 K. Two 'snap-shots' of the cross-section showing the distribution of the atoms are depicted after a time interval of 0.5 ns. for Pb, in the molten state. The *cap-shaped* Pb mass extends an atomically thin precursor film that wets the underlying Cu surface. The contact angle, ψ, in this case diminishes as the spreading advances at a constant nuclear volume. The details of contact-angle relaxation dynamics during heterogeneous nucleation remain unknown at present, but atomic-scale events similar to those depicted in these simulations are doubtless involved (adapted from [28])

12.6.3 Accommodation Strains

Careful selection of inoculants can provide both compatible atomic bonding for good chemical affinity between solid and substrate, and reduce, but not eliminate, some of the lattice mismatch. The solid nuclei, particularly if they are strongly bonded to the substrate, must accommodate any residual lattice mismatch by elastic straining, and introduction of misfit dislocations. The free (strain) energy density, E_ϵ, associated with isothermal linear-elastic straining of a crystalline nucleus (as given by Hooke's law) is

$$E_\epsilon = \frac{1}{2}C_{ijkl}\{\epsilon_{ij}\cdot\epsilon_{kl}\}, \qquad (12.39)$$

where the energy coefficient C_{ijkl} is the linear elastic stiffness matrix relating stress and strain, and the contraction $\{\epsilon_{ij}\cdot\epsilon_{kl}\}$ is the square of the strain tensor. The Einstein summation notation for repeated indices, $ijkl$ is used in Eq. (12.39). Fundamentally, Eq. (12.39) shows that strain energy is always positive, and proportional to the square of the strains.[6]

The strains, in turn, are proportional to the lattice mismatch between the solid nucleus and the substrate. Thus, for a given lattice mismatch the total elastic energy may be written as a slight modification of Eq. (12.20),

$$\Delta G_{het} = \frac{\pi R^3}{3}\left(2+\cos^3\psi - 3\cos\psi\right)(\Delta G_V + E_\epsilon)$$
$$+2\pi R^2 (1-\cos\psi)\gamma_{s\ell} + \pi(R^2)\sin^2\psi\,(\gamma_{sa}-\gamma_{\ell a})\,, \qquad (12.40)$$

where the strain energy density, Eq. (12.39), is now added to the 'volume' term that previously only contained the Gibbs free energy change for the $\ell \to s$ transformation from the supercooled melt. The two remaining terms are both area terms involving the relevant interfacial tensions. Given that the volume of the nucleus is proportional to the number of atoms, n, and its areas are proportional to $n^{\frac{2}{3}}$, it is possible to represent the total free energy for heterogeneous nucleation of a *strained* spherical cap of solid as

$$\Delta G_{het} = nF_V + n^{\frac{2}{3}}F_A. \qquad (12.41)$$

where

$$F_V = \frac{\Omega}{N_A}f(\psi^*)(G_v+E_\epsilon)\,, \qquad (12.42)$$

---

[6] Calculation of the nuclear strain energy density, E_ϵ, in the free energy expressions, Eqs. (12.40) and (12.42), by applying linear elasticity requires stipulation of the shape of the nuclear volume, as well as knowing the elastic stiffnesses of both the crystalline nucleus and the substrate. Such data are difficult to obtain.

and where

$$F_A = \left(\frac{3\Omega}{4N_A}\right)^{\frac{2}{3}} (8\pi)^{\frac{1}{3}} \left(1 - \cos^3 \psi^{\star}\right) \gamma_{s\ell}. \tag{12.43}$$

The effect on the catalytic efficiency of an inoculant by lowering the strain energy through lattice matching is suggested in Fig. 12.10. In the example shown by plotting Eq. (12.41) the strain energy exerts a deleterious influence on the efficacy of the inoculant when its magnitude becomes more than about 50% of the melt's free energy density. Good lattice matching, i.e., low nuclear strain energy, is clearly essential for efficient nucleation catalysis and effective grain refinement.

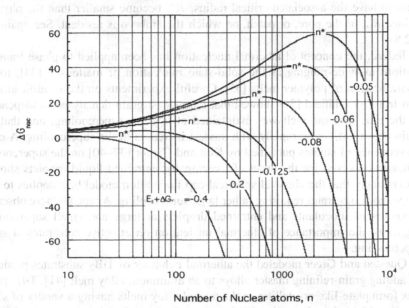

Fig. 12.10 Semi-logarithmic plot of Eq. (12.41) for various total volume free energies, $G_v + E_\epsilon$, versus the number of atoms in a nucleus, n. As the strain energy E_ϵ, decreases, with improving lattice match between the solid and the substrate, the number of atoms (at constant supercooling) to form a critical fluctuation, $n^{\star}$, rapidly diminishes. This demonstrates the effectiveness of good lattice matching when selecting an inoculant

12.6.4 Athermal Heterogeneous Nucleation

In contrast to the stochastic aspects of thermally-controlled nucleation, where random fluctuations over time control the process, there also exists *athermal* nucleation, for which sub-critical embryonic clusters can become activated by additional supercooling. Activation of athermal nuclei occurs in a time-independent manner [33].

The first suggestion of athermal nuclei in melts was argued by Turbull, in his account on the interesting observation that the maximum supercooling sustained by melts prior to their spontaneous nucleation increased, if the melt was previously superheated above its liquidus temperature [12]. Moreover, experiments showed that the higher the superheat, the greater was the tendency of the melt to supercool subsequently. Turnbull's explanation for this curious relationship between prior superheating and subsequent supercooling was that so-called athermal nuclei of the solid were present that could persist up to some level of superheating well above the melt's liquidus temperature! Their resistance to melting was caused by their being located within microscopic pores, cracks, and fissures, located here and there within the crucible walls. Turnbull pointed out that the activation of an athermal embryo to elicit crystal growth required achieving a supercooling level sufficient to have the associated critical radius, $R^\star$, become smaller than the physical radius, R_p, of the pore, or patch, on which the embryo is resident. See again Eq. (12.8).

Indeed, the concept of athermal nucleation has been applied to phase transformations as wide-ranging as the solid-state nucleation of martensite [34], to the crystallization of polymer melts [35]. Careful experiments on the solidification of cast irons by Oldfield [36] showed that the as-cast grain density was independent of the time that the melt was maintained at a given supercooling, and that the grain density also rose rapidly with modest increases in the supercooling. A careful sequence of studies published by Kim and Cantor [37–40] on the supercooling behavior observed for the onset of freezing for entrained liquid droplets showed convincingly that the classical spherical-cap fluctuation model best applies to heterogeneous substrates requiring rather large supercooling. Again, all these observed behaviors of inoculants and entrained droplets at large and small supercooling suggested the importance of athermal nucleation on effective substrates at small supercooling.

Quested and Greer modeled the athermal behavior of TiB_2 substrates produced by adding grain-refining master alloys to an aluminum alloy melt [41]. TiB_2 particles form plate-like crystallites in the treated alloy melts having a variety of sizes. See Fig. 12.11 that shows the development of athermal embryos resident on a flat circular TiB_2 substrate, each having a contact area $A = \pi(R_p)^2$. At some small supercooling such TiB_2 substrates catalyze a layer of crystalline Al by the usual process of heterogeneous nucleation. These athermal embryos develop the critical radius of curvature, $R^\star$, appropriate to the supercooling. Lowering the temperature of the melt slightly then allows $R^\star$ to decrease slightly. However, so long as $R^\star \geq R_P$, the athermal embryo remains inactive, as any additional increase in its volume would further reduce its radius below that dictated by the supercooling. A reduced radius is thermodynamically unstable, so the embryo is forced to shrink back to its prior larger critical radius, which is permitted by the supercooling.

If, however, the melt temperature is reduced further, then the embryo's critical radius can become sufficiently small that it becomes less than its base radius on the substrate, R_P. At that point the spherical cap can continue to grow spontaneously without restriction, which is designated as 'free growth'.

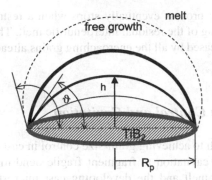

Fig. 12.11 Quested and Greer's model of an athermal crystalline nucleus resident on a surface of TiB$_2$ inoculant. Depending on the supercooling, an *inactive* athermal nucleus can have the height, h, of its *cap-shaped* layer of solid vary from near zero (for a monolayer of solid with repose angle $\vartheta \approx 0$), up to R_p, which is the limiting radius of the patch or substrate, having a repose angle $\vartheta = \pi/2$. If the supercooling exceeds a value where the critical radius $R^\star \leq R_p$, then the athermal embryo becomes *activated* and 'free growth' ensues (*dashed spherical cap*, exhibiting an increasing unstable repose angle $\vartheta \geq \pi/2$). Athermal nucleation as shown here accounts for grain refinement catalysis at the small supercoolings (several Kelvins) typically encountered in commercial casting practice where inoculants are added to the melt

The critical supercooling, ΔT_{fg} required for activation of an athermal embryo of base radius, R_P, and the initiation of its 'free growth' condition is easily found by setting the critical radius, $R^\star$, given by Eq. (12.8) equal to the embryo's base radius, namely,

$$R^\star = \frac{2\gamma_{s\ell}\Omega}{\Delta S_f \cdot \Delta T_{fg}} = R_P, \tag{12.44}$$

and so the supercooling for activating a pre-existing athermal nucleus is

$$\Delta T_{fg} = \frac{2\gamma_{s\ell}\Omega}{\Delta S_f \cdot R_P}. \tag{12.45}$$

It is interesting to note that achieving efficient catalysis for grain refining inoculants such as TiB$_2$ requires *both* classical heterogeneous nucleation—a *stochastic* process—to form an initial distribution of athermal (non-active) embryos on substrates of varying sizes (typically base radii in the range 0.1–1 μm), followed by *deterministic* activation of the largest of these athermal embryos remaining in the melt. As the temperature is gradually lowered during casting, smaller (but yet the largest remaining) athermal embryos become activated sequentially. For *inactive* athermal nuclei with a given base radius, the angle of repose may vary between $0 \leq \vartheta < \pi/2$, depending on the supercooling, $\Delta T \leq \Delta T_{fg}$. When finally the condition is reached that $\Delta T > \Delta T_{fg}$, the athermal embryo activates, the angle of repose $\vartheta > \pi/2$, and the barrier against free growth is removed.

The grain refining process eventually stops when a restriction is imposed preventing further cooling of the residual interdendritic melt. This restriction is caused by the latent heat released by all the encroaching grains already growing throughout the melt.

12.6.5 Melt Flow Control and Cavitation

An alternate approach to achieving grain-size control in cast structures is to employ melt convection and cavitation to fragment fragile dendritic crystals. Flow interactions between the melt and the developing cast microstructure are extremely complicated, and involve modification of the thermal and solutal fields adjacent to the advancing interface [42]. In situ experiments performed with transparent alloys show that both forced and natural convection can effectively 'fragment'[7] growing alloy dendrites and then disperse numerous particles back into the melt, some of which 'seed' new crystals that resume equiaxed growth and help to refine the final grain size [43–45]. This type of flow-mediated fragmentation is generally *not* associated with the fluid mechanical stresses developed between the growing dendrites and the flowing melt. Instead the observations suggest that the fragmentation mechanism is a type of capillary-induced melting, where portions of a dendrite are selectively melted, and branches detach from the primary dendritic stem. Figure 12.12 shows the as-cast grain structure of three castings of Sn-3wt%Zn alloy. The conventionally (static) casting exhibits columnar dendritic crystals that abruptly stop growing and are replaced by coarse equiaxed dendritic grains. The centrifugally cast sample, causing higher levels of melt convection, exhibits near elimination of the equiaxed grains plus a slight refinement of the columnar grains. Finally, the casting which was oscillated continually during solidification experienced the strongest melt convection that effectively fragmented the dendrites, eliminating all the columnar grains, and producing a uniform population of well-refined grains.

Cavitation of the melt can be induced by high intensity sonic and ultrasonic waves coupled to the melt through a buffer rod from a powerful transducer or hydrophone stack. See Fig. 12.13. If the acoustic impedance between the melt and the transducer is reduced, the emitted sound waves induce strong rarefactions (negative pressure regions), which, if powerful enough, can rupture the melt at selective 'weak' spots, where either the gas concentration is high, or where poorly wetted areas exist on particles or external surfaces. When these ultrasonically-induced cavities eventually collapse, their high closing velocities generate pressure shocks that propagates and fragment dendrites into a shower of fine crystallites. Some crystalline fragments survive and 'seed' the formation of new grains. Besides the relatively high costs of setting up powerful transducers to induce cavitation in a

---

[7] The verb 'fragment', used here is not meant to imply merely the mechanical breakage, or disruption, of the advancing dendrites, but also includes an array of thermo-chemical processes including dissolution, coarsening, and local re-melting of already solidified crystallites.

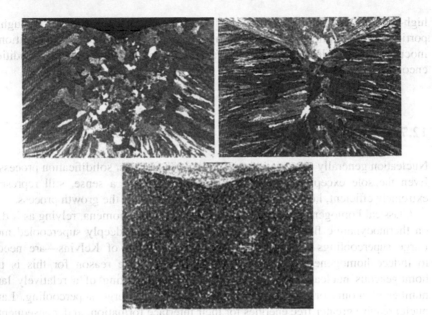

Fig. 12.12 As-cast microstructures Sn-3wt%Zn alloy. *Top left:* Statically cast, displaying the classical formation of columnar grains growing inwards, with an abrupt transition to a zone of equiaxed grains. *Top right:* Centrifugally spun during casting, showing finer columnar grains lacking the transition to equiaxed grains. *Bottom:* Oscillated, exhibiting efficient grain refinement by flow-induced 'fragmentation', with elimination of all columnar dendritic grains

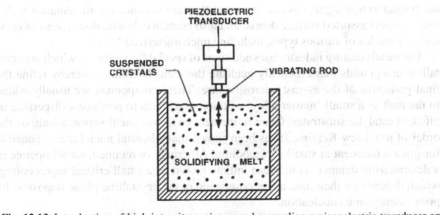

Fig. 12.13 Introduction of high intensity sonic waves by coupling a piezoelectric transducer and the melt. Intense sound waves can produce rarefactions of sufficient negative pressure to locally rupture the melt. Upon collapse of these cavities high pressure shocks can form that fragment fine dendritic crystals. Dendritic fragments can act as 'seed' crystals to refine the as-cast grain size

high-temperature melt, this process is difficult to implement uniformly through all portions of a complicated shaped casting or ingot. Grain refinement by addition of inoculants is much less expensive, and far easier to apply under practical conditions encountered in the cast shop.

12.7 Summary

Nucleation generally comprises the initiating step for most solidification processes. Even the sole exception of 'seeded' crystallization, in a sense, still represents extremely efficient, i.e. near-perfect, epitaxial catalysis of the growth process.

Classical homogeneous nucleation is a stochastic phenomena, relying as it does on thermodynamic fluctuations occurring throughout a deeply supercooled melt. Large supercoolings—on the order of some hundreds of Kelvins—are needed to induce homogeneous nucleation in such melts. The reason for this is that homogeneous nucleation requires critical nuclei consisting of a relatively large number of atoms or molecules, except at extremely large supercooling. Large nuclei require greater free energies for their interface formation, and, consequently, develop a high energy barrier that reduces the probability of successful fluctuations occurring.

By contrast, heterogeneous nucleation involves the assistance of extrinsic substrates, which catalyze the nucleation event by reducing the number of atoms in a critical nucleus. The efficacy of such substrates depends on factors such their total area and orientation, crystallographic match, and similarity of their bonding (metallic, ionic, covalent) to that of the crystallizing solid. Normally, most melts are subject to heterogeneous nucleation, as they are in contact with container walls, often support assorted surface debris, and also contain colloidal dispersions of crystalline particles of various types, including uncharacterized 'dirt'.

The metals casting industry uses additions of special 'inoculants', which are crystalline compounds that efficiently nucleate the melt. Inoculants thereby refine the final grain-size of the as-cast microstructure. These compounds are usually added to the melt as a small 'master alloy' addition that reacts to produce a dispersion of efficient catalytic substrates. Grain refinement ensues at small supercooling, on the order of just a few Kelvins. It has been shown that athermal nuclei are responsible for grain refinement at small supercooling. Athermal, or retained, nuclei operate in a deterministic manner, as they can initiate growth at a small critical supercooling, which depends on their area and repose angle of the crystalline phase deposited by prior heterogenous nucleation.

Grain refinement can also be accomplished by dendritic fragmentation accomplished by using melt flow or ultrasonic vibration. These methods are of interest, as they can control grain size without the addition of inoculants, which add foreign particulates to the as-cast structure. Unfortunately, melt flow control and ultrasonics are expensive and technically difficult to implement in the cast shop, and are not uniformly effective in grain refining complicated castings.

References

1. V.K. LaMer, *Ind. Eng. Chem.*, **44** (1952) 1270.
2. K.C. Russell, *J. Chem. Phys.*, **50** (1969) 1809.
3. J.L. Katz, *J. Stat. Phys.*, **2** (1970) 137.
4. M. Volmer, *Z. Elektrochem.*, **35** (1929) 555.
5. M. Volmer and A. Weber, *Z. Phys. Chem.*, **119**, (1925) 277.
6. M. Volmer and H. Flood, *Z. Phys. Chem.*, **170A** (1934) 273.
7. L. Farkas, *Z. Phys. Chem.*, **125**, (1927) 239.
8. R. Becker and W. Döring, *Ann. Phys.*, **24** (1935) 719.
9. J.B. Zeldovich, *Acta Physicochim.*, **18** (1943) 1.
10. J. Frenkel, *Kinetic Theory of Liquids*, Dover Publications, Inc., New York, NY, 1955, p. 176.
11. D. Turnbull, *J. Chem. Phys.*, **20** (1952) 411.
12. D. Turnbull, *J. Phys. Chem.* **18** (1950) 198.
13. D. Turnbull and J.C. Fisher, *J. Phys. Chem.* **17** (1949) 71.
14. K.F. Kelton and A.L. Greer, *Nucleation in Condensed Matter*, **15**, Pergamon Materials Series, Elsevier BV, Holland, 2010.
15. G.S. Heffelfinger et al., *Mol. Simul.*, **2** (1989) 393.
16. A. Schmidt and V. Smirnov, *TopCatal.*, **32** (2005) 71.
17. M.K. Harbola, *Proc. Natl. Acad. Sci. USA*, **89** (1992) 1036.
18. D. Turnbull, *J. Chem. Phys.*, **18** (1950) 198.
19. M.E. Glicksman and W.J. Childs, *Acta Metall.*, **10** (1962) 925.
20. J.H. Perepezko and W.S. Tong, *Phil. Trans. R. Soc. Lond. A*, **361** (2003) 447.
21. W. Yourgrau, A. van der Merwe and G. Raw, *Treatise on Irreversible and Statistical Thermophysics*, Dover Publications, New York, NY, 1982.
22. E.H. Kennard, *Kinetic Theory of Gases*, McGraw-Hill Book Company, New York, NY, 1938.
23. M.C. Flemings et al., *AFS Trans.*, **70** (1962) 1029.
24. C. Vivès, *Metall. Trans. B*, **27** (1996) 445.
25. M. Easton and D. St. John, *Metall. Mater. Trans. A*, **30** (2007) 1625.
26. J.P. Perepezko, 'Nucleation kinetics and grain refinement', *ASM Handbook*, **15**, American Society for Materials, Int., Materials Park, OH, 2008, p. 276.
27. C. Limmaneevichitr and W. Eidhed, *Mater. Sci. Eng. A*, **349** (2003) 197.
28. E.B. Webb III, G.S. Grest, and D.R. Heine, *Phys. Rev. Lett.*, **91** (2003) 236103.
29. N.H. Fletcher, *J. Chem Phys.*, **29** (1958) 572.
30. N.H. Fletcher, *J. Chem Phys.*, **31** (1958) 1136.
31. N.H. Fletcher, *J. Chem Phys.*, **38** (1958) 237.
32. T.B. Blake and J.M. Haynes, *J. Coll. Inter. Sci.*, **30** (1969) 421.
33. J.C. Fisher, J.H. Holloman and D. Turnbull, *J. Appl. Phys.*, **19** (1948) 775.
34. G. Ghosh and G.B. Olsen, *Acta Metall. Mater.*, **42** (1994) 3361.
35. P. Sajkiewicz, *J. Polym. Sci. B*, **41** (2003) 68.
36. W. Oldfield, *Trans. ASM*, **59** (1966) 945.
37. W.T. Kim, D.L. Zhang and B. Cantor, *Metall. Trans. A*, **22A** (1991) 2487.
38. W.T. Kim and B. Cantor, *J. Mater. Sci.*, **26** (1991) 2868.
39. W.T. Kim and B. Cantor, *Acta Metall.*, **40** (1992) 3339.
40. W.T. Kim and B. Cantor, *Acta Metall.*, **42** (1994) 3045.
41. T.E. Quested and A.L. Greer, *Acta Mater.*, **53** (2005) 2683.
42. J. Szekely, *Fluid Flow Phenomena in Metals Processing*, Academic Press Ltd., New York, NY, 1979, 305.
43. M.E. Glicksman, C.J. Paradies, G.T. Kim and R.N. Smith, 'The effects of varying melt flow velocity on the grain fragmentation in a mushy zone', *Light Metals 1993*, S.K. Das, Ed., The Minerals, Metals & Materials Society, Warrendale, PA, 1993, p. 779.
44. M.E. Glicksman, R.N. Smith, C.J. Paradies and P. Rao, 'Grain refinement of castings via melt-flow interaction with the mushy zone', *Near-Net Shape Casting in the Minimills*, I.V. Samarasekera, Ed., The University of British Columbia, Vancouver, BC, 1995, p. 217.
45. C.J. Paradies, M.E. Glicksman and R.N. Smith, *Metall. Mater. Trans. A*, **28A** (1997) 875.

Part IV
Microstructure Evolution

Chapter 13
Dendritic Growth

13.1 Introduction

In Chapter 10 it was shown that the onset of *linear* morphological instability initiates exponential growth of local perturbations on the interface. A modal analysis using sine-wave ripples disclosed a particular wavenumber, $k^\star$, that was found to be 'most dangerous' with respect to inducing interface instability. High-amplitude features develop further on the interface by non-linear processes to form deep cells, which evolve into yet more complex solidification structures[1] including dendrites, which are non-equilibrium structures that persist throughout the entire subsequent process of solidification.

Dendrites (which derives from the Greek word 'dendron' ($\delta\epsilon\nu\delta\rho\omega\nu$—a tree) are ramified (branched) single crystals, exhibiting morphological features that display crystallographic directionality, such as straight primary stems, secondary side arms, and even tertiary branches. These branches bear special angular relations between each other, and also exhibit appropriate symmetries reflecting the underlying crystal structure of the material. For example, most cubic materials develop dendrites with each higher order of branching developing orthogonally to the previous order along one of the six ⟨100⟩ cube-edge directions.

Dendrites, in a sense, represent the fully evolved end-state of an unstable crystal-melt interface. In this respect, dendrites form solidification microstructures that have passed through enough morphogenic processes to have totally 'forgotten' their origin as some initially small fluctuation, or persistent disturbance, on an otherwise smooth and featureless, crystal-melt interface.

Dendrites gain their importance in cast materials as they establish the scale of chemical microsegregation and the casting's solidification texture and grain size, both of which influence mechanical properties and post-casting processing.

---

[1] See again the experimental observations presented in Chapter 9. There, in situ micrographs are shown of the evolution of unstable structures, including dendrites that develop near grain boundary grooves and other persistent disturbances on dilute alloy crystal-melt interfaces.

M.E. Glicksman, *Principles of Solidification*, DOI 10.1007/978-1-4419-7344-3_13,
© Springer Science+Business Media, LLC 2011

13.1.1 Some Early History

Snow flakes and frost patterns provide the most common and obvious examples of dendritic growth in nature. Snow flakes grow from miniscule ice grains falling and tumbling through a water-supersaturated atmosphere, whereas frost patterns form from supercooled moisture condensing and eventually freezing dendritically on cold surfaces, such as a window pane. Snow flakes, the iconic symbol of winter in the northern hemisphere, are well known for their distinctive sixfold symmetry, reflecting their underlying arrangement of H_2O molecules in the rhombohedral crystal structure of ice, as well as for their variable and beautiful branching patterns [1, 2]. See Fig. 13.1.

Fig. 13.1 Magnified images of snow flakes (H_2O ice) as viewed on the {0001} basal plane. The sixfold symmetry arises from ice's rhombohedral crystal structure. Inspection of these snow flake patterns reveals three levels of dendritic branching. One notes, however, that despite the apparent—but only approximate—symmetry, each main branch, or primary stem, is unique in its detailed form, with each stem exhibiting fine-scale features not mirrored by the other five arms. Photos adopted from Libbrecht [3]

Indeed, dendritic snow-flake patterns over the ages have caught the attention of such astute observers of nature as Johannes Kepler—the great German mathematician and astronomer—who in 1611, wrote the essay, 'Strena Seu De Nive Sexangula',[2] and later Henry David Thoreau—the well-known nineteenth century American philosopher and naturalist—who added in his Journal, 'How full of the creative genius in the air in which these [snow flakes] are generated!'

13.2 Dendrites in Metals Processing

Dendrites comprise a ubiquitous solidification morphology, appearing as they do in most microstructures in metal castings. They remain important in a wide variety

---

[2] Kepler's title, 'Strena Seu De Nive Sexangula', which translates to 'A New Year's gift of hexagonal snow', is his commentary on how complex and beautiful patterns seemingly emerge mysteriously out of 'thin air' [4].

of metals processing technology. Specifically, dendritic morphologies occur commonly in as-solidified alloy crystals and polycrystalline castings produced by many industrial metal molding methods, including shape casting, semi-finished ingot practice, directional casting, die casting, and in all fusion weldments.

13.2.1 Observations and Simulation of Dendritic Growth

Metallic dendrites, as a class, are capable of growing relatively rapidly (0.1–1 mm/s) even under weak thermal gradients and up to speeds as great as tens of m/s in highly supercooled metallic melts [5, 6].[3] An early photo sequence taken *in situ* during dendritic growth in a metallic melt is shown in Fig. 13.2. The four photos were taken over uniform time intervals of a Sn dendrite growing over the surface of slightly supercooled molten tin. The Sn surfaces were protected from oxidation by a layer of a transparent flux (molten anhydrous $SnCl_4$). This sequence of micrographs clearly shows a propagating metallic dendrite developing small, wake-like branches trailing behind its steadily advancing tip.

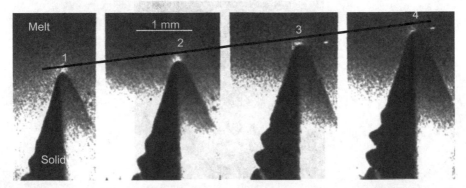

Fig. 13.2 In situ photo-sequence of dendrites growing on the surface of supercooled pure molten Sn. The molten Sn was protected from oxidation by a transparent (invisible)) flux layer. The four micrographs, taken at equal time intervals, demonstrate the steady *tip shape* and the uniform rate of advance of thermal dendrites. Side branches continually develop behind the growing tip (adopted from [8])

Impressive strides have been made over the past four decades in understanding the fundamentals of dendritic growth, allowing advanced engineering applications of dendritic growth in castings. One may glimpse a bit of this technological advancement by comparing the above experimental observation of a free-growing Sn den-

---

[3] These high-speed dendrites never encounter absolute stability because the thermal gradients responsible for their growth are *negative* in supercooled melts. Absolute stability requires positive thermal gradients. See Section 10.5.

drite growing in its low-temperature single-component melt, with the predictive power of combining computational thermodynamics with multicomponent diffusion and simulated microstructure evolution. Today, dendritic growth may be simulated with reasonable accuracy in complex, high-temperature alloys. For example, Fig. 13.3 shows the computed image of a Ni-dendrite and its surrounding thermal field while growing from a 5-component alloy melt. Commercial data bases now exist similarly to allow simulation of the solidification of tool steels, superalloys, semiconductors, ceramics, as well as numerous non-ferrous casting alloys based on Al, Ti, Mg, and Zr [7]. Dendritic growth occurs in most supercooled pure melts or supersaturated alloy melts. As shown in Section 2.3, the Gibbs free energy per unit mass of a supercooled melt is greater than that that of the solid at its equilibrium transition temperature. Nucleation of the solid phase in such a metastable melt leads to prompt, spontaneous dendritic crystal growth. The fraction of solid, X_s, that spontaneously forms is eventually limited by the supercooling, or thermal Stefan number. As long as a melt remains supercooled, there is no need to remove the latent heat of freezing to an external environment, as freezing would cease only at the point of adding sufficient latent heat to raise the melt's temperature back to its equilibrium liquidus.

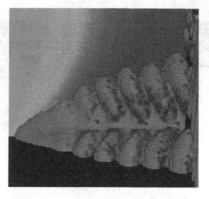

Fig. 13.3 Modern computer simulation of a Ni-alloy dendrite growing from a casting wall in a 5-component (Ni-Al-Cr-Ta-W) melt. *Color scale* suggests the thermal field developed in the surrounding melt into which the dendrite advances. This simulation combines commercially-available software packages developed for computational thermodynamics (THERMO-CALC®), multicomponent diffusion simulations (DICTRA®), and a microstructure evolution code (MICRESS®). Simulation snapshot adapted from [7]

13.2.2 Initiation of Dendritic Growth

The metastability of supercooled melts allows the initiation of dendritic growth by either spontaneous nucleation at some favorable site within the melt itself, on the container wall, or through a controlled 'triggering' event induced by contacting the melt with a seed crystal, usually in the form of a fine wire.[4] Figure 13.4-left shows an example of a random nucleation event occurring within the volume of a transparent supercooled melt, initiating the early stages of an equiaxed, dendritic, cubic crystal of succinonitrile (SCN).[5] This micrograph was taken normal to a {100} plane, with two of the dendrite's ⟨100⟩ growth directions oriented normal to the page. Again, hints of side-branches can be observed developing near the base of the four project-ing tips. Figure 13.4-right shows the later stage of the downward pointing primary stem. Well developed side arms grow outward in orthogonal branching sheets, which are fin-like extensions of the {100} planes of this cubic crystal.

Fig. 13.4 *Left:* Photomicrograph of the nucleation and early stages in the growth of an equiaxed dendritric crystal from a supercooled melt of pure, transparent, SCN. This crystal has the BCC crystal structure, and exhibits early development of primary dendritic stems extending in the six, orthogonal, cube-edge ⟨100⟩ directions—with four stems growing in the plane of the photograph, and two growing normal to the page. The six dendrite tips locate the vertices of an octahedron in the melt space, which is the cubic analog to the hexagonal snow flake shown in Fig. 13.1. Evidence for the development of side branches is just beginning on each stem behind the advancing tips. *Right:* Dendritic SCN crystal at a more advanced stage, showing development of nearly periodic side arms along the downward-pointing primary stem. These arms protrude in the four orthogonal {100} planes, forming branching sheets

13.2.3 Directional Solidification

Alloy dendrites, by contrast can develop and grow spontaneously during directional solidification (DS) of a melt. Figure 13.5-left shows aligned dendrites of a dilute

---

[4] The topic of nucleation catalysis of solid phases from metastable melts is developed in Chapter 12.

[5] Succinonitrile is a transparent organic 'plastic crystal' with the molecular formula 'N$_2$–(CH$_4$)$_2$–N$_2$'. This compound freezes at approximately 58 C. The solid phase allows hindered molecular rotation of two geometric isomers, which permits formation of a ductile BCC crystalline phase. Plastic crystalline behavior, although common in metals, is relatively rare among organic compounds. It allows such special substances to be used as 'metallic' analogs in studies that include dendritic crystal growth and plasticity [9].

organic alloy (SCN + acetone) growing steadily with a distinctive uniform tip shape and a well-developed, regular, primary stem spacing. Here the heat of solidification was extracted through the solid. Figure 13.5-right is a video frame of directionally solidifying low-carbon steel, freezing relatively quickly under a steep applied thermal gradient. Alloy dendrites are non-equilibrium structures, as their finely branched forms cause their interfacial area per unit mass to be relatively large, with the excess stored free energy ultimately residing in the as-cast microstructure. Further discussion of surface energy-driven phenomena in cast dendritic microstructures, beyond those already mentioned in Section 8.3, will be deferred until later.

Fig. 13.5 *Left:* An early in situ micrograph of self-aligned organic alloy (SCN-acetone) dendrites growing by slow directional solidification (DS). The heat of fusion is extracted through the solid phase below the *bottom* of the photo, so the melt is not thermally supercooled, just constitutionally supercooled. Note the spacing of the primary stems, the well-developed secondary arm spacing, and the uniform *tip shapes.* Adapted from [10]. Size scale not available from source. *Right:* Frame from a video of solidifying low-carbon steel (Fe-0.11wt.%C) growing at 1 cm/s with an applied thermal gradient of 100 K/mm. Two grains compete in the field of view with about a 17° misorientation in their [100] dendritic growth directions. Note the fine scale of the tips and primary stems of these steel dendrites. Size scale not available from source (adapted from http://www.msm.cam.ac.uk/phase-trans/phase.field.models/movies2.html)

Figure 13.6-left depicts some of the major milestones in the technological development of superalloys cast as near net-shape turbine blades and disks. Today, directionally solidified (DS) single-crystal components for jet engines provide the critical, high-temperature components in high-efficiency turbo-machinery, namely, blades and disks. In addition, because of increasing demand for efficient electrical energy production, large, land-based, gas-fired turbo-generators that use single crystal superalloy components[6] are being used for electric power generation.

Despite their single-crystal nature, DS alloy castings are comprised microstructurally of a dense 'forest' of dendrites, all crystallographically aligned in their pre-

---

[6] The Pratt & Whitney JT9D-7R4 jet engine used single-crystal superalloy technology. It was the first 'single-crystal' aircraft engine, and was flight-certified for use in civilian aircraft in 1982. Today jet aircraft engines from Rolls Royce, General Electric, and Pratt & Whitney all use some variant of solidification technologies that eliminate all grain boundaries in the turbine blade castings, further improving the jet engine's operating temperature, fuel efficiency, safety, reliability, and service life.

Fig. 13.6 *Left:* Cast superalloy turbine blades, each about 10 cm long, with inserts of macrophotographs indicating their cast macrostructures. The three blades, from *left* to *right*, typify the major milestones in superalloy casting technology: (1) Prior to 1960. Polycrystalline blade comprised of random crystallites. The superalloy melt was poured into a shaped mold, and heat was removed omnidirectionally. (2) Circa 1966. Directionally solidified (DS) multi-grain blade, as invented by VerSnyder at Pratt & Whitney Corp., eliminating transverse grain boundaries. (3) After 1982. Single crystal jet engine blade. Random nucleation of multi-grains near the tang is eliminated, and directional single-crystal growth is controlled using a 'seed' crystal, and an orientation selector. Adapted from [11]. *Right:* Scanning electron micrograph of a dense forest of nickel-base superalloy dendrites aligned during DS processing of a single-crystal turbine blade. Note the near-perfect alignment of the growth orientation of every dendrite, and the parallelism of their secondary branches, or sidearms. Scanning electron microscopy specimen prepared by sudden accidental decanting of a superalloy melt. Courtesy of S. David and L. Boatner, Oak Ridge National Laboratory (Private Communication, 1995)

ferred growth direction—typically $\langle 100 \rangle$ for the cubic metallic phases in Ni-based superalloys. As shown in Fig. 13.6-right, the strict alignment of all the dendrites freezing across a section of the DS mold, precludes the formation of any high-angle grain boundaries throughout the casting. Eliminating high-angle grain boundaries confers exceptional creep, rupture life, and fatigue properties. Unavoidable thermo-mechanical strains occur within the dendrite forest, causing some low-angle 'polygonization' boundaries to form. Low-angle boundaries with misorientations less than 1° are not considered deleterious to the performance of DS single-crystal blades. A particularly harmful casting defect termed 'freckle grains', however, can also develop during DS processing, particularly where sharp section changes occur in the blade mold. Freckle grains are believed to be caused by fragmentation (melting off) of dendrite arms [12] or by spurious nucleation. Considerable engineering effort and expense is warranted to avoid the appearance of freckle grains by both judicious alloy selection and appropriate heat-flow control during DS.

13.3 Dendrites in Castings

As suggested in Chapter 9 (See again Fig. 9.11.) alloy castings often solidify under conditions of low thermal gradients at relatively fast rates (several mm/s). Under

such circumstances dendrites are expected to form in the microstructure without
the assistance of directional freezing imposed by direct heat transfer. These dendrites are more randomly oriented in space, although each is again a single crystal,
exhibiting a well-developed primary stem crystallography usually developing along
a $\langle 100 \rangle$ zone axis in cubic materials. Figure 13.7 shows a good example of alloy
dendrites in a casting imaged using scanning electron microscopy (SEM).

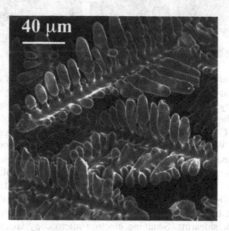

Fig. 13.7 SEM image of dendrites from a Al-Cu alloy casting. Note how the proximity of these
randomly growing crystals causes foreshortened side branches to develop on the upper dendrite.
This effect is caused by melt convection. Here latent heat released from the lower dendrites causes
the melt (removed by decanting) to be buoyant and rise, thereby slowing the growth of some of
the branches on the upper dendrite. These complex heat-transfer effects profoundly influence the
microstructures of conventionally cast alloys

13.4 Dendritic Growth Theory

Kinetic and transport theories were developed to predict the solidification characteristics of dendrites as a consequence of their widespread commercial importance
in castings, ingots, weldments, and shaped crystals, as discussed above. These theories provide predictions (estimates) that include dendritic growth direction, speed,
dendrite size, and morphology. All these growth characteristics, in turn, help determine properties often considered crucial in many cast materials applications, especially their rupture strength and ductility, texture, fatigue and fracture properties.
Even the chemical and electrical characteristics of cast alloys, including their corrosion resistance and electrical conductivity are influenced strongly by the dendritic
microstructure. The interested reader should see references [13, 14] for a comprehensive review of the extensive research literature of the subject of dendritic
growth.

An outline of the standard theoretical treatments for predicting dendritic
behavior using fundamental principles follows. The theory consists of two distinct

components: (1) transport theory for heat conduction and mass diffusion, developed for dendrites in the 1940s, and (2) interfacial physics, considered by physicists some thirty years later, who applied aspects of linear morphological stability to dendritic growth. See again Chapter 10. Dendritic growth theory was tested experimentally in the 1990s, and provides reasonable engineering estimates of dendritic character- istics through so-called 'scaling laws', which will be fully developed subsequently. A few additional contemporary ideas regarding the fundamental basis for dendrite formation, introduced only recently, are provided toward the end of this chapter in Section 13.9.

13.4.1 Ivantsov's Transport Model

Russian mathematician G.P. Ivantsov developed an exact mathematical solution to the temperature distribution in the melt surrounding an isothermal steady-state dendrite [15]. Ivantsov, in 1947, solved the case of a needle-like, or branchless dendrite, the steady-state form of which—a paraboloid—was initially suggested through observations of dendritic crystals published in 1935 by Achille Papapetrou [16].

Figure 13.8-top shows a dendrite of SCN. The vertical line drawn on this photomicrograph separates the 'steady-state' tip region from the time-dependent branching region. The paraboloidal, branchless form used in Ivantsov's theory is suggested in Fig. 13.8-bottom, including its associated co-moving coordinate sys- tem. The lines of heat flow and isotherms in the surrounding melt form a set of orthogonal surfaces described by confocal paraboloidal coordinates [17, 18], a pla- nar projection of which is also included on the lower sketch.

The solid and the melt are both assumed to be pure (one-component) phases, with a melting point, T_m, which is the prescribed temperature at the steadily growing solid-melt interface. The effect of curvature on the melting point (Gibbs–Thomson effect) is not included in Ivantsov's transport analysis.

Of course, a real dendrite, see again Fig. 13.8-top, displays a far more complex, time-dependent morphology than is suggested by the smooth, branchless, 'steady- state' shape, which is well-described as a paraboloid of revolution. The branchless shape is compared in Fig. 13.8-bottom. Despite overlooking the time-dependent complexities of real dendrites, Ivantsov's steady-state heat flow analysis captures several important aspects of the thermal conduction process surrounding a growing dendrite tip. Thermal conduction, or heat diffusion, in fact, controls the dendrite's axial propagation speed, and its tip radius.

Time-dependent features, such as the trailing wake of periodic side branches, are of great practical importance in casting processes, because side branches induce lat- eral solute segregation, and are responsible for the occurrence and scale of chemical microsegregation in dendritically solidified alloys. The presence of side branches, fortunately, has only a modest effect on either the thermal or solute diffusion pro- cesses around the smoother tip region. The tip regions of some dendrites, as shown

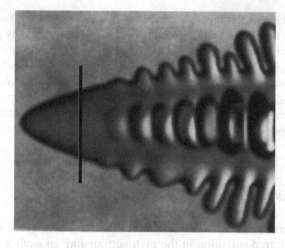

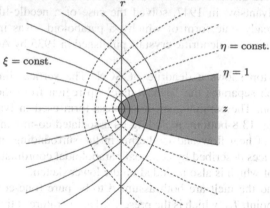

Fig. 13.8 *Top:* In situ photomicrograph of a dendrite of SCN. The *vertical bar* separates the 'steady-state' tip region, which is nearly a body of revolution, from the time-dependent side branches that protrude in the four ⟨100⟩ directions flanking the primary stem. *Bottom:* Steady-state branchless paraboloid, as suggested by Papapetrou, used to formulate Ivantsov's dendritic transport model. Heat flow lines, $\xi =$ const., and isotherms, $\eta =$ const., are indicated in this projection onto the plane of the diagram of the three-dimensional thermal field, $T_\ell(r, \xi, \eta)$, where the (r, ξ, z) coordinate system is co-moving with the tip. The solid–liquid interface shape in this moving coordinate system is the surface $\eta = 1$, and the extent of the dendrite's diffusion boundary layer is suggested by different values of $\eta =$ const. (adapted from [19])

in Fig. 13.8-top, seem well described by the needle-like form of a paraboloid of revolution.

13.4.1.1 Boundary Conditions

For pure materials, solute diffusion is, of course, inoperative, and the conditions for seeking Ivantsov's heat transport solution involve only the steady-state conduction of latent heat away from the moving dendritic interface into the surrounding supercooled melt. The following mathematical assumptions and boundary conditions establish the basis of Ivantsov's transport solution:

1. The thermal field is expressed in confocal paraboloidal coordinates, or, equivalently, in cylindrical coordinates, that remain embedded in a frame co-moving at the speed, v, of the dendrite tip growing along the z-axis. See again Fig. 13.8-bottom. Ivantsov's thermal field solution is conveniently expressed in dimensionless moving cylindrical coordinates, using the (unknown) tip radius as the scale factor, viz., z/R_{tip}, and r/R_{tip}.
2. The temperature of the paraboloidal solid-melt interface is everywhere constant and fixed at the freezing point, T_m. The influences on the equilibrium temperature of the interfacial curvature encountered near the tip and the solid–liquid interfacial energy are ignored in Ivantsov's analysis. This is tantamount to assuming that the interfacial energy is zero.
3. A supercooled temperature below the melting point exists at all points in the supercooled melt that are far away from the solid–liquid interface. During dendritic growth, the melt far from the advancing dendrite ($r \rightarrow \infty$) has a uniform supercooled temperature given as $T(\infty, z) = T_\infty$. The melt, as described above in Section 13.2.1, is initially supercooled by an amount $\Delta T = T_m - T_\infty$, permitting dendritic growth to occur by steadily absorbing the latent heat generated at the interface via thermal conduction into the surrounding cooler melt, locally raising the melt's temperature.
4. Ivantsov correctly suggested that the steady-state solution to the energy equation, Eq. (4.38), could be obtained in the co-moving coordinate frame. Many materials with cubic crystal structures have their primary growth direction ($+z$ in Ivantsov's theory) aligned with the crystal's $\langle 100 \rangle$. Moreover, dendrites of cubic materials also exhibit fourfold zone-axis symmetry—not the circular rotational symmetry assumed in the Ivantsov analysis. See again the photomicrograph of a cubic dendrite in Fig. 13.8-top, which shows the four branch sheets in a BCC material. Fortunately, many cubic dendrites show nearly circular symmetry extremely close to their tips.

13.4.1.2 Governing Equations

Ivantsov formulated two steady-state energy equations: one in the melt phase as

$$\frac{\partial T_\ell}{\partial t} = \nabla^2 T_\ell + \frac{v}{\alpha_\ell} \nabla T_\ell = 0, \tag{13.1}$$

and one in the solid as

$$\frac{\partial T_s}{\partial t} = \nabla^2 T_s + \frac{v}{\alpha_s} \nabla T_s = 0. \tag{13.2}$$

The heat flow solution that Ivantsov chose to represent the interior temperature field of the dendrite is the trivial isothermal field,

$$T_s(z, r) = T_m.$$ (13.3)

This interior solution relies on the fact that the temperature within an enclosed slender body with a fixed boundary temperature, T_m, such as a dendritic needle, reaches an isothermal condition almost instantaneously.

13.4.1.3 Dimensionless Parameters

The following dimensionless parameters and equations prove useful both in developing and interpreting Ivantsov's heat conduction solution:

- $\vartheta(z, r)$ is the dimensionless temperature field defined as

$$\vartheta(z, r) \equiv \frac{T(z, r)}{\Delta H_f / C_p}.$$ (13.4)

- The growth Péclet number, Pe, is a dimensionless group defined as

$$Pe \equiv \frac{v R_{tip}}{2\alpha_\ell}.$$ (13.5)

The Péclet number may be thought of as the ratio of the tip radius, R_{tip}, to the thermal conduction length, $2\alpha_\ell/v$.
- The steady-state energy equation for the melt, Eq. (13.1), may be non-dimensionalized using the variables defined above as

$$\nabla^2 \vartheta + 2Pe \nabla \vartheta = 0,$$ (13.6)

where the Laplacian operator, ∇^2, and the gradient operator, ∇, are each taken with respect to the non-dimensional coordinates, z/R_{tip} and r/R_{tip}.
- Ivantsov chose boundary conditions on the solid and melt, and assumed an isothermal solid–liquid interface at the melting point, and a uniform far-field temperature far from the dendrite in the melt. In dimensionless variables, these boundary conditions are given as the interface temperature,

$$\hat{\vartheta} = \frac{T_m}{\Delta H_f / C_p},$$ (13.7)

and the far-field temperature,

$$\vartheta_\infty = \frac{T_\infty}{\Delta H_f / C_p}.$$ (13.8)

13.4.2 Ivantsov's Transport Solution

A clever insight that Ivantsov added to his analysis is that the paraboloidal shape of the dendrite tip allows the use of a separable coordinate system (confocal paraboloids). Ivantsov's solution has since been generalized to other shapes that are not bodies of revolution. These other dendritic shapes also grow at constant speed, and provide a complete family of shape-preserving elliptical paraboloidal interfaces [20].

Ivantsov obtained the dimensionless temperature solution surrounding the growing dendrite, $\vartheta(z/R_{tip}, r/R_{tip})$, for the paraboloid of revolution, i.e., for the 'needle' dendrite. His transport solution may be used to relate the dimensionless supercooling, $\Delta\vartheta$, to the growth Péclet number, Pe. The so-called *characteristic* equation connecting these transport parameters is given by

$$\Delta\vartheta = Pe\, e^{Pe} E_1(Pe) \equiv \mathcal{I}v(Pe), \tag{13.9}$$

where $E_1(Pe)$, is the 1st exponential integral, a tabulated function which is defined as the definite integral

$$E_1(Pe) \equiv \int_{Pe}^{\infty} \frac{e^{-\zeta}}{\zeta} d\zeta. \tag{13.10}$$

Ivantsov's steady-state dendrite solution, Eq. (13.9), implies that the dimensionless supercooling can be expressed as function of the growth Péclet number, namely, $\Delta\vartheta = \mathcal{I}v(Pe)$. In most solidification processes, however, the preferred independent variable is the supercooling or supersaturation, viz., $\Delta\vartheta$, and the growth Péclet number usually becomes the dependent variable. Thus, the inverse to Eq. (13.9) would offer a more convenient mathematical solution of the transport problem. An analytic inverse to Eq. (13.9), unfortunately, cannot be found as a 'closed form', as is often the case when inverting transcendental equations. That is, a combination of elementary functions to represent the inverse cannot be found. Instead, one may think of the inverse to Eq. (13.9), and symbolically express it as $Pe = \mathcal{I}v^{-1}(\Delta\vartheta)$. Here the symbol $\mathcal{I}v^{-1}$ represents the formal inverse of Ivantsov's result, Eq. (13.9), which is easily presented as a graph of the Ivantsov solution, with supercooling chosen as the independent variable, or abscissa.

In any event, a specified value of the supercooling, $\Delta\vartheta$, fixes the magnitude of the growth Péclet number. If the melt supercooling is specified, then the following relationship holds between the dendritic growth speed and the tip radius,

$$vR_{tip} = 2\alpha_\ell \cdot \mathcal{I}v^{-1}(\Delta\vartheta). \tag{13.11}$$

Equation (13.10) is plotted on double logarithmic coordinates in Fig. 13.9. At large supercoolings the growth Péclet number increases rapidly, and as $\Delta\vartheta \to 1$, it diverges, and $Pe \to \infty$. The growth Péclet number becomes nearly linear on log-log coordinates in the range of small supercooling, indicating that the Ivantsov solution can be expressed approximately as the power law,

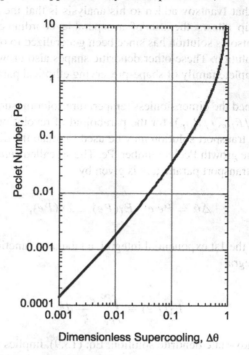

Fig. 13.9 Plot on double logarithmic coordinates of Ivantsov's transport solution, Eq. (13.10). The growth Péclet number becomes singular ($Pe \rightarrow \infty$), diverging as the dimensionless supercooling, $\Delta\vartheta$, approaches unity. Over part of the range of dimensionless supercooling shown here, viz., $0.001 < \Delta\vartheta < 0.05$, the (log) Péclet number appears nearly linear with (log)supercooling, and may be approximated as a power law

$$Pe \approx 0.647 \Delta\vartheta^{1.27} \quad (0.001 < \Delta\vartheta < 0.05). \tag{13.12}$$

Equation (13.12) provides a convenient approximation for making estimates of dendrite behavior in situations where the thermal supercooling is not large; specifically where it falls in the range $0.001 < \Delta\vartheta < 0.05$. Again, although Ivantsov's treatment yields a useful relationship connecting the observable dendrite parameters, viz., $v = v(R_{tip})$, the *unique* transport solution, or 'operating state' of the dendrite, (v_{op}, R_{op}), for a specified supercooling is not provided. The reason, of course, is that Eq. (13.11) contains only the product of the two unknowns, so it is incapable of predicting separately the unique velocity and radius of the dendrite. Thus, an additional independent equation—to be found later in this discussion— must be sought in order to solve the steady-state dendritic growth problem uniquely, and to be able to test *individual* its predictions against observable crystal growth behavior.

Figure 13.10 shows the hyperbolic relationship established between v and R_{tip} by Ivantsov's solution, Eq. (13.11), for several supercoolings at a fixed value of the melt's thermal diffusivity, α_ℓ. These curves show how the velocity of a thermal dendrite must vary with its corresponding tip radius, at constant supercooling. Thick dendrites grow slowly, relative to slender ones that grow quickly. The tendency for slowly growing dendrites to have large blunt tips, and rapidly growing ones to have small sharp tips, is referred to as the 'point effect' of diffusion—a well-known phenomenon in diffusion and electrostatics.

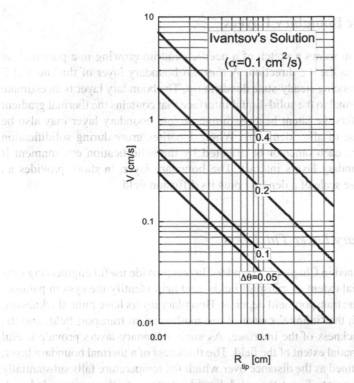

Fig. 13.10 Double logarithmic coordinates of Ivantsov's transport solution, displaying the hyperbolic relationships (*straight lines*) between growth velocity, v, and tip radius, R_{tip} for several values of the dimensionless supercoolings, of heat or solute from a dendrite tip is more efficient at smaller tip radii, which increase the local gradient field. Thus, fast growing dendrites usually exhibit finer scale features, and slower dendrites are coarser. This inverse relationship between speed and feature scale is borne out in the dendrites shown on the LHS and RHS of Fig. 13.5

The Ivantsov solution thus fails to predict the unique combination of variables, (v_{op}, R_{op}), that are displayed if one sets the supercooling and conducts a solidification experiment in a specific melt. The dendrite's 'operating state' in that melt remains indeterminate within the predictive power of Ivantsov's transport analysis.

This, of course, is not a surprising outcome, because the energy transport solution, Eq. (13.11), provides just a single equation in two unknowns, viz., v and R_{tip}. Some additional (interfacial) physics, which would add an independent equation between these variables, will be discussed in Section 13.6 to find the specific operating state of a steady-state dendrite.

Before adding interfacial physics, some engineering concepts of thermal and solutal boundary layers will be discussed to make clear how growing dendrites interact during casting.

13.5 Dendritic Boundary Layers

Figure 13.8-bottom shows a sketch of a needle dendrite growing in a pure melt at a steady speed v in the $+z$ direction. A *thermal* boundary layer of thickness $\Delta \mathcal{Z}$ surrounds the advancing steady-state dendrite tip. The boundary layer is an estimate of the distance normal to the solid–liquid interface that contains the thermal gradient field which transfers the latent heat. A thinner *solute* boundary layer may also be present in the case of alloy dendrites. When dendrites grow during solidification they can influence each other or be affected by the solidification environment if their thermal boundary layers interact. The boundary layer, in short, provides an estimate of the size scale of a dendrite plus its diffusion field.

13.5.1 Boundary Layer Thickness

As already explained in Chapter 6, boundary layers provide useful engineering estimates of the spatial extent of transport fields, and help identify the system parameters upon which the transport field depends. Boundary layers have finite thicknesses, as contrasted with the 'infinite' extent of the mathematical transport field, and the 'infinitesimal' thickness of the interface. As such, boundary layers provide useful estimates of the spatial extent of the field. The thickness of a thermal boundary layer, $\Delta \mathcal{Z}_{th}$, may be defined as the distance over which the temperature falls substantially from the melting point, T_m, at the solid–liquid interface, to the supercooled melt temperature, T_∞, far away from the interface. Similarly, a solute boundary layer thickness may be defined as the distance, $\Delta \mathcal{Z}_{sol}$, over which the solute concentration in the melt, $\hat{C}_\ell$, adjacent to the interface, falls (or rises) to the nominal melt composition, C_0, ahead of the interface.

First, consider thermal boundary layers. A thermal flux balance holds at any arbitrary point on the solid-melt interface, as the latent heat released at each location is conducted away into the nearby supercooled melt via the local normal temperature gradient. This is the local form of Stefan's energy balance, and may be expressed as

$$\frac{\Delta H_f}{\Omega} \mathbf{v} \cdot \mathbf{n} = -k_\ell \nabla T \cdot \mathbf{n}, \qquad (13.13)$$

where **n** is the unit normal vector on the solid–liquid interface. If the RHS of Eq. (13.13) is applied to conditions at the dendrite tip, $r = 0$, the normal vector, **n**, becomes parallel to the z-axis. The magnitude of the thermal gradient vector, $|\nabla T|$, may be estimated as the total temperature drop ahead of the tip, $T_\infty - T_m$, divided by the thermal boundary layer thickness, $\Delta \mathcal{Z}_{th}$, so the steady-state heat flux flowing out from the dendrite tip is,

$$\frac{\Delta H_f}{\Omega} v \approx k_\ell \left(\frac{T_m - T_\infty}{\Delta \mathcal{Z}_{th}} \right). \tag{13.14}$$

Solving Eq. (13.14) for the boundary layer thickness, $\Delta \mathcal{Z}_{th}$, and recalling the definition of the dimensionless supercooling, $\Delta \vartheta \equiv (T_m - T_\infty) C_p / \Delta H_f$, one finds that the thermal boundary layer thickness at a dendrite tip is

$$\Delta \mathcal{Z}_{th} = \frac{k_\ell \Omega}{v} \left(\frac{T_m - T_\infty}{\Delta H_f} \right) = \frac{k_\ell \Omega}{C_p} \frac{\Delta \vartheta}{v}. \tag{13.15}$$

Dividing both sides of Eq. (13.15) by the dendritic tip radius, R_{tip}, multiplying the numerator and denominator of the right-hand side by 2, and recognizing the definitions of the thermal diffusivity of the melt, $\alpha_\ell \equiv k_\ell \Omega / C_p$, and the Péclet number, $Pe \equiv v R_{tip} / 2\alpha_\ell$, combine to yield the scaled form of the boundary layer thickness,

$$\frac{\Delta \mathcal{Z}_{th}}{R_{tip}} = \frac{2\alpha_\ell}{v R_{tip}} \frac{\Delta \vartheta}{2} = \frac{\Delta \vartheta}{2Pe}. \tag{13.16}$$

If the Ivantsov solution, Eq. (13.9) is substituted for the dimensionless supercooling, $\Delta \vartheta$, on the RHS of Eq. (13.16), the exact expression for this boundary layer thickness results,

$$\frac{\Delta \mathcal{Z}_{th}}{R_{tip}} = \frac{1}{2} e^{Pe} E_1(Pe). \tag{13.17}$$

13.5.2 Boundary Layers at Small Péclet Numbers

One may employ the following series expansions for each of the terms appearing in Ivantsov's characteristic equation, $\Delta \vartheta = Pe \exp(Pe) E_1(Pe)$, to simplify the estimates of dendritic boundary layers under limiting solidification conditions at low Péclet numbers:

i. $e^{Pe} = 1 + Pe + (Pe)^2/2 + \cdots$,
ii. $E_1(Pe) = -(\ln Pe + \gamma_E) + Pe - (Pe)^2/4 + \cdots$, where $\gamma_E = 0.57721 \ldots$ is Euler's constant.

Substituting these expansions into Ivantsov's transport solution, Eq. (13.9), yields the following first-order approximation for the dimensionless supercooling when $Pe \ll 1$:

$$\Delta\vartheta \approx Pe(1 + Pe)(Pe - \ln Pe - \gamma_E). \tag{13.18}$$

Substituting the expansion developed for $\Delta\vartheta$ in Eq. (13.18), and dropping higher-order terms yields

$$\frac{\Delta Z_{th}}{R_{tip}} \approx \frac{1}{2}\left[\ln\left(\frac{1}{Pe}\right) - \gamma_E\right] \quad (Pe \ll 1). \tag{13.19}$$

13.5.3 Boundary Layers at Large Péclet Numbers

Ivantsov's solution is also easily approximated in the limit of large Péclet numbers, where $Pe \gg 1$. Large Péclet numbers are encountered frequently in the transport solution for alloy dendrites, where $\Delta\vartheta$ represents the dimensionless solute supersaturation of the melt, defined as the solute concentration, C, divided by the width of the tie-line, $\hat{C}_\ell - \hat{C}_s$, at the growth temperature. By straightforward analogy with Ivantsov's thermal transport solution, the solutal Péclet number is defined as $Pe_{sol} = vR_{tip}/2D_\ell$, where D_ℓ is the solute diffusivity in the melt. Ivantsov's transport solution at high Péclet numbers becomes

$$\Delta\vartheta = Pe\, e^{Pe} E_1(Pe) \approx 1 - \frac{1}{Pe}. \tag{13.20}$$

If a flux balance is applied to the tip of the dendrite, and the procedures outlined for Eqs. (13.14) through (13.19) are followed—excepting the use of expansion Eq. (13.20) instead of Eq. (13.18)—one finds the scaled boundary layer thickness in the limit $Pe \gg 1$ as

$$\frac{\Delta Z}{R_{tip}} = \frac{1}{2Pe}\left(1 - \frac{1}{Pe}\right). \tag{13.21}$$

Given that Eq. (13.21) is valid when $Pe \gg 1$, so $1/Pe \ll 1$, Eq. (13.21) may be simplified as

$$\frac{\Delta Z}{R_{tip}} \approx \frac{1}{2Pe}. \tag{13.22}$$

The dimensional form of Eq. (13.22) is

$$\Delta Z \approx \frac{R_{tip}}{2Pe} = \frac{\alpha_\ell}{v}. \tag{13.23}$$

It is interesting to point out that at high dimensionless supersaturations the boundary thickness at the tip of a dendrite is the same as the diffusion length, α_ℓ/v, ahead of a planar front moving at the same speed. Note that such boundary layers are thinner than the tip radius, and would appear similar to the boundary layer portrayed in Fig. 13.8. Finally, Eq. (13.19) confirms that the scaled boundary layer thickness depends only on the Péclet number. Because the thermal and solute transport solutions given by Ivantsov's transport theory are mathematically identical, excepting the definition of their respective thermal and solute Péclet numbers and using either supercooling, ΔT, or supersaturation, $\Delta\theta$, Eq. (13.19) serves to predict *both* thermal and solute boundary layers in dendritic growth. Solute Péclet numbers are usually much larger than thermal Péclet numbers, because of the large disparity (about 4-orders of magnitude) between the the thermal and solutal diffusivities in melts. Consequently, the associated solute boundary layers surrounding dendrites tend to be far thinner than are thermal boundary layers. Most importantly, boundary layer analysis suggest that dendrites in most solidification processes interact with each other and with their solidification environments only at relatively close proximity, i.e., at separation distances of, at most, a few tip radii. Thus, dendritic crystals, while growing, tend to act nearly independently of their environments, and exhibit surprising uniformity of their tip shapes and branching patterns. In the next section, solidification dendrites—like snow flakes—will be shown to develop unique branching patterns, as environmental noise and fluctuations play at least some role in their development, even long after the initial interfacial stability that may have led to dendritic growth is 'forgotten' (Fig. 13.11).

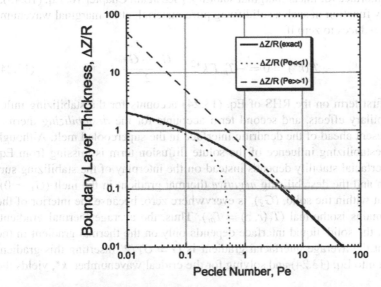

Fig. 13.11 Dendritic boundary layer thickness (scaled to the tip radius) versus Péclet number. Approximations for the boundary layer thickness at low and high Péclet numbers are included (*dotted* and *dashed curves*) to show their respective ranges of validity. Note that the exact transcendental relationship (*solid curve*), Eq. (13.17), is required over the intermediate range $0.1 < Pe < 10$

13.6 Dendritic Capillarity

The introduction of capillarity physics (Gibbs-Thomson effect) on the solid-liquid interface of a dendrite along with steady-state conduction of latent heat into the supercooled melt proved to be difficult, because the shape of such a non-isothermal dendrite was unknown. Capillarity and heat conduction were successfully merged using a self-consistent dendrite shape found by Nash and Glicksman [21, 22]. These investigators derived the self-consistent dendrite shape as a 'correction' to Ivantsov's (isothermal) paraboloid of revolution by adding *isotropic* capillarity. Later, this Nash-Glicksman analysis was extended to include *anisotropic* capillarity by Kessler and Levine [23–25] and by Barbieri and Langer [26]. The selection of a self-consistent steady-state dendrite solution using advanced analytical methods is termed 'solvability' theory, and is considered as the correct analysis of dendritic growth.

J.S. Langer and H. Müller-Krumbhaar [27–29] introduced the concept that morphological stability, or some variant of it, provides the additional physical principle needed to find the operating state of a steadily growing dendrite, namely $v = v_{op}(\Delta\vartheta)$ and $R_{op} = R_{tip}(\Delta\vartheta)$. Their suggestion was based, in part, on experimental observations by Huang and Glicksman [30, 31] that the side branches and the tip scale of dendrites are not independent manifestations of dendritic growth, despite the fact that dendritic side branches are time-dependent features, whereas the tip region 'appears' to be a steady-state feature.[7]

Consider a pure crystal-melt system, so that $C_0 = 0$, and recall the general conditions at an interface for linear marginal stability. [See again Chapter 10, Eq. (10.43).] The stability function in such a solidifying pure material, at its marginal wavenumber, $\mathcal{F}(k^*)$, reduces to zero if

$$\mathcal{F}(k^*) = 0 = -T_m \Gamma k^{*2} - \frac{G_s + G_\ell}{2}, \tag{13.24}$$

where the first term on the RHS of Eq. (13.24) accounts for the stabilizing influence of capillary effects, and second term accounts for the *destabilizing* thermal gradient present ahead of the dendritic interface in the supercooled melt. Although the usual destabilizing influence of the solute diffusion term is missing from Eq. (13.24), interfacial stability depends instead on the interplay of the stabilizing surface tension and the destabilizing *negative* thermal gradient in the melt ($G_\ell < 0$). The gradient within the solid, (G_s), is everywhere zero, because the interior of the dendrite remains isothermal ($T_s(r, z) = T_m$). Thus, the average thermal gradient, $\hat{G}$, acting at the solid–liquid interface depends only on the thermal gradient in the melt, so that the average interfacial gradient is $\hat{G} = G_\ell/2$. Inserting this gradient information into Eq. (13.24) and solving for the critical wavenumber, k^*, yields the

---

[7]As will be discussed later in Section 13.9, the tip region of a dendrite is not growing in a steady-state, but undergoes slight oscillations in its motion that are so faint that they remain almost undetectable.

result,

$$k^\star = \sqrt{\frac{-G_\ell}{2T_m\Gamma}}, \tag{13.25}$$

where previously the solid-melt capillary length was defined as $\Gamma \equiv \gamma_{s\ell}\Omega/\Delta H_f$. The thermal energy balance at the dendrite tip may be found from Ivantsov's transport solution, which yields the non-dimensional thermal gradient in the melt, if evaluated just ahead of the dendrite's tip,

$$\hat{G}_\ell = -2Pe. \tag{13.26}$$

The dimensional thermal gradient is found by rescaling Eq. (13.26) by the characteristic thermal gradient $\Delta H_f/(C_p R_{tip})$, and substituting the definition of the growth Péclet number, $Pe \equiv vR_{tip}/(2\alpha_\ell)$. These steps lead to an explicit formula for the steady-state thermal gradient in the melt at the dendrite tip,

$$G_\ell = -\left(\frac{\Delta H_f}{\alpha_\ell C_p}\right)v. \tag{13.27}$$

Inserting Eq. (13.27) into Eq. (13.25), provides an important relationship defining the marginal dendritic wavenumber and wavelength,

$$k^\star = \frac{2\pi}{\lambda^\star} = \sqrt{\frac{v\Delta H_f}{2\alpha_\ell C_p T_m \Gamma}}. \tag{13.28}$$

Solving for $\lambda^\star$, and grouping some of the materials constants, one obtains the central result of this theory that

$$\lambda^\star = 2\pi\sqrt{\frac{2\alpha_\ell d_0}{v}}, \tag{13.29}$$

where a microscopic length scale, d_0, has been introduced that is proportional to the capillary length, Γ. The so-called 'capillary length', d_0, is typically $\approx 10^{-7}$ cm for many materials,[8] and is defined as,

$$d_0 \equiv \frac{C_p T_m \Gamma}{\Delta H_f} = \frac{C_p}{\Delta S_f}\Gamma. \tag{13.30}$$

---

[8] The small variation of the magnitude of d_0 among different materials (metals, semiconductors, ceramics, and polymers) suggests that certain features of dendritic crystals are nearly 'universal' among different materials classes.

If both sides of Eq. (13.29) are squared, and the terms rearranged slightly, a testable result is obtained from stability analysis of dendrites, namely, that

$$v\lambda^{*2} = 4\pi^2(2\alpha_\ell d_0) = \text{const.} \tag{13.31}$$

13.7 Marginal Stability Hypothesis

Oldfield [32] is credited with performing the first computer calculations of dendritic interface behavior in the early 1970s. Oldfield's results suggested that dendrites grew in a state where there was a kinetic 'balance' between the stabilizing influence of interface capillarity and the destabilizing effect of diffusion or thermal conduction. Oldfield's work was strictly numerical, and was not widely accepted at the time. Not until Langer and Müller-Krumbhaar [27, 28] followed up his interesting suggestion a few years later, adding the critical physical assumption that the steady dendritic tip radius, R_{tip}, scales with the marginal wave length, λ^*, were Oldfield's ideas given recognition. The marginal stability hypothesis of Langer and Müller-Krumbhaar posits that the dendrite tip operates within the margin of stability, whereas the remainder of the dendritic interface grows in an unstable, time-dependent manner, producing quasi-periodic waves that amplify into side arms, or branches.

Equation (13.31) represents an important, and surprising dendritic 'scaling law'. It predicts that the speed of the dendrite tip multiplied by the square of the marginally stable wavelength is a constant, which is independent of the super-cooling. Langer and Müller-Krumbhaar's qualitative suggestion that the tip of a dendrite operates at the margin of stability implies that the dendrite tip becomes as large as possible, but avoids becoming large enough to go unstable, and split. Their hypothesis for the dendritic operating state as one of noise-mediated marginal stability was a reasonable 'guess', which postulated $R_{tip} \approx \lambda^*$. The approximation symbol used here between the *observable* tip radius, R_{tip}, and the *unobservable* theoretical wavelength, λ^*, underscores the fact that linear stability analysis, which is responsible for Eq. (13.29), was developed for a planar crystal front that becomes cellularly unstable, and not for a paraboloidal needle crystal that becomes oscillatory and dendritic. The veracity of the latter application of linear stability remains to be proven, and is probably incorrect.

Nevertheless, accepting tentatively that Langer and Müller-Krumbhaar's suggestion for selecting the operating state of a dendrite is correct, allows the scaling law for a steady-state dendrite to be quantified as,

$$\frac{2\alpha_\ell d_0}{vR_{tip}^2} = \frac{1}{4\pi^2} \cong 0.025. \tag{13.32}$$

The constant, $1/4\pi^2$ is usually designated σ^* in the literature. Its approximate value in real systems was confirmed qualitatively by Glicksman et al. [33, 34] using precision experimental measurements of steady-state dendritic growth carried out under microgravity conditions.[9]

The dendritic scaling law, $vR^2 = \text{const.}$, derived from Langer and Müller-Krumbhaar's stability hypothesis, has a number of important applications. It permits, purely from the perspective of foundry engineering, qualitative estimates to be made of cast microstructures and their responses to several key solidification variables, including pouring temperature, cooling rate, and alloy composition.

13.7.1 Estimating v and R_{tip}

Ivantsov's transport solution, derived from energy conservation, Eq. (13.11), and Eq. (13.32), derived using linear marginal stability theory, are sufficient, when taken together to determine the operating state of a steady-state dendrite. One can summarize these equations as follows:

Transport theory:

$$vR_{tip} = 2\alpha_\ell \mathcal{I}v^{-1}(\Delta\vartheta),\tag{13.33}$$

Linear stability theory:

$$vR_{tip}^2 = \sigma^\star(2\alpha_\ell d_0).\tag{13.34}$$

Solving Eqs. (13.33) and (13.34) simultaneously for the values of the dendrite's velocity and tip radius gives

$$v = \frac{2\alpha_\ell\sigma^\star}{d_0}\mathcal{I}v^{-2}(\Delta\vartheta),\tag{13.35}$$

and

$$R_{tip} = \frac{d_0}{\sigma^\star}\mathcal{I}v^{-1}(\Delta\vartheta).\tag{13.36}$$

The operating states of a dendrite can be expressed as numerical approximations that are convenient for checking theoretical predictions against experimental observations.

$$v \cong 0.018\frac{\alpha_\ell T_m \Delta S_f^2}{C_p \gamma_{s\ell}\Omega}\cdot\Delta\vartheta^{2.5}\ \text{cm}\cdot\text{s}^{-1},\tag{13.37}$$

[9] Experimental verification of the scaling law given by Eq. (13.32) does not, however, test all the important elements of dendritic growth theory that are based on selective noise amplification. The scaling law instead only verifies that thermal diffusion and interface capillarity are significant aspects of the physics of dendrite formation, which is correct.

and

$$R_{tip} \cong 55 \frac{C_p \gamma_{s\ell} \Omega}{T_m \Delta S_f^2} \cdot \Delta \vartheta^{-1.25} \text{ cm}. \tag{13.38}$$

The expressions given as Eqs. (13.37) and (13.38) remain valid for moderate values of the dimensionless supercooling, $\Delta \vartheta$, which for typical free dendritic growth experiments fall in the approximate range of dimensionless supercooling, $0.01 \leq \Delta \vartheta \leq 0.3$. At smaller values of $\Delta \vartheta$, the theory predicts that $v \approx \Delta \vartheta^2$, and $R_{tip} \approx \Delta \vartheta^{-1}$. As a practical matter, however, the thermal boundary layer surrounding a slowly growing dendrite at extremely small supercoolings becomes sufficiently wide that the dendrite becomes sensitive to convective flows in the melt. Such dendrites may no longer be considered as freely growing and 'isolated' from their environment. At large values of $\Delta \vartheta$—i.e., for values of $\Delta \vartheta$ approaching unity—the assumption of local equilibrium at the crystal-melt interface—inherent in formulating this transport and stability analysis—breaks down, and the basic theory is rendered inaccurate.

13.8 Dendritic Growth Experiments

13.8.1 Velocity Experiments

A careful series of dendritic velocity experiments conducted in the 1980s by Glicksman et al. [9], Huang and Glicksman [30, 31], and by Fujioka and Sekerka [35], all qualitatively support the theoretical predictions given in Eqs. (13.37) and (13.38). Growth speeds for freely growing dendritic crystals in pure material systems were measured as a function of the supercooling. The data for these systems exhibit reasonable precision when sufficient care was taken to eliminate the influence of container walls and other environmental factors that interfere with 'free' dendritic growth.

Growth speed data as a function of supercooling are shown in Fig. 13.12 for the systems succinonitrile and ice/water. These data roughly confirmed the power-law relationship given in Eq. (13.37).

13.8.2 Gravitational Effects

A subtle but significant environmental factor not accounted for in the experiments reported in Fig. 13.12 is the presence of gravitational convection. It is now known that convection of the melt around growing dendrites influences the local thermal (and solute) transport fields surrounding the crystal, and modifies the size, shape and growth rate of the dendrite. See Fig. 13.13 that illustrates how gravity modifies the form and speed of dendritic crystals. The theory outlined above treats the transport

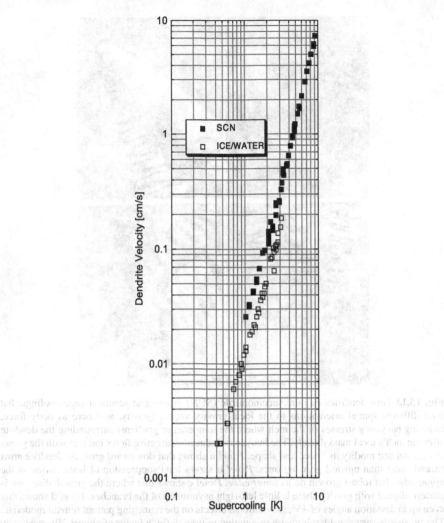

Fig. 13.12 Free dendritic velocity measurements versus supercooling measured in high-purity succinonitrile [9] and water/ice [35]. These data are affected by gravitationally-induced melt convection, and, therefore, only exhibit qualitative agreement with theoretical dendritic velocity predictions based on Eqs. (13.37) and (13.38)

process as pure thermal conduction, so convective effects are excluded. This experimental difficulty was eventually resolved through a series of three microgravity experiments conducted in the mid-1990s using NASA's remotely controlled space laboratory aboard the shuttle *Columbia* [36].

Figure 13.14 shows the observed growth Péclet number versus supercooling for succinonitrile derived from both terrestrial dendritic growth experiments and microgravity experiments carried out in low-Earth orbit. Only the convection-free microgravity data in Fig. 13.14 show agreement with predictions based on Ivantsov's

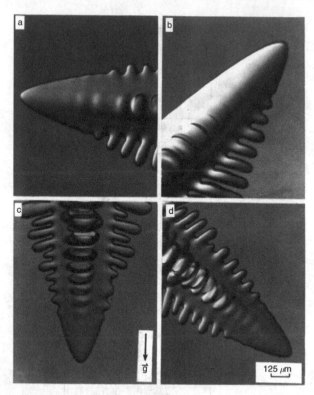

Fig. 13.13 Four dendrites of pure succinonitrile (SCN) growing at identical supercoolings, but with different spatial orientations to the local gravity vector. Gravity, acts here as body force, inducing buoyancy stresses in the melt where the temperature gradients surrounding the dendrite alter the melt's local mass density. The buoyancy-induced convective flows interact with the growing crystal and modify its speed and shape. *Panel* **a** shows that downward growing dendrite arms extend faster than upward growing ones. *Panel* **b** shows total suppression of instabilities on the *upper edge*, but robust growth on its *lower edge*. *Panel* **c** shows that where the growth direction is closely aligned with gravity, there is little left-right asymmetry of the branches. *Panel* **d** shows that even up to deviation angles of 45°, gravitational effects on the branching pattern remain moderate. The tip speeds vary considerably with orientation to gravity (by a factor of almost 20), especially in materials such as SCN that have large Prandtl numbers ($Pr = 27$). Thermal interactions from gravity are weaker in metallic melts, which have relatively small Prandtl numbers ($Pr \approx 0.1$)

analysis, Eq. (13.9) [33]. The agreement with transport theory is especially gratifying given that there are not any adjustable parameters used in the comparison. The microgravity data presented in Fig. 13.14 constitute quantitative verification of the transport theory component of dendritic growth theory.

13.8.3 Stability Constants for Dendritic Growth

Analyses based on several noise-amplification models [34, 37–46] for dendritic growth show that the stability constant $\sigma^\star \approx 0.02$; however, its precise value

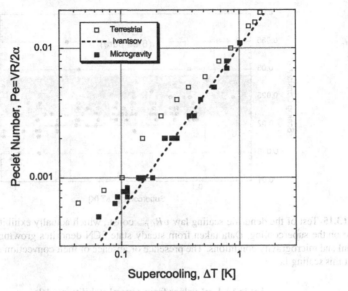

Fig. 13.14 Experimental verification of Ivantsov's transport solution. *Open data symbols* derive from terrestrial measurements of the growth Péclet numbers, where gravitational convection influences the crystal growth, especially at small supercoolings. *Filled symbols* were derived from velocities and tip radii observed under microgravity conditions, where latent heat transport occurs only by thermal conduction. *Dashed line* is Ivantsov's theoretical transport solution, Eq. (13.9), which does not contain any 'adjustable' parameters. Deviations from theory occur at small supercooling ($\Delta T < 0.1$ K) because the thermal boundary layer thickness increases so much that it exceeds the size of the experimental chamber enclosing the melt, allowing latent heat to escape into the stirred thermostat

depends weakly on the geometrical details of the stability model and the melt supercooling [47]. (Cf. Fig. 13.15.)

Table 13.1 lists values of $\sigma^\star$ obtained from several analyses published since the numerical study by Oldfield, who was the first to recognize that interface stability, *per se*, might be a factor in dendritic growth. Dendritic growth kinetic experiments in microgravity were conducted with NASA's Isothermal Dendritic Growth Experiment [33, 34]. v and R_{tip} were measured simultaneously on steady-state dendrites growing in terrestrial (Earth gravity) and in microgravity conditions. These data show that $\sigma^\star$ is indeed nearly constant over a wide range (one decade) of supercooling. As shown in Fig. 13.15 the dendritic scaling law, $vR_{tip}^2 = \text{const.}$, is only approximated in a real system, *both* under terrestrial and microgravity conditions, and the value of $\sigma^\star \cong 0.022$. Although the microgravity data for $\sigma^\star$ supports the dendritic scaling law, it cannot be taken as 'proof' of any particular kinetic theory of dendritic growth. As discussed next in Section 13.9, a fundamentally different theory of dendritic pattern formation and growth leads to similar scaling laws and predictions, as does the marginal stability approach suggested by Langer and Müller-Krumbhaar.

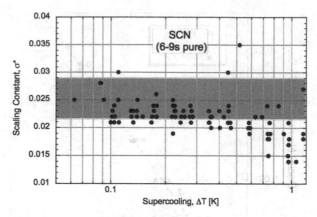

Fig. 13.15 Test of the dendritic scaling law $vR_{tip}=$ const., which actually exhibits a weak dependence on the supercooling. Data taken from steady-state SCN dendrites growing under both terrestrial and microgravity conditions. The presence or absence of melt convection does not seem to affect this scaling law.

Table 13.1 σ^* values from several stability models

Stability model	σ^*
Planar front (analytical)	0.0253
Parabolic eigenstates (numerical)	0.025±20%
Spherical harmonic (analytical)	0.0192
Oldfield's "force balance" (numerical)	0.02

13.9 Stochastics and Determinism

13.9.1 Background

This chapter has been based, up to this point, on the widely-accepted premise that heat flow, capillarity, and selective noise amplification combine to form the physical structure of an instability mechanism that generates the patterns and scaling laws for dendritic crystal growth. It would be unfortunate, however, to allow this approach—its considerable successes notwithstanding—to remain as the sole basis for dendrite formation, even in an introductory treatment of this subject.

The appeal of the marginal stability hypothesis, with its reliance on selective noise amplification, is certainly understandable. As dendrites grow under conditions determined by thermal and/or solutal diffusion, the possibility of natural interface oscillations, or eigenfrequencies, was disregarded.[10] If solid–liquid interfaces really

---

[10] Heat and mass diffusion are described by parabolic differential equations, whereas acoustic waves are described by hyperbolic differential equations. The latter commonly result in wave propagation solutions, whereas the former provide marginally oscillatory solutions that are usually highly damped in space and time. Not surprisingly therefore, self-sustaining oscillations were not expected for diffusion-mediated phenomena such as dendritic growth.

lacked the ability to sustain self-oscillations, then extrinsic, stochastic sources, such as thermodynamic fluctuations and environmental noise, might be expected to provide them.

The exception to this trend of thought were a series of mathematical studies published by J.J. Xu and collaborators [38, 39], who developed a framework for so-called 'trapped waves', as a type of global interfacial instability, which provided an alternative description of dendritic side-branch formation. Xu's analyses are clearly supportive of considering sustained eigenfrequencies as the source of dendritic side-branching [40–42].

A potentially acceptable alternative *physical* basis for producing periodic interface patterns was published recently [43, 44]. The alternative length scales uncovered in those studies, curiously, also devolve from Ivantsov's transport theory, and interface anisotropy; but, in contradistinction to the usual explanation, noise amplification does not influence the fundamental pattern-forming dynamics. Instead, new length scales are connected through the physics of capillarity and heat or solute diffusion, but now emerge directly from the classical *anisotropic* interface boundary condition. This boundary condition, it has been found, interacts non-linearly with the growth dynamics to produce periodic, oscillations in the interface temperature and self-sustained tip oscillations through a 'limit cycle' mechanism, which, in turn, induces the growth of dendritic side branches. Noise, per se, plays no direct role.

13.9.2 Anisotropic Capillarity

The *anisotropic* version of the Gibbs-Thomson capillary relationship, Eq. (8.29), called the Gibbs-Thomson-Herring (GTH) equation [48], also connects the chemical potentials at a one-component crystal-melt interface with its curvature, κ. The equilibrium temperature at any point along such a curved interface in two dimensions [11] given by Herring is

$$T_e(\kappa) = T_m \left(1 - \frac{\Omega}{\Delta H_f} (\gamma_{s\ell} + \gamma_{\theta\theta}) \kappa(\theta) \right), \tag{13.39}$$

where the second-derivative term, $\gamma_{\theta\theta} = \frac{\partial^2 \gamma_{sl}}{\partial \theta^2}$, is the torque, or interfacial 'stiffness', and θ is the normal angle on the interface.

---

[11] Although dendrites are three-dimensional objects, the two-dimensional case is more easily discussed with little loss of generality. A few details for three-dimensional dendrites are provided for the interested reader in Appendix C.

13.9.3 Energy Anisotropy

The crystal-melt interfacial energy densities, $\gamma_{s\ell}$, of cubic crystals (in this example a *two-dimensional* 'square lattice') always reflect a small variation with the normal orientation of the interface, expressing the fourfold symmetry about each of the four equivalent $\langle 10 \rangle$ zone axes. Higher-order anisotropies of $\gamma_{s\ell}$ may also occur, adding symmetries about other important crystallographic directions. Consider for the present just the major fourfold anisotropy of the interface energy around a $\langle 10 \rangle$ axis.

If a fourfold symmetric 'dendrite' grows in the x-y plane along one of its $\langle 10 \rangle$-type axes, equivalent growth directions occur at $\theta = 0$, π, and $\pm\pi/2$. The anisotropic free energy per unit length, $\gamma_{s\ell}$, along the interface may be written as the fourfold harmonic expression at *any* orientation, θ,

$$\gamma_{s\ell} = \gamma_0 \left(1 + \eta \cos 4\theta\right), \tag{13.40}$$

where γ_0 is the modulus of the free energy, and η is a material constant measuring the strength of this fourfold anisotropy. Substituting Eq. (13.40) into Eq. (13.39) and evaluating the 'stiffness' (second derivative) term, yields an expression for the dimensionless temperature shift caused by interfacial curvature in two dimensions,

$$\frac{T_m - T_e(\kappa)}{T_m} = \frac{\gamma_0 \Omega}{\Delta H_f} \left(1 - 15\eta \cos 4\theta\right) \kappa. \tag{13.41}$$

Implementation of the GTH boundary condition for thermodynamic equilibrium in two spatial dimensions requires specification of Eq. (13.41) over the solid-melt interface of a crystallite with fixed area. The resultant closed, equilibrium shape—called the Wulff shape—is essentially the 'pedal' coordinates of the polar plots of $\gamma_{s\ell}$. Polar plots of $\gamma_{s\ell}(\theta)/\gamma_0$ for two cubic substances are shown in Fig. 13.16. These plots also give the *equilibrium* profiles projected along a $\langle 100 \rangle$ axis. Each profile represents a small fixed volume of melt reaching its equilibrium shape within the surrounding anisotropic crystal. The melt droplets were subject to uniform temperature and external pressure. These shapes—and only these shapes—result in a uniform chemical potential at all points along the crystal-melt interface. The data for these polar plots were derived from experimentally observed, three-dimensional shapes of both crystalline substances containing tiny equilibrated liquid droplets (negative crystals) [9, 49].

13.9.4 Shape Anisotropy

If the GTH boundary condition is applied locally to the interface of any *static* form—other than the closed Wulff shape itself—the equilibrium interfacial temperature distribution would no longer remain strictly isothermal. Consequently, some

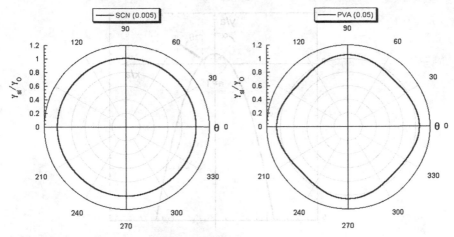

Fig. 13.16 Polar plots of $\gamma_{s\ell}(\theta)/\gamma_0$ versus normal angle, θ for two different test substances. *Left:* Polar plot for SCN, a BCC crystal, showing the polar coordinate projection of its Wulff shape onto a {100} plane. The anisotropy parameter, $\eta = 0.005$, is relatively small in SCN, making the nearly circular projection of SCN's interfacial energy correspond with its equilibrium shape (almost a sphere). *Right:* Plot for pivalic anhydride, PVA, an organic FCC crystal, showing the polar projection of its Wulff shape onto a {100} plane. The anisotropy parameter for PVA, $\eta = 0.05$, is almost 10 times larger than that for SCN, making PVA's equilibrium shape a rounded octahedron, which is markedly different from a sphere, with distinctive bulges exhibited in each of its six $\langle 100 \rangle$ directions

heat flow would result, and subsequent shape changes, if allowed to act over a sufficiently long time, would eventually recapture the Wulff shape.

If one applies the GTH interface condition to some curved convex shape, the interfacial equilibrium temperature is also found to decrease smoothly with slowly increasing curvature, similar to that expected with the isotropic Gibbs-Thomson effect. A surprise occurs, however, if the GTH boundary condition is applied to an interface with strong 'shape anisotropy',[12] which couples with the crystal's energy anisotropy. The two anisotropies—shape and surface energy—interact non-linearly, and produce an interfacial temperature that may vary in a *non-monotonic* manner, even on a smooth, convex interface. The implication that interface geometry and its energy can interact through the GTH equilibrium condition and induce a non-monotone interface temperature distribution is both surprising and of profound importance to dendritic growth theory. Nevertheless, this coupling has, until recently, been overlooked. As will be shown, it is the coupling of surface curvature

---

[12] Shape anisotropy is just a non-uniform distribution, or surface gradient, of the curvature along some direction on an interface. In two dimensions, for example, only a circle represents an isotropic curve, whereas any other curve, such as an ellipse, is anisotropic in its curvature distribution. In three dimensions, similarly, spheres are isotropic, whereas other shapes, such as ellipsoids, are anisotropic along most directions.

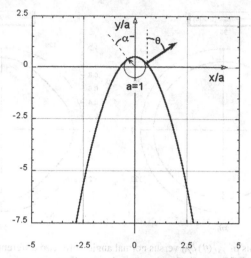

Fig. 13.17 Parabolic needle crystal, prescribed by Eq. (13.42), with a tip radius of $a/2$. Positions along this interface can be parameterized by the orientation of the local normal vector, given by the angle θ, or by the coordinate polar angle, $\alpha \equiv \arctan(y/x)$. The tip of this interface is located at $\theta = \alpha = 0$, and it extends to $y/a \to -\infty$ as $\theta \to \pm\pi/2$

gradients and the energy anisotropy that provides a *deterministic* mechanism for interface oscillation and dendritic eigenfrequencies.

To demonstrate the surprising interaction between shape and energy anisotropies, a static parabolic interface is chosen on which to apply the GTH condition. The parabola, shown in Fig. 13.17, is selected as a smooth convex shape similar in form to the tip region of a two-dimensional 'dendrite'. A parabola, as would any 'finger'-shaped curve, provides shape anisotropy, insofar as the distribution of its curvature changes markedly with the direction of the local normal as the tip is approached.

Any parabolic curve may be conveniently expressed in standard Cartesian form as,

$$\frac{y}{a} = \frac{1}{2} - \frac{x^2}{a^2}. \tag{13.42}$$

Here the size scale of the entire parabola in Fig. 13.17 is set by its tip radius, $a/2$. Its curvature, κ, at any point in the x-y plane is also parameterized conveniently here by the normal angle, θ, which is easily shown, using elementary calculus, to be

$$\kappa = \left| \frac{2}{a} \left(1 + \frac{4x^2}{a^2} \right)^{-\frac{3}{2}} \right| = \left| \frac{2}{a} \cos^3 \theta \right|. \tag{13.43}$$

The curvature distribution, $\kappa(x)$, Eq. (13.43), with a scaled to unity, is plotted in Fig. 13.18. The shape anisotropy of the parabola is indicated by its concentration

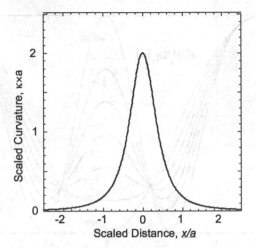

Fig. 13.18 Scaled curvature, $\kappa \times a$, for the parabola in Fig. 13.17, choosing $a = 1$. Plot shows that large curvature anisotropy is concentrated near the tip, $x/a = 0$, where the scaled curvature rises to a relatively high value ($\kappa = 2$)

of large curvatures near its tip at $x/a = 0$. Substituting the parabola's curvature, Eq. (13.43), into the GTH boundary condition, Eq. (13.41), yields the interfacial (anisotropic) equilibrium temperature distribution,

$$T_e(\theta) = T_m - \frac{2T_m \Omega \gamma_0}{a \Delta H_f}(1 - 15\eta \cos 4\theta)\cos^3 \theta, \qquad (13.44)$$

or, equivalently, the dimensionless temperature shift caused by the anisotropic distribution of its curvature,

$$\frac{T_e(\theta) - T_m}{T_m} = -\frac{\gamma_0 \Omega}{\Delta H_f} \frac{2}{a}(1 - 15\eta \cos 4\theta)\cos^3 \theta. \qquad (13.45)$$

The normalized interfacial temperature distributions corresponding to several values of η for a static parabola are displayed in Fig. 13.19. Equation (13.45) shows that the temperature distribution along the crystal-melt interface depends on the strength of the interfacial energy anisotropy, η, as well as on the detailed tip shape and size. Thus, if a two-dimensional crystal were parabolic, then its equilibrium interface temperature would become non-monotonic as the tip is approached ($\theta \to 0$). This non-monotonicity occurs providing that, at least in this case, the energy anisotropy is slightly greater than 1%, i.e., $\eta \geq 1/95$. Moreover, as the energy anisotropy increases, and $\eta \to 1/15 \approx 0.067$, the tip becomes increasingly warmer than the immediately surrounding region of the crystal-melt interface, and actually approaches the melting temperature of a *flat* interface, T_m. This is unexpected.

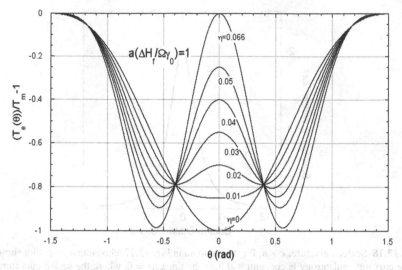

Fig. 13.19 Relative temperature shift along the parabolic interface parameterized by the normal angle, θ. The actual boundary temperature shift induced by capillarity is proportional to the melting temperature, T_m, and inversely proportional to the parabola's length scale, a. The data are plotted here for the dimensionless parameter $\frac{2\gamma_0 \Omega}{a \Delta H_f} = 1$. When the energy anisotropy is small ($\eta \leq 1/95$) the interface temperature remains monotone down as the tip is approached. For larger anisotropy coefficients the temperature becomes non-monotonic, with a distinct maximum developing at the tip, and symmetrical minima positioned about $\pm a/4$ from the parabolic axis, $x/a = 0$. Minimum tip temperature occurs for isotropic surface energy, $\eta = 0$. The non-monotonic temperature behavior shown here is limited to a small region close to the tip, $\theta = 0$

The implications for the occurrence of non-monotonicity in the interfacial temperature will be discussed in the next section, where solidification dynamics and interface motions are introduced.

13.9.5 *Deterministic Dynamics: Two Dimensions*

The example of a static parabolic interface mainly serves to show that anisotropies of the shape and interface energy can couple to provide an unusual (non-monotone) temperature distribution on an otherwise smooth, convex, static interface.

The combined influence of shape and energy anisotropies on growth dynamics were simulated recently by Li and Lowengrub (Private Communication, 2009) in both two- and three-dimensions. They programmed their dynamic solver to implement front tracking from arbitrary initial shapes. Interface dynamics and front tracking relied on solutions to the quasi-static heat equation, subject to anisotropic boundary conditions, and computed accurately as Greens function integrals. These simulations included both the GTH anisotropic capillary condition just described, and an anisotropic interface attachment mobility. Complete details of the dynamic solver are available, especially regarding its freedom from numerical noise [43, 44].

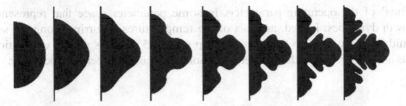

Fig. 13.20 Time sequence of interface patterns showing evolution of an initially circular interface with attachment kinetics in the form of an anisotropic mobility with four-fold symmetry. The interface energy here is isotropic. Only the *right half-plane* is shown, and the branching patterns are re-scaled for ease of comparison. Adapted from S. Li and J. Lowengrub (Private Communication, 2009)

Figure 13.20 shows the development of branch-like features in two dimensions, using an anisotropic mobility condition applied to a circular starting shape with isotropic interface energy. Neither interface perturbations (noise) nor any initial shape anisotropy were present, so the sequence of interfacial patterns follow a strictly deterministic path set by the kinetics.

The influence on two-dimensional pattern formation of anisotropic surface energy via the GTH condition, now with isotropic mobility, is portrayed in Fig. 13.21. There appears to be little doubt that these patterns show that a dendritic morphology develops once the starting interface deviates sufficiently from a circle to exhibit adequate shape anisotropy. Given that noise is completely absent in these simulations, the wake of periodic side branches trailing behind the tip suggests strongly the onset of a 'limit cycle'. Limit cycles, in general, are repeating, closed

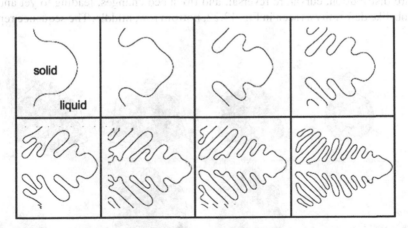

Fig. 13.21 Time sequence of interface patterns showing dendrite evolution from an initially circular interface, with isotropic attachment kinetics and anisotropic surface energy with four-fold symmetry. Only the right extension of the interface is shown, and the patterns are re-scaled for ease of comparison. Results derived from these dynamic solver simulations present evidence for a deterministic mechanism of dendritic growth, and the operation of a noise-free limit cycle that generates a self-sustained eigenfrequency. Adapted from (S. Li and J. Lowengrub, Private Communication, 2009)

'orbits' of the operating parameters in some parameter space that represent the growth dynamics. Indeed, analysis of the temperatures occurring along the solid–liquid interface close to the advancing tip ($\alpha = \pm 5°$) shows perfect correlation of side-arm generation with time-periodic, non-monotone temperature behavior.

13.9.6 Deterministic Dynamics: Three Dimensions

This chapter concludes with presentation of a few deterministic simulation results computed by Lowengrub and Li in three space dimensions (Private Communication, 2009). Here the starting interface shape is spherical, so shape anisotropy is not present initially. Weak anisotropies in both the attachment kinetics and the interface energy were both specified with cubic symmetry. Temporal development of the dendritic patterns is shown in Figs. 13.22 and 13.23, in which the pattern is viewed along the growth direction selected by the anisotropy in the kinetics, viz., $\langle 111 \rangle$, and along a cube-edge, or $\langle 100 \rangle$ direction, respectively.

The results show for this selection of interface energy anisotropy and attachment kinetics that the fastest growth occurs initially along the $\langle 111 \rangle$ cubic body diagonal, causing bulges to protrude in the 8 octahedral directions. See Fig. 13.22, top row, middle. When sufficient shape anisotropy has developed, non-monotone temperature changes appear that cause the mean curvature to reverse sign, and the tip speed to decrease then increase. A knob-like deformity behind the tip forms, as seen in Fig. 13.22, top row, right. Further growth smoothes the deformity into a protrusion shown in Fig. 13.22, bottom row, left, that triggers another non-monotone temperature distribution, curvature reversal, and tip speed changes, leading to yet another knob-like deformity as seen in Fig. 13.22, bottom row, middle. The sequence repeats

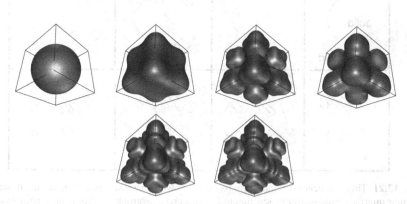

Fig. 13.22 Time sequence of deterministic interface patterns showing dendrite evolution in three dimensions from an initially spherical interface. Isotropic attachment kinetics, but anisotropic interface energy with cubic symmetry. Interface viewed along a $\langle 111 \rangle$ cube-edge direction, shows that the preferred growth directions for this choice of anisotropy lie along the body diagonals. Adapted from S. Li and J. Lowengrub (Private Communication, 2009)

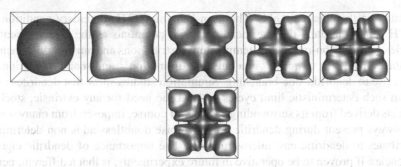

Fig. 13.23 Same time sequence of deterministic interface patterns as shown in Fig. 13.22, with interface viewed along a ⟨100⟩ cube-edge direction. Each pulsation on the interface is synchronized with the appearance of a non-monotone 'hot spot' in the interface temperature, and a slight oscillation in the tip speed. Adapted from S. Li and J. Lowengrub (Private Communication, 2009)

again and again, Fig. 13.22, bottom row, right, and leads to a dendritic stem extending in the octahedral directions. Figure 13.23 shows this deterministic sequence as viewed along a cube-edge direction, and provides some additional shape details more easily observed from this view.

13.10 Summary

Dendritic growth is a ubiquitous crystallization form found in both nature and industrial materials processing. Its importance in casting and welding technology derives from the effects of dendritic crystal growth on the microstructure, particularly its manifold influences on microsegregation, solidification texture, deformation and ductility, mechanical strength, and corrosion resistance.

Dendritic growth is widely accepted as the diffusion-limited solidification morphologies resulting from the 'end-stages' of instabilities on interfaces in supercooled—thermally or constitutionally—or supersaturated alloy melts. As such, the conventional theoretical treatments of dendritic growth kinetics apply marginal stability concepts by relating the tip radius of a dendrite to the most dangerous wavenumber calculated from stability analysis. In essence, the characteristic dendritic form is considered as a stochastic, noise-mediated, unstable morphology, for which side-branching is caused by selective amplification of the environmental noise spectrum imposed near the tip. Stability treatments lead to dendritic scaling laws, such as $vR_{tip}^2 = $ const., that have proved both qualitatively, and semi-quantitatively, correct for estimating the influence of casting parameters and alloy characteristics on dendrite size and speed.

Recently, however, a deterministic source of self-sustained tip oscillations, or eigenfrequencies, has been discovered. Periodic oscillations were shown to result from a dynamic limit cycle that occurs as the shape anisotropy (curvature distribution) interacts non-linearly with the interface energy anisotropy. This interaction occurs via the classical Gibbs-Thomson-Herring (GTH) boundary condition. The GTH condition alters the interfacial chemical potentials as the crystal orientation varies, causing non-monotonic temperature distributions and anisotropic attachment mobility. These in turn result in a pulsatile tip motion[13] and shape oscillations that give rise to dendritic side-branching. Simulation studies show that dendrites evolve from such deterministic limit cycles without the need for any extrinsic, stochastic effects derived from its surroundings. Noise, of course, imposed from many sources, is always present during dendritic growth; noise doubtless adds non-deterministic attributes to dendritic cast microstructures. The importance of dendritic eigenfre-quencies, if proven to be operative in future experiments, is that a different, perhaps more fundamental, physical basis for dendritic growth phenomena could be devel-oped, helping to advance this important aspect of solidification science.

References

1. U. Nakaya, *Snow Crystals: Natural and Artificial*, Harvard University Press, Cambridge, MA, 1954.
2. W.A. Bentley, *Snowflakes in Photographs*, Dover Publications, Mineola, NY, 2000. Originally published as *Snowcrystals*, McGraw-Hill, New York, NY, 1931.
3. K.G. Libbrecht, *The Snowflake: Winters Secret Beauty*, Voyageur Press, Stillwater, MN, 2003, p. 112.
4. A. Janner, *Acta Cryst.*, **53** (1997) 615.
5. G.A. Colligan and B.J. Bayles, *Acta Metall.*, **8** (1962) 895.
6. Y. Wu et al., *Metall. Mater. Trans. A*, **18** (1987) 915.
7. *THERMO-CALC*® Software, a product of Thermo-Calc AB, Stockholm, Sweden. Applica-tions brochure, TMS Annual Meeting, Seattle WA, 2010.
8. M.E. Glicksman and R.J. Schaefer, *Acta Metall.*, **14** (1966) 1226.
9. M.E. Glicksman, R.J. Schaefer and J.D. Ayers, *Metall. Trans. A*, **7** (1976) 1747.
10. H. Esaka and W. Kurz, *Z. Metall.*, **76** (1985) 127.
11. J.Blackford, http://www.cmse.ed.ac.uk/AdvMat45/superEng.pdf, *Engineering of Superalloys*, 2002.
12. M.E. Glicksman, A. Lupulescu and M.B. Koss, *J. Thermophys. Heat Transf.*, **17** (2003) 69.
13. M.E. Glicksman and S.P. Marsh, 'The dendrite', *Handbook of Crystal Growth*. Hurle D.T.J. Ed., Elsevier Science Publishers, Amsterdam, 1993, p. 1075.
14. K.A. Jackson, *Kinetic Processes: Crystal Growth, Diffusion, and Phase Transitions in Mate-rials*, Wiley-VCH, Weinheim, 2004.
15. G.P. Ivantsov, *Dokl. Akad. Nauk. USSR*, **558** (1947) 567.
16. A. Papapetrou, *Z. Kristallographie*, **92** (1935) 89.
17. P.M. Morse and H. Feshbach, *Methods of Theoretical Physics, Part I*, McGraw-Hill, New York, NY, 1953, p. 664.
18. P. Moon and D.E. Spencer, *Field Theory Handbook, Including Coordinate Systems, Differ-ential Equations, and Their Solutions*, 2nd Ed., Springer, New York, NY and , Berlin, 1988, p. 47.
19. Y-Q Chen, X-X Tang, J-J Xu, *Chinese Phys. B*, **18** (2009) 671.

---

[13] See Appendix C for some additional details on tip speed and shape oscillations, non-monotone temperature distributions, and limit cycles in dendritic crystal growth. A summary is also provided of the computational approach used to study the dynamics of anisotropic interface behavior.

20. G. Horvay and J.C. Cahn, *Acta Metall.*, **9** (1961) 695.
21. G.E. Nash and M.E. Glicksman, *Acta Metall.*, **22** (1974) 1283.
22. G.E. Nash and M.E. Glicksman, *Acta Metall.*, **22** (1974) 1291.
23. D. Kessler and H. Levine, *Phys. Rev. A*, **36**, (1987) 4123.
24. D. Kessler and H. Levine, *Phys. Rev. Lett.*, **57** (1986) 3069.
25. D. Kessler and H. Levine, *Acta Metall.*, **36** (1987) 2693.
26. A. Barbieri and J.S. Langer, *Phys. Rev. A.*, **39**, (1989) 5314.
27. J.S. Langer and H. Müller-Krumbhaar, *J. Cryst. Growth*, **42** (1977) 11.
28. J.S. Langer and H. Muller-Krumbhaar, *Acta Metall.*, **26** (1978) 1681.
29. J.S. Langer, *Rev. Mod. Phys.*, **52** (1980) 1.
30. S.C. Huang and M.E. Glicksman, *Acta Metall.*, **11** (1981) 701.
31. S.C. Huang and M.E. Glicksman, *Acta Metall.*, **11** (1981) 717.
32. W. Oldfield, *Mater. Sci. Eng.*, **11** (1973) 211.
33. M.E. Glicksman, M.B. Koss, L.T. Bushnell, J.C. LaCombe and E.A. Winsa, *Adv. Space Res.*, **16** (1995) 181.
34. M.E. Glicksman, M.B. Koss, L.T. Bushnell, J.C. Lacombe and E.A. Winsa, *Iron Steel Inst. Jpn Int.*, **35** (1995) 604.
35. T. Fujioka and R.F. Sekerka, *J. Cryst. Growth*, **24–25** (1974) 84.
36. M.E. Glicksman, M.B. Koss and E.A. Winsa, *Phys. Rev. Lett.*, **73** (1994) 573.
37. D.A. Kessler and H. Levine, *Phys. Rev. B*, **33** (1986) 7867.
38. X.J. Chen, Y.Q. Chen, J.P. Xu and J.J. Xu, *Front. Phys. China*, **3** (2008) 1.
39. Y.Q. Chen, X.X. Tang and J.J. Xu, *Chinese Phys. B* **18** (2009) 686.
40. J.J. Xu, 'Dynamics of dendritic growth interacting with convective flow—global instabilities and limiting state selection', *Advances in Mechanics and Mathematics* **1**, Springer, New York, NY, 2002, p. 113.
41. J.J. Xu, 'Dynamical heory of dendritic growth in convective flow', *Advances in Mechanics and Mathematics* Springer, New York, NY, 2004.
42. J.J. Xu, *Introduction of Dynamical Theory of Solidification Interfacial Stability*, Chinese Academy Press, Beijing, 2007.
43. M. Glicksman, J. Lowengrub and S. Li, *Proceedings in Modelling of Casting, Welding and Advanced Solidification Processes XI*, C.A. Gandin and M. Bellet, Eds., The Metallurgical Soc., Warrendale, OH, (2006) p. 521.
44. M.E. Glicksman, J. Lowengrub and S. Li, *J. Metals*, **59** (2007) 27.
45. R. Pieters and J. Langer, *Phys. Rev. Lett.*, **56** (1986) 1948.
46. P. Pelce, *Dynamics of Curved Fronts*, Academic Press, New York, NY, 1988.
47. D.A. Kessler and H. Levine, *Phys. Rev. A*, **34** (1986) 4908.
48. Herring, C., *Physics of Powder Metallurgy*, W.E. Kingston, Ed., McGraw-Hill, New York, NY, 1951, 143.
49. N.B. Singh and M.E. Glicksman, *Thermochim. Acta*, **159** (1990) 93.

Chapter 14
Microsegregation

An important issue connected to the stability of solid–liquid interfaces is that unstable interfaces, which give rise to cast microstructures that are cellular or, ultimately, even dendritic, also remain the basic cause of chemical microsegregation. Microsegregation, a chemical separation and concentration of alloy elements or impurities on small scales, has the identical thermodynamic origin as does macrosegregation, discussed in Chapter 5. The essential differences between micro- and macrosegregation are the length scales over which these phenomena operate. As a general rule, microsegregation may be removed, or at least reduced significantly, by annealing a cast sample in the solid state. By contrast, macrosegregation cannot be influenced much by annealing in the solid state.

More specifically, macrosegregation is associated with *stable* solidifying interfaces. Macrosegregation affects solute concentrations over a characteristic length scales in the melt of D_ℓ/v, where D_ℓ is the solute diffusivity in the melt, and v is the rate of advance of the interface. When an interface is stable, its associated macrosegregation—particularly the initial transient solute distribution discussed in Section 6.2.1—can alter a solidified crystal's composition over its entire length. Microsegregation, by contrast, increases and decreases the chemical composition *locally*, often in a fluctuating manner, in virtually all dendritic cast microstructures. Microsegregation develops as solute is rejected and trapped among microstructural elements formed by *unstable* interfaces, such as cells, grooves, and dendrites. For example, the primary stem of a dendrite may be relatively low in solute compared to nearby interdendritic regions surrounded by the side branches that were the last to solidify.

Other forms of segregation that have been recognized, include 'inverse' segregation, which occurs as localized solute enrichment near the surface of castings. Inverse segregation is caused by interdendritic flows of enriched melt exuded near a casting's surface [1, 2]. Inverse segregation, which is based on long-range melt flows through the microstructure, will not be discussed in this chapter.

M.E. Glicksman, *Principles of Solidification*, DOI 10.1007/978-1-4419-7344-3_14,
© Springer Science+Business Media, LLC 2011

14.1 Introduction

Practical solidification processes are operated at sufficiently high freezing rates and low thermal gradients that cells and dendrites are normally expected as at least part of the cast microstructure. Cells and dendrites, as discussed in Chapters 9 and 13, are three dimensional solidification structures that form at length scales typically in the range 10^{-5} to 0.1 cm. These unstable microstructural features exhibit characteristic forms of chemical microsegregation associated with their formation and growth.

Consider the nature of microsegregation that occurs on the smaller size scales that characterize the last melt regions to freeze between interfacial cells and secondary dendritic side branches. As mentioned, microsegregation is not different, in principle, from earlier considerations of solute redistribution during freezing developed in Chapters 5 and 6, excepting the much shorter diffusion times and far smaller length scales over which solute transport occurs in microsegregation.

14.2 Cellular Microsegregation

In dilute alloys undergoing relatively slow freezing rates, cells may develop from an initially planar solid–liquid interface and persist almost indefinitely. The presence of deep trailing grooves behind these cells allows strong lateral solute rejection to occur. Numerical calculations of microsegregation in moderately deep cells were illustrated in Chapter 11, which clearly showed that rejected solute becomes trapped within the deep, steep-sided grooves formed between the cell walls. As solidification progresses, the concentration of solute within the encroaching groove walls keeps rising, and quickly reaches the solubility limit of the primary phase. At that point during solidification, second-phase precipitates can form via some type of invariant reaction, such as a eutectic or peritectic polyphase freezing. If the trailing end of a cellular groove gets slender enough, it may 'pinch' off periodically under the destabilizing action of capillary forces that round off the cell roots and promote coalescence of opposing cell walls. The shedding of solute-rich droplets from the base of the cell roots within the microstructure gives rise to a trail of molten droplets behind the main solidification front. Precipitates that form between cell walls and droplet chains are frequently observed in slowly frozen dilute alloys.

14.2.1 Key Assumptions

Flemings and co-workers [3] developed a general model of cellular microsegregation. The key physical assumptions required in developing their theory of cellular microsegregation are listed below. These assumptions are also suggested by the sketches provided in Fig. 14.1. The results from this theory have been verified, at least semi-quantitatively, by direct microprobe measurements of solute distributions in cellular microstructures that were directionally solidified in dilute alloys.

   i. No intercellular supercooling occurs.
   ii. Local equilibrium is maintained between the solid and melt.
  iii. Motion of the interface is steady.
  iv. Solute diffusion does not occur in the solid.
   v. Linear temperature and concentration gradients develop between cells.
  vi. Solute transport is limited to diffusion in melt, i.e., no convection.
 vii. Phase diagram simplified by a constant liquidus slope and distribution
     coefficient.

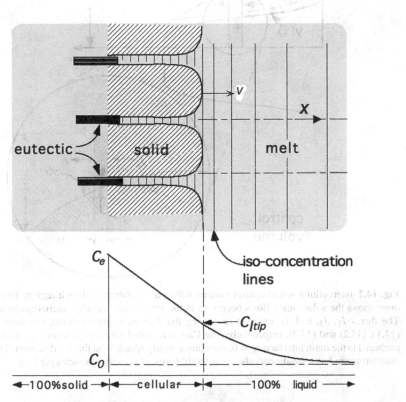

Fig. 14.1 Steady-state solute distribution at a cellular interface solidifying at a velocity, v. *Upper:* Physical configuration of a cellular channel showing *vertical lines* of isoconcentration. *Lower:* Schematic of the spatial variation of solute concentration, $C(x)$. The concentration field within the intercellular region, located behind the main interface, $x < \hat{x}$, is assumed to be linear (note equi-spaced isoconcentration lines between cells). The extracellular concentration field, $x > \hat{x}$, is also assumed to decay as an exponential in x back towards the melt's initial concentration, C_0 (note unevenly spaced isoconcentration lines in the melt ahead of the cellular interface)

14.2.2 Intercellular Solute Mass Balance

Consider the small composite (solid plus liquid) parallelepiped, sketched in Fig. 14.2, as the differential control volume for intercellular microsegregation. The

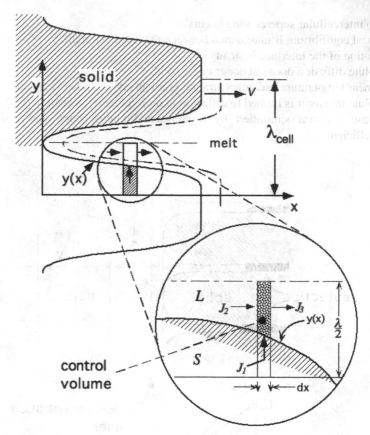

Fig. 14.2 Intercellular solute control volume within a cellulated solid-melt region. The expanded *inset* shows the solute mass flows occurring in the steady-state cellular microsegregation model. The fluxes J_1, J_2, and J_3, entering or leaving the 2-phase control volume, are defined in Eqs. (14.1), (14.2) and (14.3), respectively. The Cartesian coordinates, x, y, form a co-moving frame anchored to the main interface, $x = 0$, traveling a steady speed, v, in the $+x$ direction. The cellular microstructure has a steady periodic 'gear tooth' form, $y(x)$, with a wavelength λ_{cell}

control volume has a height $y = \lambda_{cell}/2$, a width of dx, and a depth of $z = 1$ unit into the plane of the sketch. Three distinct contributions are recognized that comprise the total solute mass changes affecting the concentration within this control volume:

1. The solute mass rejected in the y-direction into the molten portion of the control volume, per unit time, from the element of the solid–liquid interface located at $y(x)$ is

$$J_1 = \left(\hat{C}_\ell - \hat{C}_s\right) \frac{\partial y}{\partial t} dx. \qquad (14.1)$$

2. The solute mass diffusing per unit time along the x-direction. This flux is parallel to the growth direction of the cellular tips and enters the control volume according to Fick's 1st law as

$$J_2 = -D_\ell \frac{\partial C_\ell}{\partial t} \left(\frac{\lambda_{cell}}{2} - y \right). \tag{14.2}$$

3. The solute mass diffusing in the x-direction, per unit time, along the cellular channel at the point $y(x + dx)$, and leaving the 2-phase control volume, is

$$J_3 = -D_\ell \frac{\partial C_\ell}{\partial t} \left(\frac{\lambda_{cell}}{2} - (y + dy) \right). \tag{14.3}$$

Note that the amount of solute flowing out, Eq. (14.3), differs from the amount flowing in, Eq. (14.2); this difference is caused by the changing width for the available diffusion flow between the cell walls. Specifically, the area within the melt channel for diffusion transport increases with x as the cell tips are approached. See again Fig. 14.2. The total steady-state mass balance for the cellular control volume is obtained by adding algebraically each of the mass-flow contributions specified in Eqs. (14.1), (14.2) and (14.3). The balance of these three contributions during steady-state advance of the interface is shown on the LHS of Eq. (14.4),

$$\left(\hat{C}_\ell - \hat{C}_s \right) \frac{\partial y}{\partial t} dx - D_\ell \frac{\partial C_\ell}{\partial t} \left(\frac{\lambda_{cell}}{2} - y \right) + D_\ell \frac{\partial C_\ell}{\partial t} \left(\frac{\lambda_{cell}}{2} - y - dy \right)$$
$$= \frac{\partial C_\ell}{\partial t} \left(\frac{\lambda_{cell}}{2} - y \right). \tag{14.4}$$

The RHS of Eq. (14.4) gives the time rate of change of the solute concentration within the intercellar channel itself. The assumption of local equilibrium at the solid-melt interface permits the following substitution for the first term on the LHS of Eq. (14.4),

$$\hat{C}_\ell - \hat{C}_s = \hat{C}_\ell (1 - k_0). \tag{14.5}$$

The differential volume, dV, for a unit thickness of the interface is

$$dV = \frac{\lambda_{cell}}{2} dx \times 1. \tag{14.6}$$

The fraction of solid, f_s, developed at the control volume location, x, is

$$f_s(x) = \frac{2}{\lambda_{cell}} y(x), \tag{14.7}$$

and the corresponding fraction of melt at that point is

$$f_\ell(x) = 1 - f_s(x). \tag{14.8}$$

Dividing Eq. (14.4) through by the control volume, dV, and then substituting the relationships provided in Eqs. (14.6), (14.7) and (14.8) yields, after some additional rearrangement,

$$\frac{2}{dx\lambda_{cell}} \frac{\partial C_\ell}{\partial t} \left(\frac{\lambda_{cell}}{2} - y \right) dx = \frac{2}{dx\lambda_{cell}} C_\ell(1 - k_0) \frac{\partial y}{\partial t} dx - \frac{2}{dx\lambda_{cell}} D_\ell \frac{\partial C_\ell}{\partial x} dy, \tag{14.9}$$

or, equivalently,

$$\frac{\partial C_\ell}{\partial t} f_\ell = C_\ell(1 - k_0) \frac{\partial f_s}{\partial t} - D_\ell \frac{\partial C_\ell}{\partial x} \frac{\partial f_s}{\partial x}. \tag{14.10}$$

Continuity of the phases requires that the following partial differential relations hold between the changing phase fractions,

$$\frac{\partial f_s}{\partial t} = -\frac{\partial f_\ell}{\partial t}, \tag{14.11}$$

and

$$\frac{\partial f_s}{\partial x} = -\frac{\partial f_\ell}{\partial x}. \tag{14.12}$$

The steady-state assumption also requires that the total differentials of the phase fractions vanish, so

$$df_\ell(x, t) = \frac{\partial f_\ell}{\partial x} dx + \frac{\partial f_\ell}{\partial t} dt = 0. \tag{14.13}$$

Dividing both sides of Eq. (14.13) by dx yields the steady-state requirement that

$$\frac{df_\ell}{dx} = \frac{\partial f_\ell}{\partial x} + \frac{\partial f_\ell}{\partial t} \frac{dt}{dx} = 0. \tag{14.14}$$

The steady-state velocity of the cellular solidification system being considered is v, defined as

$$v = \frac{dx}{dt}, \tag{14.15}$$

and substituting the inverse of Eq. (14.15) into Eq. (14.14) yields the steady-state relationship describing how the melt fraction, f_ℓ, within the cellular channel must vary with time and distance, namely,

$$\frac{1}{v}\frac{\partial f_\ell}{\partial t} = -\frac{\partial f_\ell}{\partial x}. \tag{14.16}$$

Substituting the continuity and steady-state conditions, Eqs. (14.11), (14.12), and (14.16), into Eq. (14.10) provides the following differential mass balance for a steady-state cellular interface,

$$\frac{\partial C_\ell}{\partial t} f_\ell = -C_\ell(1 - k_0)\frac{\partial f_\ell}{\partial t} - \frac{D_\ell}{v}\left(\frac{\partial C_\ell}{\partial x}\right)\frac{\partial f_\ell}{\partial t}. \tag{14.17}$$

Dividing Eq. (14.17) by C_0, yields

$$\frac{\partial C_\ell}{\partial t}\frac{f_\ell}{C_0} = -\left[\frac{C_\ell}{C_0}(1 - k_0) + \frac{D_\ell}{vC_0}\left(\frac{\partial C_\ell}{\partial x}\right)\right]\frac{\partial f_\ell}{\partial t}. \tag{14.18}$$

A linear temperature distribution was assumed earlier to exist along the cellular channel, which is tantamount to assuming (via local equilibrium) that a constant solute concentration gradient exists along the cell channel. The ratio of the applied thermal gradient, G, to the liquidus slope, m_ℓ, may be substituted for the concentration gradient using the following logic,

$$\frac{G}{m_\ell} = \frac{\partial T_\ell}{\partial x} \cdot \left(\frac{\partial T_\ell}{\partial C_\ell}\right)^{-1} = \frac{\partial C_\ell}{\partial x}. \tag{14.19}$$

Substitution of Eq. (14.19) into Eq. (14.18) eliminates the explicit presence of the spatial variable, x, providing an ODE for the steady-state solute mass balance during steady-state cellular microsegregation. This differential equation relates the remaining melt fraction, f_ℓ, within the cellular channel, with the local melt composition, C_ℓ. Specifically,

$$\frac{dC_\ell}{dt}\frac{f_\ell}{C_0} = -\left[\frac{C_\ell}{C_0}(1 - k_0) + \frac{D_\ell G}{vm_\ell C_0}\right]\frac{df_\ell}{dt}. \tag{14.20}$$

14.2.2.1 Cellular Microsegregation Parameter

Equation (14.20) can be solved by first introducing the cellular microsegregation parameter, a, defined here as the lumped system/materials parameter[1]

$$a \equiv \frac{D_\ell G}{vm_\ell C_0}, \tag{14.21}$$

---

[1] Note that the cellular microsegregation parameter, a, is inversely proportional to the product of the interface speed, v, times the alloy concentration, C_0. So for highly dilute alloys undergoing slow solidification rates this parameter could become large in magnitude; moreover, at sufficiently slow freezing speeds and low solute concentrations, however, the interface will re-establish marginal stability, so the interfacial cells would disappear and chemical microsegregation would cease.

to yield

$$\frac{dC_\ell}{dt}\frac{f_\ell}{C_0} = -\left[\frac{C_\ell}{C_0}(1-k_0)+a\right]\frac{df_\ell}{dt}. \tag{14.22}$$

The variables appearing in Eq. (14.22) may now be separated and integrated, according to the assumptions stipulated as boundary conditions in Section 14.2.1,

$$\frac{1}{C_0}\int_{C_\ell^{tip}}^{C_\ell}\frac{dC_\ell}{(C_\ell(x)/C_0)(k_0-1)-a} = \int_1^{C_\ell}\frac{df_\ell}{f_\ell}. \tag{14.23}$$

Evaluation of the solute mass balance integrals, Eq.(14.23), gives the result

$$\frac{1}{k_0-1}\ln\left(\frac{\frac{C_\ell(f_\ell)}{C_0}(k_0-1)-a}{\frac{C_\ell^{tip}}{C_0}(k_0-1)-a}\right) = \ln f_\ell, \tag{14.24}$$

which can be rewritten as

$$f_\ell^{(k_0-1)} = \left(\frac{(k_0-1)C_\ell(f_\ell)/C_0-a}{(k_0-1)C_\ell^{tip}/C_0-a}\right). \tag{14.25}$$

14.2.2.2 Cellular Stefan Solute Mass Balance

All points along the channel boundary contribute to the process of cellular microsegregation. Solute conservation within the channel may be written as a Stefan mass balance, although the interface in this case consists of cell tips advancing at a speed v. Nevertheless, a Stefan balance may be written for a cellular interface as

$$v(C_\ell^{tip}-C_0) = -D_\ell\left(\frac{\partial C_\ell}{\partial x}\right)_{x_{tip}} = -D_\ell\frac{G}{m_\ell}. \tag{14.26}$$

The second equality shown in Eq. (14.26) substitutes the ratio of the thermal gradient to the liquidus slope, G/m_ℓ, for the concentration gradient. Equation (14.26) can now be solved for the concentration ratio at the cell tips,

$$\frac{C_\ell^{tip}}{C_0} = 1 - \frac{D_\ell G}{vm_\ell C_0}. \tag{14.27}$$

Inserting the definition of the microsegregation parameter, a, provided in Eq. (14.21), into Eq. (14.27), allows a simple relationship to be established between the parameter a and the melt concentration ratio just ahead of the cell tips,

$$\frac{C_\ell^{tip}}{C_0} = 1 - a. \tag{14.28}$$

If Eq. (14.28) is substituted into the denominator of Eq. (14.25) one obtains the result,

$$\frac{C_\ell}{C_0} = \frac{a}{k_0 - 1} + \left(1 - \frac{ak_0}{k_0 - 1}\right) f_\ell^{(k_0-1)}. \tag{14.29}$$

Equation (14.29) predicts the steady-state solute concentration in the melt within a cellular channel at any melt fraction, from the base of a cell groove, where $f_\ell = 0$, to the cell tip, where $f_\ell = 1$.

If preferred, cellular microsegregation may be expressed instead in terms of the solid-state concentration, C_s, at any solid fraction, $f_s = 1 - f_\ell$. To accomplish this requires that local equilibrium between solid and melt holds along the channel, allowing application of the definition of the equilibrium distribution coefficient, $k_0 \equiv \hat{C}_s / \hat{C}_\ell$ to the LHS of Eq. (14.29). Applying these substitutions yields the cellular microsegregation expressed in terms of the solid-state concentration,

$$C_s = k_0 C_0 \left[\frac{a}{k_0 - 1} + \left(1 - \frac{ak_0}{k_0 - 1}\right)(1 - f_s)^{(k_0-1)}\right]. \tag{14.30}$$

Equations (14.29) and (14.30) comprise Flemings's cellular microsegregation equations.

Note, that if the cellular segregation parameter, a, were to attain a magnitude greater than $a = (k_0 - 1)/k_0$, then the second term on the RHS of Eqs. (14.29) and (14.30) would vanish. Under such a circumstance, (low solute concentration, slow freezing rate) the solute concentration, C_s, would become independent of f_s, achieving a consistent steady-state value of $C_s = C_0$. This, of course, is precisely the result expected for steady plane-front growth *without* cell formation or microsegregation. These behaviors are demonstrated in Fig. 14.3, for the case of an Al-Si alloy. The microsegregation parameter, a, which is normally negative (because usually $m_\ell < 0$) cannot become more negative than $(k_0 - 1)/k_0$, or else cell formation would cease, and the solid-melt interface would remain stable and planar. We note that this condition for plane-front stability may be expressed through the inequality

$$a \equiv \frac{D_\ell G}{v m_\ell C_0} \geq \frac{k_0 - 1}{k_0}. \tag{14.31}$$

It is clear that this inequality is equivalent to the constitutional supercooling criterion for achieving marginal stability of a planar solid–liquid interface, developed previously as Eq. (9.13) in Chapter 9. The mass balance model for cellular microsegregation is, therefore, consistent with the concept of interfacial stability and constitutional supercooling.

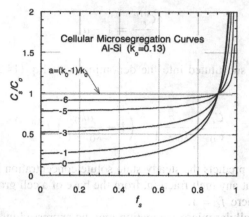

Fig. 14.3 Cellular microsegregation curves for a dilute Al-Si alloy undergoing steady directional solidification. As the cellular microsegregation parameter, a, gets less negative and approaches zero, microsegregation along the cell channels becomes increasingly severe. When the parameter becomes highly negative, $a = k_0 - 1/k_0 \cong -6.9$, the solidification interface regains stability, so the both the interfacial cells and the solute microsegregation vanish

A cellular distribution coefficient, k_{cell}, may be defined to relate the concentration in the solid freezing at the cell tips relative to the average melt concentration, C_0, namely

$$k_{cell} \equiv \frac{C_s^{tip}}{C_0} = k_0 \frac{C_\ell^{tip}}{C_0}. \tag{14.32}$$

Inserting Eq. (14.28) into the RHS of the second equality shown in Eq. (14.32) gives

$$k_{cell} = k_0(1 - a). \tag{14.33}$$

Substituting Eq. (14.33) into the LHS of Eq. (14.32) yields the result

$$C_s^{tip} = k_0(1 - a)C_0, \tag{14.34}$$

which just restates that local equilibrium holds at the cell tips. The concentration of the solid freezing at the base of the cell grooves may be found similarly by inserting the value $f_s = 1$ into Eq. (14.30). This gives the result

$$C_s^{base} = \frac{k_0}{k_0 - 1} aC_0. \tag{14.35}$$

Dividing Eq. (14.35) into Eq. (14.34) yields an expression for the *maximum* microsegregation ratio encountered in a cast structure solidified under steady-state cellular growth conditions,

$$\frac{C_s^{tip}}{C_s^{base}} = (k_0 - 1)\frac{1 - a}{a}. \tag{14.36}$$

Equation (14.36) is plotted in Fig. 14.4 for the case of various alloys undergoing directional cellular solidification. The cellular segregation ratio diminishes to unity for values of $a \leq (k_0 - 1)/k_0$, because the interface becomes planar. Also, as shown in Fig. 14.4, the maximum cellular segregation ratio rises rapidly in the range $-100 \leq a \leq -1$, where cellular interfaces develop deep grooves trailing behind the interface with almost parallel sides. Extreme lateral segregation occurs under these conditions.

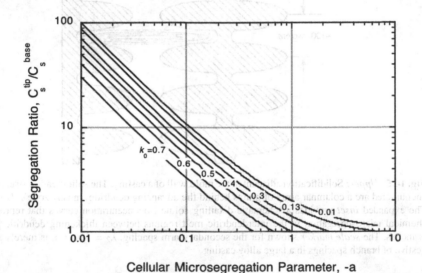

Cellular Microsegregation Parameter, -a

Fig. 14.4 Ratio of solid-phase solute concentration, C_s^{tip}/C_s^{base}, for directionally solidified, cellulated alloys versus (minus) the microsegregation parameter, $a = D_\ell G / v m_\ell C_0$. Microsegregation and cellulation vanish when the magnitude of $|a| \geq (k_0 - 1)/k_0$.

14.3 Dendritic Microsegregation

The model of the solute mass balance occurring within a cellular region can be extended to the often encountered situation of microsegregation in alloy castings and ingots. Solidification in castings, often carried out with low thermal gradients transferring the latent heat to the mold walls, almost invariably occurs by the advance of a dendritic interface, that is referred to as a 'mushy zone'. See Fig. 14.5 for a schematic of these casting structures.

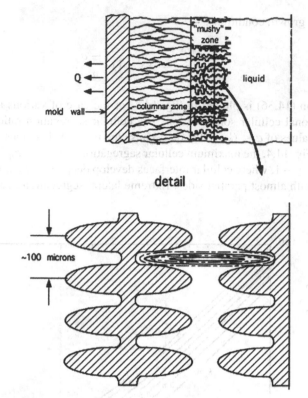

detail

Fig. 14.5 *Upper:* Solidification adjacent to the mold wall of a casting. The solidification structures encountered are a columnar array of grains behind the advancing dendritic 'mushy zone'. *Lower:* The expanded *inset* provides a schematic detailing solute isoconcentration curves that represent chemical microsegregation in the interdendritic melt regions between thickening dendritic side branches. The *scale marker* shown for the secondary arm spacing, $\lambda_2 = 100\,\mu$m, is merely suggestive of branch spacings in a large alloy casting

14.3.1 Microsegregation in Mushy Zones

The name mushy zone implies a fine-scale intimate mixture of thickening dendritic crystals and melt. Mushy zones form in castings because the ratio of the thermal gradient, G, to the local interdendritic freezing rate, v is extremely small. The criterion for 'small' is that $G/v \ll m_\ell C_0/D_\ell$, which is equivalent to the segregation parameter $|a| \ll 1$. Note that these solidification conditions are far beyond the marginal stability condition predicted from constitutional supercooling theory.

If the casting condition at the solid-melt interface is that of a low thermal gradient and a fast freezing rate, then the microsegregation parameter $a \approx 0$. Substituting the value $a = 0$ back into the cellular segregation equation, Eq. (14.30), yields the limiting form of the cellular microsegregation equation that is appropriate to the case of solute rejection in and among the thickening dendritic side-branches in a mushy zone.

$$C_s = k_0 C_0 (1 - f_s)^{k_0 - 1} \qquad (14.37)$$

Surprisingly, this limiting form, Eq. (14.37), is identical to the Gulliver–Scheil *macrosegregation* equation derived earlier in Section 5.3. The Gulliver–Scheil segregation relationship, however, applies only in cases of directional solidification at stable planar interfaces, where solid-state diffusion of the solute can be ignored, and vigorous convective mixing of rejected solute occurs in the melt leading to 'perfect' homogenization of the solute. Figure 14.6 shows the application of Eq. (14.37) to the solidification of a mushy zone in Al-4.5wt.%Cu casting alloy. The microsegregation curve shows how the concentration of Cu rises in the interdendritic region as freezing progresses until the eutectic composition is reached. Solid-state diffusion is ignored in this instance.

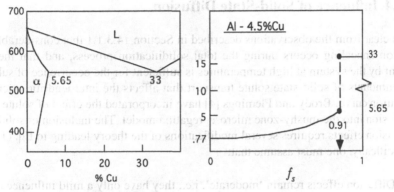

Fig. 14.6 *Left:* Partial Al-Cu phase diagram. *Right:* Interdendritic segregation curve calculated from Eq. (14.37). As indicated in the *graph*, about 91% of the alloy has solidified at the point where the solute concentration has risen sufficiently to induce eutectic solidification

The more general situation of chemical segregation in mushy zones, however, is substantially different from that at a stable planar interface. Mushy zones, as mentioned above, consist of an intimate mixture of fine dendritic crystallites and melt, often held for extended time at the liquidus temperature, which gradually falls as solidification within the mushy zone progresses. Little convective mixing of the melt is possible within an alloy mushy zone because of the tiny length scales among nearby dendritic interfaces that magnify the viscous forces resisting flow. Consequently, however, liquid state diffusion between such closely spaced dendritic branches is rapid—compensating for the diminished convective mixing—and solid-state diffusion may become important during interdendritic solidification, depending on the cooling rate of the ingot or casting. The parameter needed in deciding the relative importance of solid-state diffusion on dendritic microsegregation is the solutal Fourier number, Fo—a dimensionless diffusion-time parameter already introduced in the Brody–Flemings macrosegregation theory discussed in Eq. (5.31), Chapter 5.4.

14.3.1.1 Application to Al-Cu Alloys

The mushy-zone microsegregation equation, Eq. (14.37), tends to overestimate the amount of eutectic formed in the final cast structure of Al-4.5 wt.% Cu. This binary alloy provides an approximation to the behavior of a commercial Al-Cu casting alloy, and was studied extensively [3, 4]. The experiments indicate errors of about 50% in the amount of eutectic predicted without correction for solid-state diffusion. The overestimate predicted with Eq. (14.37) occurs because solid-state diffusion through the already solidified portions of the dendritic mushy zone reduces the build-up of Cu in the interdendritic interstices. In addition, the minimum solute concentrations actually observed in solidified mushy zones are typically about twice those calculated from Gulliver-Scheil theory.

14.4 Influence of Solid-State Diffusion

It is clear from the observations described in Section 14.3.1.1 that considerable diffusion annealing occurs during the total solidification process, and that the time spent by the system at high temperatures is sufficient for the occurrence of substantial amounts of solid-state solute transport that affects the interdendritic microsegregation curve. Brody and Flemings [4] have incorporated the effect of solute back-diffusion into the mushy-zone microsegregation model. The inclusion of solid-state diffusion effects requires several modifications of the theory leading to Eq. (14.37). Specifically, one must assume that:

i. Diffusion effects remain 'moderate', i.e., they have only a mild influence on the final microsegregation process.
ii. The rate of solidification within the entire mushy zone is steady and uniform.
iii. A plate-like one-dimensional interdendritic cell is chosen for calculating the mass balance, as sketched in the lower part of Fig. 14.5, with an average side-branch spacing λ_2.
iv. Local interfacial equilibrium applies.
v. The 'pockets' of interdendritic melt are sufficiently small that they may be assumed to be uniform, but changing, in concentration with time. The homogenization of solute in the small interdendritic spaces occurs by relatively rapid diffusion in the melt.
vi. Convective transport within the melt is ignored. Long-range flows of the melt through the permeable mushy zone are not permitted in this model.

14.4.1 Solute Mass Balances in Mushy Zones

The solute mass rejected per unit time and unit thickness by the downward freezing interdendritic interface into the liquid portion of the control volume is detailed on the right-hand side of Fig. 14.7. The flux of solute rejected at the solid-melt interface into the interdendritic melt may be expressed through a Stefan mass balance as,

$$J_{rej} = -(\hat{C}_s - \hat{C}_\ell)\frac{d\hat{y}}{dt}dx, \tag{14.38}$$

with the height of the interdendritic interstice, $\hat{y}$, decreasing with time.

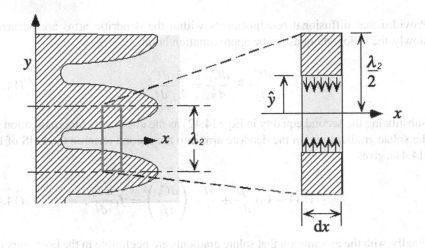

Fig. 14.7 Control volume (*gray rectangle* on *left*) for determining the solute mass balance during interdendritic solidification with microsegregation occurring within a mushy zone. The average secondary side-branch spacing is λ_2. The detailed *inset* of the control volume shows the coordinate, $\hat{y}$, of the solid-melt interface. As the branch thickens with increasing solidification time, $\hat{y} \rightarrow 0$, and $d\hat{y}/dt < 0$

The solute flux diffusing back through the solid phase toward the center of the thickening dendrite arms, per unit depth normal to the plane of the diagram, is given by Fick's 1st law as

$$J_s = -D_s \left(\frac{dC_s}{dy}\right)_{\hat{y}} dy. \tag{14.39}$$

The signs for the mass fluxes in Eqs. (14.38) and (14.39) are selected to be consistent with the coordinates chosen in Fig. 14.7. The mass balance for the upper half of the control volume is obtained by algebraically summing these fluxes and equating that sum to one-half the net increase in solute mass accumulating between the side arms. The rate at which solute mass accumulates in the melt must equal the time rate of its concentration change. Thus, the total mass balance becomes

$$-(\hat{C}_s - \hat{C}_\ell)\frac{d\hat{y}}{dt}dx + D_s \left(\frac{dC_s}{dy}\right)_{\hat{y}} dy = \frac{dC_\ell}{dt}\hat{y}dx. \tag{14.40}$$

Local equilibrium between the phases may now be applied to the dendritic interfaces between the side branches. If Eq. (14.40) is divided by half the control volume, $dV/2 = (\lambda_2/2)dx$, one obtains a differential equation with the melt fraction, f_ℓ, as the independent variable, namely,

$$- C_\ell(1 - k_0)\frac{df_\ell}{dt} + \frac{2D_s}{\lambda_2}\left(\frac{\partial C_s}{\partial y}\right) = f_\ell\frac{dC_\ell}{dt}. \qquad (14.41)$$

Provided that diffusional re-adjustments within the dendritic arms are occurring slowly, the following quasi-static approximation holds,

$$\frac{dC_s}{dy} \cong \frac{d\hat{C}_s}{d\hat{y}} = \frac{2}{\lambda_2}\frac{d\hat{C}_s}{df_\ell}. \qquad (14.42)$$

Substituting the second equality in Eq. (14.42) as the quasi-static approximation for the solute gradient within the dendrite arm into the second term on the LHS of Eq. (14.41), gives

$$- C_\ell(1 - k_0)\frac{df_\ell}{dt} + \frac{4D_s}{\lambda_2^2}\left(\frac{d\hat{C}_s}{df_\ell}\right) = f_\ell\frac{dC_\ell}{dt}. \qquad (14.43)$$

Finally, with the assumption that solute gradients are negligible in the homogeneous interdendritic melt itself, which would be true for slow thickening of the dendritic branches, then it is also true that

$$C_\ell \cong \hat{C}_\ell. \qquad (14.44)$$

Substituting the approximation Eq. (14.44) into both sides of Eq. (14.43) yields a differential expression for the solute mass balance between dendrite branches in terms of the interface compositions, time, and fraction liquid:

$$\hat{C}_\ell(k_0 - 1)\frac{df_\ell}{dt} + \frac{4D_s}{\lambda_2^2}\left(\frac{d\hat{C}_s}{df_\ell}\right) = f_\ell\frac{d\hat{C}_\ell}{dt}. \qquad (14.45)$$

The assumption of *uniform* heat extraction from the mushy zone over the total period of freezing, t_f, permits the use of an approximation for the fractional rate of side-branch freezing,

$$\frac{\partial f_s}{\partial t} = -\frac{\partial f_\ell}{\partial t} \approx \frac{1}{t_f}. \qquad (14.46)$$

The quantity t_f is the *local* freezing time within the mushy zone. The local freezing time is the average dendrite side-branch spacing divided by twice the average side-branch thickening rate, so

$$t_f \equiv -\frac{1}{2} \frac{\lambda_2}{\left(\frac{\partial \hat{y}}{\partial t}\right)} = -\frac{\partial t}{\partial f_\ell} = \frac{\partial t}{\partial f_s}. \tag{14.47}$$

The local freezing time defined in Eq. (14.47) may be substituted into the first term on the left-hand side of the solute mass balance, Eq. (14.45). In addition, constraints on the phase fractions, viz., $f_s + f_\ell = 1$, and $df_s = -df_\ell$, are applied to the remaining two terms in Eq. (14.45) to give

$$-\frac{\hat{C}_\ell(k_0 - 1)}{t_f} - \frac{4D_s}{\lambda_2^2} \frac{\partial \hat{C}_s}{\partial f_s} = (1 - f_s) \frac{\partial \hat{C}_\ell}{dt}. \tag{14.48}$$

Now, one may introduce the definition of a solute diffusion Fourier number that can be defined for the mushy zone solidification process occurring over the interdendritic length scale, λ_2,

$$Fo \equiv \frac{4D_s t_f}{\lambda_2^2}. \tag{14.49}$$

This Fourier number provides the dimensionless 'control parameter' needed to estimate the amount of solid-state diffusion that occurs during the period, t_f, of interdendritic freezing at the position of the control volume within the mushy zone. If Eq. (14.48) is now multiplied through by t_f, and the definition of the interdendritic solutal Fourier number given in Eq. (14.49) is inserted into the differential mass balance, one obtains

$$\hat{C}_\ell(1 - k_0) - Fo \frac{\partial \hat{C}_s}{\partial f_s} = (1 - f_s) \frac{\partial \hat{C}_\ell}{\partial t} t_f. \tag{14.50}$$

Applying the local interfacial equilibrium relation to the second term on the LHS of Eq. (14.50), that $d\hat{C}_s = k_0 d\hat{C}_\ell$, gives

$$\hat{C}_\ell(1 - k_0) - k_0 Fo \frac{\partial \hat{C}_\ell}{\partial f_s} = (1 - f_s) \frac{\partial \hat{C}_\ell}{\partial t} t_f. \tag{14.51}$$

In view of Eq. (14.47), Eq. (14.51) may be converted to an ODE for the interdendritic solutal mass balance,

$$\hat{C}_\ell(1 - k_0) - k_0 Fo \frac{d\hat{C}_\ell}{df_s} = (1 - f_s) \frac{d\hat{C}_\ell}{df_s}. \tag{14.52}$$

The variables in Eq. (14.52) can be separated, and the resulting expression integrated as

$$\int_0^{f_s} \left(\frac{1-k_0}{1-f_s+k_0 Fo} \right) df_s = \int_{C_0}^{C_\ell} \frac{\hat{d}C_\ell}{\hat{C}_\ell}. \tag{14.53}$$

The general limits of integration appearing on the integrals in Eq. (14.53) correspond to the initial condition within the interdendritic spaces that $C_\ell = C_0$ when $f_s = 0$. Carrying out the indicated integrations yields

$$\ln \frac{C_\ell}{C_0} = (k_0 - 1) \ln \left(\frac{1 - f_s + k_0 Fo}{1 + k_0 Fo} \right), \tag{14.54}$$

or, equivalently

$$C_\ell = C_0 \left(1 - \frac{f_s}{1 + k_0 Fo} \right)^{k_0 - 1}. \tag{14.55}$$

The companion relationship for the dendritic microsegregation in the solid phase is easily found by multiplying Eq. (14.55) through by the distribution coefficient, k_0, and again assuming that local equilibrium holds, one finds that

$$C_s = k_0 C_0 \left(1 - \frac{f_s}{1 + k_0 Fo} \right)^{k_0 - 1}. \tag{14.56}$$

A normalized plot of Eq. (14.56) for a range of solutal Fourier numbers is provided in Fig. 14.8.

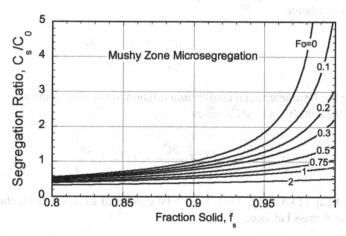

Fig. 14.8 Interdendritic microsegregation in a binary Al-4.5wt%Cu alloy mushy zone for various solutal Fourier numbers. The equilibrium distribution coefficient $k_0 = 0.14$. The preponderant amount of microsegregation occurs in this cast alloy over the last 15% of its freezing range

If the parameter $k_0 \times Fo \ll 1$, (i.e., corresponding to the case of a solute with small solid-state diffusivity, and a casting subject to a brief local freezing time, then, not surprisingly, the Gulliver–Scheil segregation equation is recovered. In this context, however, the Gulliver–Scheil mass balance applies to microsegregation occurring at the localized scale between dendritic branches. A typical interdendritic microsegregation curve is predicted in Fig. 14.9 for an Al-4.5 wt% Cu alloy with an interdendritic arm spacing of 20 μm. A segregation ratio (maximum-to-minimum concentrations) of about 7 results from the solidification conditions chosen. When the Cu content of the aluminum-rich phase reaches about 5.7 wt% Cu, eutectic microconstituent forms within the interdendritic spaces. For the particular case shown in Fig. 14.9, about 5 vol% (non-equilibrium) eutectic would be expected to form in this as-cast microstructure. Thus, as a practical matter, one cannot avoid eutectic formation during solidification of 4.5 wt% Al-Cu by just reducing the cooling rate, and thereby increasing the solute diffusion Fourier number. Post-solidification annealing would be required to remove the non-equilibrium eutectic by solid-state diffusion.

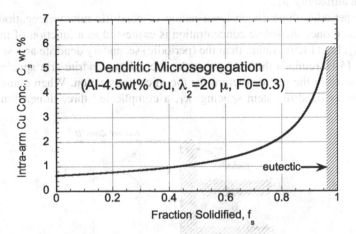

Fig. 14.9 Predicted distribution of Cu concentration versus distance resulting from interdendritic microsegregation in cast Al-4.5 wt% Cu alloy. The average sidebranch spacing is $\lambda_2 = 20\,\mu$m, which sets the length scale of the segregation profile. See also Fig. 14.10, which shows how the inter-arm solute distribution establishes an overall repeating pattern of microsegregation. The eutectic reaction initiates over the last 5% of solidification

One finds as a practical matter that solid-state diffusion in a dendritic mushy zone is relatively unimportant provided that $k_0 \times Fo < 0.1$. More rigorous theories of dendritic microsegregation were developed over the twenty-year period following the seminal work by Brody and Flemings that account more fully for the solute mass balance during freezing [5–7]. The details of these more accurate microsegregation models, however interesting, will not be explored in this chapter.

To assure compliance with the rule that solid-state diffusion may be ignored if $k_0 \times Fo < 0.1$, one must be able to estimate the 'local' freezing time, t_f, which may be approximated as the freezing range—liquidus temperature minus the eutectic temperature—divided by the thermal gradient times the dendritic branch thickening rate,

$$t_f \cong \frac{T_\ell - T_{eu}}{Gv}. \tag{14.57}$$

Substituting this result into Eq. (14.49) yields an estimate for the relevant Fourier number, namely

$$Fo \cong \frac{D_s}{\lambda_2^2} \frac{(T_\ell - T_{eu})}{Gv}. \tag{14.58}$$

The secondary side-branch spacing, λ_2, needed in Eq. (14.58) may be approximated using the dendritic scaling laws developed earlier in Chapter 13, keeping mindful of the fact that when dealing with the solidification of an alloy the growth Péclet number should be based on the solute diffusivity in the melt, D_ℓ, rather than on the thermal diffusivity, α_ℓ.

The periodic, three-dimensional nature of dendritic microsegregation is easily overlooked once the solute concentration is expressed as a function of the fraction solid, f_s, Eq. (14.56), rather than the (periodic) secondary dendrite arm spacing. λ_2. Figure 14.10 reminds the reader that the secondary dendrite arm spacing sets the length scale in the microstructure for the microsegregation. When combined with the primary dendritic stem spacing, λ_1, a complicated three-dimensional pattern

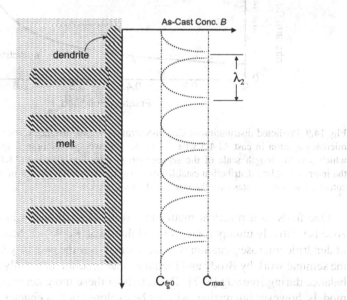

Fig. 14.10 *Left:* Sketch of a dendrite within the partially solidified mushy zone of a solidifying alloy casting. *Right:* The relationship of the inter-arm spacing, λ_2, set by the dendrite to the periodic microsegregation that develops later within the *fully-solidified* as-cast microstructure. In castings, both the primary dendrite stem spacing, λ_1, and the secondary dendrite arm spacing, λ_2, determine the 3-dimensional microsegregation pattern

of microsegregation in the as-cast structure results. Fortunately, numerical models are now available that comprehensively include the effects of solute microsegregation, phase coarsening, and interface temperature on the predictions of as-cast microstructures in multicomponent alloys [8].

Removal of dendritic microsegregation, or at least its amelioration by diffusion annealing, is both time consuming and expensive. Smaller average length scales of microsegregation, $\langle \lambda \rangle$, are usually desired to hasten the post-casting annealing process, as diffusion theory shows that $t_{diff} \cong \langle \lambda \rangle^2 / D_s$, where D_s is the diffusivity of the solute in the solid state. As suggested by this diffusion scaling law, a modest refinement of the dendritic arm spacing can greatly reduce the time and temperature needed to anneal an alloy casting and reduce its chemical microsegregation significantly.

14.5 Structure of Castings

Earlier chapters have dealt with specific phenomena occurring during solidification, with citations of numerous experiments that were relevant to each topic. Here, some practical applications of solidification science to casting technology are explored, drawing on general principles developed in earlier chapters.

Traditional foundry methods of producing shaped and ingot castings involve pouring molten metal into permanent, refractory-lined ingot molds or into sand molds. Many alloy systems cast by these so-called 'static' methods produce, with some variations, the classical grain structures depicted in Fig. 14.11 left, for cast Cu, and impressively predicted in Fig. 14.11, right, using simulation techniques based on cellular automaton and finite element numerical methods. Readers interested in the details of extended numerical modeling of cast microstructure evolution in alloys should refer to the comprehensive monograph written on this subject by Nastac [9].

Three distinct regions may be discerned in Fig. 14.11 left, which together comprise the classic cast structure of metals [10], namely:

1. *Chill zone*: The chill zone forms where rapid local supercooling occurs in the melt along the cold mold wall, which stimulates rapid heterogeneous nucleation of a thin layer of randomly oriented, fine grains.
2. *Columnar zone*: As the layer of chill zone grains rapidly thickens by competitive grain growth from the melt, steep thermal gradients acting near the mold walls allow only a few grains the advantage of faster growth. Most of the dendritic grains are not aligned well with respect to the local thermal gradients, and they are soon 'squeezed' out by their more favorably oriented neighbors. (See Section 13.3 for details on preferred dendritic growth directions in cubic crystals.) The changing directions of the thermal gradients as solidification progresses keeps reducing the remaining subset of competing columnar grains for additional growth. Note the thickening of the most favorably oriented grains as they grew toward the central region of the casting.

Fig. 14.11 *Left:* Classical structural zones in a Cu rod casting. As shown, immediately adjacent to the mold walls (*black borders*) a 'chill zone' of fine-grained polycrystals is nucleated where the molten alloys contacts the cold walls. These grains grow and compete, eliminating all but those favorably oriented grains, which extend inward as a dendritic 'columnar zone'. Columnar grains tend to track the local thermal gradients as solidification progresses. When some thermal supercooling of the melt interior to the casting finally occurs, an 'equiaxed zone' of new dendrites forms, either by heterogeneous nucleation or grain fragment growth ahead of the columnar zone. The equiaxed grains block further inward growth of the columnar grains. *Right:* Simulation of a similar as-cast structure using cellular automaton-finite element (CAFE) computer simulation. Adapted from Rappaz [11]

3. *Equiaxed zone*: The thermal gradients weaken near the center of castings, and the temperature of the melt there cools over time. In some locations the melt will supercool slightly, stimulating the growth of free-floating nuclei or grain fragments. A theory developed by Lipton et al. [12] to explain the appearance of such equiaxed grains in the central regions of castings shows that both thermal and constitutional supercooling must be present simultaneously to allow the growth, *de novo*, of fresh equi-axed dendrites near the completion of solidification. These equi-axed grains in turn release additional latent heat, which inhibits further columnar growth. See again the micrographs in Fig. 12.12 for Sn-Zn alloys that exhibit the classic three structural zones for cast grain structures.

14.6 Summary

The development of fine-scale solidification structures, including cellular and dendritic microstructures, allows solute segregation to occur on comparably small length scales. In this sense, microsegregation and macrosegregation certainly share similar thermodynamic origins, but they are treated as separate phenomena. It is sometimes stated that macrosegregation is the crystal growers problem, whereas microsegregation is the foundryman's problem. Indeed, the latter arises ubiquitously in almost casting scenarios, as the temperature gradients are usually smaller, the rates of freezing are larger, and often the alloys undergoing solidification are more concentrated with respect to solute additions. Fusion welds suffer from microsegregation as well, because their much higher thermal gradients in the melt are still overwhelmed by the rapidity of the weldment's solidification, leading to fine dendritic structures with complicated textures induced by the weld geometry and heat input.

The finer scales of chemical segregation developed during casting and welding also permit significant amounts of solid-state diffusion to occur. Post solidification annealing is an effective processing step for reducing the extent of microsegregation, whereas macrosegregation in single crystals or that caused by interdendritic melt motion will seldom respond to such heat treatments. The early analysis by Brody and Flemings contributed much to our basic understanding of this important aspect of casting, albeit for binary alloys cast in simplified solidification geometries. Modern simulation techniques, such as the development of 'CAFE' models, using cellular automaton and finite elements, continue to improve our capability for computing both the macroscopic (thermal gradients) and microscopic (nucleation and interface motion) aspects of cast microstructures. Adding further the associated multi-dimensional diffusion of solutes, greatly expands one's ability to determine the nature of the residual chemical segregation patterns and grain structures in engineering materials cast into complex forms. This type of 'information engineering', capable of including the panoply of solidification fundamentals, is contributing toward improvements needed in the overall efficiency of metal casting and prediction downstream processing responses of cast materials [13–16].

References

1. V.R. Voller and S. Sundarraj, *Int. J. Heat Mass Transf.*, **38** (1995) 1009.
2. B. Prabhakar and F. Weinberg, *Metall. Mater. Trans. B*, **9** (1978) 150.
3. M.C. Flemings, *Solidification Processing*, McGraw-Hill, New York, NY, 1974.
4. H.D. Brody and M.C. Flemings, *Trans. AIME*, **236** (1966) 615.
5. T.W. Clyne and W. Kurz, *Metall. Trans.*, **12A** (1981) 965.
6. I. Ohnaka, *Trans. Iron Steel Inst. Jpn*, **26**, (1986) 1045.
7. S. Kobayashi, *Trans. Iron Steel Inst. Jpn*, **28** (1988) 725.
8. T. Kraft, M. Rettenmayr and H.E. Exner, *Mater. Sci. Eng.*, **4** (1996) 161.
9. L. Nastac, *Modelling and Simulation of Microstructure Evolution in Solidifying Alloys*, Kluwer Academic Publishers, New York, NY, 2004.

10. B. Chalmers, *Principles of Solidication*, Ch. 8, 253, John Wiley & Sons, Inc., New York, 1964.
11. M. Rappaz, *EPFL Supercomput. Rev.*, **8**, (1996). http://sawww.epfl.ch/SIC/SA/publications/SCR96/scr8-page11.html
12. J. Lipton, M.E. Glicksman and W. Kurz, *Mater. Sci. Eng.*, **65** (1984) 57.
13. M. Rappaz, Ch.-A. Gandin, *Acta Metall. Mater.*, **41** (1993) 345.
14. Ch.-A. Gandin, M. Rappaz and R. Tintillier, *Metall. Trans. A*, **25A** (1994) 629.
15. M. Rappaz, Ch.-A. Gandin and J.-L. Desbiolles, *Metall. Mater. Trans. A*, **27A** (1996) 695.
16. Ch.-A. Gandin, R.J. Schaefer and M. Rappaz, *Acta Metall. Mater.*, **44** (1996) 3339.

Chapter 15
Interface Structure and Growth Kinetics

15.1 Introduction

Implicit in our discussions of solid–liquid interfaces are the related concepts of local equilibrium and *molecular accommodation*. The former was carefully detailed in several earlier chapters, whereas the latter, thus far, has not been explored critically. Chalmers has discussed molecular accommodation at solid–liquid interfaces, and describes the relative ease of individual and small groups of atoms or molecules making energetic and configurational transitions between the crystalline and molten phases [1].

As described in Chapter 8, a narrow transition zone always exists between an equilibrium solid and its melt phase. An energy barrier accompanies this transition zone, reflecting the fact that atoms or molecules comprising the transition zone are 'trapped' in higher energy states than are those positioned away from the interface in the bulk phases. This situation remains true even at a static interface, where the Gibbs energies of the bulk phases are equal. Labile excursions of the atoms from the solid to liquid, or vice versa, which must occur during freezing or melting, require an interface that accommodates significant changes in local order. As will be shown, a sufficiently high density of vacant interface sites and 'ad-atoms' are needed to achieve high mobility of the interface. Highly mobile interfaces are termed 'atomically rough', 'diffuse', or 'continuous'. The kinetics of atomically rough interfaces are reasonably well understood.

Interfaces in pure materials that lack solute drag, or those in substances with smooth order parameter transitions, cannot easily accommodate a net positive or negative flux of atoms and advance or recede normal to itself without an adequate density of vacant and ad-atom sites, respectively. Interfaces lacking these characteristics must instead resort to alternative—and more energetically costly—atomic attachment mechanisms. These include occasional nucleation events occurring epitaxial on the interface that allow the initiation and lateral spreading of one or more monolayers of solid. Kinetically hindered interfaces are variously deemed as 'faceted', 'singular', or 'atomically smooth'. Alternatively, screw dislocations threading through faceted solid–liquid interfaces may appear, which permit spiral hillocks to grow, thereby promoting interface motion, layer by layer, through cyclic lateral spreading of atomic steps and ledges surrounding the dislocations.

M.E. Glicksman, *Principles of Solidification*, DOI 10.1007/978-1-4419-7344-3_15,
© Springer Science+Business Media, LLC 2011

Theories that attempt to relate the growth rate and morphologies of crystal melt interfaces to basic structural and thermodynamic parameters remain incomplete. Several of the classical modeling approaches will be discussed. Some experimental data relevant to the kinetics of crystal-melt interface motion will be presented at the end of this chapter to provide the reader with a glimpse of the difficulties involved in performing and analyzing critical kinetic experiments.

15.1.1 Faceted Interfaces

Perhaps the easiest solid–liquid interfaces to visualize are atomically flat, featureless crystallographic surfaces. Such interfaces are termed 'singular', or faceted, because they exhibit unique crystallographic orientations called facet planes. Approximations to such idealized 'singular' interfaces usually contain in addition certain distinctive elements such as those depicted in Fig. 15.1. Here one sees a $\langle 100 \rangle$ plane on a growing cubic Kossel crystal [2] exposed to its melt or vapor phase. The crystal-fluid interface formed by the $\langle 100 \rangle$ plane is mostly atomically smooth and nearly featureless. The finescale structures not resolved here are the individual atomic sites themselves. Several typical interfacial features occur commonly, as is schematically illustrated in Fig. 15.1. Locations A and B denote the positions of so-called repeating 'kink-sites', where at B, an atom, or molecule—if it attaches—extends the growing monolayer, or step edge, by one additional unit. At A, a persistent kink, or empty bonding site remains available for the next ad-atom that might arrive. The kink site provides an energetically favorable, repeating location for the accommodation of more molecules. By continuously adding atoms to the next available kink site, the trailing ledge moves laterally, eventually completing an additional line of atoms, to advance the step location on the interface. This configuration allows the continued *lateral* growth of the solid, as the repeating kink site is able to accommodate a sequence of atoms onto the solid from the fluid phase. Note also that at B-type sites, three nearest-neighbor bonds, out of a total of six (based on simple-cubic packing of the Kossel crystal) would be satisfied by the arrangement of nearest neighbors. After incorporating an atom at B, the kink site at A is shifted and re-formed and ready to accept another atom, and so on. The series of atoms accommodated at such kink sites develop long trailing ledges, or interface steps, where additional kink sites form by chance fluctuations. An advancing solid–liquid interface is gradually covered over by the spreading ledges, steps, and kink-sites. As these singular interfaces become completely covered by a new atomic layer, the interface itself advances toward the fluid phase by one atomic distance. This mechanism is called crystal growth by lateral step and ledge motion. This type of interfacial behavior is typical of that found in many covalently bonded semiconductor and organic crystals, where the bond orientation between atoms or molecules is important. Faceted interfaces in melt crystallization undergoing lateral crystal growth are not commonly found in materials that possess either ionic or metallic bonding, both of which are non-directional. By contrast, faceted interfaces are common for almost all crystals, irrespective of bonding type, when grown from aqueous solutions, where interface reactions such as dehydration or ion exchange must occur to advance the interface.

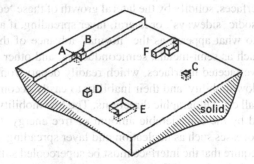

Fig. 15.1 Schematic representation of a ⟨100⟩ faceted solid-melt interface. The labeled interface locations represent additional elements considered to be important during crystal growth: (*A*) an available repeating kink-site; (*B*) an accommodated atom or molecule adding to the spreading layer; (*C*) an interfacial vacancy; (*D*) an ad-atom; (*E*) exposed underlying lattice plane; (*F*) a cluster of ad atoms

Position *D* shown in Fig. 15.1 denotes an isolated ad-atom site, which is a relatively unfavorable interfacial configuration for starting a new crystallized layer arranged by epitaxy with the underlying layer of the solid. At sites such as *D* only one nearest-neighbor bond (directed normally into the interface) is satisfied. The ad-atom, were it thermally activated and vibrating with sufficient amplitude, could easily detach from the interface and return to the surrounding melt or vapor phase. One should think of ad-atoms, or ad-molecules, as a dynamic population of units joining the solid temporarily. Some ad-atoms will diffuse rapidly along the interface and either encounter, by chance, a suitable step edge, or kink-site, that allows the ad-atom to bond more securely to the solid. Failing that, a loosely bound ad-atom may become activated energetically, de-bond from the interface, and then return to the fluid phase.

Site *C* represents an isolated interface vacancy, an especially favorable—but non-repeating—interface location for accommodating one ad-atom that happens to migrate over it. Such isolated interface sites form five nearest-neighbor bonds in a cubic Kossel crystal, leaving only a single unsatisfied interfacial bond pointing out into the melt or vapor. Ad-atoms accommodated at isolated vacant interface sites, such as configuration *C*, are less likely to leave the solid again. The energy reduction caused by satisfying five nearest neighbor bonds tends to be too great to be overcome by normal thermal fluctuations, provided that the interface remains below its thermodynamic melting or sublimation point.

Location *F* designates an ad-atom cluster, or 'island', which occurs when a single ad-atom accretes around itself some additional neighbors. The island, as an entity, becomes more and more stable as it grows laterally on the interface and builds up more and more bonds with the underlying crystal.

Region E on Fig. 15.1 is, in a sense, the reciprocal configuration to that depicted at location F. Configuration F represents the residual of the underlying layer almost totally covered over excepting patches such as F. Eventually, kink-sites and islands merge, and completely cover the preceding layer. Materials with faceted, or atomically smooth, interfaces, solidify by the lateral growth of these 'colliding' sequential layers. Thus, episodic 'sideways', or lateral, layer spreading, if averaged over time, leads indirectly to what appears as the 'normal' advance of the solid-melt interface. Materials such as semi-metals, semiconductors, and other covalently-bonded solids tend to have faceted interfaces, which readily depart from local equilibrium because of their low mobility, and their inability to easily accommodate molecular attachment from all crystallographic directions. The low mobility of such interfaces is actually caused by the considerable amount of free energy dissipated per layer by irreversible processes such as nucleation and layer spreading. Lateral attachment processes often require that the interface must be supercooled sufficiently to provide the requisite large free energy difference between the melt and the growing crystal. By contrast, normal crystal growth on atomically rough interfaces requires relatively little free energy dissipation, and can occur close to thermodynamic equilibrium.

Singular, or faceted interfaces, exhibit the following characteristics:

- small interfacial ad-atom density
- small interfacial vacant site density
- narrow interfacial transition thickness, ca. 2–5 molecular spacings
- large free energy dissipation per layer added

The features listed just above are incorporated in Fig. 15.2, left, which shows an idealized 2-layer representation of an atomically smooth, solid–liquid interface. Figure 15.2 left schematically suggests that at faceted interfaces the populations of ad-atoms (black squares) added from the melt and the available interfacial vacancies (gray squares) are both dilute. To advance such an atomically smooth interface in a direction normal to itself during crystal growth requires lateral growth of new layers, containing steps and kink-sites plus other interfacial features already mentioned. Some features, such as 'islands' or clusters are transient arrangements of atoms and interfacial vacancies, similar to those depicted in Fig. 15.1. Others include more persistent features, such as screw dislocations, and twin-plane doublets that may assist in advancing faceted interfaces by lowering the attachment barrier for molecules.

15.1.2 Non-faceted Interfaces

Metals with body-centered cubic (BCC), face-centered cubic (FCC), and hexagonal close-packed (HCP) crystal structures, and the numerous engineering alloys based on them, along with many other materials with centro-symmetric crystal structures, generally exhibit atomically rough solid-melt interfaces. A characteristic kinetic behavior of atomically rough interfaces is their relative ease of maintaining small departures from local equilibrium between the crystal and its melt, even when the

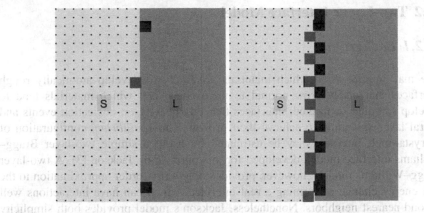

Fig. 15.2 *Left:* A two-layer representation of an atomically smooth, faceted, solid-melt interface. The *black squares* suggest a dilute concentration of ad-atoms, and the *gray square* represents a dilute population of interfacial vacancies. Such a configuration must grow by lateral completion of each atomic layer. *Right:* A two-layer representation of an atomically rough, more diffuse, solid-melt interface. The *black squares* now suggest a higher concentration of ad-atoms, and the *gray square* represents a comparable population of interfacial vacancies. Atomically rough interfaces can grow close to equilibrium by normal advance of the diffuse planes

interface advances toward the melt rapidly. Atomically rough interfaces exhibit the following general characteristics:

- higher interfacial concentration of ad-atoms
- higher interfacial population of vacant sites
- difficulty in distinguishing ad-atoms on one atomic layer from vacant sites on an adjacent layer
- relatively wide, i.e. diffuse, interfacial transition, ca. 5–20 molecular spacings
- small free energy dissipated per layer added

Atomically rough solid–liquid interfaces are easily able to accommodate the addition of new atoms, or molecules, from the melt from *all* crystallographic directions. Little free energy is dissipated in attaching ad-atoms and thereby advancing an atomically-rough interface. Thus, an atomically rough interface remains close to equilibrium, and easily adopts the shape dictated by the prevailing temperature and solute fields that define the temperature and solute concentration for local equilibrium. Kinetically-induced facets seldom form during freezing, in stark contrast with atomically smooth interfaces. Furthermore, growth and propagation of atomically rough interfaces do not require the presence of any special interfacial defects, such as described in Section 15.1.1 for atomically smooth, faceted interfaces. Thus, the lateral spreading of distinct monolayers is much less prevalent than would be observed on faceted interfaces. Instead, an atomically rough interface advances by the omnidirectional accretion of ad-atoms that attach easily to the interface at many nearby accommodation sites.

15.2 Two-Layer Interface Model

15.2.1 Background

One may inquire as to which materials are likely to develop atomically rough interfaces that easily grow normal to themselves, and which materials tend to develop singular, or faceted, interfaces that rely heavily on nucleation events and lateral layer spreading to achieve their growth. The *equilibrium* configuration of a crystal-melt interface may be estimated by using a simple two-layer Bragg–Williams interface model, developed for this purpose by Jackson [3]. A two-layer Bragg–Williams model,[1] however, provides only a first-order approximation to the free-energy change of forming a real interface, which can have interactions well beyond nearest neighbors. Nonetheless, Jackson's model provides both simplicity and considerable insight as to the thermodynamic and crystallographic factors that determine whether a crystal-melt interface for a particular material will resemble the atomically smooth configuration shown in Fig. 15.2, left, or the atomically rough configuration shown in Fig. 15.2, right.

Diffuse and singular interfaces obey radically different kinetic laws regarding their solidification response under the action of the chemical potential difference developed between the solid and liquid in the presence of supercooling or supersaturation. That is, the speed of advance of the interface depends in characteristic ways on the supersaturation or supercooling developed between the melt and the solid. The kinetics of growth—a nonequilibrium process—will not be investigated directly in developing this model. Instead, some approximate information will be extracted on the expected equilibrium populations of ad-atoms and vacancies on the interface. Jackson's thermodynamic analysis provides solid clues into whether an interface tends to be atomically rough or singular and faceted. The qualitative *kinetic* behavior of either rough or faceted interfaces may then be inferred from their estimated equilibrium behavior. A discussion of interface kinetics, per se, during irreversible freezing will be deferred until Section 15.3.

15.2.2 Energy Changes for Interfacial Configurations

Consider transferring N_i atoms from the fluid phase to a crystalline interface composed of N atomic sites. The Gibbs free energy change associated with this ad-atom transfer is as follows,

---

[1] The physics of many-body interactions, such as are applicable to crystal-melt and crystal-vapor interface behavior, has been treated by a variety of so-called 'mean-field' theories. Simply stated, the complex combinatoric problem of each particle interacting with all its neighbors is replaced by a suitable potential, or 'mean-field', representing these difficult-to-calculate interactions. The Bragg–Williams approximation is one such method developed for accomplishing this, although other theoretical schemes also exist.

$$\Delta G_{trans} = \Delta E - T\Delta S + P\Delta V, \tag{15.1}$$

where ΔE is the sum of the internal energy changes; ΔS is the sum of the entropy changes; ΔV is the total volume change for the liquid-to-solid transformation. The internal energy change, ΔE, for the ad-atom transfer from the melt to the interface may be divided into two contributions:

$$\Delta E = \Delta E_0 + \Delta E_1, \tag{15.2}$$

where ΔE_0 represents the bond interaction energy between the ad-atoms and the underlying interface, and ΔE_1 represents the in-plane interaction energy among nearest-neighbor ad-atoms.

The entropy changes for the ad-atom transfer onto the interface are

$$\Delta S_{trans} = \Delta S_0 + \Delta S_{mix}, \tag{15.3}$$

where ΔS_0 is the entropy change upon transferring the atoms from their initial average configurations in, say, the melt, to their 'solid' configurations in the crystal, and ΔS_{mix} equals any additional 'mixing' entropy created among the ad-atoms scattered over the interface, due to different interfacial configurations, or microstates, arising statistically from their presence.

15.2.2.1 Add-Atom Bond Counting on Interfaces

The two-layer, solid-on-solid (SOS) interface model, distinguishes only between those bonds formed among nearest-neighbor ad-atoms (the 'in-plane' bonds) and those formed between the interface and the ad-atoms (the 'normal' bonds). Consider first the case shown in Fig. 15.3, left, as an example of this bond counting scheme for an SOS 2-layer interface with an isolated ad-atom residing on a $\langle 100 \rangle$ plane of a cubic Kossel crystal. Cubic Kossel crystals contain 6 bonds per atom in the bulk state. At a $\langle 100 \rangle$ SOS interface, a maximum of four in-plane bonds and only one normal interface bond may develop for an isolated ad-atom.

The second example of bond counting in the Bragg–Williams model for an interfacial ad-atom is shown in Fig. 15.3, right. Here an ad-atom resides on a close-packed $\langle 111 \rangle$ plane in a face-centered cubic (FCC) crystal containing twelve bonds per atom within the bulk crystal. A maximum of six in-plane and three normal interface half-bonds may form. The number of nearest-neighbor half-bonds, $\mathcal{Z}$, formed within the bulk crystal structure (basically, just the crystal's coordination number) is the sum of the maximum number of in-plane bonds, η_1, established between pairs of neighboring ad-atoms, plus twice the number of half-bonds directed toward the interface, η_0, developed between the ad-atom and the underlying atomic plane. This bond counting scheme accounts for the fact that half the normal interface bonds are not 'satisfied' when an atom leaves the melt and attaches to the interface, because half of the normal bonds still extend outwards toward atoms in the melt phase. Thus, the coordination number, $\mathcal{Z}$, and the half-bond counts, η_0 and η_1, may be related

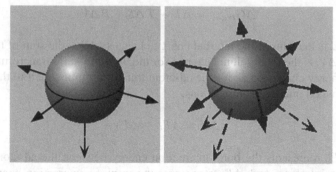

Fig. 15.3 Bond counting diagrams for Jackson's Bragg–Williams interface model. In-plane bonds (*solid arrows*) are distinguished among nearest neighbors on the solid-on-solid (SOS) interface, and normal bonds (*dashed arrows*) are shown for available nearest neighbors residing on the atomic layer below. *Left:* Half-bond directions for surface ad-atoms residing on cubic Kossel lattice, where a total of six bonds are allowed within the bulk crystal. *Right:* Half-bond directions for ad-atoms residing on a ⟨111⟩ plane on an FCC lattice, where a total of 12 bonds are allowed within the bulk crystal

through the sum rule,

$$\mathcal{Z} = 2\eta_0 + \eta_1. \tag{15.4}$$

15.2.2.2   Internal Energy Change

Expressions for the energy terms appearing in Eq. (15.2) may now be developed. The total change in internal energy from completing η_0 additional 'normal' interface bonds per atom, when transferring N_A atoms from the melt to the interface, is given by

$$\Delta E_0 = -L_0 \frac{2\eta_0}{\mathcal{Z}} N_i, \tag{15.5}$$

where L_0 is defined here as the internal energy change per ad-atom. The factor of two appearing on the right-hand-side of Eq. (15.5) is present because the total number of interatomic bonds in a bulk crystal containing N_i atoms is only $(1/2)\mathcal{Z}N_i$, so the energy per completed 'normal' interfacial bond is $2L_0 N_A / \mathcal{Z}$.

The energy change associated with completing all the in-plane bonds formed among $N_i/2$ pairs of neighboring ad-atoms that transferred onto the interface is

$$\Delta E_1 = -L_0 \frac{2\eta_1}{\mathcal{Z}} \frac{N_i}{N} \frac{N_i}{2}. \tag{15.6}$$

Here N is the total number of available interface sites, so that N_i/N is the probability that an in-plane interface bond forms by chance between neighboring ad-atom pairs.

15.2.2.3 Entropy Changes

The entropy terms appearing in Eq. (15.3) can also be found. The change in entropy associated with transferring N_i ad-atoms from the bulk melt to the bulk crystal is

$$\Delta S_0 = -\frac{\Delta H_f}{T_e} N_i \cong -\frac{L_0}{T_e} N_i. \tag{15.7}$$

The term ΔH_f in Eq. (15.7) denotes the latent heat per atom, which may be taken approximately as equal to the internal energy per atom, L_0, because at ordinary atmospheric pressure the pressure-volume term in the enthalpy change is negligible compared to the internal energy. See again the detailed discussion on pressure effects in solidification provided in Chapter 2.

The mixing entropy term in Eq. (15.3), ΔS_{mix}, arises from the number of ways of arranging the N_i ad-atoms among the N interface sites. The entropy change based on the number of ways, or independent interfacial configurations, W, can be determined using combinatorics and statistical mechanics by applying Boltzmann's entropy formula [4], namely,

$$\Delta S_{mix} = k_B \ln W = k_B \ln \frac{N!}{N_i!(N - N_i)!}, \tag{15.8}$$

where k_B is Boltzmann's constant (8.617×10^{-5} ev/K). For large numbers of interfacial sites, N, and ad-atoms, N_i, Stirling's approximation may be employed for estimating the factorial functions appearing in Eq. (15.8). Sterling's approximation may be written for any large factorial, $n!$, as

$$\ln n! \cong n \ln n - n \quad (n >> 1). \tag{15.9}$$

If the RHS of Eq. (15.8) is expanded

$$\Delta S_{mix} = k_B[\ln N! - \ln N_i! - \ln(N - N_i)!], \tag{15.10}$$

and Stirling's approximation, Eq. (15.9), is substituted for the factorials appearing in Eq. (15.10), one finds after a few steps of algebra that

$$\Delta S_{mix} = k_B[N \ln N - N - N_i \ln N_i + N_i - (N - N_i) \ln(N - N_i) + N - N_i], \tag{15.11}$$

and after combining logarithms

$$\Delta S_{mix} = k_B N \ln \left(\frac{N}{N - N_i}\right) + k_B N_i \ln \left(\frac{N - N_i}{N_i}\right). \tag{15.12}$$

Substitution of the terms developed in Eqs. (15.5), (15.6), (15.7), and (15.12) back into the Gibbs free energy function, Eq. (15.1), yields the free energy change for the ad-atom transfer,

$$\Delta G_{trans} = L_0 \frac{2\eta_0}{Z} N_i + L_0 \frac{\eta_1}{Z} \frac{N_i^2}{N} +$$

$$T \left[\frac{L_0}{T_e} N_i - k_B N \ln \frac{N}{N - N_i} - k_B N_i \ln \frac{N - N_i}{N_i} \right]. \tag{15.13}$$

Consistent with the low pressures normally encountered in conventional solidification and crystal growth processes, the pressure-volume work term will again be ignored in Eq. (15.1).

Equation (15.13) can be conveniently normalized by the Boltzmann energy factor, $Nk_B T_e$, to yield the dimensionless form of the free energy change for the interfacial ad-atom transfer to the interface at the equilibrium temperature, namely,

$$\frac{\Delta G_{trans}}{N k_B T_e} = \frac{L_0 N_i}{k_B T_e N} \left(\frac{2\eta_0}{Z} + \frac{\eta_1}{Z} \frac{N_i}{N} \right) +$$

$$\frac{T N_i}{T_e N} \left(\frac{L_0}{k_B T_e} - \frac{N}{N_i} \ln \left(\frac{N}{N - N_i} \right) - \ln \left(\frac{N - N_i}{N_i} \right) \right). \tag{15.14}$$

15.2.2.4 Interfacial α-Factor

The interface and its ad-atom population achieve equilibrium when $T = T_e$. A parameter, known in the crystal growth literature as the interfacial 'α-factor', may be defined to assist in the description the *equilibrium* configuration expected for the mixture of interfacial ad-atoms and vacancies, namely,

$$\alpha \equiv \frac{L_0}{k_B T_e} \frac{\eta_1}{Z} \cong \frac{\Delta S_f}{R_g} \cdot \frac{\eta_1}{Z}. \tag{15.15}$$

Substituting this definition of the α-factor, Eq. (15.15), into Eq. (15.14) yields

$$\frac{\Delta G_{trans}}{N k_B T_e} = -\alpha \frac{N_i}{N} \left(\frac{N_i}{N} + \frac{2\eta_0}{\eta_1} \right) + \alpha \frac{N_i}{N} \frac{Z}{\eta_1} - \ln \frac{N}{N - N_i} - \frac{N_i}{N} \ln \frac{N - N_i}{N_i}. \tag{15.16}$$

For the 2-layer Bragg–Williams interface model under consideration, Eq. (15.4) shows that $Z = 2\eta_0 + \eta_1$, so the following bond-counting constraint applies to the first two terms on the right-hand side of Eq. (15.16),

$$\frac{Z}{\eta_1} - \frac{2\eta_0}{\eta_1} = 1. \tag{15.17}$$

Substitution of the bond counting constraint, Eq. (15.17), into Eq. (15.16), gives the required result after several steps of algebra that

$$\frac{\Delta G_{trans}}{N k_B T_e} = \alpha \frac{N_i}{N} \left(1 - \frac{N_i}{N} \right) + \left(1 - \frac{N_i}{N} \right) \ln \left(1 - \frac{N_i}{N} \right) + \frac{N_i}{N} \ln \frac{N_i}{N}. \tag{15.18}$$

As is shown by the structure of Eq. (15.18), the normalized change in the Gibbs free energy is a function only of the site fraction coverage of ad-atoms, N_i/N, which, in turn, depends solely on the material's interface parameter, α. A plot of Eq. (15.18), shown in Fig. 15.4, demonstrates the central results of Jackson's theoretical treatment of the 2-layer, or SOS, interface model: namely, that the free energy change for the ad-atom transfer either exhibits a maximum or a minimum at a site coverage of $\xi = N_i/N = 0.5$. When a maximum occurs, for cases where the parameter $\alpha \geq 2$, it is flanked by two widely-spaced, shallow minima occurring near $\xi = 0$ and $\xi = 1$, respectively. Study of these free energy extremals, shown in Fig. 15.4, indicates that the qualitative behavior of the ad-atom population on the interface changes abruptly above and below the 'critical' value of $\alpha = 2$.

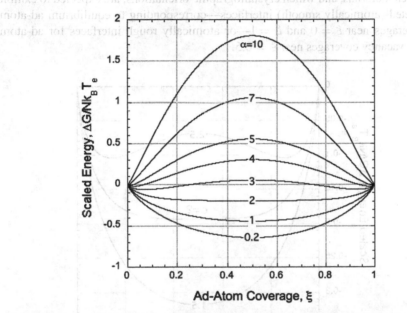

Fig. 15.4 Plot of Eq. (15.18), showing the variation of the scaled free energy change, $\Delta G_{trans}/Nk_BT_e$, for the ad-atom addition against site fraction coverage, $\xi \equiv N_i/N$. The α-factor acts as the interface parameter, causing either a maximum or a minimum at $\xi = 0.5$. For $\alpha \geq 2$ a maximum forms, with two shallow minima occurring near $\xi = 0$ and $\xi = 1$. For $\alpha \leq 2$ only a single minimum forms near $\xi = 0.5$

An interpretation of the single energy minimum ($\alpha < 2$), or the dual energy minima ($\alpha > 2$), in the ad-atom free energy change follows:

1. When energy minima occur near $\xi = 0$ and $\xi = 1$, for materials with $\alpha > 2$, one expects that the *equilibrium* population of ad-atoms residing on the interface would be sparse. That is, an equilibrium site fraction coverage near $\xi = 0$ implies, by definition, few ad-atoms. Similarly, an equilibrium site fraction coverage near $\xi = 1$ implies the presence of only a few vacant sites remaining on

an otherwise almost fully-packed layer of ad-atoms. Note, however, that both of
these interfacial configurations really describe the same condition, viz., atomic
smoothness! That is, few ad-atoms or few vacant sites provide an identical situation, save for the addition, or removal, of one monolayer of the crystal. In a
two-layer SOS model, these interfacial states are indistinguishable.

2. By contrast, Figs. 15.4 and 15.5 clearly show the equilibrium site coverage for
   materials with $\alpha < 2$ exhibits a single minimum in the free energy change at
   $\xi = 0.5$. In such cases one expects an equal density of ad-atoms and vacant interfacial sites. Such an interface, modeled as just two layers, if covered with about
   equal concentrations of atoms and vacancies, would be an atomically rough interface. Cf. Fig. 15.2, left and right.

3. Thus, the alpha-factor is a convenient interfacial parameter to help one judge
   which materials, and which crystallographic orientations, are expected to exhibit
   faceted (atomically smooth) interfaces—corresponding to equilibrium ad-atom
   coverages near $\xi = 0$ and $\xi = 1$—or atomically rough interfaces for ad-atom
   and vacancy coverages near $\xi = 0.5$.

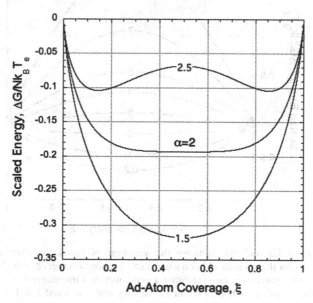

Fig. 15.5 Plot of Eq. (15.18), showing the extreme sensitivity of the scaled free energy change,
$\Delta G_{trans}/Nk_B T_e$, versus the ad-atom fractional coverage, ξ, when the alpha-factor is near 2. The
change in interfacial configurations with the alpha-factor value—from a single value near $\xi = 0.5$
to two values near 0 and 1—occurs so swiftly over a narrow range of alpha-factors near $\alpha = 2$, that
the behavior may be treated as a kind of two-dimensional phase transition

Of course, the 2-layer interface model used by Jackson limits the treatment of
the interface structure to just adjacent lattice planes. Within the framework of this
model the interface is really not any thicker when $\alpha < 2$, merely rougher. In real
solid-melt interfaces, atomically rough interfaces usually exhibit a more gradual

transition from solid to melt, whereas atomically smooth interfaces show more abrupt transitions.

The alpha-factor itself correlates strongly with the complexity of the underlying crystal structure. For example, high-symmetry, centrosymmetric crystals tend to have relatively low alpha-factors ($\alpha \ll 2$). By contrast, materials with either complex unit cells, or low-symmetry crystal structures usually exhibit high values ($\alpha \gg 2$).

The interfacial alpha-factor, $\alpha \equiv \frac{\Delta S_f}{R_g} \cdot \frac{\eta_1}{\mathcal{Z}}$, as defined by the second equality in Eq. (15.15), consists of the product of two terms arising from distinctly different physical origins:

1) The factor $\Delta S_f / R_g$ is a thermodynamic term equal to the ratio of the molar entropy of fusion to the universal gas constant. This entropy ratio for typical crystal/melt transitions varies from values close to unity, typical of the metals, to somewhat higher values for semi-metals and semiconductors, to values as large as 10–100 for macromolecular crystals, such as proteins and long-chain polymers.

2) The geometric factor, $\eta_1 / \mathcal{Z}$, is of crystallographic origin, and equals the ratio of the in-plane half-bond count to the crystal's coordination number (number of nearest neighbors in the bulk crystal). This second factor depends both on the crystal structure itself and on the orientation of the interface. It is always less than unity, but greater than 1/4.

15.2.2.5 Verification

Values of the thermodynamic term in the alpha-factor for a few materials with non-faceting, atomically rough, interfaces are listed below in Table 15.1. All the values for the entropy ratios of the crystals listed in Table 15.1 are less than 2, thereby insuring that $\alpha \leq 2$. The materials listed here are a sampling of pure metallic elements which crystallize in centro-symmetric crystal structures, including FCC, BCC, HCP, and BCT. Predictions of their interface structures based on applying the α-factor criterion have been verified by direct studies of their solid-melt interfaces. None of these low α-factor materials tends to form kinetically-induced interfacial facets upon solidification from their melts. The next series of materials to be discussed develop faceted (atomically smooth) solid–liquid interfaces on at least one crystallographic plane exposed to their melt phase. A selection of these materials are grouped in Table 15.2. The substances listed in this table comprise a diverse mixture of non-metallic, semi-metallic, semiconductor, and organic materials. These materials also exhibit a variety of bonding types, including covalent, metallic, hydrogen, and van der Waals, and it should be noted that all solidify from their melts into more complex crystal structures. The one obvious exception that 'proves the rule' is the inclusion of molecular P_4, white phosphorus. This substance is indeed unusual, insofar as it exhibits an extremely small entropy decrease upon solidification, because its P_4

molecules retain hindered rotations in its solid-state α-manganese cubic structure, which consists of 56 P_4 molecules per unit cell! Unsurprisingly, the 2-layer SOS interfacial model being considered here is too limited to capture the subtle bonding interactions that occur in a complex molecular solid such as cubic α-P_4. Nevertheless, most materials that form facets during crystal growth will have alpha-factors greater than 2.

Table 15.1 Materials without facets on freezing

Material	K	Pb	Cu	Ag	Zn	Al	Sn
$\Delta S_f / R_g$	0.82	0.93	1.1	1.1	1.3	1.4	1.6

Table 15.2 Materials with facets on freezing

Material	P_4	Ga	Bi	Ge	Si	Salol
$\Delta S_f / R_g$	1.0	2.2	2.4	3.2	3.6	7.0

The corresponding values of the α-factor's 'crystallographic' term for several common crystal structures on low-index planes are listed in Table 15.3.[2] As seen here, the bonding terms are always less than unity. This insures that if the ratio of a material's molar entropy to the gas constant, R_g, is less than 2, it should not form facets on freezing from its melt.

Table 15.3 Bonding factors

$\eta_1 / \mathcal{Z}$	Lattice	Interface
2/3	SC	$\langle 100 \rangle$
1/3	FCC	$\langle 111 \rangle$
1/3	BCC	$\langle 110 \rangle$

15.3 Kinetic Theories

An important distinction between atomically smooth (faceted and constrained) and atomically rough (non-faceted and compliant) interfaces is their response to the thermodynamic 'driving force' provided by supercooling or supersaturating the melt. Atomically rough interfaces easily accommodate the transfer of ad-atoms in all directions, whereas faceted interfaces grow by constrained lateral step edge motion of independently nucleated layers on crystallographically defined planes. Several recommended references providing major reviews on the kinetics of melting and

---

[2] The symbol SC denotes simple, or primitive cubic.

freezing, which are based on informed, but highly divergent opinions, may be found in [5–7].

15.3.1 Atomically Rough Interfaces

Rough interfaces consist of atoms, or molecules, with high intrinsic mobilities, defined here with reference to their liquid-state diffusivities as,

$$\mathcal{M} \equiv \frac{D_\ell}{k_B T}, \tag{15.19}$$

where D_ℓ is the melt's self-diffusion coefficient [8]. The kinetics of freezing on such a rough interface may be estimated from the atomic diffusion speed, v_{diff}, which may be found from the product of the atomic mobility and the thermodynamic driving force, **F**, acting across the interface. The driving force for crystal growth is defined as the (negative) gradient of the chemical potential across the crystal/melt transition zone, as suggested in Fig. 15.6.

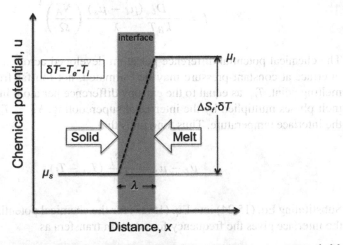

Fig. 15.6 Schematic representation of a crystal-melt interface, of thickness λ, supercooled by a small amount $\delta T = T_e - T_i$, where T_e is the equilibrium melting temperature, and T_i is the (supercooled) interface temperature. The slope of the *dashed line* is the chemical potential gradient, $\nabla \mu = (\mu_\ell - \mu_s)/\lambda$, acting across the interface

Formally, one may write this generalized interfacial 'force' as the negative gradient of some related potential, as is done in physics for classical (conservative) electrostatic, magnetic, or gravitational forces, namely,

$$\mathbf{F} = -\nabla \mu \approx -\frac{\mu_\ell - \mu_s}{\lambda} \mathbf{u}_x, \tag{15.20}$$

where λ is the width of the solid-melt transition zone, and $\mathbf{u}_x$ is a unit vector normal to the interface. Thus, the speed, v_d, of diffusive ad-atom transfers across the transition zone may be found by combining Eqs. (15.19) and (15.20), with the definition of the atomic diffusion speed,

$$v_d \equiv \mathcal{M} \cdot |\mathbf{F}| = -\frac{D_\ell}{k_B T} \frac{(\mu_\ell - \mu_s)}{\lambda}. \tag{15.21}$$

If the average diffusion speed is divided by the atomic jumping distance, $\delta \approx (\Omega/N_A)^{1/3}$, where Ω is the molar volume of the atoms and N_A is the Avogadro number, we obtain the average frequency, f, of attaching ad-atoms on the atomically rough solid-melt interface.

$$f = v_d \left(\frac{N_A}{\Omega}\right)^{\frac{1}{3}} \tag{15.22}$$

Substituting Eq. (15.21) into Eq. (15.22) yields

$$f = -\frac{D_\ell}{k_B T} \frac{(\mu_\ell - \mu_s)}{\lambda} \left(\frac{N_A}{\Omega}\right)^{\frac{1}{3}}. \tag{15.23}$$

The chemical potential difference per atom developed across a planar crystal-melt interface at constant pressure may be estimated, not too far from the equilibrium melting point, T_m, as equal to the entropy difference per atom in the crystalline and melt phases multiplied by the interfacial supercooling, $\delta T = T_m - T_i$, where T_i is the interface temperature. Thus, one finds that

$$\mu_\ell - \mu_s = \frac{\Delta S_f}{N_A} (T_m - T_i). \tag{15.24}$$

Substituting Eq. (15.24) into Eq. (15.23) for the chemical potential difference across the interface gives the frequency for ad-atom transfers as

$$f = -\frac{D_\ell}{k_B T_i} \frac{\Delta S_f}{N_A} \frac{(T_m - T_i)}{\lambda} \left(\frac{N_A}{\Omega}\right)^{\frac{1}{3}}. \tag{15.25}$$

Now, for the case of crystallization in which the densities of the crystal and its melt are comparable, the rate of adding atomic layers with spacing d_{hkl} is also given by f, since an atomically rough crystal-melt interface can accept atoms equally in all crystallographic directions. The interface velocity, $\mathbf{v}$, which results from the ad-atom flux occurs in a direction *opposite* to the atomic flux, which remains normal to the crystallographic plane (hkl). The interface speed is therefore,

$$v_n = -f \times d_{hkl} = -f \zeta_{hkl} \left(\frac{\Omega}{N_A} \right)^{\frac{1}{3}}, \qquad (15.26)$$

where the factor ζ_{hkl} is a pure number of unit order that varies slightly with the interface's growth direction. Combining Eqs. (15.25) and (15.26) gives

$$v_n = \frac{D_\ell}{k_B T} \frac{\Delta S_f}{N_A} \frac{(T_e - T_i)}{\lambda} \zeta_{hkl}. \qquad (15.27)$$

The component of the diffusive motions responsible for transporting ad-atoms from the supercooled melt onto the crystalline solid is normal to the interface. In fact, this type of solidification at atomically rough interfaces is referred to as *normal* growth. The diffusion coefficient, $\hat{D}_\ell$ for atomic motions which are *normal* to the crystal-melt interface is given by the classical Einstein formula in one spatial dimension [8],

$$\hat{D}_\ell = \frac{1}{2} v \delta^2 \cong \frac{1}{2} v \left(\frac{\Omega}{N_A} \right)^{\frac{2}{3}}, \qquad (15.28)$$

where v is the jumping frequency of atoms making normal jumps to the interface, and δ is their mean jump distance. Again, one assumes that the appropriate jumping distance through the interfacial region is about one atomic diameter. Substituting Eq. (15.28) into Eq. (15.27) gives the normal interface speed as

$$v_n = \frac{v \zeta_{hkl}}{2\lambda} \left(\frac{\Omega}{N_A} \right)^{\frac{2}{3}} \frac{\Delta S_f}{R_g} \left(\frac{T_m - T_i}{T_i} \right). \qquad (15.29)$$

Equation (15.29) was obtained using the fact that the universal gas constant $R_g = k_B N_A$. The width of the crystal/melt transition zone, λ, in a given crystallographic direction $\langle hkl \rangle$ can also be expressed in terms of the atomic diameter as

$$\lambda = \Lambda_{hkl} \left(\frac{\Omega}{N_A} \right)^{\frac{1}{3}}, \qquad (15.30)$$

where Λ_{hkl} is a pure number of the order of 10 or less. Substituting Eq. (15.30) into Eq. (15.29) yields the desired result for the speed of an atomically rough interface as a function of its supercooling, δT.

$$v_n = \frac{v}{2} \frac{\zeta_{hkl}}{\Lambda_{hkl}} \left(\frac{\Omega}{N_A} \right)^{\frac{1}{3}} \frac{\Delta S_f}{R_g T} \cdot \delta T. \qquad (15.31)$$

The temperature T used in Eq. (15.31) is close to both the interface temperature and the bulk melt temperature, as the interface supercooling is usually extremely small in the case of atomically rough interfaces. Consequently Eq. (15.31) can be written as a general *linear* kinetic expression for atomically rough interfaces,

$$v_n = K_n \delta T. \tag{15.32}$$

Equation (15.32) is a form of Wilson–Frenkel theory, which predicts the speed of crystal growth occurring via normal (continuous) atomic attachment. The growth speed, v_n, is proportional to the interfacial supercooling [9, 10]. The constant K_n that appears in Eq. (15.32) is known as the 'kinetic attachment coefficient', which is proportional to the interface mobility. In many metallic systems that typically exhibit atomically rough interfaces, K_n is likely have a value between about 10–100 $(\text{cm-s}^{-1})/\text{K}$.[3]

Consequently in a typical metal alloy casting process, with interface speeds of about 0.1 cm/s, the interfacial supercooling would be less than 0.01 K. Thus, ad-atom attachment in the case of solidification of materials with atomically rough interfaces has little influence on most metallurgical solidification processes. Instead, transfer of the latent heat and diffusion of solute near the interface are by far the dominant (and slowest) kinetic processes, not the microscopic kinetics of ad-atom attachment from the melt. As shown next, the opposite is true in many cases of faceted, atomically smooth, crystal growth, where interfacial kinetics usually exerts a significant influence on the overall crystal growth process. Transport of heat and solute may impose relatively minor effects on the morphology of crystals growing with faceted interfaces. Figs. 15.7 and 15.8 provide, respectively, a comparison of a molecularly rough, low alpha-factor interface of CBr_4 held in a thermal gradient, with that for a high alpha-factor, faceted interface in salol ($C_{13}H_{10}O_3$).

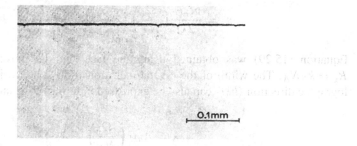

0.1mm

Fig. 15.7 Solid-melt interface in CBr_4. This low alpha-factor material aligns its molecularly rough interface with a horizontal isotherm, excepting the locations around a few grain boundary grooves. See also Appendix B for an explanation of the behavior of grain boundary grooves in temperature gradients. In situ micrograph taken through parallel glass slides spaced 25 μm apart. Adapted from [6]

---

[3] The interface attachment coefficient, K_n, derived here for normal crystallization at atomically rough interfaces is estimated in units of $(\text{cm-s}^{-1})/\text{K}$, where the degree difference, in K, refers to *interfacial* supercooling, not melt supercooling. Interfacial supercooling during crystal growth is notoriously difficult to measure accurately because of the large temperature gradients associated with conduction of the latent heat.

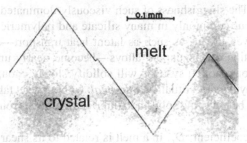

Fig. 15.8 Solid-melt interface in salol. This high α-factor material forms sharp interfacial facets that 'ignore' the horizontal thermal gradients that are present here, and consequently become non-isothermal. Here the *top portion* of this image is slightly hotter than the *lower portion*. In situ micrograph taken through parallel glass slides spaced 25 μm apart. Adapted from [6]

15.3.2 Molecularly Smooth Interfaces

Molecular attachment kinetics occurring in crystallizing systems with smooth, faceted interfaces results in an interface speed that, in contrast with non-faceted interfaces, increases non-linearly with the applied supercooling or supersaturation of the melt. In fact, as will be shown, the kinetic resistance encountered in adding new layers to a faceted solid–liquid interface can cause the interface itself to super-cool significantly (by at least several Kelvins), and depart appreciably from local thermodynamic equilibrium, especially at higher crystal growth rates.

Examples of crystallization processes that operate far from equilibrium on faceted crystal-melt interfaces include nucleation of fresh monolayers, interface dehydration of ions from aqueous solutions, activated ad-atom attachment, layer spreading, step bunching, and spiral ramp growth around screw dislocations thread-ing through the interface. As discussed in the following section, activated ad-atom attachment, and nucleation, lead to growth speeds that theoretically increase exponentially with supercooling [11], whereas screw dislocation-enhanced crystal growth yields theoretically a quadratic dependence on the supercooling [12].

15.3.2.1 Attachment-Limited Growth: Viscous Effects

Some systems with very large α-factors, such as complex silicates, polymers, and macromolecular systems exhibit extremely slow crystallization rates, limited by either nucleation of new interfacial layers, or by the sluggish transfer and incor-poration of the large molecules from their highly viscous melts to the solid-melt interface. The rates of solidification in such systems are kinetically hindered, thus, the crystal-melt interface must supercool appreciably to activate ad-atom attach-ment, and, finally, the melt viscosity itself can increase rapidly at high supercool-ing. The increasing viscosity makes molecular transport ever more difficult, fur-ther driving the crystallizing system away from equilibrium, even eventually reduc-ing its transformation rate sufficiently for glass formation to occur. Thus, crystal-lization can cease despite the increase in supercooling and the associated avail-

able free energy. The sluggishness of such viscously-dominated crystallization processes, which occur commonly in many silicate and polymeric melts, can become
so marked that even processes such as latent heat transport—so important in the
case of solidification of metals and alloys—become nearly undetectable. In fact,
some highly viscous molten systems will solidify slowly enough that during their
crystallization they remain virtually *isothermal*, without detectable temperature gradients developing between the solid-melt interface and its surrounding highly supercooled melt.

The diffusion coefficient, D_ℓ, in a melt is related to its shear viscosity, η, by the
Stokes–Einstein relationship [13],

$$D_\ell = \frac{k_B T}{6\pi \eta R},$$ (15.33)

where R is the radius of gyration of the diffusing molecule. If Eq. (15.33) is substituted into Eq. (15.27), used here as the general kinetic expression for molecular
attachment, one obtains the interfacial speed as function of its shear viscosity, shown
as a function of the melt temperature.

$$v_n = \frac{\Delta S_f \Delta T \zeta_{hkl}}{6\pi \eta(T) R\lambda}.$$ (15.34)

The shear viscosity is a macroscopic measure of the relative resistance of molecular motions under the action of an applied shear stress. In addition, in such melts
the shear viscosity commonly exhibits an Arrhenius correlation with the absolute
temperature, so that

$$\eta(T) = \eta_0 e^{-\Delta Q/k_B T}.$$ (15.35)

Here ΔQ appearing in the Boltzmann exponential in Eq. (15.35) is the activation
enthalpy for viscous flow, and the term η_0 is the pre-exponential coefficient that car-
ries the units of viscosity. Inserting Eq. (15.35) into Eq. (15.34) gives the viscosity-
dependent growth kinetics as

$$v_n = \frac{\Delta S_f}{6\pi \eta_0 R} \frac{\zeta_{hkl}}{\Lambda_{hkl}} \delta T e^{-\Delta Q/k_B(T_e - \delta T)}.$$ (15.36)

The interfacial growth rate behavior predicted according to Eq. (15.36) is plotted
in Fig. 15.9. Attachment-limited crystallizing systems, such as shown here, exhibit
an initially rising interface speed with melt supercooling, followed by a slowing
down as the viscosity of the melt increases at an exponential rate. Here the opposing
effects of increasing kinetic activation for molecular attachment, and the decreasing transport of molecules from the melt as its viscosity rises, compensate at a
melt supercooling of almost $\Delta T = 200$ K. Beyond this point of kinetic compen-
sation, the overall rate of solidification begins to fall with increased supercooling.

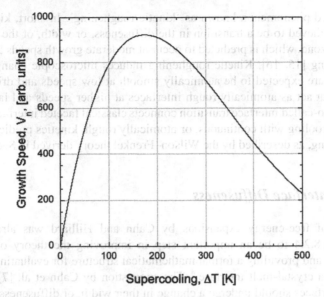

Fig. 15.9 Growth rate dependence for attachment-limited crystal growth from a viscous melt. As the melt supercooling, ΔT, initially increases, the interfacial supercooling, δT, also increases, stimulating more rapid attachment of ad-atoms, and the solidification rate rises. However, with further increases in melt supercooling, the exponential increase in the melt's viscosity eventually slows the transport of molecules to the interface, reducing the overall rate of crystallization. A material's kinetic crystallization curve may exhibit other effects at small supercooling that are not included here in Eq. (15.36)

Crystallization kinetics and glass formation in complex silicate and organic glass-forming melts is discussed by Doremus, along with specific examples of experimental data and observations drawn from a number of high-viscosity systems. [14].

15.4 Kinetic Roughening

15.4.1 Introduction

In the foregoing discussion of interface kinetics, crystallizing systems were divided for convenience into those exhibiting atomically rough or atomically smooth interfaces. The alpha-factor was shown to discriminate between these two classes of interfacial behavior based on a simple equilibrium criterion of a material's thermodynamic and crystallographic properties. The α-factor method, however, is based on reasoning that only applies to solidification taking place close to equilibrium, and which was derived from a highly simplified Bragg–Williams approximation for a two-layer crystal-melt interface.

There is ample experimental evidence and theoretical support that the nature of a solid–liquid interface should also depend on its speed—caused perhaps by a complicated phenomenon known as 'kinetic roughening'. In short, kinetic roughening is believed to be a transition in the diffuseness, or width, of the solid–liquid transition zone, which is predicted to occur at moderate growth speeds and interface supercooling [15, 16]. Kinetic roughening induces microscopic changes in interfaces that are expected to be atomically smooth at low speeds and driving forces, to ones that act as atomically rough interfaces at higher speeds and larger driving forces. A so-called interface transition connects classical faceted interface kinetics at low supercooling with continuous, or atomically rough, kinetics predicted at larger supercooling, as described by the Wilson–Frenkel theory derived in Section 15.3.1.

15.4.2 Interface Diffuseness

The use of free-energy expansions by Cahn and Hilliard was already shown in Section 8.2.4 to be an important step in improving the theory of interphase interfaces, and providing a formal mathematical structure for evaluating the excess energy of a crystal-melt interface.[4] The suggestion by Cahn et al. [7] that solid–liquid interfaces should undergo a change in their width, or diffuseness, as they are driven by higher kinetic driving forces is also an extremely interesting one, notwithstanding the long-standing controversy that still surrounds this idea.

In short, an advancing (faceted) crystal-melt interface, undergoing repeated lateral layer formations, is thought to undergo periodic changes in its interface energy, $\gamma_{s\ell}$. See Fig. 15.10. This conjecture is based on the idea that the minimum free energy of a static interface would be raised by microscopic changes in its structure. As a faceted interface attempts to advance in a direction normal to itself, it must develop non-equilibrium, energetically costly features such as shown schematically in Fig. 15.1.

15.4.3 Roughening Transition

The theory of kinetic roughening predicts that faceted growth occurs on relatively sharp (non-diffuse) interfaces at low interface supercooling, δT, specifically where

$$\delta T < \frac{g_m \gamma_{s\ell} \Omega}{a_0 \Delta S_f}. \tag{15.37}$$

Here $\Delta S_f / \Omega$ is the volumetric entropy of fusion, and a_0 is the interplanar spacing normal to the interface. A transition toward continuous growth is predicted as the interfaces becomes more diffuse. Continuous growth is expected for all crystal-melt

---

[4] Indeed, the related free-energy expansion methods known as Cahn–Hilliard, and Allan–Cahn, theories form parts of the underlying mathematical basis for the successful modeling approach known as phase-field methods [17].

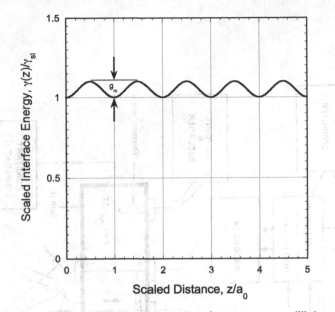

Fig. 15.10 Interface energy scaled to its minimum interface energy at equilibrium, $\gamma_{s\ell}$, versus scaled distance of advance, z/a_0, as suggested by Cahn [16]. The relative increase in interface energy defines the 'diffuseness' parameter, g_m, which provides a measure of the kinetic resistance encountered in advancing a faceted interface from its equilibrium configuration and energy, $\gamma_{s\ell}$, through various non-equilibrium configurations and higher interface energies. Here the maximum interface energy occurs periodically at each location, $z = n - 1 + a_0/2$ ($n = 1, 2, 3, \ldots$), where the interface has advanced to a half-integer position of the interplanar spacing

interfaces when the interface supercooling becomes sufficiently large, as specified by the inequality,

$$\delta T > \frac{\pi g_m \gamma_{s\ell} \Omega}{a_0 \Delta S_f}. \tag{15.38}$$

15.4.4 Experiments

Although crystallization experiments measuring growth rates and supercooling have been performed for more than 75 years [11], suffice it to say the evidence for, or against, kinetic roughening remains controversial [6]. Major difficulties in interpreting experiments on complex viscous melts, including alkali metal silicates [18–23] and organic compounds, such as glycerine and salol [24–27], are the lack of accuracy measuring their interface temperatures, and uncertain corrections to the interface velocity to account for melt viscosity increases at large supercoolings.

There are, however, classes of elemental materials for which it is possible to make highly accurate interface temperature measurements. These include semimetals (Bi, Sb, As, Ga) and elemental semiconductors (Si, Ge). These particular

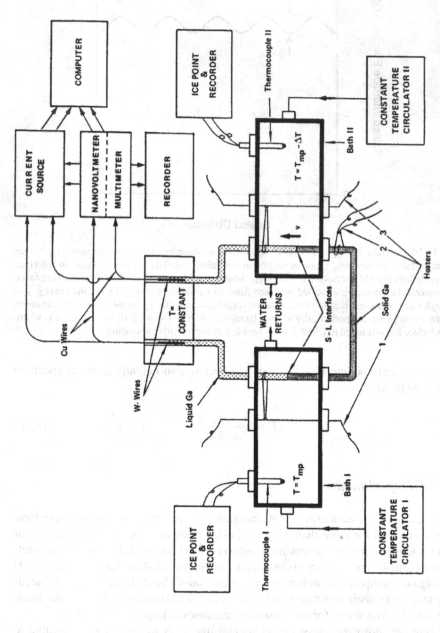

Fig. 15.11 Configuration of the interface kinetic experiment reported by Peteves and Abbaschian. This experiment used Seebeck voltages generated at a moving crystal-melt interface in pure Ga to resolve its interface temperature to ±0.003 K

substances exhibit large thermoelectric power, including their Peltier (heating and cooling) effects, and their Seebeck (junction voltage) coefficients [28].

Peteves and Abbaschian [29, 30] carried out a critical investigation of interface temperatures and velocity measurements in high-purity (7–9s) Ga. Two Ga solid–liquid interfaces were formed in capillary tubes, and arranged in the circuit shown schematically in Fig. 15.11. The left-hand solid–liquid interface was maintained at rest in its equilibrium state, whereas the right-hand interface grew at a specified speed, v. A differential Seebeck voltage developed between the two solid–liquid interfaces that was proportional to the interfacial supercooling. This experimental arrangement provided impressive resolution (± 0.003 K) in the interface temperature resolution, which set a new standard for precision interface kinetic measurements in a crystallizing system.[5]

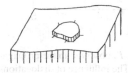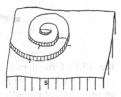

Fig. 15.12 Lateral growth mechanisms studied in high-purity Ga using Seebeck voltage measurements to resolve the interface supercooling. *Left Panel:* Mononuclear 'island' formation; *Middle Panel:* Polynuclear 'islands'; *Right Panel:* Spiral ramp spreading around a screw dislocation threading through the interface. Adapted from [30]

Several aspects of lateral growth behavior in pure Ga metal were studied and reported by Peteves and Abbaschian, including mononuclear and polynuclear 'island' formations, and spiral ramp spreading across an interface threaded by a screw dislocation. These lateral growth mechanisms are portrayed in Fig. 15.12. The type of high-quality kinetic data obtained by these investigators is shown in Fig. 15.13.

15.5 Summary

The structure and behavior of interfaces comprise a range of basic research subjects in physics, chemistry, and engineering. Characterizing the structure and kinetic behavior of crystal-melt interfaces is not an exception to that vast enterprise known as interface science, inasmuch as the interface controls the deepest aspects of crystallization and solidification processes.

---

[5] Peteves and Abbaschian's experiments on Ga formed the basis for the French development of the major microgravity experiment named MEPHISTO (Material pour l'Etude des Phenomenes Interessant la Solidification sur Terre et en Orbite), which was flown in low-Earth orbit several times on the space shuttle *Columbia*, as part of NASA's United States Microgravity Program (USMP) in the late 1990s [31].

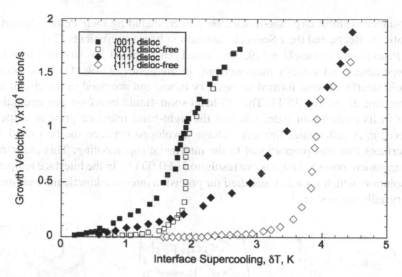

Fig. 15.13 Growth rate, v, versus measured interfacial supercooling, δT, observed in the propagation of {100} and {111} type crystal-melt interfaces in 7–9s pure Ga. The influence of dislocations on the interface mobility is striking. When the interfaces were free of dislocations, they remained virtually immobile, despite the application of more than 1 K supercooling on the {100} interfaces, or more than 3 K on the {111} interfaces. The sudden increase in velocity above these supercooling values might be described as a form of 'kinetic roughening', perhaps due to the onset of mononuclear or polynuclear processes. Data adapted from [29]

Crystal-melt interfaces form the boundaries between condensed phases, and consequently share the experimental and theoretical difficulties attending all such condensed matter interphase transitions. These interfaces may be characterized by macroscopic observation rather simplistically as being either atomically 'rough' or 'smooth', depending on the width of the equilibrium phase transition zone. These characteristics devolve from the equilibrium density of atomic-scale features such as kink sites, ledges, vacancies, and ad-atoms.

The original two-layer interface theory put forth by Jackson is developed. This theory partitions solid–liquid interfaces of different materials according to their α-factor—basically just the ratio of a material's molar entropy of fusion to the universal gas constant, $\Delta S_f / R_g$, multiplied by a bonding ratio of unit order that reflects the crystallography of the plane exposed to its melt. If $\alpha < 2$, the material during freezing is likely to exhibit an atomically rough, kinetically mobile solid–liquid interface. If $\alpha > 2$, the material tends to form interface facets, which exhibit much lower mobilities. This easily applied criterion distinguishes these two types of interface behavior, despite the fact that the model it is based upon is an equilibrium model that is restricted to a two-layer description of the interfacial transition zone.

More sophisticated approaches by Cahn et al. include 'diffuseness' as an important interface parameter. As discussed earlier in Chapter 8, the Cahn–Hilliard free energy expansion introduces additional factors relevant to interfaces that depend on spatial gradients of order and composition that change rapidly through the interfacial transition. A major prediction of such models is that the atomic smoothness of a moving crystal-melt interface should depend on its rate of motion, which is known as 'kinetic roughening'. At present, the experimental status of confirming kinetic roughening remains subject to interpretation.

Quantitative experiments, using differential Seebeck voltage measurements performed by Peteves and Abbaschian in high-purity Ga, clearly expose the influence of dislocations and interface speed on the interfacial mobility. These early experiments led to an important series of space-flight experiments by Abbaschian and French co-workers, named MEPHISTO, that were operated in low-Earth orbit in the 1990s. These studies measured interface kinetic phenomena on pure Bi crystals and Bi alloys using thermoelectric techniques.

References

1. B. Chalmers, *Trans. AIME*, **200** (1956) 519.
2. H.M. Cuppen, H. Meekes, W.J.P. van Enckevort and E. Vlieg, *J. Cryst. Growth*, **286** (2006)188.
3. K.A. Jackson, *Acta. Metall.*, **7** (1959) 148.
4. R.T. DeHoff, *Thermodynamics in Materials Science*, Chap. 3, McGraw-Hill, Inc. New York, NY, 1993.
5. K.A. Jackson and B. Chalmers, *Can. J. Phys.*, **34**, (1956) 473.
6. K.A. Jackson, D.R. Uhlmann and J.D. Hunt, *J. Cryst. Growth*, **1** (1967) 1.
7. J.W. Cahn, W.B. Hillig and G.W. Sears, *Acta Metall.*, **12** (1964) 1421.
8. M.E. Glicksman, *Diffusion in Solids: Field Theory, Solid-State from an and Applications*, Wiley Interscience Publishers, New York, NY, 2000.
9. H.A. Wilson, *Philos. Mag.*, **50** (1900) 238.
10. J. Frenkel, *Physik. Z. Sovjetunion*, **1** (1932) 498.
11. M. Volmer and M. Marder, *Z. Phys. Chem.*, **154** (1931) 97.
12. W.B. Hillig and D. Turnbull, *J. Chem. Phys.*, **24** (1956) 914.
13. S. Kou, *Transport Phenomena and Materials Processing*, Wiley, New York, NY, 1996.
14. Robert H. Doremus, *Glass Science*, Chap. 5, Wiley, New York, NY, 1973.
15. J. Cahn and J. Hilliard, *J. Chem. Phys.*, **28** (1958) 258.
16. J.W. Cahn, *Acta Metall.*, **8** (1960) 554.
17. H. Emmerich, *Phase Field Methods*, Springer, Berlin, 2003.
18. H.R. Lillie, *J. Am. Ceram. Soc.*, **22** (1939) 367.
19. J.P. Poole, *J. Am. Ceram. Soc.*, **32** (1949) 230.
20. L. Shartsis, S. Spinner and W. Capps, *J. Am. Ceram. Soc.*, **35** (1952) 155.
21. W. Kost, *Z. Elektrochem.* **57** (1953) 431.
22. J.O'M. Bockris, J.D. Mackenzie and J.A. Kitchener, *Trans. Faradat Soc.*, **51** (1955) 1734.
23. W.D. Scott and J.A. Pask, *J. Am. Ceram. Soc.*, **44** (1961) 181.
24. H. Pollatschek, *Z. Phys. Chem.*, **142** (1929) 289.
25. L. Neumann and G. Micus, *Z. Phys. Chem.*, **2** (1954) 245.
26. V.I. Malkin, *Zh. Fiz. Khim.*, **28** (1954) 1966.
27. O. Jantsch, *Z. Krist.*, **108** (1956) 185.
28. M.E. Glicksman and R.J. Schaefer, *Acta Metall.*, **16** (1968) 1009.
29. S.D. Peteves and R. Abbaschian, *Metall. Trans. A*, **22A** (1991) 1259.
30. S.D. Peteves and R. Abbaschian, *Metall. Trans. A*, **22A** (1991) 1271.
31. M.E. Glicksman, 'Solidification Research in Microgravity', *ASM Handbook*, **15**, Casting, ASM International, Materials Park, OH, 2008, p. 398.

Chapter 16
Polyphase Solidification

16.1 Introduction

Solidification has been treated up to this point as a two-phase transformation process, wherein a melt is rearranged at the molecular scale to form a crystalline phase. Local molecular configurations and long-range atomic order undergo profound changes at the crystal-melt interface, as do the concentrations of the species present. The system's atomic density and energy states are affected less dramatically, but they too are altered in important ways. Heterogenous phase equilibria, even for a binary system, admit up to three-phases in *simultaneous* thermodynamic equilibrium, although the Gibbs phase rule dictates that the temperature and pressure associated with binary, three-phase equilibria must remain invariant. Three commonly encountered invariant reactions involving solidification and melting are the eutectic, monotectic, and peritectic. These particular three-phase solidification reactions, to be elaborated throughout this chapter, may be represented as reversible transformations:

- eutectic: $\ell \rightleftharpoons \alpha + \beta$.
- monotectic: $\ell_1 \rightleftharpoons \alpha + \ell_2$.
- peritectic: $\ell + \alpha \rightleftharpoons \beta$.

In addition to the invariance of the temperature and pressure, each of the phases participating in these reactions are of fixed composition determined in the phase diagram by the end-points of their tie lines where they contact each of the adjoining single-phase fields. Each of these three invariant reactions, as indicated, are driven toward the right by removing latent heat from the system, and toward the left by adding heat. The idealized symmetric binary eutectic system shown in Fig. 16.1 supports two phases, α and β, with different crystal structures, each in equilibrium with the melt, and exhibiting complete miscibility of the system components, A and B.

Another practically important polyphase solidification reaction occurs when a gas is released as the third phase during freezing at an advancing solid–liquid interface. In industrial casting and ingot practice the release of gases during solidification is of practical concern because trapped gas bubbles in the cast microstructure can

M.E. Glicksman, *Principles of Solidification*, DOI 10.1007/978-1-4419-7344-3_16,

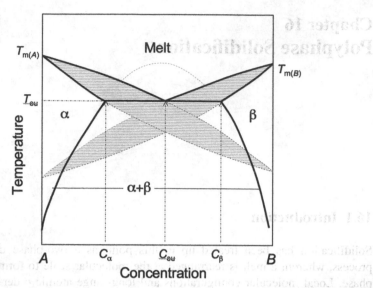

Fig. 16.1 Eutectic phase equilibrium. Two crystalline phases, α and β, each based respectively on the differing crystal structures of pure components A and B, respectively, crystallize from a common melt. Their individual liquidus and solidus curves intersect at an unique composition, C_{eu}, called the 'eutectic point' that represents this system's lowest equilibrium melting point, T_{eu}. The heavy solidus/liquidus lines represent the individual stable equilibria with the melt. The heavy solvus curves delineate the immiscibility gap between the crystalline phases below T_{eu}, whereas the *dashed gray lines* above and below T_{eu} exhibit various metastable extensions. The metastable extensions below T_{eu} play important roles in understanding the kinetics of eutectic solidification

undermine the integrity and strength of the solidified material. In the case of the build-up during freezing of chemically active solutes, such as oxygen and nitrogen, their concentrations in the melt need not even reach levels where bubbles are produced to be harmful to the cast product. High concentrations of oxygen and nitrogen, developed through interdendritic microsegregation, can react with metallic components in the alloy to produce oxide and nitride inclusions that are brittle phases detrimental to the material's fatigue life and stress-rupture properties.

16.2 Eutectics

16.2.1 Thermodynamics: Polyphase Solidification

The three-phase equilibria implied on the phase diagram in Fig. 16.1 are established at the 'fixed' eutectic reaction temperature, T_{eu}, at some fixed pressure, P. The corresponding free energy relations among the three participating eutectic phases are shown schematically in Fig. 16.2.

The Gibbs phase rule states:

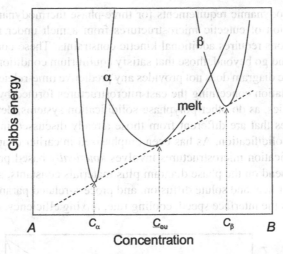

Fig. 16.2 Gibbs free energy versus composition at the eutectic isotherm, $T = T_{eu}$. The 'common tangent' construction insures that the chemical potentials of component species A and B (μ_A and μ_B) are independently pairwise equal in all three phases. The common *tangent* represents the unique combination of temperature, pressure, and phase compositions where eutectic equilibria hold. The *binary eutectic* represents an invariant 3-phase state, as the temperature, pressure, and all phase compositions are fixed quantities

$$f = C - \varphi + 1, \tag{16.1}$$

where f is the degree of freedom; C is the number of components; and φ is the number of phases present. For a binary system, $C = 2$, with three conjugate phases, $\varphi = 3$, Eq. (16.1) shows that $f = 0$. This is the formal statement that eutectic equilibrium is *invariant*, and occurs at a fixed temperature, pressure, and composition for each of the three participating phases. The thermodynamic invariance requires that the free-energy versus composition curves for the phases align in such a way that there is a common tangent to all three curves. Figure 16.2 shows the alignment of the free energies needed to satisfy this invariance. The common tangent construction guarantees that at the fixed eutectic reaction temperature (and pressure), the thermodynamic activities and chemical potentials of the components A and B are pairwise identical in each of the three phases:

i. $\mu_A^\alpha = \mu_A^\beta = \mu_A^\ell$
ii. $\mu_B^\alpha = \mu_B^\beta = \mu_B^\ell$

Imposition of these thermodynamic constraints at $T = T_{eu}$, and adding some additional features as shown in Fig. 16.1—such as decreasing solid solubility with temperature—and then eliminating the metastable portions of this phase diagram leads to the 'generic' binary eutectic phase diagram drawn in Fig. 16.3. Although the equilibrium phase diagram is consistent with, and satisfies the general thermodynamic requirements for three-phase thermodynamic equilibrium, the solidification of eutectic microstructures from a melt under realistic solidification conditions requires additional kinetic constraints. These constraints are different from, and go beyond, those that satisfy equilibrium conditions. More specifically, the phase diagram does not provides any predictive time-related or length scale related information concerning the cast microstructures formed by eutectic alloys. Indeed, eutectics, as do other polyphase solidification systems, develop distinctive microstructures that are different from those already discussed in some detail for single-phase solidification. As has been emphasized in earlier chapters, the formation of solidification microstructures involves *kinetically* based processes, and, as such, they depend on the phase diagram plus materials constants, such as transport coefficients for heat and solute diffusion, and process-related parameters, including such details as the interface speed, cooling rate, mixing efficiency, etc.

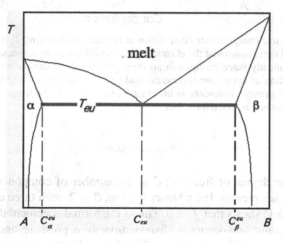

Fig. 16.3 Generic binary eutectic phase diagram. The compositions shown for the three phases are their equilibrium compositions only at $T = T_{eu}$. Although the equilibrium eutectic invariant does not provide any information about the microstructure produced by eutectic freezing, it supplies an important basis for the kinetic analyses that follow

16.2.2 Classification of Eutectics

The microstructures of eutectics have been studied experimentally and classified since the 1960s [1]. Chadwick published an excellent review of the nature of eutectic microstructures [2]. Although the classification schemes and nomenclature for eutectic types vary, the following system based on the entropy of fusion of the participating phases [3] is adopted for the purposes of subsequent discussions:

i. *Type I, 'regular' eutectics.* Here the two solid phases solidify simultaneously from the melt in a cooperative manner, with each phase participating in the development of the local diffusion fields along their common solid-melt interfaces.

The microstructures observed in type-I regular eutectics are typically lamellar (plate-like) or rod form. That is, the phases either alternate as broad plates or lamellae, or if one crystalline phase predominates in volume fraction relative to the second, the minor phase can form as long rods or needles. It is generally thought that type-I eutectics have interfacial alpha-factors that are less than two for *both* the α-phase/melt interface and the β-phase/melt interface. See again Section 15.2.2.4. This statement is consistent with the requirement that the two interfaces must be virtually isothermal and aligned along a local isotherm that is reasonably close (within a few Kelvins) to the equilibrium eutectic temperature, T_{eu}. If the interfaces were, instead, faceted, (alpha-factors greater than two) one phase could lead or lag the other by a distance large enough to preclude the coupling of their interfacial diffusion fields. This insistence on small alpha-factors for both phases usually limits type-I eutectics to alloys with centrosymmetric, i.e., highly symmetric crystal structures, such as BCC, FCC, HCP, and BCT. Thus, type-I eutectics occur most often, but not exclusively, in metallic alloy systems, although type-II eutectics, to be described next, are also commonly found.

i. *Type II, 'irregular' eutectics.* In this case, the α-phase/melt interface and the β-phase/melt interface both grow in a more loosely coordinated manner. As explained above, this situation occurs when either one or both of the eutectic crystalline solids have an interfacial α-factor greater than two. The cast microstructures produced by type-II eutectics vary greatly. If the major phase has a low α-factor, and the minority phases has an α-factor only slightly greater than two, the eutectic can form a rod-type microstructure, provided that it is directionally solidified under a sufficiently steep temperature gradient. Usually, however, type-II eutectics fail to couple the solidification of their crystalline phases, and an irregular eutectic microstructure forms. The microstructures of irregular eutectics

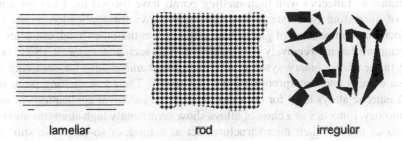

lamellar   rod   irregular

Fig. 16.4 Microstructural appearance (schematic) of resolved eutectic constituents in several directionally-solidified eutectic alloys: (1) Lamellar eutectic. Lamellar eutectics require type-I behavior—i.e., small interfacial alpha-factors for both of its crystalline phases. (2) Rod-form eutectic. Rod eutectics are usually type-I, but may even be type-II, provided that the majority phase has a low alpha-factor, and the minority phase has an alpha-factor not much above two. (3) Irregular eutectic. Irregular eutectics have large alpha-factors in one or both phases. These systems, by contrast with lamellar and rod-form eutectics, fail to exhibit any appreciable crystallographic coupling between the phases as they solidify from the melt. The morphology of irregular eutectics depends sensitively on the crystallography of the phases and their growth rate anisotropy in various directions. Typical as-cast eutectic phase spacings are several micrometers

vary greatly, depending in detail upon the crystallographic properties of the two solid phases, as well as the differences of their interfacial speed.

Figure 16.4 shows schematically several typical forms of eutectic microconstituents after directional solidification. The ability to recognize these microconstituents requires that they be resolved at an appropriate magnification and resolution. In fact, most eutectic microconstituents in ordinary castings cannot be resolved, that is, observed as distinct phases arrangements, unless the magnification is at least 1,000, or greater. Some eutectics produced by rapid freezing require much higher levels of magnification than would be convenient, or even practicable, with a conventional optical microscope. These ultra fine-scale eutectics require the use of scanning electron microscopy (SEM) to resolve the individual phases present in the eutectic microconstituent.

16.2.3 Importance of Eutectics

Many eutectics yield polyphase solidification structures that form from a melt at relatively high temperatures, that is at temperatures comparable to the melting points of the pure binary components, and under conditions removed, but usually not too far, from interfacial thermodynamic equilibrium. In this limited sense, eutectics as a class, form highly stable microstructures that do not revert, or coarsen, easily at elevated temperatures. Stated simply, phases produced at high temperatures retain the lattice stability to be used at high temperatures.

Type-I eutectics, which couple their two crystalline phases, also naturally favor the development of low-energy lamellar and rod-form boundary structures, which also adds to their high thermodynamic stability and kinetic resistance to thermal degradation. Eutectics with high-melting points have formed the basis for a number of interesting candidate high-temperature alloys for application to the high-temperature components of gas turbine engines. On the other hand, eutectics with components having relatively low melting points, such as the one at 183 C occurring in the Pb-Sn binary system, are used almost universally for soldering electronic circuits and waterproofing container seams. Their low melting point make such eutectic alloys ideal for both mechanical and electrical applications of solder technology. Eutectics as a class of alloys show surprisingly high strengths and creep resistance because their microstructures act as natural, or so-called 'in situ' composite materials. Their responsiveness to solidification processing and their superior mechanical properties have stimulated large amounts of research on eutectics. As will be shown in subsequent sections of this chapter, eutectics have their phases distributed on relatively small scales, even under ordinary solidification conditions. Sub-micrometer spacings are not uncommon under ordinary freezing conditions. This unusual feature in cast microstructures provides enhancement of their mechanical properties, because the fine-scale mixture of phases interfere with the mobility of slip dislocations during plastic deformation. Indeed, at more rapid solidification speeds, eutectics have been prepared as nanostructures, with lamellar spacings verified to be as small as several hundred Ångstroms. When a eutectic microstructure is refined to such extremely small length scales, however, its stability, even at moderately elevated temperatures, can be compromised because of its high density of stored interfacial energy.

Further enumerating the useful features of eutectic solidification structures: (1) near-equilibrium microstructures that resist change at temperatures as high as their reaction temperature; (2) low-energy phase boundaries; (3) controllable microstructures; (4) high rupture strength; (5) stable defect structures; (6) good high-temperature creep resistance; (7) interesting and sometimes unusual electrical, magnetic, and optical properties. For a comprehensive review of eutectic solidification, the interested reader is referred to [4].

16.2.4 Nucleation of Eutectics

The subject of nucleation of eutectic phases from the melt was studied in detail by Sundquist and Mondolfo [5]. Sundquist and Mondolfo's main findings are summarized below.

- *Non-reciprocal nucleation catalysis*: Generally, if phase α nucleates phase β effectively (i.e., with little melt supercooling), then phase β tends not to nucleate phase α easily. Specifically, Sundquist and Mondolfo observed in common Pb-Sn eutectic solder, that BCT Sn-phase efficiently nucleated FCC Pb-phase from the eutectic melt. That is, only a small supercooling of the eutectic melt occurred before the FCC Pb-phase starts to overgrow the BCT Sn-phase, initiating the eutectic reaction where both phases grew simultaneous from the melt. However, the molten Pb-Sn eutectic alloy supercooled considerably when in the sole presence of FCC Pb-phase before any BCT Sn-phase nucleates, allowing the eutectic reaction to initiate. The observation of non-reciprocity in catalytic behavior could be caused by significant differences in the energy barrier to form the nuclei of each solid phase, as the crystal-melt energies for the two phases could differ substantially.
- Phases that act as effective nucleation catalysts tend to have 'complex', non-centrosymmetric, crystal structures. Again, referring to their experiments on Pb-Sn eutectic, the lower-symmetry BCT Sn-phase acts more effectively as a nucleant for the symmetric FCC Pb-phase than the reverse.
- The crystalline phase that exhibits difficulty in being initially nucleated from the eutectic melt provides an efficient substrate for the subsequent nucleation of the more symmetric conjugate phase. For example, in the case of Pb-Sn eutectic, Sn-phase nucleates with difficulty from the melt, and then easily nucleates Pb-phase. By contrast, Pb-phase nucleates from the eutectic melt with little supercooling, but it does not nucleate Sn-phase without considerable additional supercooling.

The findings of Sundquist and Mondolfo listed above were challenged initially on the basis that their experiments might not have provided observation of the nucleation process, per se, but, rather, just the inadvertent growth of the eutectic from retained, or pre-existing athermal 'nuclei'. See again the earlier discussion in Section 12.6.4 of the role of athermal nuclei in grain refinement. The concept of retained athermal eutectic phase nuclei is sketched in Fig. 16.5. As shown if nanocrystals of β-phase were retained in microscopic fissures or pores in the α-phase (even at temperatures somewhat above the eutectic isotherm), then the eventual appearance of a macroscopic crystal of β-phase when the liquid is cooled to a lower temperature might be considered more a matter of 'growth' and not 'nucleation'. The retention of microscopic crystals caused by their large negative curvatures, as described in Fig. 16.5, is also believed to be behind the oft-heard claim that re-used molds nucleate the freezing of an alloy more efficiently than does a freshly prepared mold, the surfaces of which would not contain any stray trapped nanocrystals. Whether or not such foundry lore is indeed based on scientific fact remains to be seen. In view of the success of describing the microtopography of solid-melt interfaces using Jackson's alpha-factor criterion, discussed in Section 15.2.2.4, the 'complexity' argument put forth by Sundquist and Mondolfo to explain non-reciprocal nucleation behavior in eutectics seems reasonable.

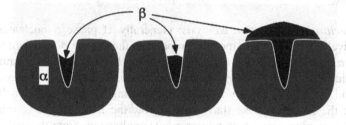

Fig. 16.5 Retained nanocrystals of β-phase growing within a micropore in the α-phase. The strong negative interfacial curvature stabilizes the β-crystal, even at temperatures greater than T_{eu}. This phenomenon was thought originally to explain some of the curious observations reported by Sundquist and Mondolfo [5] on the non-reciprocal nature of eutectic nucleation. See Section 16.2.4. Their correlations between crystal 'complexity' and eutectic nucleation seem justified today on the basis of Jackson's α-factor to represent the equilibrium microtopography of solid-melt interfaces

16.2.5 Growth of Eutectics

The growth of eutectics has provided a controversial subject when traced back through the early metallurgical literature. The difficulty in interpreting the growth of eutectics, as mentioned earlier, was that eutectics often solidify as a fine-phase mixture, or microconstituent, which, without good resolution optical or electron microscopes, was extremely difficult to resolve. Eutectics also form in a bewildering variety of microstructures. Vögel [6] was the first investigator to claim that eutectic growth occurred by simultaneous growth of both phases. Vögel also observed qualitatively that the phase spacing clearly depended on the growth speed. Tammann challenged Vögel's view of eutectic solidification in 1925, and suggested instead that the phases of a eutectic grow as an alternating, transverse sequence of individual phase layers [7]. Figure 16.6 shows the gist of Tammann's counter-hypothesis to Vögel's original idea of simultaneous, longitudinal growth of the eutectic phases.

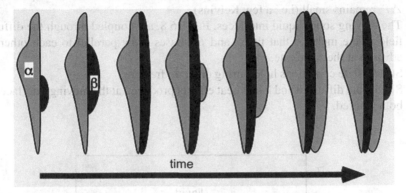

Fig. 16.6 Tammann's model of lamellar eutectic growth by alternating, *transverse* growth of phase layers. Each phase layer allegedly builds up a supersaturated zone of the rejected component that is relieved by the nucleation and growth of the conjugate phase. This eutectic solidification model, accepted for many years, was a counter-hypothesis to Vögel's original—and now known to be correct—idea that eutectics solidify by the simultaneous, *parallel* growth of both phases

It remained until Straumanis and Braaks [8] carried out a series of experiments that proved Tammann—who was usually right about most matters metallurgical—was actually wrong, and that Vögel, who had suggested that the eutectic growth mechanism occurs by *simultaneous* edgewise motion of the phases, was right. Descriptions regarding the kinetics of eutectic growth remained basically qualitative for another 25 years, until the diffusion model of eutectic growth was presented by Tiller [9] in 1957. This model was the first description of eutectic solidification and microstructure formation that clearly, and quantitatively, embraced the concept of simultaneous, coupled phase growth. Tiller's ideas for eutectic freezing, in fact, follow closely those of Zener, who in the early 1950s analyzed coupled-diffusion as the fundamental mechanism responsible for solid-state eutectoid transformations in steel. Thus, theoretical concepts, originally developed for eutectoids, were adapted by Tiller for eutectics, and used to provide testable kinetic predictions. Tiller's adaptation of Zener's coupled diffusion concept led to quantitative 'scaling laws' for eutectic microstructures that guided experimental developments in this field for many years, and stimulated more advanced theoretical work.

16.2.6 Tiller's Theory of Eutectic Growth

Their are five major assumptions needed in developing Tiller's theory of eutectic solidification:

1. The eutectic diagram, the thermal properties, and the eutectic microstructure itself are symmetrical. See Fig. 16.7 to observe how the symmetrical phase diagram results in certain simplifications of the thermodynamic parameters, and the eutectic structure itself.
2. The eutectic phases form close to equilibrium, so their interfacial supercooling, ΔT, remains small (i.e., a few Kelvins).
3. The moving solid–liquid interfaces, Fig. 16.8, are coupled through the diffusion field in the melt, so that the α and β phases grow parallel to each other and advance at the same rate.
4. Steady-state conditions hold during eutectic freezing.
5. Solid-state diffusion and latent heat effects produced at the moving interfaces are both ignored.

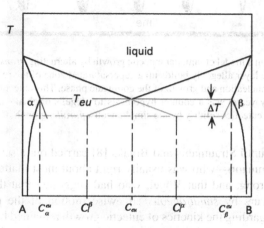

Fig. 16.7 Symmetrical eutectic phase diagram needed when developing Tiller's theory of coupled eutectics. The two liquidus slopes remain constant, equal and opposite, so that $m_\ell^\alpha = -m_\ell^\beta$; the eutectic composition, $C_{eu} = 0.5$; the supercooling of the eutectic interface, $\Delta T \equiv T_{eu} - \hat{T}$, is small, so that $\hat{T} \cong T_{eu}$

Tiller envisioned a steady oscillatory solute diffusion field, varying in the lateral x-direction, ahead of the co-moving α-liquid and β-liquid interfaces, as indicated schematically in Fig. 16.8. The triple-lines for all three phases, spaced apart by a distance $\lambda/2$, allow mutual contact along a single line normal to the page. The melt composition at these periodic triple-lines is chosen on the basis of local equilibrium to be C_{eu}. Away from the triple points, out ahead of either the α-melt or β-melt interfaces, the composition rises with respect to the rejected component (either A ahead of the β-phase, or B ahead of the α-phase). Specifically, the phase diagram demands that just ahead of the α-melt interface the melt composition at steady-state is enriched in component B to satisfy the tie-line equilibrium between them, whereas just ahead of the β-melt interface the melt is correspondingly enriched in component A to satisfy the tie-line equilibrium between those two

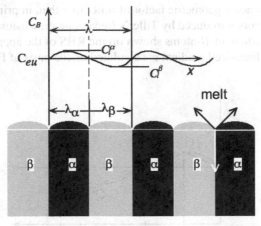

Fig. 16.8 Diffusion coupling in the melt at an eutectic-melt interface. The microstructure at the interface is symmetric, so that $\lambda_\alpha = \lambda_\beta = \lambda/2$. The solute concentration field in the melt just ahead of the advancing α-β-ℓ interface is shown oscillating along the x-direction. The peak (and valley) concentration differences about the average eutectic composition, C_{eu}, are taken to be equal and opposite ahead of adjacent lamellae, so that $C_{eu} - C_\ell^\alpha = -(C_{eu} - C_\ell^\beta)$. The triplet of vectors on the right-most pair of lamellae shows the surface tension balance among the three phases at their triple points

phases. Figure 16.8 suggests the oscillatory shape of a steady-state diffusion field that satisfies both of these interface conditions.

16.2.6.1 Diffusion at Lamellar Eutectic Interfaces

Tiller used an approximate solute mass balance at the freezing eutectic interface to formulate his model. As suggested in Fig. 16.8, the concentration field, $C_\ell(x)$ was assumed to be sinusoidal in x. This field is expressed in terms of the concentration of B-atoms, which accumulate ahead of the α-lamellae, and which is depleted ahead of the β-lamellae. These variations in the melt's solute concentration reflect rejection of B-atoms from the advancing α-phase, which grows by accepting mostly A-atoms, and their 'cooperative' incorporation of these 'rejected' B-atoms that permit the simultaneous growth of the β-phase, which grows by accepting mostly B-atoms. Stated more simply: the α/melt portions of the eutectic interface represents 'sources' of B-atoms, whereas the β/melt portions of the interface represents 'sinks' of B-atoms. The transported fluxes of B-atoms diffusing in the $\pm x$-directions arriving through the melt just ahead of the β-phase lamellae, and the compensating flux of A-atoms arriving through the melt just ahead of the α-phase lamellae, was approximated by applying Ficks 1st law of diffusion,

$$J_B = -D_\ell \frac{\partial C_\ell}{\partial x} \cong \pm D_\ell \left(\frac{C_\ell^\alpha - C_\ell^\beta}{\lambda/2} \right) \cdot \xi, \qquad (16.2)$$

where ξ is an unknown geometric factor of unit order that, in principle, corrects the solute flux for errors introduced by Tiller's linear, one-dimensional estimate for the concentration gradient of B-atoms shown on the RHS of the approximation used in Eq. (16.2). This linearized gradient is calculated on the basis of Fig. 16.9.

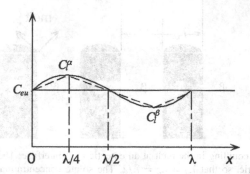

Fig. 16.9 Linearized approximation of the concentration field, $C_\ell(x)$, ahead of an eutectic interface. The *vertical axis* represents the concentration of component B at any point, x, along the advancing solid-melt interface, which is chosen here as the x-axis. The local concentration in the melt is raised above C_{eu} ahead of the α-lamellae, and depressed ahead of the β-lamellae

Tiller's main contribution here is to point out that the advance of a coupled eutectic front normal to the solid–liquid interface is controlled primarily by the *lateral* flux of solute atoms diffusing from neighboring lamellae. Each lamella acts simultaneously as a continuous source for the A- or B-atoms it rejects, and as a continuous sink for the atoms it needs for growth. The atoms required by each lamella are precisely those rejected and made available from its adjacent, neighboring lamellae. Thus, Tiller's model of lamellar eutectic solidification provides a physical mechanism that allows the β-phase lamellae to accept an excess number of B-atoms from adjacent α-phase lamellae; the β-phase grows and rejects an excess number of A-atoms, which diffuse away and allow the growth of α-phase lamellae, which in turn grow and reject more excess B-atoms that diffuse to the β-phase, and so forth. This process of continuous steady exchange, sorting, and separation of A- and B-atoms that were mixed in the melt, although surprising at first, is enabled through cooperative lateral diffusion along a lamellar interface. As we shall now see, this diffusive sorting of atoms from an initially uniform melt represents an irreversible, non-equilibrium process that dissipates free energy.

16.2.6.2 Mass Balances at Lamellar Eutectic Interfaces

The mass of B-atoms rejected per unit time per unit area of the α/melt interface may be expressed using a simple linear expression for the left-hand liquidus on the phase diagram. (See again Fig. 16.7.)

$$v_{eu}\left(C_\ell^\alpha - C_s^\alpha\right) = v_{eu}\left(1 - k_0^\alpha\right)C_\ell^\alpha, \tag{16.3}$$

where v_{eu} is the speed of steady-state advance of the eutectic interface, and k_0^α is the distribution coefficient for component B at the α/melt interface. Equations (16.3) balances the steady-state rates of the B-atom solute mass rejected by the α-phase, to their transport to the β-phase via the diffusion flux, Eq. (16.2). This conservation of mass is the local Stefan solute balance ahead of each α-lamella.

$$v_{eu}\left(1 - k_0^\alpha\right) C_\ell^\alpha = D_\ell \frac{C_\ell^\alpha - C_\ell^\beta}{\lambda/2} \cdot \xi. \tag{16.4}$$

Equation (16.4) may be solved for the concentration difference as a function of growth speed, lamellar spacing, λ, and thermophysical parameters, such as melt diffusivity, and the equilibrium distribution coefficient.

$$C_\ell^\alpha - C_\ell^\beta = \frac{v_{eu}}{2D_\ell}\left(1 - k_0^\alpha\right) C_\ell^\alpha \lambda \cdot \xi. \tag{16.5}$$

One may further approximate the concentration in the melt ahead of the α-lamellae, for small interfacial supercoolings below the eutectic temperature, as $C_\ell^\alpha \approx C_{eu}$. One finds using this approximation on the right-hand side of Eq. (16.5) that the steady-state concentration difference developed between adjacent lamellae is

$$C_\ell^\alpha - C_\ell^\beta = \frac{v_{eu}}{2D_\ell}\left(1 - k_0^\alpha\right) C_{eu} \lambda \cdot \xi. \tag{16.6}$$

This small composition difference along the interface causes the interface temperature, $\hat{T}$, to be slightly depressed below the equilibrium eutectic temperature, T_{eu}. Thus, one may determine the small interfacial supercooling, $\Delta \hat{T}_{diff} = T_{eu} - \hat{T}$, which acts at the eutectic-melt interface and provides the necessary free energy that 'drives' the solute diffusion process, namely

$$\Delta \hat{T}_{diff} = -m_\alpha \left(C_\ell^\alpha - C_{eu}\right) = +m_\beta \left(1 - k_0^\alpha\right) C_{eu}, \tag{16.7}$$

where m_α and m_β are the corresponding liquidus slopes for the α- and β-phases, respectively. Equations (16.7) are equivalent to the coupled pair of equations given next as Eqs. (16.8) and (16.9):

$$\frac{\Delta \hat{T}_{diff}}{m_\alpha} = C_{eu} - C_\ell^\alpha, \tag{16.8}$$

and

$$\frac{\Delta \hat{T}_{diff}}{m_\beta} = C_{eu} - C_\ell^\beta. \tag{16.9}$$

Subtracting Eq. (16.8) from Eq. (16.9) provides an expression for the amount of interfacial supercooling required to 'unmix' the melt into its component A- and

B-atoms. This undercooling is proportional[1] to the Gibbs free energy being dissipated for diffusion and 'unmixing' of the component atoms as eutectic solidification proceeds.

$$\Delta\hat{T}_{diff}\left(\frac{1}{m_\beta}-\frac{1}{m_\alpha}\right)=C_\ell^\alpha-C_\ell^\beta. \tag{16.10}$$

Substituting the right-hand side of Eq. (16.6) into the right-hand side of Eq. (16.10) yields Tiller's expression for the interfacial eutectic supercooling required by diffusion in the melt, namely

$$\Delta\hat{T}_{diff}=\frac{v_{eu}}{2D_\ell}\left(1-k_0^\alpha\right)\left(\frac{m_\alpha m_\beta}{m_\alpha-m_\beta}\right)C_{eu}. \tag{16.11}$$

In view of the assumed symmetrical eutectic diagram, where $m_\alpha=-m_\beta$, the supercooling caused by solute diffusion, Eq. (16.11), simplifies to

$$\Delta\hat{T}_{diff}=-\frac{v_{eu}}{4D_\ell}\left(1-k_0\right)m_\ell C_{eu}\lambda\cdot\xi. \tag{16.12}$$

Equation (16.12) shows that the interfacial supercooling required to provide the free energy needed to 'drive' the lateral solute diffusion process and 'unmix' the melt for a symmetric eutectic interface is proportional to the interface speed, v_{eu}, the average shift in melt composition, $C_{eu}(1-k_0)$, and the steady-state lamellar spacing, λ. The supercooling is, however, inversely related to the diffusion coefficient in the melt, D_ℓ.

16.2.6.3 Stored Free Energy (Capillarity)

The other major energetic requirement in solidifying an eutectic alloy is providing the free energy needed to form the lamellar microstructure itself. Specifically, one must address the fact that the eutectic constituent being formed behind the crystal/melt interface contains a dense array of parallel α/β interfaces. These interfaces were shown in many cases studied experimentally to consist of special crystallographic boundaries established between the conjugate crystalline phases. These boundaries provide an α/β eutectic with relatively low interfacial free energy. Nevertheless, the total area of these solid-state boundaries per unit volume of eutectic constituent is considerable, and, therefore, the free energy required to form them must be included in assessing the free energy requirement for steady-state eutectic

---

[1] The Gibbs energy per mole of eutectic associated with this component of the interfacial supercooling is $\hat{T}_{diff}\Delta S_f$, where ΔS_f is the molar entropy of fusion, which may be considered the same for each of the eutectic's crystalline phases, as the system's phase diagram is assumed to be symmetric.

solidification.[2] This energy resides within the eutectic microstructure as the 'stored' free energy of its crystalline interfaces. These interfaces are created continually during the solidification process and impose an additional free energy 'cost', causing yet a further lowering of the interfacial temperature, $\hat{T}$, in addition to that consumed for diffusion transport. Thus, the requirement to store copious amounts of interfacial energy among the α/β lamellae comprising the fine eutectic microstructure causes the total interfacial supercooling to increase beyond $\Delta \hat{T}_{diff}$, given by Eq. (16.12). The free energy stored in a unit volume of lamellar eutectic may be estimated using the sketch provided in Fig. 16.10. Here one views one square centimeter of the crystal-melt eutectic interface advancing steadily to the right at a speed v_{eu} [cm/s]. The advancing square centimeter of interface produces after a time interval of $1/v_{eu}$ s one cubic centimeter of the eutectic microconstituent.[3] The microstructure depicted in Fig. 16.10 contains many internal α/β interfaces of specific energy $\gamma_{\alpha\beta}$. The free energy that must be created and stored per unit time when producing this eutectic at a speed v_{eu} is

$$\Delta \dot{G}_{\alpha\beta} = \frac{2 v_{eu}}{\lambda} \gamma_{\alpha\beta}, \qquad (16.13)$$

where the over-dot notation implies the total time derivative. The free energy storage rate, $\Delta \dot{G}_{\alpha\beta}$, given in Eq. (16.13) is again provided by the appearance of an

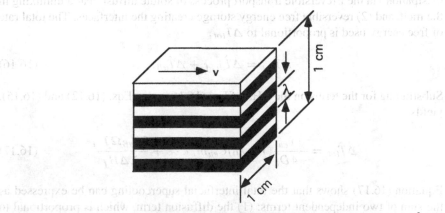

Fig. 16.10 Unit volume of a lamellar eutectic microstructure produced by advancing 1 cm² interface through 1 cm of additional growth. Each α/β interface stores free energy in the amount $\gamma_{\alpha\beta}$ per unit area

---

[2] For example, a typical metallic eutectic with a lamellar spacing of $\lambda = 1 \mu$ m, and an average α-β interfacial energy of 100 mJ/m², has an equivalent stored energy of the order of 0.1 J/cm³.

[3] The term 'microconstituent' is used to denote the combination of two crystalline phases that are recognizable as being produced by the eutectic reaction.

additional interfacial supercooling, $\Delta \hat{T}_{\alpha\beta}$,

$$\Delta \dot{G}_{\alpha\beta} = \Delta \hat{T}_{\alpha\beta} \frac{\Delta S_f}{\Omega} v_{eu}. \tag{16.14}$$

Equating the RHS of Eqs. (16.13) and (16.14) allows the additional capillary-induced supercooling needed at the solid–liquid interface to create free energy at the required rate of storage in the α/β interfaces. Thus, one finds

$$\Delta \hat{T}_{\alpha\beta} = \frac{2\gamma_{\alpha\beta}\Omega}{\lambda\Delta S_f} = \frac{2\gamma_{\alpha\beta}\Omega T_{eu}}{\lambda\Delta H_f}. \tag{16.15}$$

Equation (16.15) indicates that the additional capillary supercooling, $\Delta \hat{T}_{\alpha\beta}$, accounting for the rate of free energy storage in the α/β interfaces, is inversely proportional to the lamellar spacing, λ, selected by the sytem.

16.2.7 Eutectic Phase Spacing

One can now appreciate the fact that the total interfacial supercooling, $\Delta \hat{T}_{tot}$, needed to solidify a moving eutectic front entails two coupled processes: (1) free energy dissipation via the irreversible transport process of solute diffusion and unmixing in the melt, and (2) reversible free energy storage creating the interfaces. The total rate of free energy used is proportional to $\Delta \hat{T}_{tot}$,

$$\Delta \hat{T}_{tot} = \Delta \hat{T}_{diff} + \Delta \hat{T}_{\alpha\beta}. \tag{16.16}$$

Substituting for the terms on the RHS of Eq. (16.16) using Eqs. (16.12) and (16.15), yields

$$\Delta \hat{T}_{tot} = -\frac{v_{eu}}{4D_\ell}(1 - k_0)C_{eu}m_\ell\lambda \cdot \xi + \frac{2\gamma_{\alpha\beta}\Omega T_{eu}}{\lambda\Delta H_f}. \tag{16.17}$$

Equation (16.17) shows that the total interfacial supercooling can be expressed as the sum of two independent terms: (1) the diffusion term, which is proportional to the lamellar spacing, and (2) the interfacial term, which depends inversely on the lamellar spacing. Tiller suggested that in the course of solidification type-I eutectics minimize their interfacial supercooling by 'selecting' an optimal spacing for its microstructure, λ^*. If one views the lamellar spacing, λ, as a 'free' parameter of the solidification process, then the functional dependence of the total interfacial supercooling on λ may be written as

$$\Delta \hat{T}_{tot} = a_1\lambda + a_2\lambda^{-1}, \tag{16.18}$$

where the coefficients a_1 and a_2 in Eq. (16.18) appear explicitly in Eq. (16.17).

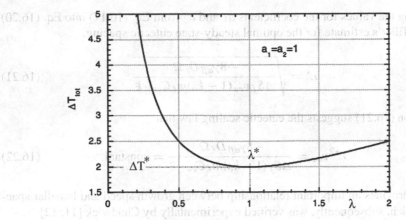

Fig. 16.11 Dependence of total interfacial supercooling on the eutectic lamellar spacing, λ, according to Eq. (16.18). The values of the coefficients a_1 and a_2 are arbitrarily set equal to unity. The minimum in the supercooling represents the minimum rate of free energy use, where λ^* balances the amount of free energy stored as interfacial free energy against the free energy dissipated for solute diffusion through the melt

Equation (16.18) is plotted in Fig. 16.11, which shows that at small lamellar spacings the interfacial supercooling falls rapidly with λ, because the stored free energy in the eutectic decreases as the spacing increases. By contrast, the interfacial supercooling increases gradually at large lamellar spacings, because solute transport becomes increasingly difficult over the larger length scale for solute diffusion. Between the limits of extremely fine and coarse lamellar spacings, an optimal spacing, λ^*, occurs, for which the interfacial supercooling achieves a minimum. Tiller assumed that the physical 'operating state' for the steady-state growth of a lamellar type-I eutectic also occurs where the total interfacial supercooling, $\Delta \hat{T}_{tot}$, is minimized.

Although not a fundamental principle of nature, minimization of the total free energy consumption is equivalent to minimum entropy production [10]—a condition that is approximated by at least some steady-state kinetic processes that operate not too far from equilibrium. Minimum interfacial supercooling in eutectics has also been confirmed experimentally by Hunt. This optimum spacing may be found analytically through the necessary condition of setting the first derivative of Eq. (16.18) with respect to λ equal to zero.

$$\frac{d\Delta \hat{T}_{tot}}{d\lambda} = 0 = a_1 - \frac{a_2}{\lambda^{*2}}. \tag{16.19}$$

Solving Eq. (16.19) for λ^* gives the result,

$$\lambda^* = \sqrt{\frac{a_2}{a_1}}. \tag{16.20}$$

Inserting the values for the coefficients a_1 and a_2 from Eq. (16.17) into Eq. (16.20) yields Tiller's estimate for the optimal steady-state eutectic spacing,

$$\lambda^{\star} = \sqrt{\frac{-8\gamma_{\alpha\beta}D_{\ell}\Omega}{\Delta S_f v_{eu}(1-k_0)m_{\ell}C_{eu} \cdot \xi}}.$$ (16.21)

Equation (16.21) suggests the eutectic scaling law that

$$\lambda^{\star 2}v_{eu} = \frac{-8\gamma_{\alpha\beta}D_{\ell}\Omega}{\Delta S_f(1-k_0)m_{\ell}C_{eu} \cdot \xi} = \text{constant},$$ (16.22)

which provides an important relationship between growth speed and lamellar spacing, which, subsequently, was verified experimentally by Chadwick [11, 12].

Equation (16.17) can be combined with Eq. (16.21) to obtain an expression for the *minimum* interfacial supercooling, $\Delta\hat{T}_{tot}^{\star}$.

$$\Delta\hat{T}_{tot}^{\star} = 2\sqrt{a_1 a_2}.$$ (16.23)

Again Eqs. (16.17) and (16.21) may be substituted for the two coefficients appearing in Eq. (16.23) to yield an explicit value for the minimum interfacial supercooling,

$$\Delta\hat{T}_{tot}^{\star} = \sqrt{\frac{-2v_{eu}(1-k_0)C_{eu}m_{\ell}\gamma_{\alpha\beta}\Omega \cdot \xi}{D_{\ell}\Delta S_f}}.$$ (16.24)

The total interfacial supercooling for typical metallic systems solidifying at modest rates (ca. 1 mm/s) is in the range 1–10 K. Equation (16.24) also suggests another interesting scaling law between the interfacial supercooling and the growth rate of the eutectic, namely,

$$\frac{\Delta\hat{T}_{tot}^{\star 2}}{v_{eu}} = \frac{2m_{\ell}(k_0-1)C_{eu}\gamma_{\alpha\beta}\Omega \cdot \xi}{D_{\ell}\Delta S_f} = \text{constant}.$$ (16.25)

Hunt and Chilton qualitatively verified Eq. (16.25) by making precise temperature measurements of a directionally solidifed eutectic interface [13].

16.3 Hunt–Jackson Theory of Eutectics

The scaling laws derived initially by Tiller, using approximate (mass-balance) methods, were expanded in 1966 by Hunt and Jackson [14], who solved the coupled diffusion problem at a regular eutectic interface. These investigators also generalized the description of aligned eutectics to cover more realistic cases of non-symmetric phase diagrams, and were able to provide predictions and observations on polyphase interface forms.

The interested reader will find this theory developed in Appendix D. Of specific interest here is determining the value of the unknown diffusion parameter, ξ, that enters Tiller's eutectic scaling laws, Eqs. (16.22) and (16.25). Hunt and Jackson determined that $\xi \approx 0.5$.

16.4 Eutectic Microstructures

An interesting aspect of regular (type-I) eutectic solidification is the phase morphology selected by the reaction. Regular eutectics, as already discussed in Section 16.2.2, directionally solidify either in the form of alternating periodic lamellae of α and β phases, or as slender rods of the minor phase embedded in the major phase. See again Fig. 16.4. We now explore the reasons why some eutectic alloys preferentially solidify in one morphology rather than the other.

16.4.1 Lamellar Eutectics

Lamellar eutectics usually adopt special crystallographic orientations, known as 'habits', forming α/β interfaces that exhibit low interfacial energies. Lamellar eutectic structures tend to exhibit extraordinary thermal stability at high temperatures, resisting the tendency to coarsen. The extraordinary stability of lamellar eutectics arises because:

i. Most of the lamellar eutectic interfaces are planar, and thereby limit the number of locations, such as lamellar faults and boundaries, at which capillary-induced coarsening can initiate.
ii. As pointed out in Section 16.2.6, the stored free energy density within a volume of eutectic microconstituent is significant enough to be accounted in the eutectic freezing process. The interfacial free energy stored among eutectic lamellae sharing special crystallographic 'habit' planes, however, is usually much smaller than that found in typical interphase interfaces with arbitrary crystallographic misorientations. Thus, by selecting special low-energy habit planes during solidification, aligned eutectics avoid any energetic advantage to either decouple their cooperative diffusion, or break-up and form a microstructure consisting of a coarser, disconnected dispersion of phase particles known as a 'divorced' eutectic [15, 16].

First consider a unit area cross-section of a directionally solidified lamellar eutectic structure, as already depicted in Fig. 16.10. We now generalize the lamellar structure beyond that used to develop Tiller's theory of symmetric eutectics (Section 2.6) by allowing the eutectic to be composed of alternating lamellae of phases with different volume fractions. In the context of Fig. 16.10 this allows the width of the α-lamellae, λ_α, to be different from the width of the β-lamellae, λ_β. The relationship of these lamellar widths to the eutectic's average lamellar spacing, λ, is just

$$\lambda_\alpha + \lambda_\beta = \lambda. \tag{16.26}$$

The total amount of interfacial area, $A_{\alpha\beta}$, per unit volume of the eutectic microstructure, V_{eu}, is S_v, defined as

$$S_v \equiv \frac{A_{\alpha\beta}}{V_{eu}} = \frac{2 \times 1 \times 1}{\lambda \times 1 \times 1} \left[\frac{cm^2}{cm^3} \right]. \tag{16.27}$$

The factor of 2 appearing in the numerator of Eq. (16.27) accounts for the fact that two α/β interfaces form per β-phase lamella. Equation (16.27) also indicates that the interphase area density of a lamellar eutectic microconstituent, S_v, (measured in units of cm^{-1}) does not depend on the volume fraction of the β-phase, V_V. In the case of a lamellar eutectic, V_V is equal to the lamellar width ratio

$$V_V = \frac{\lambda_\beta}{\lambda}. \tag{16.28}$$

Thus, the *volumetric* free-energy density of the eutectic microconstituent, $\mathcal{E}_{lam}^{\alpha\beta}$, of a lamellar eutectic is easily found using Eq. (16.27) as

$$\mathcal{E}_{lam}^{\alpha\beta} = S_v \cdot \gamma_{\alpha\beta} = \frac{2\gamma_{\alpha\beta}}{\lambda}. \tag{16.29}$$

Equation (16.29) confirms that the free-energy density of the α/β interfaces comprising lamellar eutectics is just inversely proportional to the average eutectic spacing, λ.

16.4.2 Rod Eutectics

Now consider a rod eutectic, where the microconstituent consists of long needle-like rods of β-phase regularly distributed on a square lattice through a continuous α-matrix. Figure 16.12 shows a transverse cross-section of such a regular square array of edge length λ, containing cylindrical rods with a uniform diameter, D_β.

The amount of α/β interfacial area per unit volume, S_v, of this rod eutectic microstructure, is

$$S_v = \frac{\pi D_\beta \times 1}{\lambda^2 \times 1} \ [cm^{-1}]. \tag{16.30}$$

The interface energy per unit volume, $\mathcal{E}_{rod}^{\alpha\beta}$, of this rod eutectic microstructure is therefore

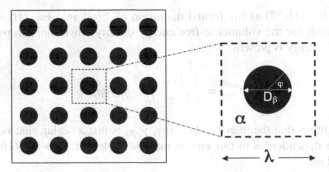

Fig. 16.12 *Left:* Geometry of an ideal transverse cross section of a regular (*square lattice*) cylindrical rod eutectic, with β shown as the minority phase. *Right:* Expanded unit cell of a β-rod within the continuous α phase matrix. The inter-rod spacing is λ, and the rod diameter is D_β. The angle φ may be used to describe any anisotropy of the α-β interfacial energy, $\gamma_{\alpha\beta}$

$$\mathcal{E}_{rod}^{\alpha\beta} = \frac{\pi D_\beta}{\lambda^2} \langle \gamma_{\alpha\beta} \rangle, \tag{16.31}$$

where the coefficient $\langle \gamma_{\alpha\beta} \rangle$ is the angularly averaged value of the α-β interfacial free energy around the circumference of a rod. This energy average is defined specifically as

$$\langle \gamma_{\alpha\beta} \rangle \equiv \frac{1}{2\pi} \int_0^{2\pi} \gamma_{\alpha\beta}(\varphi) d\varphi, \tag{16.32}$$

where φ, a dummy variable, is the azimuthal angle of the β-rod's interface normal with the α—matrix. The volume fraction of the β-phase for the rod eutectic sketched in Fig. 16.12 is

$$V_V = \frac{\pi D_\beta^2}{4\lambda^2}. \tag{16.33}$$

Solving Eq. (16.33) for the diameter of the β-rods in terms of the volume fraction and rod spacing yields an expression for their diameter,

$$D_\beta = 2\lambda \sqrt{\frac{V_V}{\pi}}. \tag{16.34}$$

Substituting Eq. (16.34) into the expression for the α-β interfacial energy, Eq. (16.31), yields the volumetric energy density of a rod eutectic, $\mathcal{E}_{rod}^{\alpha\beta}$, consisting of a square array of cylindrical β rods in a continuous α matrix,

$$\mathcal{E}_{rod}^{\alpha\beta} = \frac{2\pi \langle \gamma_{\alpha\beta} \rangle}{\lambda} \sqrt{\frac{V_V}{\pi}}. \tag{16.35}$$

Inserting Eq. (16.32) as the formal definition of $\langle\gamma_{\alpha\beta}\rangle$ into Eq. (16.35) gives the general result for the volumetric free energy density where anisotropy of the α-β interfacial energy is present,

$$\mathcal{E}_{rod}^{\alpha\beta} = \frac{1}{\lambda}\sqrt{\frac{V_V}{\pi}} \int_0^{2\pi} \gamma_{\alpha\beta}(\varphi)d\varphi. \tag{16.36}$$

If one assumes that the interfacial energy, $\gamma_{\alpha\beta}$, is just a scalar, that is, anisotropy, or angular dependences of this energy may be neglected, then Eq. (16.36) may be simplified as,

$$\mathcal{E}_{rod}^{\alpha\beta} = \sqrt{4\pi V_V} \, \frac{\gamma_{\alpha\beta}}{\lambda}. \tag{16.37}$$

If similar calculations are made for rod eutectics composed of an hexagonal array of cylindrical rods in a continuous matrix, the volumetric energy density stored in the microconstituent is increased slightly (by about 7%) to,

$$\mathcal{E}_{rod}^{\alpha\beta} = \frac{\sqrt{8\pi V_V}}{3^{\frac{1}{4}}} \frac{\gamma_{\alpha\beta}}{\lambda}. \tag{16.38}$$

16.4.3 Lamellar-to-Rod Transition

The morphology of a type-I eutectic microconstituent can be changed from lamellar to rod-form as the volume fraction of the β-phase is varied. Quite generally, the location of the eutectic composition, C_{eu}, and its associated tie-line terminal compositions for the individual phases, viz., C_α^{eu}, and C_β^{eu}, determine, via the equilibrium lever rule, what the relative mass fractions of the phases are at equilibrium. The phase mass fractions will approximate the corresponding volume fractions so long as their mass densities are similar. See again Fig. 16.7 for the symmetric situation, where the mass and volume fractions are equal to 1/2, and the directionally solidified eutectic structure is lamellar.

Now, if a eutectic of a specific phase ratio can have its stored free energy density reduced by a change in its morphology, then the thermodynamically 'preferred' form of the microstructure can be predicted from system-to-system, or, even within one system if the composition and volume fractions of phases are varied. Thus, one expects a lamellar-to-rod or, conversely, a rod-to-lamellar microstructural transition to occur in aligned regular eutectics at some critical volume fraction of the β-phase, $V_V^\star$, and, by symmetry, at the equivalent opposite phase ratio that occurs at $1 - V_V^\star$ for the α-phase.

The critical free energy at which these dramatic microstructural transitions occur may be found easily. One merely explores the ratio, $\mathcal{R}$, of the specific interfacial free energy densities for rod eutectics, given by Eqs. (16.37) and (16.38), for square and hexagonal arrays, respectively, to that for lamellar eutectics shown in Eq. (16.29).

These ratios are

$$\mathcal{R} \equiv \frac{\mathcal{E}_{rod}^{\alpha\beta}}{\mathcal{E}_{lam}^{\alpha\beta}} = 2\sqrt{\pi V_V}, \tag{16.39}$$

for square lattice arrays, and

$$\mathcal{R} \equiv \frac{\mathcal{E}_{rod}^{\alpha\beta}}{\mathcal{E}_{lam}^{\alpha\beta}} = \frac{2^{\frac{1}{2}}}{3^{\frac{1}{4}}}\sqrt{\pi V_V}, \tag{16.40}$$

for hexagonal arrays of rods.

Equations (16.39) and (16.40) show that the critical volume fractions, $V_V^\star$, at which the lamellar-to-rod or rod-to-lamellar transitions are expected in isotropic eutectics, occur for square lattice rod arrays when

$$V_V^\star = \frac{1}{\pi} \approx 0.32, \tag{16.41}$$

and for hexagonal arrays of rods when

$$V_V^\star = \frac{\sqrt{3}}{2\pi} \approx 0.28. \tag{16.42}$$

In the case of eutectic systems where α becomes the minority phase, one could expect a microstructure of α-rods through continuous β-phase. Therefore, symmetry in the energy density argument dictates that the critical transition volume fraction of β for the lamellar-to-rod transition of α-rods in a square arrangement is

$$V_V^\star = 1 - \frac{1}{\pi} \approx 0.68, \tag{16.43}$$

and, similarly, for the case of hexagonal arrays of α-rods in continuous β-phase,

$$V_V^\star = 1 - \frac{\sqrt{3}}{2\pi} \approx 0.72. \tag{16.44}$$

Figure 16.13, provides a plot of the volumetric energy ratios of rod-to-lamellar eutectics, $\mathcal{R}$, given in Eqs. (16.39) and (16.40), versus volume fraction of the β-phase, V_V, for square and hexagonal arrangements of rods, respectively. When the β-phase volume fraction is below approximately 0.3, rods of β in an α-matrix are expected. Conversely, rods of α are expected in a matrix of β when the volume fraction of the β-phase rises above about 0.7. The plot shows only a small effect caused by the precise arrangement of the rods—with square, hexagonal, and randomly dispersed rod arrays being most commonly observed in many directionally solidified alloy systems. Indeed, experiments performed on many metallic eutectics confirm that rod eutectics, in fact, form over the volume fraction range of $0.3 \leq V_V \leq 0.7$. The β-phase volume fraction range accepting the formation of *lamellar* eutectics is found to fall in the range $V_V = 0.5 \pm 0.2$, which agrees well with the energy density arguments applied in the discussion given above. Note, that these lamellar eutectic alloys are generally not too far from the symmetric case assumed in Tiller's original theory, developed in Section 2.6.

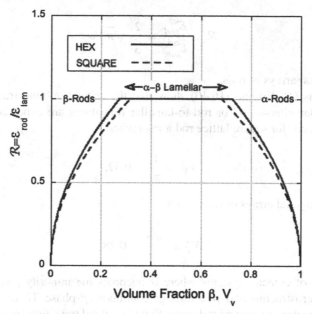

Fig. 16.13 Plot of the stored energy density ratios, $\mathcal{R}$, for aligned eutectics with square (*dashed*) and hexagonal (*solid*) arrays of rods, to that of alternating lamellae, versus volume fraction of the β-phase. Rod eutectics are expected in both high and low volume fraction eutectics until $\mathcal{R} = 1$, at which point a microstructure transition occurs. Over the mid-range of phase fractions, $0.3 \leq V_V \leq 0.7$, aligned lamellar are thermodynamically preferred. The geometrical arrangement of the rods (*square* or *hexagonal*) has little influence on the critical volume fractions for either rod-to-lamellar or lamellar-to-rod transitions of the eutectic microstructure

16.4.4 Crystallography of Lamellar Eutectics

The α/β interface remains nearly macroscopically planar in the case of directionally solidified lamellar eutectics—excepting the extremely fine-scale ripples of the individual lamellar tips—and the alternating phase lamellae crystallize from the melt in parallel. Such a two-phase microstructure is capable of developing low-energy 'habit planes'. A crystallographic habit plane, or family of such planes, provides preferred low-energy orientations that develop between the coupled, conjugate phases as they solidify simultaneously from the melt. Once a habit plane orientation is established between the eutectic phases, the system tends to 'lock' onto that orientation. The mechanism that 'locks' the solidification to these specific low-energy orientations is the energy anisotropy, or 'torque' associated with the habit-plane. Conyers Herring showed on the basis of general energy minimization for equilibrium of anisotropic interfaces that the chemical potential of a habit plane is related to its interfacial free energy, $\gamma_{\alpha\beta}$, plus the second angular derivative (or so-called torque) of this energy with respect to the orientation angle of this plane [17]. A frequently encountered habit plane of minimum chemical potential that has been observed in metallic eutectics is: $[100]\alpha//[110]\beta$, with common (parallel) $\{100\}$ crystallographic directions in the two phases aligned with the eutectic growth direction. It is generally accepted that the following facts hold true:

1. The preferred crystallographic habit plane for a lamellar eutectic is selected by the system seeking a minimum in the chemical potential of the α-β interface, per Herring's criterion.
2. Despite the presence of habit planes, however, both lamellar and rod eutectics are frequently composed of many groups of independently nucleated eutectic 'colonies', each displaying an unique orientation of the crystallographic habit with respect to the local growth direction.
3. Serrations, or jogs, also appear in lamellar eutectics because of the crystallographic habit plane. The structure of well-aligned eutectics is often interrupted by narrow regions, called 'faults', where the regular microstructure is locally disrupted. The details of the processes leading to these eutectic faults are complicated, and remain incompletely understood.

16.5   Computation of Eutectic Microstructures

Numerical software packages for computing eutectic alloy thermodynamic properties, solving the diffusion fields developed in the melt ahead of polyphase interfaces, and, finally, simulating the eutectic microstructure itself, are all available as commercial computer codes. (See again Section 13.2.1 for their description.) These codes are collectively capable of providing realistic visualization of the interesting melt-component 'unmixing' process that occurs during coupled directional solidification. Figure 16.14 clearly renders the complex phenomenon of eutectic melt unmixing, by showing how adjacent α-β lamellae simultaneously reject one component, while accepting the other. The numerically derived diffusion field in Fig. 16.14 also provides an accurate perspective of the periodic nature of the concentration field, and the manner with which it damps away toward the homogeneous eutectic melt over a small distance normal to the interface that is less than λ.

16.6   Directional Freezing of Polyphase Alloys

Hunt and Jackson also carried out experiments with low α-factor transparent organic alloys to observe eutectic morphologies and compare them to theoretically

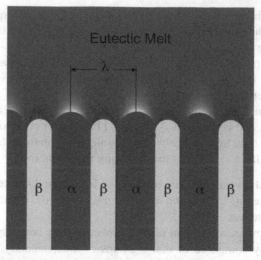

Fig. 16.14 Numerical visualization of diffusive 'unmixing' in the melt occurring near an advancing non-symmetric lamellar eutectic interface. The homogeneous melt is shown near its eutectic composition in red. The lateral unmixing process occurs by rejection of A-rich melt (*shades of yellow*) ahead of the α-lamellae, and rejection of B-rich melt (*shades of blue*) ahead of the β-lamellae. Inspection of this diffusion field indicates periodicity of the concentration field parallel to the eutectic interface, and rapid damping of the periodic diffusion field over a distance normal to the interface of approximately $\lambda/2$. Cf. Fig. E.3, showing a contour plot of the diffusion field ahead of a *symmetric* eutectic. Constructed using MICRESS® software

calculated interface shapes. Figure 16.15 shows their convincing comparison of how a steady-state eutectic interface responds to changes in the shapes of the lamellar tips and their spacing.

Directional solidification (DS)[4] was carried out in many laboratories for a wide range of metallic alloys that solidify with polyphase microstructures. These include, of course, regular, type-I eutectics, for which the phase coupling and scaling laws, as described here, are now well developed, but also other less well-investigated polyphase alloy systems. The latter include those with monotectic, and peritectic reactions [18, 19]. As in eutectic solidification, monotectic and peritectics produce two (or more) phases, which, more or less, tend to solidify simultaneously. The phase couplings established during directional freezing in these polyphase systems are, however, limited, at least when compared to those observed typically in regular eutectics.

Specifically, peritectic reactions ($\alpha + \ell \rightarrow \beta$) tend to produce crystalline phases that are self-blocking upon freezing. This peculiarity of their solidification sequence greatly inhibits diffusive transport from occurring between the melt, the reac-

---

[4] See Appendix C for brief comparative descriptions of directional casting and several important crystal growth processes.

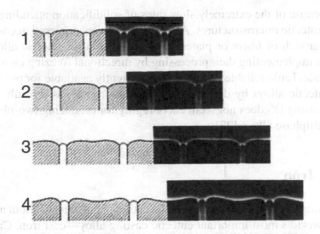

Fig. 16.15 Comparison of theoretically calculated lamellar eutectic interface shapes and their wavelengths, λ, with direct microscopic observations of a directionally solidified, low α-factor, transparent, organic eutectic alloy. Each successive row (1 to 4) was solidified at a progressively slower rate, v_{eu}, causing the wavelength at steady state to increase, in accord with the scaling law that $\lambda^2 v_{eu} = const$. In addition, on rows 3 and 4 the tips of the majority phase develop a pronounced concavity as the lamellar width of the majority phase increases. Adapted from [21]

tant phase, and the peritectic phase.[5] Consequently, responsive, well-controlled peritectic microstructures are not possible. Nevertheless, peritectic solidification proves to be important in the melt crystallization processing of high-temperature yttrium-barium oxide superconductors, where texture and directionality of the phases do play important roles in the performance of these interesting cast superconductors [20].

Monotectic alloys can couple more easily during directional freezing, but the polyphase reaction in that case $(\ell_1 \rightarrow \ell_2 + \alpha)$ involves two melt phases and one crystalline phase [15]. Some interest in directional freezing of Fe-S monotectics focussed on producing ultra-fine porous filters by etching out the fibers of the sulphide phase from the iron sulfur solid solution.

Eutectics have been investigated extensively in attempts to produce so-called 'in situ', or DS composite materials with both plate or fiber morphologies as the strengthening phase for commercial applications. The spacings of regular eutectic phases decreases in a controlled manner with increasing cooling rate and growth speed. DS eutectic composites, which provide near-equilibrium, ultra-stable microstructures, were at one time considered seriously as alloys with optimized fiber characteristics for improved high-temperature applications. A well-known example is the (Co,Cr)-$(Co,Cr)_7(C)_3$ eutectic, which exhibits lower creep rates than do conventionally cast high-temperature alloys. Excessive production costs are incurred,

---

[5] The technical term used to describe the reluctance of peritectic reactions to proceed toward completion is 'peritectic walling'.

however, because of the extremely slow rates of solidification attending the growth
of uniform eutectic microstructures. Also, difficulties encountered in controlling the
continuous growth of fibers or plates in high-temperature eutectic alloys have all
but ruled out implementing their processing by directional freezing on a meaningful
industrial scale. Little reliable information is currently available for producing commercial peritectic alloys by directional solidification, and, as a result, their future
exploitation using DS does not seem encouraging for producing two-phase or more
complex multiphase alloys [22].

16.7 Cast Iron

It would be impossible to discuss the solidification of eutectics without at least men-
tioning the world's most important eutectic casting alloy—cast iron. Cast iron was
the premier commercial casting alloy for making utilitarian and decorative objects
through the nineteenth and into the early twentieth centuries. Even industrial build-
ings were erected in that period using cast iron curtain walls.[6]

The cast irons as a group are multicomponent iron-carbon-X alloys that solidify
with a considerable amount of the eutectic microconstituent, composed of graphite
and iron. The melt composition and cooling rate during casting also determine
whether the as-cast structure is gray iron, white cast iron, or a mixture called mot-
tled cast iron. More sophisticated variants called spheroidal and ductile cast irons
were developed that make this material less brittle, and more crack tolerant. Ductile
irons show enhanced mechanical properties and have wide-ranging applications.
The interested reader should consult the excellent monograph written on this won-
derful material by Minkoff [23].

16.8 Summary

Polyphase solidification is an important aspect of the many complex crystallization
reactions that can occur in solute-rich homogeneous melts. The thermodynamics of
eutectics and their phase diagrams provide an example of suitable polyphase crystallization conditions under which a binary melt can 'unmix' its components and form
two different crystalline phases. The nucleation and growth of eutectics is reviewed
briefly, as these topics caused considerable confusion among experts studying eutectics through the early twentieth century. Despite these scientific uncertainties, a
range of eutectic materials—from common Pb-Sn solders to the cast irons—became
dominant products of metallurgical commerce.

The solidification kinetics of eutectic alloys was not understood quantitatively
until 1957, when Tiller explained how coupled liquid-state diffusion leads to the

---

[6] A famous example of a 'cast iron building' that remains standing today is one located at the US
Army's Watervliet Arsenal, in Watervliet, NY.

diffusive unmixing of a eutectic melt and the simultaneous production of two crystalline phases. Tiller's ideas were refined further by Hunt and Jackson, who developed the mathematical solution that describes the solute diffusion field for eutectic growth, thereby unlocking the detailed scaling laws for eutectic microstructures. See Appendix D for details of that theory and how it confirms Tiller's original findings.

Today, regular eutectics, i.e., those that solidify with non-faceting solid-melt interfaces, can be directionally solidified to yield in situ composites consisting of either alternating plates of the crystalline phases, or aligned fibers of one crystalline phase in a matrix of the other. The selection of either fibrous or plate-like morphologies is discussed in this chapter as an energetic outcome decided by the volume fractions and compositions of the phases.

The microstructures resulting from other polyphase solidification reactions, including peritectics and monotectics are mentioned briefly. These polyphase systems prove to be less responsive to microstructure controls imposed by directional solidification processing. Finally, the subject of cast iron is touched on, if only to remind readers of the great economic and engineering importance of this ubiquitous, and flexible class of casting alloys.

References

1. R.W. Kraft and D.L. Allbright, *Trans. Metall. Soc. AIME*, **224** (1962) 1176.
2. G.A. Chadwick, *Prog. Mater. Sci.*, **10** (1963) 97.
3. J.W. Hunt and K.A. Jackson, *Trans. Metall. Soc. AIME*, **236** (1966) 843.
4. *Solidification Processing of Eutectic Alloys*, Eds. D.M. Stefanescu, G.J. Abbaschian and R.J. Bayuzick, TMS, Warrendale, PA, 1988.
5. B. Sundquist and L.F. Mondolfo, *Metall. Trans.*, **221** (1961) 157.
6. R. Vögel, *Z. Anorg. Chem.*, **76**, (1912) 425.
7. G. Tammann, *Textbook of Metallography*, Chemical Catalogue Co., New York, NY, 1925 p. 182.
8. W. Straumanis and N. Braaks, *Z. Phys. Chem.*, **29** (1935) 30.
9. W. Tiller, *Liquid Metals and Solidification*, ASM, Cleveland, OH, 1958, p. 276.
10. H.B. Callen, *Phys. Rev.*, **105** (1957) 360.
11. G.A. Chadwick, *J. Inst. Metals*, **91** (1963) 169.
12. G.A. Chadwick, *J. Inst. Metals*, **92** (1963) 18.
13. J.D. Hunt and J.P. Chilton, *J. Inst. Metals*, **92** (1964) 2214.
14. J.D. Hunt and K.A. Jackson, *Trans. Metall. Soc.*, **236** (1966) 1129.
15. D.M. Stefanescu, *Science and Engineering of Casting Solidification*, 2nd Ed., Chap. 11, Springer, New York, NY, 2009.
16. S. Wang, T. Akatsu, Y. Tanabe and E. Yasuda, *J. Mater. Sci.*, **35** (2000) 2757.
17. C. Herring, *The Physics of Powder Metallurgy*, R. Gomer and C.S. Smith, Eds., University of Chicago Press, Chicago, IL, 1953, p. 5.
18. W. Kurz and P.R. Sahm, *Gerichtet erstarrte eutecktische Werkstoffe* [*Directionally Solidified Eutectic Materials*], Springer, Berlin (1975).
19. H. Biloni and W.J. Boettinger, 'Solidification', *Physical Metallurgy*, 4th Ed., Chap. 8, North Holland, Amsterdam, 1996, p. 669.
20. Y. Nakamura and Y. Shiohara, *J. Mater. Res.*, **11** (1996) 2450.

21. K.A. Jackson, *50 Years Progress in Crystal Growth*, R.S. Feigelson, Ed., Elsevier, Amsterdam, 2004, p. 81.
22. M. Rettenmayr and H.E. Exner, 'Directional Solidification of Crystals', in *Encyclopedia of Materials: Science and Technology* , Elsevier Science Ltd., Amsterdam, ISBN: 0-08-0431526, 2000.
23. I. Minkoff, *The Physical Metallurgy of Cast Iron*, Wiley, New York, NY, 1983.

Chapter 17
Rapid Solidification Processing

17.1 Introduction

This book has developed the subject of solidification kinetics and interfacial processes at crystal-melt interfaces by considering near-equilibrium phenomena occurring over a few layers of atoms, some temporally associated with the crystal, and some temporally associated with the melt. This thin microscopic zone of transition was usually approximated mathematically as a 'sharp' interface of zero thickness. Also, the rate-limiting kinetic processes that were considered thus far acted both separately and simultaneously, and consisted of heat conduction and solute diffusion. Both of these transport mechanisms normally develop fields over macroscopic distances, as compared to the near-atomic interface thickness. Moreover, the dynamic interchanges among the host and solute atoms at sharp interfaces were considered to occur rapidly enough, and to repeat frequently enough, to approximate closely the condition of *local* interfacial thermodynamic equilibrium. Global, i.e. *total*, thermodynamic equilibrium, was shown to be inoperative, as it is precluded for virtually all crystal-melt processes on the basis that the relevant length scales over which heat transfer and solute diffusion operate on practical time scales for the motion of the interface are much too small to allow total equilibrium to be achieved.

As was mentioned in earlier chapters, most crystal growth and conventional solidification processes operate at rates of interfacial advance somewhere between micrometers per second up to centimeters per second. Powerful 'directed energy' sources, and novel ultra-fast heat transfer methods (up to 10^9 K/s), however, have become readily available and economical for accomplishing ultra-fast partial melting, followed by rapid re-solidification.

Rapid solidification techniques devolved from the invention of optical lasers in 1959, and innovative developments in the early 1960s of 'splat quenching', by Duwez and co-workers at Caltech. Duwez et al. produced a remarkable range of rapidly solidified alloys [1, 2]. In parallel, the steady technological refinement of reliable, exquisitely controllable, high-intensity electron beam equipment for welding and surface processing of materials continually broadened options for increasing the speed of both melting and freezing processes at virtually any temperature. Today, directed energy sources now allow rapid melting of layers as thin as several

M.E. Glicksman, *Principles of Solidification*, DOI 10.1007/978-1-4419-7344-3_17,

nanometers on many types of materials, with re-freezing occurring on a time scale of nanoseconds. The equivalent interfacial speeds, in both low- and high-thermal conductivity materials, for these rapid solidification processes rise to impressively high rates measured in meters per second.

At such extreme rates of freezing, the time-scale available for accomplishing even short-range diffusion, such as solute partitioning at the interface, becomes inadequate. As a consequence, phenomena not yet discussed in prior chapters, such as 'solute trapping', metastable phase formation, nanocrystal formation, and even metallic glass formation, all occur, each often accompanied by severe departures from the dictates of the equilibrium phase diagram. Even quantities normally requiring only local interfacial equilibrium, such as the operative alloy distribution coefficient, k_0, or a phase diagram's liquidus slope, m_ℓ, begin to change values as the interface speed is ramped up to and beyond 1 m/s. The practical interests in rapid solidification processing, known as RSP, are linked toward the controlled production of unusual microstructures, extended phase chemistries, and, most interestingly, novel properties. Two comprehensive books that review the materials science and applications developed around RSP through the early 1990s may be found in [3, 4].

17.2 Background

17.2.1 Splat Quenching of Alloys

In his review of metallurgical developments evolving from applications of RSP, entitled, *Perspective on the development of rapid solidification and nonequilibrium processing and its future*, Professor Howard Jones stated [5],

> The current status and potential of rapid solidification and nonequilibrium processing is assessed from the perspective of its early origins and rapid development that followed publication in 1960 of Duwez' remarkable findings [on splat quenching]. A multiplicity of routes to a wide range of products is now available with many of these routes in commercial use. Evident priorities for the future include replacement of empirical approaches wherever possible with fully tested physically based predictive models and intensification of the search for viable applications for the products of these routes.

As splat quenching and related techniques [6] that produced gram-sized novel RSP materials matured, production of commercially interesting RSP materials were scaled up to much higher volumes. Laboratory methods such as spray quenching, and industrial-scale processes such as Osprey casting [7] demonstrated that kilograms-to-tons of a wide-range of RSP alloys could be produced economically [8]. Large-scale production of a range of RSP alloys stimulated both commercial interest and additional applications for these materials. Developments in industry became fully commercialized into such wide-ranging products exemplified by the MetGlas® series of continuously melt-spun alloy ribbons, sold by Allied Chemical Corp., for applications as diverse as brazing foils, electrical transformer cores, magnetic tags, medical devices, and anti-shoplifting detectors [9].

17.3 Early Research in Rapid Solidification

17.3.1 Non-equilibrium Phase Diagrams

In addition to the pioneering studies by Duwez and co-workers using splat quenching, already cited in Section 17.2.1, important insights were added toward developing a thermodynamically-based explanation of kinetic phenomena during rapid
crystal growth by Russian crystallographers during the same time frame. Research
studies published by by A.A. Chernov and his collaborators during the 1960s
[10–12] led to the concept of a 'kinetic' phase diagram, which showed how well-
separated equilibrium solidus and liquid lines would tend to approach each other as
the speed of the solid–liquid interface and the degree of chemical disequilibrium at
the interface both increased. If the solidus and liquidus curves approach each other
during RSP, as predicted by Chernov, the solute distribution coefficient, k, will also
increase, and rise toward unity at the so-called T_0 line. The T_0 line is the locus of
solute concentration and temperature where the solid and liquid phases have equal
free energies and identical compositions.

Chernov's approach to nonequilibrium crystal growth was borne out by different
detailed theortical analyses of RSP kinetics to be discussed later in this chapter.
Figure 17.1, for example, is representative of the predicted non-equilibrium modifications to the solidus and liquidus lines as solidification speed increases. The data
on which this kinetic phase diagram was plotted were obtained from a growth model
calculated for ideal solid and liquid solutions by Aziz and Kaplan [13]. The outer
curves are the equilibrium solidus and liquidus, applicable to a stationary, or slowly
advancing solid–liquid interface. As the interface speed increases into the RSP range
of ca. 1 m/s, one sees that the non-equilibrium liquidus curves begin to exhibit considerable supercooling below their equilibrium temperatures. The horizontal arrows
appearing on this kinetic phase diagram are 'kinetic tie-lines' that connect solid and
liquid compositions at all temperatures within the wide freezing range of this ideal
alloy system for a specified rate of freezing.

17.3.2 Hypercooling

Experimental proof that interface temperatures during high-speed solidification of
simple molecular melts depart significantly from their equilibrium melting point was
demonstrated originally by Glicksman and Schaefer [15], who 'hypercooled' pure,
molten phosphorus, cubic α-P$_4$. Hypercooling involves supercooling a melt beyond
its characteristic invariant temperature, when $\Delta T > \Delta H_f / C_p^\ell$. In the particular
case of α-P$_4$, the critical supercooling to reach the hypercooled state is unusually
small, only 25.1 K. This is a relatively modest amount of supercooling compared to
what is required to hypercool most metallic alloys (100–500 K).[1]

---

[1] The characteristic temperature, or critical supercooling, is discussed in Section 3.1.3.1. It represents the minimum supercooling needed to solidify a melt fully under adiabatic conditions, where

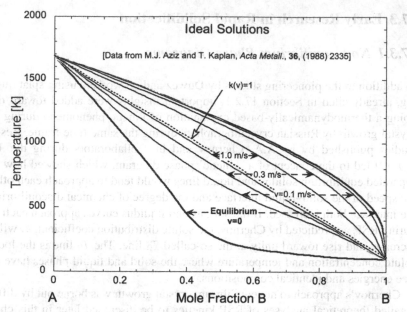

Fig. 17.1 Kinetic phase diagram showing modification of an equilibrium phase diagram calculated for an ideal binary A-B alloy system with a wide freezing range. Here pure component A melts at $T_m^A = 1,684$ K and pure component B melts at $T_m^B = 100$ K. The equilibrium solidus and liquidus are the heavier outer curves, and show the behavior expected at zero solidification rate. As the interface speed increases, the solidus and liquidus lines approach each other. Note the large supercoolings indicated by the liquidus curves as the interface speed increases into the RSP range. The dotted line, called 'T_0', represents the condition of attaining equal free energies for solid and liquid at the same composition [14]. The T_0 line demarks the *thermodynamic* limit at which the two-phase region on the phase diagram collapses

To prove the occurrence of hypercooling, Glicksman and Schaefer measured the jump in temperature immediately after rapid freezing of the melt. When ordinary *invariant* freezing occurred over a range of supercoolings (in this case for supercooling less than 25 K), the melt partially transformed to solid, and the two-phase mixture always recalesced up to its equilibrium temperature (44.3 C). The amount of solid freezing from the melt under adiabatic conditions is proportional to the initial melt supercooling. However, when the hypercooling limit ($\Delta T_{hyp} = \Delta H_f / C_p$) was finally exceeded, the P_4 melt froze completely in a mere fraction of a second. Moreover, the solid so produced responded by recalescing to a *univariant* temperature

---

external transfer of the latent heat plays no role in achieving total solidification. One-dimensional steady-state heat flow solutions describe heat transfer at hypercooled solid–liquid interfaces. Consequently, any supercooling beyond the minimum characteristic supercooling, $\Delta T_{hyp} = \Delta H_f / C_p^\ell$, appears as additional interface supercooling, ΔT. Interface supercooling reflects the occurrence of non-equilibrium processes operating at or near the interface. These non-equilibrium processes include interface attachment of molecules or atoms, irreversible heat flow, and defect generation in the solid. See temperature scheme, Fig. 17.3, lower panel.

that falls short of reaching P_4's equilibrium melting point, 44.3 C. The final temperature falls short of the melting point in proportion to $\Delta T - T_{hyp}$. These thermal measurements are plotted in Fig. 17.2.

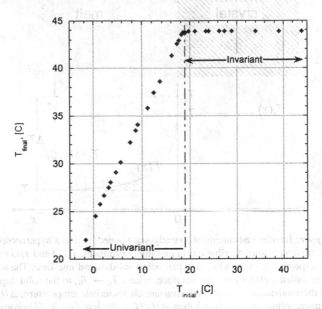

Fig. 17.2 Temperature 'jumps' following rapid solidification of supercooled and hypercooled P_4. Graph shows the initial and final temperatures of the system. At any initial melt temperature, $T_{initial}$, from just below the normal melting point, 44.3 C, down to just above the point where hypercooling begins at 19.2 C, the supercooled melt, maintained subsequently under adiabatic conditions, freezes into a two-phase mixture that rapidly recalesces to a temperature, $T_{final} = T_m = 44.3$ C. This is normal, or *invariant* freezing of a pure material. However, when the melt supercooling exceeds $\Delta T_{hyp} = 25.1$ K $= \Delta H_f / C_p$, the now hypercooled melt solidifies under *univariant* conditions, with its recalescence temperature falling below the equilibrium melting temperature. The hypercooling point is delineated clearly in these experiments by the sharp 'corner' that locates the change from invariant to univariant freezing. Data from [15]

Once hypercooling is achieved in a pure melt, a macroscopically planar interface can in principle propagate at a steady rate, v, as suggested in Fig. 17.3, upper panel. The thermal fields, $T_s(\hat{x})$ and $T_\ell(\hat{x})$ associated with the advance of such a planar interface are best described in a coordinate system, $\hat{x}$, fixed to, and co moving with, the interface at $\hat{x} = 0$. See Fig. 17.3, lower panel.

These temperature fields match on the interface ($T_s(0) = T_\ell(0) = T_i$), and satisfy the steady-state heat flow equations at all other points in the adjacent solid ($\hat{x} \leq 0$) and liquid ($\hat{x} \geq 0$), namely,

$$\frac{\partial^2 T_i}{\partial \hat{x}^2} + \frac{v}{\alpha} \frac{\partial T_i}{\partial \hat{x}} = 0, \ (i = s, \ell). \tag{17.1}$$

The solutions to Eq. (17.1) in the solid and liquid phases are, respectively,

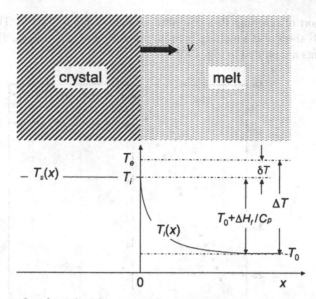

Fig. 17.3 *Upper:* Interface advancing at a steady-state speed, v, into a hypercooled melt. *Lower:* Schematic of the temperature distributions in moving coordinates, $T_s(\hat{x})$ and $T_\ell(\hat{x})$ in the solid and melt phases, respectively, surrounding a hypercooled solid–liquid interface. The temperature rise from the hypercooled melt far from the interface, where $T_\ell \to T_0$, to the solid–liquid interface at T_i is fixed by thermodynamics to equal the invariant characteristic temperature, $\Delta H_f / C_p$. Thus, if the applied supercooling, ΔT, is greater than $\Delta H_f / C_p$, the interface itself becomes supercooled by the amount δT. After Glicksman and Schaefer [15]

$$T_s(\hat{x}) = T_0, \quad (\hat{x} \leq 0), \tag{17.2}$$

and

$$T_\ell(\hat{x}) = T_0 + \frac{\Delta H_f}{C_p} e^{-\frac{v}{\alpha}\hat{x}}, \quad (\hat{x} \geq 0). \tag{17.3}$$

Glicksman and Schaefer also accurately measured the rate of advance of the solid-melt interface at various supercoolings from 5 to 50 K, exceeding almost twice the characteristic temperature, and achieved crystal growth speeds in the RSP range of 3.5 m/s. Their speed measurements are plotted in Fig. 17.4. One of the main characteristics of crystal-melt interfaces freezing during RSP is the presence of enormous thermal gradients acting to transport the latent heat. For example, during growth from hypercooled melts, the temperature gradient just ahead of the interface may be found from the melt's temperature field, Eq. (17.3). Differentiating Eq. (17.3) with respect to $\hat{x}$ yields the thermal gradient, which if evaluated at the interface, $\hat{x} = 0$, gives the interfacial gradient on the liquid side as

$$\text{Grad}(T)_{\hat{x}=0} = -\frac{v}{\alpha} \frac{\Delta H_f}{C_p}. \tag{17.4}$$

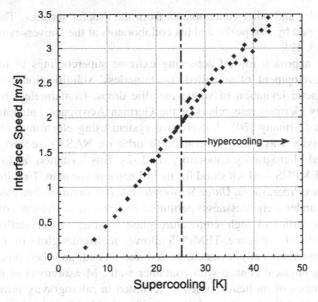

Fig. 17.4 Interface speed versus supercooling, α-P_4. The hypercooling region begins at a supercooling of 25.1 K. No discernible break in slope occurs in the interface speed curve either at or beyond the hypercooling point. Optical scattering experiments, also reported in [15], indicate that this system solidified with a dendritic interface below the hypercooling limit. By exceeding the hypercooling point, it was demonstrated that the interface, during its maximum observed crystallization speed of about 3.5 m/s, was supercooled below its equilibrium freezing temperature of 44.0 C by about 25 K

Thus, the hypercooled P_4 melt discussed above when freezing at 3 m/s experiences a thermal gradient of about 2×10^6 K/cm. The adjacent crystalline solid, by contrast, experiences a thermal gradient of zero! Such extremes often occur during RSP.

Also interesting was the observation that if Hg was dissolved into molten P_4, and the saturated molten alloy was supercooled and rapidly solidified, clear evidence of 'solute trapping' could be observed by noting the failure of the Hg solute to precipitate as a dense, opaque dispersion of small liquid Hg droplets, which normally turned the transparent melt an inky black. The sudden onset of solute trapping was noted visually in solidifying Hg-P_4. Trapping occurred when the molten alloy was initially supercooled at or beyond its hypercooling point and then remained transparent in the totally crystalline state. The crystallized alloy slowly darkened after RSP, as minute Hg droplets precipitated over time from the metastable, supersaturated solid.

RSP techniques based on deep supercooling and hypercooling of melts were extended over time to include many other materials, including metallic [6], organic [16], and ceramic melts [17]. The achievement of deep supercooling conditions to produce metastable phase formation in alloys, and improve nucleation rate measurements in fine metal drops was accelerated by improvements in alloy dispersion methods using chemical fluxes that eliminated contact with containers, and removed surface oxides and other contaminants from the droplet surfaces. These advances were developed by Perepezko and his collaborators at the University of Wisconsin, Madison [18, 19].

Another approach toward achieving extreme supercoolings in metallic melts was the development of so-called 'containerless' solidification methods, using electromagnetic levitation of single metallic drops. Containerless methods were perfected by German researchers at the German Aerospace Laboratory, DLR, in Köln-Portz, Germany [20]. A derivative system using electromagnetic levitation in microgravity was flown in near-Earth orbit on NASA's second flight of the International Microgravity Laboratory (IML-2). This levitation facility—given the acronym TEMPUS—which stood for its full name in German, *Tiegelfreies Elektromagnetisches Prozessieren Unter Schwerelosigkeit* (Container-free electromagnetic processing under weightlessness) permitted accurate measurements of the thermophysical properties of high-temperature glass forming and crystallizing metallic melts. Specifically, in space, TEMPUS allowed multigram globs of metallic alloys to be electromagnetically levitated, heated and melted, and then supercooled—all without *any* physical contact with container walls. Measurement of the static and dynamic shapes of molten samples so levitated in microgravity provided precise non-contact high-temperature calorimetry, along with determination of the molten alloy's mass density, viscosity, and surface tension. The TEMPUS facility and other levitation systems also allowed observation and quantitative assessment of such important solidification phenomena as crystal nucleation rates, glass formation temperatures, dendritic growth speeds and grain size variations arising from melts solidified at large values of thermal supercooling (hundreds of Kelvins) [21–25].

17.3.3 Thermodynamic Limits

Early attempts to establish general, quantitative thermodynamic limits on phase transformation behavior, as well as modifications to the phase diagram for solid–liquid transitions occurring far from equilibrium were successfully pursued by Hillert and Sundman [26] for solid-state metallurgical systems, and by Baker and Cahn for splat-quenched alloys [27]. Monte Carlo computer calculations developed by Gilmer [28] provided further microscopic insights into the kinetics of rapid freezing, especially regarding the types of interface configurations and motions when solidification processes were simulated under large kinetic driving forces.

17.4 Solute Trapping

17.4.1 Interface Non-equilibrium

The definitive proof of appreciable non-equilibrium occurring at a rapidly quenched solid–liquid interface was provided by Baker and Cahn, who splat quenched binary melts of Zn-Cd. This alloy exhibits a retrograde (recurving) solvus for the HCP Zn+Cd solid-solution [27, 29]. It was shown by these investigators that the metastable extension of the Zn+Cd solid-solution cooled below about 600 K resulted in retention, or 'trapping' of Cd atoms within the Zn crystal, despite their increased chemical potential when trapped in the supersaturated solid state. The metastable extension of the Cd solubility limit in a Zn solid solution, occurring over a temperature range where local equilibrium would actually demand a *decreasing* solubility, proved conclusively that they observed significant departures from interfacial equilibrium during rapid freezing, which were responsible for the unexpected solubility increase.

This phenomenon, known as 'solute trapping', is now generally taken to imply that at sufficiently high freezing rates the distribution coefficient, k, will begin to change as some function of the interface speed, v. The distribution coefficient, $k(v)$, at extremely high interface speeds can, in principle, drift away from k_0 and approach unity! Thus, ideal solute trapping would actually allow a crystallizing solid phase to retain the identical concentration of solute as is dissolved in its melt phase adjacent to the interface, in which case the non-equilibrium distribution coefficient, $k(v) \rightarrow 1$. This limit is referred to in RSP technology as 'partitionless solidification', which has only been approximated using several experimental settings of RSP, including hypercooling (Section 17.3.2), splat quenching, pulsed laser melting (Section 17.4.2), and electron beam surface glazing. Achieving the exact condition for partitionless solidification, $k(v) = 1$, requires higher supercooling than that predicted by diffusion-based RSP theories [14].

17.4.2 Laser Melting

The success of splat quenching studies conducted by many investigators through the 1970s eventually led to interesting attempts to exploit the phenomenon of RSP using alternative techniques. These later, more sophisticated experiments used powerful, albeit brief, laser pulses to melt an extremely thin layer—only a few hundred nanometers thick—on carefully oriented, doped semiconductor crystals. Here the directionality of the laser beam, its wavelength, intensity, pulse shape, and duration, were all finely tuned variables available through the technology of laser optics. The laser energy was initially deposited in the layer of molten semiconductor, which in the case of Si forms a molten metallic phase that readily adsorbs the incident laser energy in the form of photons, converting them via free electrons to heat. This brief pulse of energy, which is stored as both sensible and latent heat, rapidly diffuses out of the melt and back into the underlying semiconductor crystal. By adjusting both the temperature of the substrate semiconductor and the characteristics of the laser pulse, controlled and measurable re-solidification rates could be accomplished exceeding 1 m/s [30].

The solute, or dopant, redistribution in the thin refrozen layer, after ultra-fast solidification of the semiconductor, was monitored using advanced analytical tools, such as Rutherford Back Scattering (RBS). RBS is a nuclear-based analytical tool that is capable of measuring the energy distribution of a beam of neutral helium ions backscattered from the near-surface regions of the molten and partially refrozen Si. Remarkably, RBS is capable of resolving accurately the concentration profiles of the dopant atoms in molten or crystalline layers only a few hundred nanometers thick. The laser melting and rapid re-solidification was indeed found to suppress the equilibrium partitioning of the dopant atoms in silicon. The increase noted in the distribution coefficients caused corresponding large increases in the dopant's (non-equilibrium) solubility, which, in some instances, exceeded 4–5 orders-of-magnitude [31, 32].

17.5 Theory

17.5.1 Introduction

Earlier chapters of this book dealt with diverse solidification topics including changes of long-range order in freezing melts (Chapter 1); influences of interface speed on crystal composition (Chapters 5, 6, and 7) and phase and microstructure selection (Chapters 9, 11, 13, and 16). Much of the modeling and theoretical concepts relied on descriptions of continuum fields acting on phases that were also described as continua, meeting at 'sharp interfaces', but maintained during freezing in near-equilibrium states.

These same topics of solidification science and technology are also influenced by extreme departures from interfacial equilibria that occur at high growth speeds. Theoretical interest in RSP, therefore, became focused on modifying crystal growth models using phenomenological methods, irreversible thermodynamics, and other techniques already well established in kinetic analysis to describe how, to what extent, and under what conditions would the rejection of solute, or retention of residual disorder in crystalline solids be altered because of the rapid advance of the solid–liquid interface. In short, this section deals with theoretical estimates of how one might discriminate between 'conventional' or 'slow' solidification phenomena and RSP. Specifically, we address the key question surrounding RSP: What aspects of conventional solidification are altered when a crystal-melt interface advances rapidly?

Some examples of non-equilibrium processes were already mentioned in earlier chapters of this book, including heat and mass transport, and defect formation, all of which occurred as solid–liquid interfaces advanced and transformed a melt to a crystalline solid. The interface, however, even in these cases, always moved slowly enough to permit *local* equilibrium to prevail. The associated long-range transport processes, acting primarily through thermal and diffusion fields established in the melt, dissipated free energy and, consequently, required the presence of some slight supercooling at the interface. In Chapter 16, moreover, specific account was given of long-range diffusion in the melt ahead of a polyphase solid–liquid interface that 'unmixed' the components and allowed the cooperative growth of eutectic phases. The supercooling required for free energy dissipation during solute transport and for free energy storage associated with interface formation during polyphase solidification was examined in detail with Eq. (E.46). In practice, the total supercooling required during freezing at type-I eutectic interfaces seldom exceeds a few K. In Section 17.3.2, by contrast, rapid solidification from a hypercooled melt was shown to achieve interfacial speeds approaching almost 4 m/s in elemental α-P_4, and the associated interfacial supercooling attained values exceeding 20–30 K. As will be shown quite generally, RSP processes including, for example, hypercooling, splat quenching, and laser- and electron-beam melting, develop such large interfacial supercoolings and dissipate so much free energy that the phase diagram itself can appear to be altered significantly.

In reality, when an interface departs substantially from local equilibrium, several important aspects of solidification predicted by the phase diagram and conventional kinetics become modified. These modifications include changes in the value of the distribution coefficient from its equilibrium value, k_0 (predicted on the basis of thermodynamic equilibrium), alteration of the solute distribution measured in the rapidly frozen solid, appearance of metastable phases in the microstructure, and, finally, development of novel solidification defects, including excess monovacancies, 'trapped' solute, high dislocation densities, and reduced LRO in ordered phases.

17.5.2 Continuous Growth Models

The development of theories to predict the behavior of interfaces far from thermodynamic equilibrium presents considerable difficulties. Near-atomic scale phenomena become important in the regime of RSP, and short of attempting quantum mechanical description of such microscopic events, one is forced to rely on phenomenological approaches, i.e., comparing 'driving forces' to 'kinetic resistances', by applying the methods of non-equilibrium thermodynamics. Continuous growth models (CGM) grew basically from such an approach, led by Aziz, Boettinger and their co-workers [13]. Such theories have proved successfully to capture the phenomenology of most RSP effects, and will suffice for our purposes. Alternative theoretical approaches also yield results that confirm some CGM predictions, based, for example, on developing diffuse interface models and by making phase-field computations [33]. Figure 17.5 shows one such model of RSP, developed by Aziz, in which solute trapping occurs through lateral growth of atomic layers, bordered by (111) terraces and orthogonal step edges.[2] Solute trapping becomes more severe as the speed, v_n, increases, and as the terrace orientation becomes normal to the

---

[2] The solidification of elemental Si, with its solid–liquid interface near (111) occurs by lateral motion of monolayers separating (111) terraces—both at low interface speeds [34] and under RSP conditions [35].

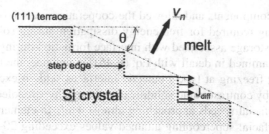

Fig. 17.5 CGM model for RSP of a re-solidifying doped Si crystal. This model uses phenomenological descriptions of solidification by lateral edge motion over terraces accompanied by solute diffusion. The model provides estimates of how the system's response functions depend on the interface growth speed, v_n, step edge solute flux, J_{diff}, terrace solute flux, and interface orientation θ. Experiments support several important aspects of CGM predictions, including interfacial orientation dependences of the solute distribution coefficient, $k(v, \theta)$, which is most sensitively influenced when the orientation is closest to (111) ($\theta \approx 0$) and the step edge velocity reaches a maximum [38]. That CGM provides such good predictions is indeed remarkable, considering that CGM itself is based on a continuum description of sharp-interface kinetics

growth direction, and $\theta \to 0$. For a given speed, v_n, the speed of step edge advance increases as $v_n/|\sin\theta|$.

Predictions for RSP are made in the form of so-called response functions. For example, in RSP alloy solidification, to connect the influence of interface speed, v, to the segregation coefficient, one requires the response function, $k(v)$, that predicts the changes of k as the speed increases, and 'solute trapping' occurs. In the case of ordered phases, including intermetallic compounds produced through RSP, one desires the analogous response function for the compound's LRO parameter. LRO decreases with interface speed because of 'disorder trapping', which is an early RSP concept proposed by Chernov [36], and then developed later using CGM theory by Boettinger and Aziz [37].

As explained in Section 17.5.1, additional response functions relating interface temperature, or supercooling, to the speed of freezing and alloy composition are also needed. The challenge posed from the experimental side, which is considerable, is to measure the speed, composition, and temperature of RSP interfaces to verify theoretical predictions of these response functions.

17.5.3 Response Functions

Some examples of RSP response functions derived from a phenomenological CGM are those found for the solute distribution function, $k(v)$, and the interface speed, $v(\hat{T}, C_s, C_\ell)$, both of which were derived by Aziz and Kaplan. These investigators found expressions for response functions that were applicable to dilute or ideal alloys, namely,

$$k(v) = \frac{k_0 + v/v_D}{1 + v/v_D}, \tag{17.5}$$

and

$$v(\hat{T}, C_s, C_\ell) = v_{col}(\hat{T}) \left[1 - e^{\Delta G(\hat{T}, C_s, C_\ell)/R_g \hat{T}} \right].$$ (17.6)

As mentioned earlier, the parameter v/v_D appearing in Eq. (17.5) is the ratio of the interface speed, v, to the 'diffusion speed', v_D, where the diffusion speed is defined in this instance as the *maximum* theoretical diffusion flux of solute atoms across the interface, divided by the solute's molar volume. This particular diffusion speed is estimated by assuming that every atomic jump contributes to the interface advance.

Equation (17.6), when applied to the case of a pure material, where $C_s = C_\ell = 0$, provides another phenomenological kinetic expression for the interface speed in terms of the maximum molecular 'collision speed', v_{col}, and the net molar free energy available for solidification, $\Delta G(\hat{T}, C_s = 0, C_\ell = 0) < 0$. Thus, if the driving free energy, $\Delta G(\hat{T}, C_s = 0, C_\ell = 0)$, becomes extremely large and negative, the interface is driven ahead by virtually every atomic jump, and advances into the melt at its maximum theoretically allowed rate. In the case of a pure metallic melt under an enormous driving force, the interface speed is theoretically proportional to the propagation speed of sound, v_c, in the melt [39]. This upper limit to interfacial advance rates, or so-called 'collision speed', can be of the order of several km/s, which is a speed far higher—by some three orders-of-magnitude—than the solute diffusion speed, v_D.

Moreover, as the solid–liquid interface speed in an alloy approaches, and eventually exceeds, its diffusion speed, the solute atoms cannot diffuse quickly enough to keep pace with the interface. Solute trapping occurs, and the value of the solute distribution coefficient, $k(v)$, (assuming that $k_0 < 1$) increases, and, in principle, approaches unity. An example of a typical set of CGM response functions predicted for alloys with various equilibrium distribution coefficients is shown in Fig. 17.6. Here the change in $k(v)$ is plotted against the interface speed as a fraction, or multiple, of the diffusion speed, namely, v/v_D.

17.5.4 Quantitative RSP Experiments

The first comprehensive test of CGM predictions, where both the interface speed and the solute concentration profiles were both measured quantitatively, and not merely estimated, was reported by Aziz et al. [40]. This remarkable experiment used brief ultra-violet laser pulses at energy densities of $1.0\,\text{J/cm}^2$ to melt thin layers of As-doped Si crystals to an initial depth of 500 nm, in approximately 100 ns. This molten layer re-solidifies in about 3,000 ns, achieving interface speeds that exceeded 10 m/s. Figure 17.7 tracks the melting and re-freezing of the interface in response to the initial laser pulse. These investigators recorded the speed of rapid re-solidification, and then measured, in situ, the instantaneous dopant concentration profiles using RBS techniques. Some of their RBS data are shown in Figs. 17.8 and 17.9. These

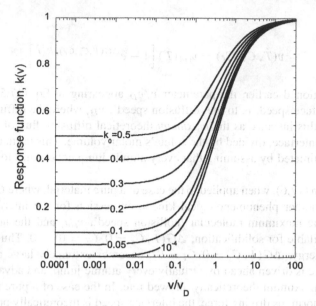

Fig. 17.6 Response function predicted with CGM theory for the solute distribution coefficient, $k(v)$, plotted against the ratio of the interface speed to the diffusion speed, v/v_D. The value of $k(v) \approx k_0$ until the interface speed exceeds about 1% of the diffusion speed, v_D. For many alloys the value of the diffusion speed is in the range 1–100 m/s. See Fig. 17.10 for some estimates of v_D for several binary systems

data greatly assist visualizing the effects of solute re-distribtution and trapping at an RSP solid-melt interface. These data sets also permit direct calculation of the non-equilibrium distribution coefficient, $k(v)$, as a function of the instantaneous interface speed, and allow critical comparison with the response function calculated from CGM.

Some typical diffusion speeds that have been determined experimentally for doped Si and binary Al alloys are plotted in Fig. 17.10 against a convenient equilibrium redistribution parameter. As shown, only a rough linear correlation was found between the value of v_D observed during RSP and the equilibrium distribution parameter, k_0, by using the function $\ln k_0 / |(k_0 - 1)|$. This approximate correlation may be expressed as the following linear regression,

$$v_D \approx 10.0 \times \ln \left[\frac{k_0}{|k_0 - 1|} \right] - 17.0 \text{ [m/s].} \qquad (17.7)$$

Unfortunately, there does not exist straightforward methods to obtain the interface diffusion speed in an alloy, either theoretically or from standard experiments. The usual method for 'guestimating' a value of the diffusion speed is to use the suggestion [42] that the diffusion speed

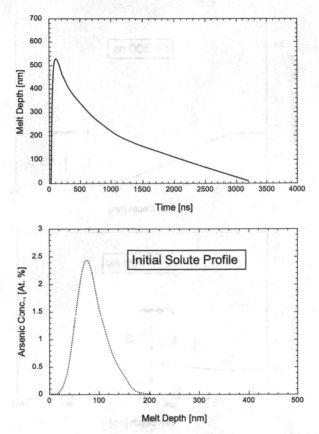

Fig. 17.7 *Upper:* Melt depth versus time recorded by Aziz et al. in As-doped Si using fast UV laser pulse melting. RSP occurred over a period of approximately 3,000 ns after the pulse ended, resulting in a maximum melt depth of about 500 nm. The average interface speed in this experiment was approximately 17 cm/s, with a maximum speed of about 65 cm/s achieved shortly after re-solidification began. *Lower:* Initial As dopant distribution implanted in the Si. Profile determined using Rutherford Back Scattering (RBS). Adapted from [41]

$$v_D = \frac{D_\ell}{\lambda}. \tag{17.8}$$

Aziz and co-workers also studied RSP of Bi-doped Si, and observed the changes in this system's response, $k(v)$, as the solidification speed increased. The diffusion velocity is estimated to be 32 m/s in Bi-doped Si. The variation found by these investigators in the non-equilibrium solute distribution coefficient, $k(v)$, is displayed as Fig. 17.11. As is shown in Fig. 17.11, at an interface speed of 2 m/s, which is already about 6% of the diffusion speed, the solute distribution coefficient has changed substantially from its equilbrium value, $k_0 \approx 0.05$. An increase of the interface speed to about 40% of v_D greatly enhances solute trapping, and further increases the value of the non-equilibrium k-value by almost 400%.

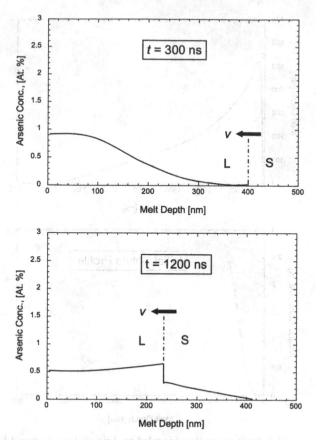

Fig. 17.8 Concentration profiles of As in re-solidified Si and in the remaining melt at 300 ns (upper RBS data) and at 1,200 ns (lower RBS data). Times (nanoseconds) denote the duration of RSP after melting ceased, and re-freezing began. The solid–liquid interface positions at each time are located by the *broken vertical lines*, with the direction of interface motion shown by the *arrows*. Adapted from [41]

These and other experiments demonstrate clearly that RSP is capable of con-trolled modification of the solidification behavior of doped semiconductors, pro-vided that the substrate temperature and the energy input via laser pulsing are care-fully controlled.

17.6 Summary

Rapid solidification processing, or RSP, represents a significant departure from most conventional casting or crystal growth methods. Even local interfacial equilibrium fails to hold as the solid–liquid interface speed is increased above thresholds deter-mined by heat transfer and solute diffusion.

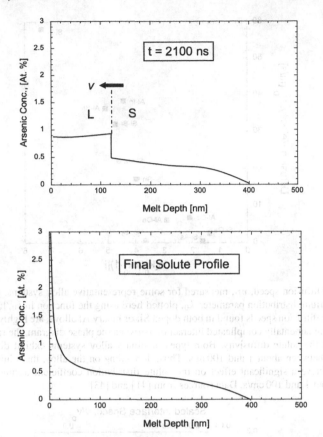

Fig. 17.9 Concentration profiles of As in re-solidified Si and in the remaining melt at 2,100 ns (*upper* RBS data) and after RSP is completed (*lower* RBS data). The solid–liquid interface positions at each time are again located by the *broken vertical lines*, with the direction of interface motion shown by the *arrow*. The solute concentration profile achieved using RSP may be compared with the initial dopant distribution shown in the *lower panel* of Fig. 17.7. Adapted from [41]

Although RSP methods can be traced back to gas quenching of liquid metals in conventional shot towers, the solidification science of RSP phenomena, including solute trapping and metastable phase formation, were discovered in the early 1960s by Duwez and co-workers, by using so-called splat-cooling methods. Significant interface supercooling, solute trapping, and RSP interface speeds exceeding 3 m/s were reported by Glicksman and Schaefer studying hypercooled P_4 and Hg-saturated P_4 melts. It was not until 1970 that Baker and Cahn provided the first clear account of thermodynamic restrictions imposed by phase diagrams for producing metastable phase extension and solute trapping by RSP. By the 1970s, an industrial-scale process, Osprey casting, proved that tonnage amounts of RSP alloys could be produced economically. The development of pulsed optical lasers and electron beam technology over the last 40 years allowed rapid expansion of the application areas

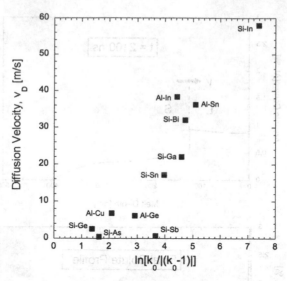

Fig. 17.10 Diffusion speed, v_D, measured for some representative alloy systems, plotted against their equilibrium distribution parameter, k_0, plotted here using the function $\ln[k_0/|(k_0-1)|]$. The values of the diffusion speeds found in both doped Si, or binary Al alloys, are highly variable. This suggests a fundamentally complicated interaction between the phase diagram, the rate of interface motion, and the solute diffusivity. Both types of binary alloy systems exhibit diffusion speeds somewhere between about 1 and 100 m/s. Thus, depending on the alloy, the minimum interface speed to achieve a significant effect on the solute distribution coefficient during RSP must be between about 1 and 100 cm/s. Data sources from [41] and [43]

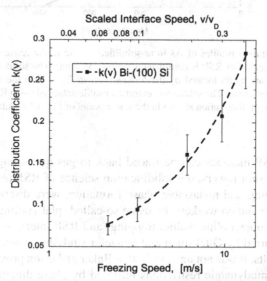

Fig. 17.11 Variation with interface speed of the $k(v)$ response function observed during RSP of Bi-doped Si. Re-solidification occurred on the (100) plane of Si. The *dashed line* is estimated from CGM theory. Values for the scaled interface speed, relative to the diffusion speed, v/v_D, are indicated along the top abscissa. Data sources drawn from [41] and [43]

of RSP technologies to sophisticated applications including surface re-melting and laser glazing of alloys.

Detailed kinetic studies of far-from-equilibrium RSP phenomena in doped semi-conductors have been carried out by several groups of researchers, led by Aziz et al., who have quantified key relationships among the interface speed, diffusion speed, with changes in the dopant distribution coefficient. Although far from being fully understood kinetically, RSP phenomena are now predictable and fall within the scope of engineering solidification processes.

References

1. Pol Duwez, R.H. Willens and W. Klement, *J. Appl. Phys.*, **31** Chap. 6 (1960) 1136.
2. D.O. Belanger, *Enabling American Innovation*, Purdue University Press, 1998, p. 165.
3. H.H. Liebermann, Ed., *Rapidly Solidified Allos: Processes, Ptructures, Properties*, CRC Press, Boca Raton, FL, 1993.
4. T.S. Srivatsan and T.S. Sudashan, Eds., *Rapid Solidification Technology: An Engineering Guide*, CRC Press, Boca Raton, FL, 1993.
5. H. Jones, *J. Mater. Sci. Eng. A*, **304** (2001) 11.
6. J.H. Perepezko et al., 'Rapid solidification of highly undercooled aluminum powders', *Rapidly Solidified Powder Aluminum Alloys: A Symposium*, M.E. Fine and E.A. Starke, Eds., ASTM International, W. Conshohocken, PA, 1986.
7. A.G. Leatham and A. Lawley, *Int. J. Powder Metall.*, **29** (1993) 321.
8. G. Thursfield and H. Jones, *J. Phys. E.*, **4** (1971) 675.
9. http://www.metglas.com/products/, *Products and Solutions for the Electronics Industry*, Allied Chemical Corp., undated.
10. A.A. Chernov, *Growth of Crystals*, **3**, A.V. Shubnikov and N.N. Sheftal, Eds., Consultants Bureau, New York, NY, 1962, 65.
11. V.V. Voronkov and A.A. Chernov, *Sov. Phys. Crystallog.*, **12** (1967) 186.
12. A.A. Chernov, *Sov. Phys. Uspekhi*, **13** (1970) 101.
13. M.J. Aziz and T. Kaplan, *Acta Metall.*, **36** (1988) 2335.
14. M. Rettenmayr and M. Buchmann, *Int. J. Mater. Res.*, **99** (2008) 961.
15. M.E. Glicksman and R.J. Schaefer, *J. Cryst. Growth*, **1** (1967) 297.
16. A. Ludwig, *Scr. Mater.*, **34** (1996) 579.
17. M. Li and K. Kuribayashi, *Acta Mater.*, **52** (2004) 3639.
18. J.H. Perepezko, *Mater. Sci. Eng.*, **65** (1984) 125.
19. J.H. Perepezko and J.S. Palik, *J. Non-Cryst. Solids*, **61–62** (1984) 113.
20. R. Willnecker, D.M. Herlach, B. Feuerbacher and H.J. Fecht, *J. Mater. Sci. Eng. A*, **133** (1991) 443.
21. A.L. Greer, *J. Less Common Metals*, **145** (1988) 131.
22. K.F. Kelton and A.L. Greer, *J. Non-Cryst. Solids*, **79** (1986) 295.
23. M.B. Robinson, R.J. Bayuzick and W.H. Hofmeister, *Adv. Space Res.*, **8** (1988) 321.
24. D.M. Herlach, R.F. Cochrane, I. Egry, H.J. Fecht and A.L. Greer, *Int. Metals Rev.*, **38** (1994) 273.
25. D.M. Herlach, *Mater. Sci. Eng. Rep.*, **11** (1994) 355.
26. M. Hillert and B. Sundman, *Acta Metall.*, **25** (1977) 11.
27. J.C. Baker and J.W. Cahn, *Acta Metall.*, **17** (1969) 575.
28. G.H. Gilmer, *Mater. Res. Soc. Symp.*, **13** (1983) 249.
29. J.C. Baker and J.W. Cahn, *Solidification*, ASM, Metals Park, OH, 1970, p. 23.
30. R.F. Wood, *Appl. Phys. Lett.*, **37** (1980) 302.
31. P. Baeri, et al., *Appl. Phys. Lett.*, **37** (1980) 912.

32. C.W. White, et al., *J. Appl. Phys.*, **51**, (1980) 738.
33. J.W. Cahn, S.R. Coriell and W.J. Boettinger, *Laser and Electron Beam Processing of Materials*, C.W. White and P.S. Peercy, Eds., Academic Press, New York, NY, 1980, p. 8.
34. T.F. Ciszek, *J. Cryst. Growth*, **10** (1971) 263.
35. D.M. Zehner, C.W. White and G.W. Ownby, *Surf. Sci.*, **92** (1980) 67.
36. A.A. Chernov, *Sov. Phys. JETP*, **26** (1968) 1182.
37. W.J. Boettinger and M.J. Aziz, *Acta Metall.*, **37** (1989) 3379.
38. L.M. Goldman and M.J. Aziz, *J. Mater. Res.*, **2** (1987) 524.
39. S.R. Coriell and D. Turnbull, *Acta Metall.*, **30** (1982) 2135.
40. M.J. Aziz, et al., *Mater. Res. Soc. Symp. Proc.*, **35** (1985) 153.
41. M.J. Aziz, *Metall. Mater. Trans. A*, **27A** (1996) 671.
42. M.J. Aziz, *J. Appl. Phys.* **53** (1982) 1158.
43. N.A. Ahmad et al., *Phys. Rev. E*, **58** (1998) 3436.

Part V
Appendices

Appendix A
Thermodynamic Functions and Legendre Transforms

Background

Chapter 2 prepared a brief thermodynamic basis for understanding transitions from crystalline solids to their conjugate melt phases, and vice versa. The internal energy, enthalpy, and Gibbs free energy played important roles in that development. Certain system parameters, such as the temperature, pressure, and component concentrations, were selected as independent intensive variables when discussing state function changes under carefully prescribed conditions. The relationships among the many thermodynamic state functions and a system's intensive and extensive parameters can seem arbitrary, and occasionally confusing. Students often ask why and how the selection of state functions relates to these variables, and most of those who do not, have merely memorized the connections. This appendix provides a rigorous yet manageable alternative.

Thermodynamic Functions and Variables

Internal energy, U, is a fundamental thermodynamic state function providing the tendency of isolated, or 'closed' material systems to approach equilibrium. Moreover, U is a linear, homogeneous function, $U(S, V, N_i, \ldots)$, of all the system's so-called 'natural' *extensive* variables, including its entropy, S, volume, V, component molar masses, N_i, etc. Intensive parameters, by contrast, such as temperature, T, pressure, P, and chemical potential, μ_i, respectively, are also system variables that often are more conveniently measured under practical solidification conditions than would the 'natural' extensive quantities. With care, as we shall show, extensive thermodynamic variables may be easily substituted by alternative independent parameters of the system, each replacing their associated extensive variable. Substitution of intensive variables for extensive variables generates different thermodynamic functions, or potentials, that can equivalently determine the state of equilibrium for solidifying systems maintained under widely varying circumstances, such as melts communicating with a heat bath, a pressure source, or exchanging mass with a reservoir of chemical potential.

M.E. Glicksman, *Principles of Solidification*, DOI 10.1007/978-1-4419-7344-3,
© Springer Science+Business Media, LLC 2011

Interestingly, all such 'practical' intensive variables, often imposed by the system's environment, can be mathematically related to conjugate extensive thermodynamic variables. One, therefore, can reformulate the internal energy function, U, by
replacing one or more of its extensive variables by intensive ones. A good example
of needing such an alternative function is recognizing that a solidifying system's
temperature, T, is both easily measured using thermometry and controlled in an
oven or furnace with a wide variety of standard laboratory instruments, such as ther-
mocouples or pyrometers. By contrast, temperature's conjugate extensive variable is
the entropy, which, unfortunately, lacks practical devices for its direct measurement
or control. How, therefore, might one transform the internal energy function, U,
to provide an alternative thermodynamic function by replacing one, or more, of its
extensive variables, viz., $(S, V, N_i, \dots)$, by its respective intensive variables, viz.,
$(T, P, \mu_i, \dots)$?

Such a substitution may be accomplished using Legendre transforms—a standard
mathematical procedure. Legendre transforms, among other things, lead directly to
important thermodynamic 'potentials' that will be used throughout this book. These
potentials provide alternatives for the internal energy. Some of these potentials
might seem more familiar than others, as they remain in common use, but the formal
procedure outlined next of how they are created remains identical, irrespective of the
variables chosen.

The Formal Math Problem

Given a function Y defined as

$$Y = Y(X_0, X_1, X_2 \dots X_n), \tag{A.1}$$

one desires to replace its independent variables, X_i, by one or more of their partial
derivatives, P_i, where

$$P_i \equiv \left(\frac{\partial Y}{\partial X_i} \right)_{1,2,\dots n \neq i}. \tag{A.2}$$

The subscripts placed outside the parentheses in Eq. (A.2) denote the variables that
are held fixed during partial differentiation. In fact, the P_is represent the geometric
'slopes' of the function Y plotted with respect to the selected variable X_i, holding
all other variables fixed.

Consider the case of a function of just one independent variable, $Y(X)$. This
function represents a curve in X-Y space, composed of Cartesian coordinate pairs,
(X, Y). The derivative, $P = \frac{dY}{dX}$, is the tangent to the curve $Y(X)$ at any point. One
may replace the independent variable, X, by P. Simply replacing the coordinate
X by the function's slope at that point to yield a new function, $Y(P)$, however,
doesn't quite work, because, as will be shown, some information is lost by this
direct substitution. Notice that for a single variable, Eq. (A.2) becomes an ordinary

differential equation, which, when integrated, yields the solution, $Y(X)$. Indeed, this result occurs, but the solution so obtained is established only to within an as yet unknown constant of integration! See Fig. A.1. Thus, by using direct substitution some information is lost.

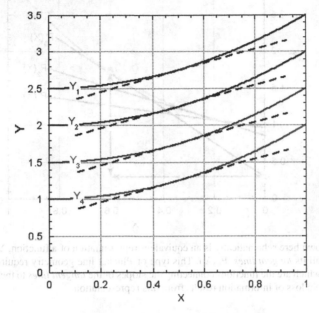

Fig. A.1 Loss of information occurs by replacing the independent variable, X, in $Y(X)$ by the slope, P, to form a new transformed function $Y(P)$. Clearly, knowing the slope at every point along the curve is *not* equivalent to knowing the coordinates, X, Y. The curves $Y_1, Y_2, \ldots$ all have identical $P(X)$ values, but are distinctly different functions

The original function $Y(X)$ may, nonetheless, be represented without loss using the substitution of slopes via Plücker line geometry. Plücker line geometry shows that a 'continuous' envelope of *tangents lines* would suffice to define the curve, Y, as suggested in Fig. A.2. Individual tangents contribute an infinitesimal line segment that join together smoothly to form the curve $Y(P)$ without any loss of information. This procedure is, therefore, equivalent to ordinary point geometry, where an infinite number of coordinate pairs connect to form a curve. The difference, however, is that each tangential segment contributing to the curve now fundamentally represents *both* a slope, or derivative, P, plus some positional information, or locus, obtained from the derived function, $P(X)$. Full specification of all the tangent lines is gained by establishing the relationship between the Y-intercepts, Ψ, and their associated slopes, P. In short, these become the transformed new independent variables. Thus, $\Psi(P)$ is mathematically equivalent to the original Y representation, namely, $Y(X)$, with all information contained in that original function fully retained in the new Ψ representation. Thus,

$$\Psi = \Psi(P). \tag{A.3}$$

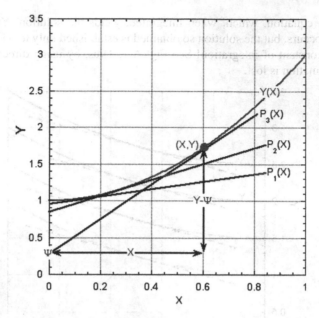

Fig. A.2 Shown here schematically is an equivalent representation of a function, $Y(X)$, using the envelope of all its *tangent lines*, $P_n(X)$. This type of Plücker line geometry requires the relationship, $\Psi(P)$, which are the functions connecting the slopes of the *tangent lines* to their Y-intercepts, i.e., the Ψs. No loss of information results from this representation

Now the formal mathematics problem reduces to how one calculates $\Psi(P)$ given $Y(X)$. A well-known mathematical manipulation called the Legendre transformation provides the appropriate procedure.

Legendre Transformations

One-Variable Transforms

As indicated in Fig. A.2, the slope of the tangent line touching the curve $Y(X)$ at the general coordinates X, Y, is

$$P = \frac{Y - \Psi}{X - 0}. \tag{A.4}$$

The linear form, Eq. (A.4), provides the definition of the new function, Ψ, which we term the *Legendre transform* of the function Y, with respect to X. Operationally, one writes $\Psi = Y[X]$, which notes that Ψ is the function Y with its independent variable X replaced by the slope $P = dY/dX$, or

$$\Psi = Y(X) - PX. \tag{A.5}$$

The inverse transform, i.e., recovering $Y(X)$ given the function $\Psi(P)$, is the symmetrical inversion,

$$d\Psi = dY(X) - PdX - XdP. \tag{A.6}$$

Recalling Eq. (A.2), which shows that $dY = PdX$, Eq. (A.6) reduces to $d\Psi = -XdP$, or, equivalently,

$$X = -\frac{d\Psi}{dP}. \tag{A.7}$$

Equations (A.5) and (A.7) together allow elimination of the variables Ψ and P from the transform function $\Psi(P)$, thereby inverting the transform and returning the original function $Y(X)$. The Legendre transform and its inverse are shown as a series of operational steps in the table below.

Transform $\rightarrow$	$\leftarrow$ Inverse transform
$Y = Y(X)$	$\Psi = \Psi(P)$
$P = \frac{dY}{dX}$	$-X = \frac{d\Psi}{dP}$
$\Psi = -PX + Y$	$Y = XP + \Psi$
Eliminate X and Y $\rightarrow$	Eliminate P and Ψ $\rightarrow$
$\Psi = \Psi(P)$	$Y = Y(X)$

Multivariate Transforms

The generalization of the Legendre transform method to multivariate functions, such as $U(S, V, N_i, \dots)$, which are commonly encountered in thermodynamics, requires a straightforward extension.

Starting with the fundamental equation $Y = Y(X_0, X_1, \dots X_n)$ portrayed in an $n + 2$ dimensional space, and represented by the coordinates $Y, X_0, X_1, X_2, \dots X_n$. The partial derivative, P_k, is defined in the usual way as

$$P_k \equiv \left(\frac{\partial Y}{\partial X_k}\right)_{0,1,2,\dots,n \neq k}, \tag{A.8}$$

and represents the partial slope of the hypersurface, Y. The total differential of the function Y is given as the usual sum,

$$dY = \sum_{k=0}^{n} P_k dX_k. \tag{A.9}$$

In an analogous manner to the one-variable case, the hypersurface can be represented either as the locus of all points satisfying the fundamental equation,

$Y = Y(X_0, X_1, \ldots X_n)$, or as the envelope of all its tangent hyperplanes. Each hyperplane may be characterized by its intercept, Ψ, which is a function of all the slopes $P_0, P_1, P_2 \ldots P_n$ and their associated independent variables, $X_0, X_1, X_2, \ldots X_n$, namely

$$\Psi = Y - \sum_{k=0}^{n} P_k X_k. \tag{A.10}$$

Taking the differential of Eq. (A.10) yields

$$d\Psi = -\sum_{k=0}^{n} X_k dP_k, \tag{A.11}$$

and thus,

$$X_k = -\frac{\partial \Psi}{\partial P_k}. \tag{A.12}$$

Thermodynamic Potentials

The general function, $Y(X_0, X_1, \ldots)$, may now be chosen as the internal energy, $U = U(S, V, N_i, \ldots)$. For simple chemical systems, such as an alloy melt, composed of n components, not subject to external fields affecting their magnetic, electric, or strain energies, the internal energy may be expressed in the linear, or Eulerian form $U(S, V, N_i) = TS - PV + \mu_i N_i$, where repeated subscripts here imply summation. The intensive conjugate parameters, T, $-P$, and μ_i, that multiply each of the extensive parameters are the temperature, (minus) pressure, and chemical potential, respectively. These 'physical' quantities correspond precisely to the partial derivatives of U, namely, $P_0, P_1, \cdots$. One tends to overlook the simple mathematical relationships between the extensive and conjugate intensive thermodynamic variables because of one's familiarity with temperature and pressure as 'physical' quantities. However, one should recall that for every extensive variable adding to a system's internal energy, there exists a companion, or conjugate, intensive quantity. The product of intensive and extensive variable pairs yields some independent contribution to the internal energy.

The Legendre transform procedure merely substitutes properly an intensive variable for its conjugate extensive variable. The functions based on $U(S, V, N_i)$ that result from taking partial Legendre transforms with respect to these variables are called thermodynamic potentials, some of which the reader will find familiar, others perhaps less so.

Helmholtz Potential

If the temperature, T, replaces the entropy, S, as an independent variable, the resulting Legendre transform is called the Helmholtz potential, or Helmholtz free energy.

Both terms occur in common useage, and are denoted by $F(T, V, N_i)$. The operational math notation used here for the partial Legendre transform, indicating the replacement of a particular extensive thermodynamic variable, (in this case replacing the entropy, S, by its conjugate, the temperature, T), is,

$$F(T, V, N_i) \equiv U[T], \tag{A.13}$$

where the square brackets enclose the 'active' transform parameter or parameters. The steps to find the partial Legendre transformation and its inverse for the Helmholtz potential are outlined the following table:

Transform $S \to T$	$S \leftarrow T$ inverse
$U = U(S, V, N_i)$	$F = F(T, V, N_i)$
$T = (\partial U / \partial S)_{V, N_i}$	$-S = (\partial F / \partial T)_{V, N_i}$
$F = U - TS$	$U = F + TS$
Eliminate S and U $\to$	Eliminate F and T $\to$
$F = F(T, V, N_i)$	$U = U(S, V, N_i)$

One concludes that the Helmholtz potential has the following total differential:

$$d\mathrm{F} = -SdT - PdV + \mu_i dN_i. \tag{A.14}$$

Enthalpy

If the pressure replaces the volume as an independent variable, the partial transform is called the 'enthalpy', or heat content. Both names remain in use, and are denoted by $H(S, P, N_i)$. The notation for the partial Legendre transform, indicating the replacement of the system's volume, V, by its conjugate, the (minus) pressure, $-P$, is,

$$H(S, P, N_i) \equiv U[-P]. \tag{A.15}$$

Steps to obtain the partial Legendre transformation and its inverse for the enthalpy are outlined the next table:

Transform $V \to P$	$V \leftarrow P$ inverse
$U = U(S, V, N_i)$	$H = H(S, P, N_i)$
$-P = (\partial U / \partial V)_{S, N_i}$	$V = (\partial H / dP)_{S, N_i}$
$H = U + PV$	$U = H - PV$
Eliminate U and V $\to$	Eliminate H and P $\to$
$H = H(S, P, N_i)$	$U = U(S, V, N_i)$

One concludes that the enthalpy has the following total differential:

$$dH = TdS + VdP + \mu_i dN_i. \tag{A.16}$$

Gibbs Free Energy

If both the temperature replaces entropy, and the pressure replaces volume, the partial transform is called the 'Gibbs potential', or chemical free energy. Both terms are in common use, and are denoted by $G(T, P, N_i)$. The notation for the partial Legendre transform, indicating replacement of both the system's entropy, S, and volume, V, by their conjugate variables, the temperature, T, and (minus) pressure, $-P$, is,

$$G(T, P, N_i) \equiv U[T, -P]. \tag{A.17}$$

The steps to obtain the partial Legendre transformation and its inverse for the Gibbs potential are outlined the next table:

Transform $S, V \rightarrow T, P$	$S, V \leftarrow T, P$ inverse
$U = U(S, V, N_i)$	$G = G(T, P, N_i)$
$T = (\partial U/\partial S)_{V,N_i}$	$-S = (\partial G/\partial T)_{P,N_i}$
$-P = (\partial U/\partial V)_{S,N_i}$	$V = (\partial G/dP)_{T,N_i}$
$G = U - TS + PV$	$U = G + TS - PV$
Elimin. U, S and $V \rightarrow$	Elimin. G, T and $P \rightarrow$
$G = G(T, P, N_i)$	$U = U(S, V, N_i)$

One concludes that the Gibbs function has the following total differential:

$$dG = -SdT + VdP + \mu_i dN_i. \tag{A.18}$$

Note, for constant T, constant P, and a closed system, $dN_i = 0, dG = 0$. Thus, it is the Gibbs potential that reaches a minimum when say thermochemical equilibrium is achieved under these typical laboratory conditions.

Grand Canonical Potential

If both the temperature replaces entropy, and the chemical potentials replace mole numbers, the partial transform is called the 'Grand Canonical potential', denoted by $\chi(T, V, \mu_i)$. The notation for the partial Legendre transform, indicating replacement of both the system's entropy, S, and mole numbers, N_i, by their conjugate variables, the temperature, T, and the chemical potentials μ_i, is,

$$\chi(T, P, \mu_i) \equiv U[T, N_i]. \tag{A.19}$$

The steps to obtain the partial Legendre transformation and its inverse for the Grand Canonical potential are outlined the next table:

Transform $S, N_i \rightarrow T, \mu_i$	$T, \mu_i \leftarrow S, N_i$ **inverse**
$U = U(S, V, N_i)$	$\chi = \chi(T, V, \mu_i)$
$T = (\partial U/\partial S)_{V,N_i}$	$-S = (\partial \chi/\partial T)_{V,\mu_i}$
$\mu_i = (\partial U/\partial N_i)_{S,V}$	$N_i = (\partial \chi/d\mu_i)_{T,V}$
$\chi = U - TS - \mu_i N_i$	$U = \chi + TS + \mu_i N_i$
Elimin. U, S and $N_i \rightarrow$	Elimin. χ, T and $\mu_i \rightarrow$
$\chi = \chi(T, V, \mu_i)$	$U = U(S, V, N_i)$

One concludes that the Grand Canonical potential has the following total differential:

$$d\chi = -SdT - PdV - N_i d\mu_i. \tag{A.20}$$

Note, for constant T, constant V, and for an open system providing constant μ_i, $d\chi = 0$. Thus, it is the Grand Canonical potential that reaches a minimum when say thermochemical equilibrium is achieved under constrained conditions where the surroundings set the chemical potentials of the grain boundaries and microscopic internal defects such as dislocations.

Appendix B
Grain Boundary Grooves

Introduction

As discussed in Chapter 9, solid–liquid interfaces normally contain 'imperfections', or persistent distortions, that provide preferred locations at which interfacial instabilities initiate. Perhaps most importantly, interfacial instabilities, when triggered by thermal or constitutional supercooling, rapidly alter the dimensionality of the solute diffusion field. Instabilities, such as cells or dendrites change an initially one-dimensional planar diffusion field into a two- or three-dimensional solute field exhibiting strong lateral microsegregation.

In order to develop a firmer understanding of interface behavior during stable and unstable freezing, quantitative descriptions of the interface morphology for various classes of materials are needed. These descriptions, moreover, fall into three major length-scale classes: (1) *macroscopic* interface shapes, including geometric description of the largest solidification features controlled by heat flow, (2) *mesoscopic* interface shapes appropriate to cellular, eutectic and dendritic crystals, and (3) *microscopic*, or molecular-scale interface configurations, that address interfacial roughness, atomic steps and layers, and facet formation. Appropriate descriptions at each length scale require knowledge of the temperature and solute distributions, the influence of these fields on the system's energetics, and the associated kinetic processes that operate during solidification. Acquiring detailed descriptions at each scale, under differing ranges of kinetic driving forces, is a daunting task, even for a single material.

This appendix develops the underlying physics of a relatively simple, commonly encountered interface distortion—the symmetric grain boundary groove. Grain boundary grooves are often present—but may not be visible—on polycrystalline solid–liquid interfaces. Indeed, the solid–liquid interface of a polycrystalline material is perturbed persistently and strongly by its intersection with grain boundaries that continue to propagate along with the interface during freezing. Thus, at moving interfaces two basic structures, the solid–liquid interface and the grain boundary, each of microscopic origin and molecular thickness, interact and produce a new mesoscopic interfacial structure, namely, a grain boundary groove.

Symmetric grain boundary grooves provide an excellent example of how heat flow, capillarity, and thermodynamics interact and distort a solid–liquid interface at

mesoscopic length scales (micrometers). Figure 9.2 provides direct observations of
how initially stable grain boundary grooves, if destabilized by constitutional supercooling, steadily evolve toward more complex three-dimensional interfacial structures during freezing of a dilute alloy. Figure 9.3 suggests schematically several of
the key intermediate steps by which a grain boundary groove initiates interfacial
instability and leads to the morphological break down of a nearly planar interface.

Symmetric Grain Boundary Grooves

Physical Model

We develop an analysis of a symmetrical grain boundary intersecting an initially
planar solid–liquid interface in a pure material. A steady, linear thermal field,
$T(y) = T_m + Gy$, where G is the constant gradient applied along the y-direction
(normal to the interface), maintains the flat portion of the interface at its equilibrium
melting point, T_m. The interface, as depicted in Fig. B.1, spontaneously forms a
grain boundary groove, and reduces the total free energy of the system.

One may think of the spontaneous changes in interface morphology as though
they were produced by the grain boundary tension 'pulling' back on the solid–liquid
interface, and forming the cusp-shaped groove. The re-configuration of the solid–
liquid system into a groove actually accomplishes several interrelated energetic
changes, which when combined, minimizes the system's total Gibbs free energy:

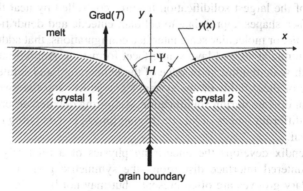

Fig. B.1 Configuration of a symmetric grain boundary groove formed between two crystals. The
x-axis coincides with the normal melting point, T_m, and locates the position of the initially planar
solid–liquid interface. A groove of depth H forms to minimize the system's total free energy under
the steady application of a constant temperature gradient, Grad(T), directed parallel to the $+y$-axis.
The dihedral angle, Ψ, at the groove root depends on the balance of interfacial tensions between
the solid–liquid interface and the intersecting grain boundary. The melt in the *upper half plane*
($y \geq 0$) is at, or above, its melting point and remains stable, whereas the melt located within the
groove ($y \leq 0$) is at, or below, its melting point, and is slightly supercooled. The profile of the
groove is given by the plane curve $y(x)$

1. The total area (i.e., total arc length per unit depth into the page) of the solid–liquid interface increases through groove formation by an amount $\Delta A_{s\ell}$, relative to its initial planar configuration. This change in length (or area per unit depth) increases the total amount of free energy stored along the interface as $\gamma_{s\ell} \times \Delta A_{s\ell}$.

2. The area (per unit depth into the page) of the symmetric grain boundary decreases after the groove develops by an amount equal to the depth of the groove, h. The change in grain boundary area decreases its energy by $\gamma_{gb} \times h$.

3. The groove itself is filled with a small volume of melt that is slightly supercooled, by virtue of the temperature field that sets the temperatures of all points $(-\infty < x < \infty, y \leq 0)$ below the melting point, T_m, by an amount equal to Gy. The melt, of course, becomes metastable in its supercooled state, and its free energy density increases at any point, $y \leq 0$, by $\Delta S_f \times Gy$ relative to its free energy density at all points along the flat portions of the interface, $(x, 0)$.

4. The solid phase surrounding the groove is stable for all points below its melting point, $(x, y \leq 0)$. However, in order to maintain local equilibrium with the supercooled melt, the interface must be curved in accord with the Gibbs–Thomson equation. See again Section 8.3. Specifically, the local curvature, $\kappa(x, y)$ of the solid–liquid interface must increase linearly with the depth, $y \leq 0$, into the groove. Thus, there is a specific shape of the groove, $y(x)$, that satisfies local equilibrium between the phases, and, equivalently, minimizes the total free energy of the system.

5. A 'natural boundary condition' of this constrained equilibrium problem may be found by using the calculus of variations. Specification of the limiting dihedral angle, ψ, as $x \to 0$ assures mechanical equilibrium. See again Fig. B.1. The geometry of the groove's triple point, where the traces of the grain boundary and the left- and right-hand sides of the solid–liquid interface intersect, shows that the equilibrium dihedral angle at the base of the groove must satisfy the vector balance of three surface forces, viz., $2\gamma_{s\ell} \cos(\psi/2) = \gamma_{gb}$.

Mathematical Description

The thermal conductivities of the solid and melt are assumed to be equal, so the temperature field, $T(y) = T_m + Gy$ has isotherms that remain parallel to the x-axis. The grain boundary groove develops a steady shape, $y(x)$, that conforms to the linear temperature distribution. Local equilibrium requires that each point on the solid–liquid interface develops a curvature, $\kappa(y)$, required by the Gibbs–Thomson equation, thus,

$$T(y) = T_m + Gy = T_m - \frac{\gamma_{s\ell}\Omega}{\Delta S_f}\kappa(y), \quad (y \leq 0). \tag{B.1}$$

Equation (B.1) expressed in the x-y Cartesian coordinates becomes,

$$Gy = \frac{\gamma_{sl}\Omega}{\Delta S_f} \frac{\frac{d^2y}{dx^2}}{\left[1+\left(\frac{dy}{dx}\right)^2\right]^{3/2}}, \qquad (B.2)$$

or, after dividing through by the temperature gradient G,

$$y = K^2 \frac{\frac{d^2y}{dx^2}}{\left[1+\left(\frac{dy}{dx}\right)^2\right]^{3/2}}, \qquad (B.3)$$

where $K^2 = \gamma_{sl}\Omega/(G\Delta S_f)$. K is a lumped parameter which defines a capillary length relevant to the scale of the groove. If both sides of Eq. (B.3) are divided by $2K$, the groove shape can be defined by two dimensionless coordinates, $\mu = x/2K$ and $\eta = y/2K$, so Eq. (B.3) becomes

$$\eta = \frac{1}{4} \frac{\frac{d^2\eta}{d\mu^2}}{\left[1+\left(\frac{d\eta}{d\mu}\right)^2\right]^{3/2}}. \qquad (B.4)$$

Equation (B.4) is a second-order non-linear ordinary differential equation. Its order can be reduced with the substitution $p = d\eta/d\mu$, which gives

$$\eta = \frac{1}{4}\left[1+(p)^2\right]^{-\frac{3}{2}}\frac{dp}{d\mu}. \qquad (B.5)$$

If both sides of Eq. (B.5) are multiplied by $d\eta$, and the substitution of $p = d\eta/d\mu$ is made on the RHS, the variables separate, and the equation can be integrated as

$$\int \eta \, d\eta = \frac{1}{4}\int \frac{p}{\left[1+p^2\right]^{\frac{3}{2}}} dp, \; (\eta \le 0). \qquad (B.6)$$

which yields

$$\frac{\eta^2}{2} = \frac{1}{4}\frac{-1}{\sqrt{1+p^2}} + C. \qquad (B.7)$$

When $\eta \to 0$, the slopes $p \to 0$ also. Thus, the constant of integration, C, introduced on the RHS of Eq. (B.7) must be $+1/4$ to satisfy the groove shape far from the root.

The second boundary condition is specified through the slope of the groove at the root, $p = \tan(\pi/2 - \Psi/2)$. The equilibrium dihedral angle of the groove, Ψ, is of course, established by the 'pull' of the grain boundary against the two solid–liquid interfaces at the root, or triple junction, $\mu = 0$. See again Fig. B.1.

Introducing the slope of the groove, p, into Eq. (B.7) yields the groove depth, $h(\Psi)$, which is the η-coordinate of the profile at the root location, $\mu = 0$. Specifically, this substitution yields,

$$h(\Psi) = \eta_0 = -\sqrt{\frac{1}{2}\left(1 - \cos\frac{\pi - \Psi}{2}\right)}. \tag{B.8}$$

If the variables are again separated by solving for p, the differential equation, now first-order and non-linear, may be integrated as,

$$\int_0^\mu d\mu' = \int_{\eta_0}^\eta \left[\left(\frac{1}{1 - 2\eta'^2}\right)^2 - 1\right]^{-\frac{1}{2}} d\eta'. \tag{B.9}$$

The lower limits placed on the integrals in Eq. (B.9) are determined by the dihedral angle boundary condition at the groove root given in Eq. (B.8). The general solution obtained for the *symmetric* groove profile in scaled coordinates is found to be

$$\mu = \frac{1}{2}\ln\left(\frac{1 - \sqrt{1 - \eta^2}}{\eta}\right) - \sqrt{1 - \eta^2} + \frac{\sqrt{2}}{2} + \frac{1}{2}\ln\left(\tan\frac{\pi - \Psi}{4}\right) + \cos\frac{\pi - \Psi}{2}. \tag{B.10}$$

This solution agrees with that found by Bolling and Tiller [1]. Grain boundary groove profiles predicted with Eq. (B.10) are plotted isometrically in Fig. B.2 for several values of the dihedral angle, Ψ. An analysis of the shape of an experimental groove profile allows determination of its physical root depth, H, which can be used to determine the interfacial energy from the expression,

$$H(\Psi) = \sqrt{\frac{2\gamma_{s\ell}\Omega}{G\Delta S_f}\left(1 - \cos\frac{\pi - \Psi}{2}\right)}. \tag{B.11}$$

Grain boundary groove depths are typically much smaller than a hundred microns—ordinarily only 0.1–1 μm, making most grooves difficult to observe and measure. However, where control of the thermal field allows sufficient reduction of the applied steady-state thermal gradient, a stationary grain boundary groove profile can attain sufficiently large size that they may be imaged and analyzed quantitatively.

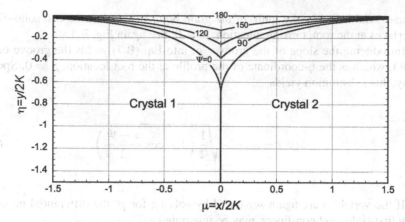

Fig. B.2 Isometric plot of Eq. (B.10) giving the true dimensionless shapes, $\mu(\eta)$, of symmetric grain boundary grooves with dihedral angles of $\psi = 0$, 90°, 120°, 150°, and 180°, the latter in the limit of a vanishingly weak grain boundary groove. The depth of a groove at $\mu = 0$ has a maximum dimensionless depth of $\sqrt{2}/2$. In general, the dimensionless groove depth is given by the relationship shown in Eq. (B.8). The corresponding real-space coordinates, x-y, and the dimensional groove depth, H, may be recovered by multiplying μ, η, and h by the capillary length scale, $2K$, which is inversely proportional to the square root of the applied temperature gradient. Thus, grain boundary grooves shrink under the influence of increasing thermal gradients

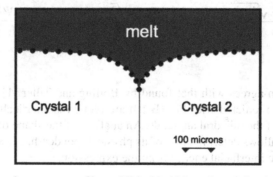

Fig. B.3 Grain boundary groove profile established in high-purity succinonitrile, under a steady temperature gradient of 0.054 K/cm. Profiles were measured for many gradient values—covering an order-of-magnitude— causing the size and depth of the groove to vary. The crystal-melt energy, $\gamma = 8.9 \times 10^{-3}$ J/m², was determined by plotting the square of the observed groove depth, H^2, shown here as about 150 μm deep, against the reciprocal of the temperature gradient, $1/G$. A straight line regressed from these data yields the ratio of the interface energy to the entropy of fusion per unit volume, in accord with Eq. (B.11). A few points from Eq. (B.10) are shown plotted along the observed interface to demonstrate their fitting of the experimental grain boundary groove profile. Adapted from [2]

Figure B.3 shows an example of symmetric high-angle grain boundary groove photographed on the solid–liquid interface in a transparent material.[1] The spatial resolution shown here was sufficient to allow determination of this material's solid–liquid energy to ±10% [2].

References

1. G.F. Bolling and W.A. Tiller, *J. Appl. Phys.*, **31** (1960) 1345.
2. R.J. Schaefer, M.E. Glicksman and J.D. Ayers, *Phil. Mag.*, **32** (1975) 725.

---

[1] This grain boundary groove occurred on the solid–liquid interface of high-purity succinonitrile (SCN). This material, and accurate knowledge of its interfacial energy, as determined from analysis of its grain boundary groove shapes (See Fig. B.3), proved to be critically important for a series of microgravity experiments on dendritic growth carried out by NASA in the 1990s.

Appendix C
Deterministic Simulation of Dendritic Growth

Background

This Appendix provides an outline of the analytic and computational techniques employed in simulating deterministic dendritic solidification kinetics, especially the generation and development of side-branches. Conventional theory for dendrite formation is fully outlined in Chapter 13.

The author, with Lowengrub and Li, recently developed a deterministic dendritic branching model that depends on non-linear interactions between the interface's shape anisotropy and anisotropic boundary conditions for the interface energy and/or its mobility [1, 2].

Lowengrub and Li had developed a sharp interface model that simulates Hele-Shaw bubble dynamics and dendritic growth in two- and three-dimensions using boundary integrals. Their numerical results reveal that both anisotropic capillarity and anisotropic interfacial attachment kinetics can induce time-periodic, non-monotone temperature distributions along the dendritic interface. When non-monotone temperatures arise, so do oscillations of the tip velocity. The consequent solidification process, operating as a dynamical 'limit cycle', generates a sequence of time-periodic protuberances near the tip. These protuberances grow and develop into secondary side-branches, thereby setting the all-important scales of thermal redistribution and chemical microsegregation.

Unlike conventional explanations of dendritic side branching, which are based on extrinsic influences, such as linear stability and selective noise-amplification [3, 4], Glicksman et al. found instead that the generation and development of side-branches during dendritic growth appears to be intrinsic to the solidification process, as they arise from deterministic dynamics at the solid–liquid interface.[1]

---

[1] A recent paper published by B. Echebarria, A. Karma and S. Gurevich, *Phys. Rev. E* **81** (2010) 021608, claims to have observed deterministic dendritic growth using phase field simulations. These authors state, "Besides noise amplified sidebranches, we have found that there is also another branch where sidebranches behave deterministically, and are not due to the influence of noise."

Stability of Dendrites and Experiments

The conventional theory for the formation of dendritic side-branches was first proposed by Langer and Müller-Krumbhaar, and developed with other collaborators over the next decade [5–7]. These theorists postulated that the origin of dendritic side-branches derives from 'selective noise amplification' occurring near the dendrite tip, which grows just below the margin of stability. Their numerical results, which were based on a linear asymptotic analysis of a two dimensional boundary-layer solidification model with kinetic crystalline anisotropy, do indeed suggest that an initially small perturbation, added via some stochastic disturbance to the tip shape, could amplify to form side-branches.

Measurements of dendritic crystallization of NH_4Br crystals from supersaturated aqueous solutions performed by Dougherty et al. [8] provided supporting data that were apparently consistent with conventional noise amplification hypotheses. Specifically, Dougherty observed that side-branch positions on opposite sides of the dendrite were imperfectly correlated. Variations clearly appeared in both the phase and amplitude of the side branches [9]. These observations helped to support the notion of stochastically-induced dendritic side branching, although the origin of the noise in Dougherty's experiment was unknown as to its intensity and spectral characteristics. It is accurate to state that dendritic side-branching is almost universally believed to arise as a kind of forced enhancement of naturally occurring disturbances acting on the solid-liquid interface.

This body of theory and experiment notwithstanding, an entirely different approach also explains the important interplay of interfacial capillarity and diffusion in dendritic pattern formation. The length scales uncovered with this new approach, curiously, also devolve from Ivantsov's transport theory, and interfacial capillarity; but, in contradistinction to convention, noise amplification plays no role in the simulations presented in Section 13.9. The new length scales connected with capillarity and kinetics emerge directly and deterministically from the thermodynamic or kinetic interface boundary condition, and then influence and interact with the growth dynamics to produce periodic oscillations through a 'limit cycle' mechanism, which, in turn, support the growth of dendritic side branches.

Branching Dynamics

Martin and Goldenfeld investigated the possible existence of a deterministic mechanism for the formation of side-branches during Hele-Shaw viscous fingering within the mathematical framework of a modal analysis of a linear stability operator [10]. They explored several possible reasons for the generation of side-branches, leading to a dendritic appearance of the fluid fingers, such as the occurrence of a limit cycle arising from a Hopf bifurcation, or by solvability-induced side-branching [10]. Moreover, their theoretical analysis emphasizes the importance of nonlinear effects, and suggests that a combination of both nonlinear dynamics and the singular nature of the dendritic steady state might be responsible for the surprising appearance of side-branching in (normally) non-branched Hele-Shaw bubble fingers [10].

Ben Jacob, in his review of non-linear pattern formation [11], mentions the peculiarity of a non-monotone temperature distribution. Almgren et al. also investigated the effect of surface tension anisotropy on the development of viscous fingerings in a Hele-Shaw cell [12]. It was, however, the clear-cut experimental results by Couder et al. that showed that nonlinear effects definitely play an important role in the generation of such fluid dynamic 'side-branches' [13, 14]. Couder attached a small air bubble to the tip of a growing fluid finger as it developed in a Hele-Shaw cell. The nonlinear interaction between the bubble and the fluid, surprisingly, gave rise to the generation of well-correlated, periodic 'side-branches' on either side of the flattened bubble finger. Couder's observations immediately suggested that noise, per se, might not be an essential physical effect, at least in his fluid dynamics experiment.

Similar time-periodic phenomena were also observed to occur in liquid crystal systems. Specifically, Borzsonyi et al. [15] demonstrated that by applying a nonlocal periodic force (e.g., time-periodic pressure or heat pulses) in the vicinity of a liquid crystal dendrite tip, the resulting tip velocity became measurably oscillatory, and the side-branches became almost perfectly periodic. Their study again suggests that noise amplification might not be an important factor here either.

Kondic et al. [16–18], experimenting with non-Newtonian fluids, found that a shear-rate dependent viscosity of the driven fluid significantly influences pattern formation in a Hele-Shaw cell. In particular, these investigators found that shear thinning suppresses tip-splitting, and produces viscous fingers that grow in an oscillatory manner, periodically developing side-branches behind their tips.

Recently, Pelce revisited the key questions surrounding fluctuations occurring near the tip of a dendrite [19]. Pelce states in his concluding remarks that,

> This problem [side branch formation] has to be considered again in connection with experiments on viscous fingering, which received explanation in a pure deterministic theory of curved interface motion.'

These independent experimental results, albeit all involving fluid dynamics, not solidification, nonetheless point towards the idea that an oscillatory tip velocity of deterministic origin might also be the hallmark of dendritic side-branches. Although considered theoretically in Pieters and Langer's work [7], branching was ascribed as having a stochastic origin.

Simulations by Lowengrub and Li

Two- and three-dimensional boundary integrals are presented in which time and space rescalings are implemented such that the total area or volume of the dendrite appears unchanged during growth. In other words, the detailed dynamics of the evolving dendrite tips, subject to anisotropic interface kinetics, were simulated, avoiding certain difficulties attending the overall net growth of the crystal. More specifically, these numerical results reveal that the tip speed and interface shape in certain parameter ranges develop periodic, non-monotone temperature distributions. These temperature distributions lead to measurable oscillations of the *scaled* tip velocity. Oscillations in tip speed would normally be undetectible in the true (unscaled) tip velocity. The underlying dynamical process uncovered by this scaling approach acts as a limit cycle, and the periodic tip speed variations generate a time-periodic sequence of interfacial protuberances located close to the dendrite tip that grow rapidly aft of the tip to form side-branches. Unlike the process of conventionally accepted noise-amplification [4], the formation of deterministic side-branches via a limit cycle is intrinsic to the interface dynamics, and can occur without the presence of any extrinsic noise.

Governing Equations

Consider a crystal growing quasi-statically in a supercooled pure melt. The interface, Σ, separates the solid phase, Ω_s, from the melt, Ω_ℓ. Assume, for simplicity, that the surface tension along the interface is isotropic, c.f. [1], but that the interfacial kinetic coefficient, $\epsilon(\mathbf{n})$ is fourfold anisotropic, i.e.,

$$\epsilon(\mathbf{n}) = \epsilon_0 \left(1 - \beta \left(3 - 4 \left(n_1^4 + n_2^4 + n_3^4\right)\right)\right), \tag{C.1}$$

where n_i denotes the components of the normal vector, $\mathbf{n}$, on the dendritic interface. In two dimensions, this kinetic form reduces to $\epsilon(\theta) = (1 + \mu \cos(4\theta))$, where θ is the angle between the normal vector and the growth axis, and μ represents the strength of the crystalline kinetic anisotropy. Again, for simplicity, the thermal diffusivities of the crystal and melt are assumed to be equal.

The length chosen for spatial re-scaling is the equivalent radius of the crystal (radius of a sphere or circle having the same volume/area) at time $t = 0$, where the time scale is the characteristic surface tension relaxation time scale [20, 21]. The following non-dimensional equations describe the growth of the dendritic crystal: The temperature field is harmonic (quasi-static) in both phases, so Laplace's equation, $\nabla^2 T_i = 0$, applies within the phase volumes ($i = s, \ell$). The interface velocity, v, is determined from the Stefan energy balance on the boundary, Σ, namely, $v = (\nabla T_s - \nabla T_\ell) \cdot \mathbf{n}$.

The temperatures of the crystal and melt match across the solid–liquid interface, respecting both capillarity and anisotropic interfacial attachment kinetics, thus,

$$T_s = T_\ell = -\kappa - \epsilon(\mathbf{n})v, \quad \text{on } \Sigma, \tag{C.2}$$

where κ is the interface's mean curvature. In addition,

$$J = \frac{1}{2\pi(N-1)} \int_\Sigma v \, d\Sigma, \tag{C.3}$$

with the quantity J representing the integral far-field heat flux that specifies the time derivative of the total volume/area of the solid phase, and $N = 2, 3$ is the dimension of space.

The flux in each space dimension is given by $J = C \cdot R(t)^{N-2}$, where R is the equivalent radius at time t, and C is a constant. The solid–liquid interface Σ evolves pointwise via the kinematic relation

$$\mathbf{n} \cdot \frac{d\mathbf{x}}{dt} = v_n \text{ imposed on } \Sigma, \tag{C.4}$$

where v_n is the normal component of the velocity of the interface, and $\mathbf{n}$ is the unit normal directed outwards towards the melt volume, Ω_ℓ.

Equation (C.2) implements the standard anisotropic kinetic boundary condition at the solid–liquid interface. The anisotropic kinetic coefficient, $\epsilon(\mathbf{n})$, reflects the underlying crystallographic orientation (fourfold cubic symmetry) and represents the rate of free energy consumption to achieve a non-zero (finite) rate of molecular attachment from the melt to the crystalline phase. An anisotropic interfacial kinetic coefficient causes the moving interface to depart from its local equilibrium temperature by an amount that is linearly proportional to its normal velocity, v_n. Dendritic growth directions are known to follow preferred crystallographic directions, such as $\langle 100 \rangle$ or $\langle 111 \rangle$ in cubic metals, to minimize the rate of free energy storage in the solid–liquid interface [22].

As the temperature fields in both solid and liquid phases are harmonic (obeying Laplace's equation), the temperature may be represented as a single-layer potential. This approach yields Fredholm integral equations of the second-kind [23] for both the velocity field, $\mathbf{v}(\mathbf{x})$, and far-field temperature in the melt, $T_\infty(t)$. Specifically, the following boundary integral representations are obtained,

$$-\kappa(\mathbf{x}) - \epsilon(\mathbf{n})v = \int_\Sigma G_N(\mathbf{x} - \mathbf{x}')v(\mathbf{x}')d\Sigma(\mathbf{x}') + T_\infty, \tag{C.5}$$

and finally,

$$J = \frac{1}{2\pi(N-1)} \int_\Sigma v(\mathbf{x}')d\Sigma(\mathbf{x}'), \tag{C.6}$$

where $G_2(\mathbf{x}) = \frac{1}{2\pi} \log |\mathbf{x}|$ and $G_3(\mathbf{x}) = \frac{1}{4\pi} 1/|\mathbf{x}|$ are the applicable Green's function kernals for Laplace's equation in two and three spatial dimensions, respectively.

Rescaling of Space and Time

In order to simulate accurately and efficiently the nonlinear dynamics of the evolving dendrites, boundary integrals as described above were computed with time and space rescaling. The combined algorithm is capable of accurately separating the dynamics of dendritic growth from the computed morphology change. This separation allows a more precise exposure of the key parameters underlying sidebranching, viz., interface velocity and temperature. Full details of the combined numerical scheme are available elsewhere, where the algorithm was originally developed to study the long-time dynamics of growing crystals [24–26], solid tumor growth [27], and Hele-Shaw viscous fingering patterns [21].

It will suffice to state, without detailed elaboration, that to evolve the interface numerically, the rescaled equations were discretized in space and solved efficiently using GMRES [28]. Rescaled field equations were discretized in two-dimensional space using spectrally accurate numerical techniques [29]. The resulting discrete system was solved in Fourier space using a diagonal preconditioner [29, 30]. In the case of three-dimensional calculations, the evolving dendrite surface was discretized using the adaptive surface-triangulated meshes developed by Cristini et al. [31]. The discretized equations for three-dimensions were then solved using the GMRES algorithm with a diagonal preconditioner. Finally, the mean curvature of the dendrite was approximated by using a least-squares parabolic fit of the enmeshed solid–liquid interface [32].

Once the re-scaled interface velocity field was obtained, the interface was advanced by applying a second-order, accurate, non-stiff updating scheme in time, with an equal arc-length parameterization in two dimensions [29, 30], or an explicit second-order Runge–Kutta method in three dimensions [26, 27].

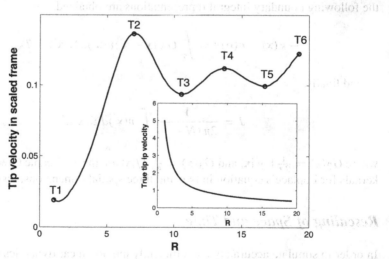

Fig. C.1 Dendrite tip velocity (scaled) versus tip radius (scaled) corresponding to simulations in 3-dimensions shown, in Figs. 13.22 and 13.23. The alternating minima and maxima in the tip speed are imperceptible in the unscaled velocity shown in the *inset*. The *inset* shows the actual tip speed in real space and time. Each of the slight slowing-down events after times labeled *T*2 and *T*4 stimulates a curvature reversal in the interface behind the tip that eventually develops into a bulge that forms a side branch. The formation of each branch is synchronized with the onset of a non-monotone interface temperature distribution, as shown in Fig. C.2

Simulation Results

The boundary integral dynamic solver provides both the interface velocity and temperature, besides revealing the noise-free interface shape evolution from a specified starting shape. A sequence of images for dendrite formation was exhibited earlier in Figs. 13.22 and 13.23, where the starting shape was a sphere (zero initial shape anisotropy) and the anisotropic mobility was chosen for a cubic crystal. As shown in Fig. C.1, the scaled velocity reveals faint oscillations 'hidden' within the actual (unscaled) tip velocity, as the dendritic shape evolves from its initial spherical form. These tip speed oscillations are not visible in the unscaled velocity inset in Fig. C.1.

If 'snap shots' of the local interface temperature distribution are taken at the extrema of the velocity plot, one observes that a pair of peaks is added when the interface slows slightly, i.e., after the tip velocity reaches and passes each maximum at times $T2$, $T4$, $T6$, etc. Comparison of these temperature distributions in Fig. C.2 clearly shows the correlation between slowing of the leading tip and the immediate appearance of a new pair of temperature peaks. The correlation of the onset of non-monotone temperature and oscillations in tip velocity and shape suggest, but do not prove, causal action. That is, when the interfacial shape anisotropy interacts nonlinearly with the anisotropy in interface energy or mobility it initiates a dendritic side-branch.

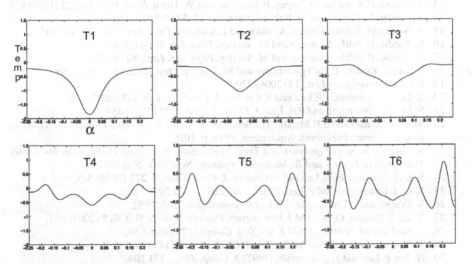

Fig. C.2 Temperature distributions corresponding to the velocity plots shown in Fig. C.1. The temperature distributions cover a small angular range, $\alpha = \pm 15°$, about the axial growth direction, $\langle 111 \rangle$. Each panel is tagged in time according to the appearance of extrema in the scaled tip velocity. The monotone temperature distribution on the *upper row*, $T1$, is that for the initial sphere, which lacks shape anisotropy. Non-monotone behavior initiates at $T3$, where two weak peaks are seen, and then again at $T4$, where four peaks become visible. Additional pairs of peaks appear periodically, each time the tip speed slows, and new branches form behind the tip. The perfect correlation of the non-monotonic interface temperature behavior with the tip velocity and shape pulsations suggests operation of a limit cycle

More study of these interesting interface interactions is clearly needed to pinpoint the fundamental cause of what appears to be a dynamic limit cycle that produces a deterministic eigenfrequency. Eventually, appropriate quantitative experiments will be needed to verify the physical mechanism that will be teased from theory. Only then will a truly robust theory be available for this important crystal growth form.

References

1. M. Glicksman, J. Lowengrub and S. Li, *Proceedings in Modelling of Casting, Welding and Advanced Solidification Processes XI*, C.A. Gandin and M. Bellet, Eds., The Minerals, Metals and Material Society, Warrendale, PA, (2006) p. 521.
2. M.E. Glicksman, J. Lowengrub and S. Li, *J. Metals*, **59** (2007) 27.
3. W.W. Mullins and R.F. Sekerka, *J. Appl. Phys.*, **34** (1963) 323.
4. J.S. Langer, *Rev. Mod. Phys.*, **52** (1980) 1.
5. J.S. Langer and H. Müller-Krumbhaar, *J. Cryst. Growth*, **42** (1977) 11.
6. J.S. Langer and H. Muller-Krumbhaar, *Acta Metall.*, **26** (1978) 1681.
7. R. Pieters and J. Langer, *Phys. Rev. Lett.*, **56** (1986) 1948.
8. A. Dougherty, P. Kaplan and J. Gollub, *Phys. Rev. Lett.* **58** (1987) 1652.
9. A. Karma and W. Rappel, *Phys. Rev. E*, **60** (1999) 3614.
10. O. Martin and N. Goldenfeld, *Phys. Rev. A* , **35** (1987)1382.
11. E. Ben Jacob and P. Garik, *Nature*, **343** (1990) 523.
12. R. Almgren, W. Dai and V. Hakim,*Phys. Rev. Lett.*, **71** (1993) 3461.
13. Y. Couder, O. Cardoso, D. Dupuy, P. Tavernier and W. Thom, *Euro. Phys. Lett.*, **2** (1986) 437.
14. Y. Couder, N. Gerard and M. Rabaud, *Phys. Rev. A*, **34** (1986) 5175.
15. T. Borzsonyi, T. Toth-Katona, A. Buka and L. Granasy, *Phys. Rev. E*, **62** (2000) 7817.
16. L. Kondic, P. Palffy-Muhoray and M. Shelley, *Phys. Rev. E*, **54** (1996) 4536.
17. L. Kondic, P. Palffy-Muhoray and M. Shelley, *Phys. Rev. Lett.*, **80** (1997) 1433.
18. P. Fast, L. Kondic, P. Palffy-Muhoray and M. Shelley, *Phys. fluid*, **13** (2001) 1191.
19. P. Pelce, *Europhysics Lett.*, **75** (2006) 220.
20. S. Li, J. Lowengrub, J.P. Leo and V. Cristini, *J. Cryst. Growth*, **267** (2004) 703.
21. S. Li, J.S. Lowengrub and P.H. Leo, *J. Comput. Phys.*, **225** (2007) 554.
22. M.E. Glicksman and S.P. Marsh, 'The dendrite', *Handbook of Crystal Growth*, Ed. D. Hurle, Elsevier Science Publishers, Amsterdam, 1993, p. 1075.
23. S. Mikhlin, *Integral Equations and Their Applications to Certain Problems in Mechanics, Mathematical Physics and Technology*, Pergamon, New York, NY, 1957.
24. S. Li, J. Lowengrub, P. Leo and V. Cristini, *J. Cryst. Growth* **277** (2005) 578.
25. S. Li, J. Lowengrub and P. Leo, *Physica D* , **208** (2005) 209.
26. V. Cristini and J. Lowengrub, *J. Crystal Growth*, **266** (2004) 552.
27. X. Li, V. Cristini, Q. Nie and J. Lowengrub, *Discrete Dynam. Syst. B*, **7** (2007) 581.
28. Y. Saad and M. Schultz, *SIAM J. Sci. Stat. Comput.*, **7** (1986) 856.
29. T. Hou, J. Lowengrub and M. Shelley, *J. Comp. Phys.*, **114** (1994) 312.
30. H. Jou, P. Leo and J. Lowengrub, (1997) *J. Comp. Phys.*, **131** 109.
31. V. Cristini, J. Blawzdziewicz and M. Loewenberg, *J. Comp. Phys.*, **168** (2001) 445.
32. A.Z. Zinchenko, M.A. Rother and R.H. Davis, *Phys. Fluids*, **9** (1997) 1493.

Appendix D
Directional Solidification Techniques

Background

This Appendix is included to provide a brief outline and overview of the most important techniques developed for applying directional solidification methods as laboratory- and industrial-scale processes. All these methods share the common feature that heat transfer, in the forms of radiation and conduction, is controlled using technical devices that limit the loss of heat from the solidifying system via some specified direction. This concept contrasts entirely with most ordinary casting methods that allow omnidirectional heat transfer to occur through the mold walls or to the environment.

The techniques listed below were invented and perfected by different individuals over many decades for the purposes of preparing crystalline materials with controlled grain textures. Moreover, the diverse skills and engineering knowledge required to construct and safely operate these devices at any scale should not be underestimated. The solidified products so produced contain crystal grains aligned to varying degrees: e.g., from vertically cast, semi-finished ingots with crystallites strongly textured by just a single zone axis still containing many residual tilt boundaries, to near-perfect, large single crystals, without so much as a single dislocation spoiling the perfection of its long-range order.

Bridgman Crystal Growth: Directional Casting

The reader should also appreciate that all directional freezing techniques, irrespective of their scale of application, will vary considerably depending on such basic details of the melt's chemical and physical attributes, such as its ease of oxidation, its solubility for, and chemical aggressiveness toward, container materials, its freezing temperature, vapor pressure, viscosity, etc. None of these practical aspects are included here, just the basic methodology for controlling heat transfer and achieving directional growth.

Vertical solidification of shaped castings uses a Bridgman crystal growth arrangement. This method was developed by Percy Bridgman, a Harvard University high-pressure physicist. Bridgman crystal growth, in its advanced industrial form, has proven economical for the manufacture of high-temperature components in turbo-machinery, including turbine blades and disks. Figure D.1 shows a shaped ceramic flask,[1] initially filled with the molten alloy, being lowered through the high-temperature zone provided by the surrounding furnace. The partially solidified casting is shown anchored either to a cooled chill plate, or is lowered into a liquid metal cooling bath to further enhance the internal temperature gradients within the casting. The casting can be directionally solidified either as an aligned multi-grain, as shown, or be fitted with a grain selector that allows only a single crystal to fill the entire mold flask from a pre-selected 'seed', oriented so as to allow freezing to advance in an easy growth direction. The twin arrow at the bottom of the figure shows the vertical heat flow direction, and the single arrow indicates the motion of the casting relative to that of the furnace.

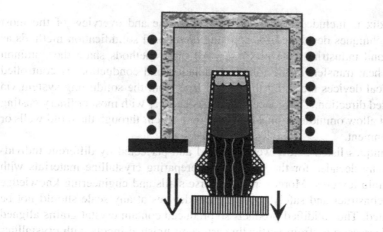

Fig. D.1 Directional solidification of a shaped casting using Bridgman crystal growth. In industry either the casting or furnace moves to 'sweep' the thermal gradients vertically through the melt. In the laboratory a small vertical furnace can also be cooled slowly to accomplish the same result. Adapted from [1]

Chalmers Method: Horizontal Directional Freezing

Professor Bruce Chalmers, while at the University of Toronto, developed a directional crystal growth method that uses a horizontal, small-bore, tubular laboratory furnace, from which a ceramic or graphite crucible, or 'boat', is slowly drawn along by a motor-driven pulley and wire system. See Fig. D.2. The twin arrow shows

---

[1] The near-net shape of directional castings, such as turbine blades, saves on finish machining costs, and reduces material waste. Those cost savings alone cover the expendable shaped ceramic mold produced by the 'lost wax' precision molding process. Directional near net-shape casting is among the most sophisticated, capital-intensive casting processes in use today.

the direction of heat flow from the hot melt (gray) towards the growing crystal (black). The horizontal orientation provokes asymmetrical convection currents in the melt, which are usually difficult to control. The single arrow shows the direction of withdrawal of the crucible from the furnace. Chalmers used this arrangement to accelerate the crystal away from its melt to decant the interface and expose the interfacial pattern. See again Section 9.3. Cooling the furnace slowly also allows directional crystal growth without relative motion between the crucible and furnace, but the growth rate would not be constant, as it could by using a gear or stepper motor to pull the crucible from the furnace mechanically.

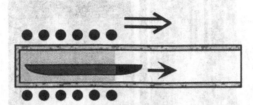

Fig. D.2 Directional solidification of a horizontal crystal using Chalmers method. This method is especially convenient in the laboratory, where small-bore tubular furnaces can be used. Heat is shown flowing to the right from the hot melt (*gray*) toward the growing crystal (*black*) Adapted from [1]

Czochralski Crystal Growth

Czochralski, or CZ, crystal growth is the basis for many industrial-scale 'bulk' crystal growth processes. The system, shown in Fig. D.3 was invented by Jan Czochralski, in 1916. The crystal grows from a suspended seed, which is slowly withdrawn

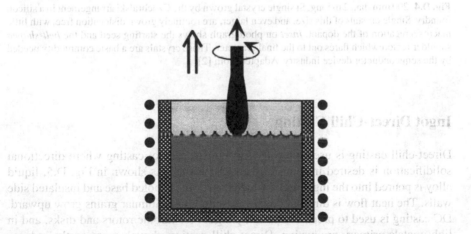

Fig. D.3 Czochralski crystal growth method. The crystal is suspended from a pull rod and a starting seed, and slowly withdrawn from the melt reservoir, which is contained in a large crucible heated by induction. Often the crystal is rotated during growth to yield a more uniform cylindrical shape and dopant concentration. Heat is withdrawn through the crystal. Adapted from [1]

from the melt, which provides a large volume of material capable of producing extremely large crystals. Figure D.4 shows a large Si crystal just removed from its Czochralski growth apparatus. Crystals of Si over 300 mm in dia. are routinely produced in silicon foundries for the electronics industry.

Fig. D.4 200 mm dia., 2 m long, Si single crystal grown by the Czochralski arrangement in a silicon foundry. Single crystals of this size, and even larger, are routinely grown dislocation free, with little macrosegregation of the dopant. *Inset* on photograph shows the starting seed and the *bell-shaped* shoulder region which flares out to the final diameter. These crystals are a basic commodity needed by the semiconductor device industry. Adapted from [2]

Ingot Direct-Chill Casting

Direct-chill casting is used typically for industrial ingot casting where directional solidification is desired in semi-finished cast forms. As shown in Fig. D.5, liquid alloy is poured into the ingot mold, which has a water-cooled base and insulated side walls. The heat flow is directed downwards, and the columnar grains grow upward. DC casting is used to prepare superalloy ingots for turbine rotors and disks, and in light metals primary production. Direct-chill casting, though more costly, reduces certain ingot defects, such as 'A-segregates' and corner cracking. These defects tend to form in conventionally cast ingots, along the join where radially inward freezing grains meet vertically freezing grains and trap excess amounts of impurities.

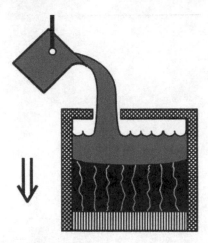

Fig. D.5 Direct-chill ingot casting into a bottom chill mold. The side walls are insulated, which directs most of the heat (*twin arrow*) towards the water-cooled chill plate at the bottom. Grains grow primarily in the vertical direction, with their rate of solidification slowing as the solid–liquid interface moves away from the chill plate. Adapted from [1]

References

1. M. Rettenmayr and H.E. Exner, 'Directional Solidification of Crystals', in *Encyclopedia of Materials: Science and Technology*, Elsevier Science Ltd., Amsterdam, ISBN: 0-08-0431526, 2000.
2. J. Page, *Smithsonian Mag.*, **30** (2000) 36–46.

Appendix E
Hunt–Jackson Theory of Eutectics

Background

Tiller's 1958 theory of coupled eutectic growth, fully discussed in Section 2.6, was generalized and improved by a more detailed, and complicated, theory developed subsequently by Hunt and Jackson [1]. Their approach, now known as 'Hunt–Jackson' eutectic theory, was to solve the 2-dimensional, coupled, solute diffusion problem ahead of a macroscopically planar eutectic front, thereby requiring fewer approximations and mathematical simplifications than did Tiller, who solved only the interfacial solute mass balance (i.e., Fick's 1st law), but not the actual eutectic diffusion field (i.e., Fick's 2nd law). In addition, Hunt and Jackson included in their theory more realistic assumptions concerning the local (microscopic) geometry of an aligned lamellar eutectic interface than did the earlier treatment. The key assumptions employed by Hunt and Jackson in developing this theory are:

- The system is a type-I, lamellar eutectic.
- Eutectic phase reactions need not be symmetrical.
- Solute transport occurs by diffusion through the melt.
- The interface remains macroscopically planar for the purposes of calculating the eutectic diffusion field, and advances at a steady state. See Fig. E.1 for geometric details, and Fig. E.2 for the nomenclature needed for the metastable extension of an eutectic phase diagram.
- The $\alpha/\beta/\ell$ interfaces are microscopically scalloped by a succession of periodic triple junctions, where the α/β boundaries intersect the solid-liquid interface and form grooves. See again Fig. 16.15 for a comparison of this theoretical structure with microscopic observations.
- Diffusion of solute is ignored in the solid-state.

Figure E.1 shows the plane-front eutectic interface geometry chosen by Hunt and Jackson for the purposes of solving the solute diffusion equation. The nomenclature used here in developing their theory is defined by the metastable extensions on the eutectic phase diagram, Fig. E.2. As discussed earlier in Section 16.2.6, some super-

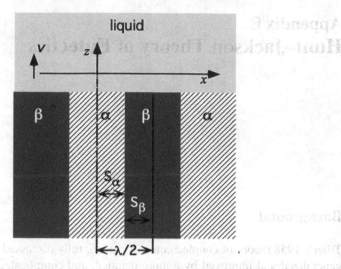

Fig. E.1 Geometry and coordinate systems at a macroscopically planar eutectic interface used by Hunt and Jackson to solve the diffusion problem in the melt

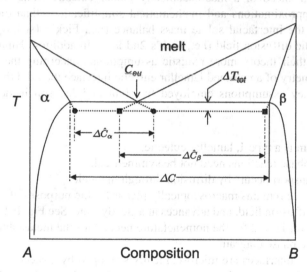

Fig. E.2 Metastable extensions for a binary eutectic, indicating the nomenclature used in developing Hunt and Jackson's diffusion solution. Here the eutectic interface is supercooled by a small amount, ΔT_{tot}, which is sufficient to provide the free energy needed for solute diffusion and interfacial energy storage

cooling is needed at the eutectic interface to provide both the free energies dissipated by the solute diffusion process and that stored in the lamellar eutectic microstructure as interfacial energy.

Diffusion Solution

A mathematical solution is required for the steady-state diffusion equation, applied here in the melt region immediately ahead of the advancing eutectic interface. The diffusion equation, written in 2-dimensional Cartesian coordinates (x, z), co-moving with the eutectic interface, is given by Fick's 2nd law, which is the the partial differential equation,

$$\nabla^2 C_\ell + \frac{v_{eu}}{D_\ell} \frac{\partial C_\ell}{\partial t} = 0. \tag{E.1}$$

The boundary conditions for the eutectic diffusion solution are as follows:

i. Solute concentration in the melt far from the solid–liquid interface: $C_\ell(z) \rightarrow C_0$, as $z \rightarrow \infty$.
ii. Zero concentration gradient (no solute diffusion) conditions apply at each mirror symmetric position: $\partial C_\ell / \partial x = 0$, at $x = \pm n\lambda/2$, $(n = 0, 1, 2, \cdots \infty)$.
iii. Stefan solute balances apply individually ahead of each of the α/melt and β/melt interfaces.

The general two-dimensional solution, $C_\ell(x, z)$, that satisfies Eq. (E.1) and the boundary conditions may be written as the following concentration field in the melt,

$$C_\ell(x, z) = C_0 + \sum_{m=0}^{\infty} B_m \cos \left(\frac{2m\pi x}{\lambda} \right) \times \tag{E.2}$$

$$\exp \left[z \cdot \left(-\frac{v_{eu}}{2D_\ell} - \sqrt{\left(\frac{v_{eu}}{2D_\ell} \right)^2 + \left(\frac{2m\pi}{\lambda} \right)^2} \right) \right], \quad z \geq 0.$$

Hunt and Jackson simplified Eq. (E.2) by assuming that the phase spacing, λ, is much smaller than the so-called 'characteristic diffusion distance', so we note that $\lambda \ll D_\ell / v_{eu}$. This assumption is based on the observation that the lamellar spacing, λ, along the x-direction, occurs on a much smaller scale than does the solute boundary layer thickness in the z-direction. It allows the first term appearing under the square-root in Eq. (E.2) to be dropped, so that

$$\left(\frac{2m\pi}{\lambda} \right)^2 \gg \left(\frac{v_{eu}}{2D_\ell} \right)^2, \quad m \neq 0. \tag{E.3}$$

The inequality expressed in Eq. (E.3) allows an approximation of the steady-state eutectic diffusion solution in the melt to be written as

$$C_\ell(x, z) \cong C_0 + B_0 e^{-\frac{v_{eu}}{D_\ell} z} + \sum_{m=1}^{\infty} B_m \cos \left(\frac{2m\pi x}{\lambda} \right) e^{-\frac{2m\pi}{\lambda} z}. \tag{E.4}$$

A two-dimensional plot of the solute concentration field in the melt, $C_\ell(x, z)$, based on Hunt and Jackson's diffusion solution, Eq. (E.4), is shown rendered in Fig. E.3. This figure shows realistically how oscillations in the solute concentration field alternate along with the different phase lamellae, but decrease in amplitude with distance away from the eutectic interface.

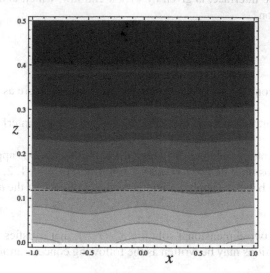

Fig. E.3 Two-dimensional field plot of Eq. (E.4), the solute concentration, $C_\ell(x, z)$, ahead of a symmetric eutectic with $\lambda = 1$. Contours show how oscillations in the solute concentration field diminish with distance from the interface, z. Beyond about $z = 0.4$ the melt is virtually homogeneous. *Lighter gray shades* represent higher solute concentration differences in the melt from its average value, C_{eu}, (*black*). Note how the *dashed horizontal line*, $z = const.$, near the interface encounters alternating *lighter* and *darker shades of gray*, depending on whether it is ahead of α-lamellae or β-lamellae, respectively

Stefan Solute Balances

The solute concentration gradient ahead of the eutectic interface is found by differentiating Eq. (E.4) with respect to z. This step yields the z-component of the concentration gradient, which is oscillatory in the x-direction, and exponentially damped in the z-direction.

$$\frac{\partial C_\ell(x, z)}{\partial z} = -\frac{v_{eu}}{D_\ell} B_0 e^{-\frac{v_{eu}}{D_\ell} z} - \sum_{m=1}^{\infty} B_m \frac{2\pi m}{\lambda} \cos \frac{2\pi m x}{\lambda} e^{-\frac{2\pi m}{\lambda} z}. \qquad \text{(E.5)}$$

The solute concentration gradient, Eq. (E.5), may be evaluated at the eutectic interface, by setting $z = 0$,

$$\left(\frac{\partial C_\ell(x, z)}{\partial z}\right)_{z=0} = -\frac{v_{eu}}{D_\ell} B_0 - \sum_{m=1}^{\infty} B_m \frac{2\pi m}{\lambda} \cos \frac{2\pi m x}{\lambda}. \tag{E.6}$$

Solute flux conservation in the melt may be applied at steady-state at the α and $\beta/melt$ interfaces by applying Fick's 1st law of diffusion. The diffusion fluxes balance the solute rejected by these advancing crystal/melt interfaces. These locally piecewise Stefan solute balances are, respectively, at the $\alpha/melt$ interface,

$$v_{eu} \Delta \hat{C}_\alpha = v_{eu} B_0 + D_\ell \sum_{m=1}^{\infty} B_m \frac{2\pi m}{\lambda} \cos \frac{2\pi m x}{\lambda}, \quad 0 \le x \le S_\alpha, \tag{E.7}$$

and at the $\beta/melt$ interface,

$$v_{eu} \Delta \hat{C}_\beta = v_{eu} B_0 + D_\ell \sum_{m=1}^{\infty} B_m \frac{2\pi m}{\lambda} \cos \frac{2\pi m x}{\lambda}, \quad S_\alpha \le x \le \lambda/2. \tag{E.8}$$

By rearranging the terms in Eqs. (E.7) and (E.8) one sees that the LHS represents the solute masses rejected by the phase lamellae, whereas the RHS are the diffusion fluxes of B-atoms flowing either from, Eq. (E.9), or to, Eq. (E.10), the advancing eutectic interface. At steady state, these solute mass flows must balance ahead of each phase as,

$$v_{eu} \left(B_0 - \Delta \hat{C}_\alpha\right) = -D_\ell \sum_{m=1}^{\infty} B_m \frac{2\pi m}{\lambda} \cos \frac{2\pi m x}{\lambda}, \quad 0 \le x \le S_\alpha, \tag{E.9}$$

and

$$v_{eu} \left(B_0 - \Delta \hat{C}_\beta\right) = -D_\ell \sum_{m=1}^{\infty} B_m \frac{2\pi m}{\lambda} \cos \frac{2\pi m x}{\lambda}, \quad S_\alpha \le x \le \lambda/2. \tag{E.10}$$

Fourier Analysis

Equations (E.9) and (E.10) contain $m + 1$ amplitudes of the solute fluxes, expressed as the unknown Fourier coefficients, $B_m s$. One applies 'orthogonality conditions' to solve for these unknown coefficients by generating a sufficient number of independent equations.[1] The formal orthogonality process involves multiplying each Fourier mode, or basis function, by an orthogonal trial function, the integral of which produce either a zero, if the trial function is orthogonal throughout the

---

[1] Orthogonality is a mathematical generalization of the projection properties of vectors into special function spaces, which include the trigonometric basis functions used in Fourier analysis.

function space, or a non-zero constant if the trial function is non-orthogonal. These orthogonality integrals generate the required $m + 1$ independent conditions. Specifically, one tests the following orthogonality integrals:

$$\frac{2\pi}{\lambda} \int_0^{\lambda/2} \cos\frac{2\pi mx}{\lambda} \cos\frac{2\pi nx}{\lambda} dx = 0, \ m \neq 0. \tag{E.11}$$

In the event that $m = n = 1$, the integral in Eq. (E.11) equals $\pi/2$, whereas if $m = n \neq 1$, then the integral equals π. Equations (E.11) represents the standard procedure used in classical Fourier analysis to solve for each of the amplitudes appearing in Eqs. (E.9) and (E.10). In principle this permits evaluating the $m + 1$ unknown B_m's. In practice, one solves for a few B_m's, and then approximates the solute diffusion solution as a finite sum of decreasing Fourier terms.

The explicit orthogonality procedure for the eutectic problem involves multiplying Eqs. (E.9) and (E.10) by trial functions, in this case $\cos(2\pi nx/\lambda)$, and then integrating the product across any *periodic* interval in x, say $\lambda/2$, which is the periodic eutectic 'half cell' chosen to represent one period of the eutectic interface shown earlier in Fig. E.1. This procedure produces the following integrals across one periodic unit of the eutectic:

$$\frac{v_{eu}}{D_\ell} \int_0^{S_\alpha} \left(B_0 - \Delta\hat{C}_\alpha\right) \cos\frac{2\pi nx}{\lambda} dx + \frac{v_{eu}}{D_\ell} \int_{S_\alpha}^{\lambda/2} \left(B_0 - \Delta\hat{C}_\beta\right) \cos\frac{2\pi nx}{\lambda} dx$$

$$\tag{E.12}$$

$$= -\int_0^{\lambda/2} \sum_{m=1}^{\infty} B_m \frac{2\pi m}{\lambda} \cos\frac{2\pi mx}{\lambda} \cos\frac{2\pi nx}{\lambda} dx.$$

The RHS of Eq. (E.12), containing the infinite Fourier series, reduces to just a single term where $m = n$. Applying this orthogonality condition gives

$$\left(B_0 - \Delta\hat{C}_\alpha\right)\frac{\lambda}{2\pi n}\left(\sin\frac{2\pi nx}{\lambda}\right)_0^{S_\alpha} + \left(B_0 - \Delta\hat{C}_\beta\right)\frac{\lambda}{2\pi n}\left(\sin\frac{2\pi nx}{\lambda}\right)_{S_\alpha}^{\lambda/2} = -\frac{\pi n}{2}B_n. \tag{E.13}$$

The concentration differences, noted on the LHS of Eq. (E.13), develop ahead of the solid phases and the melt at the moving interface (C.f. Figs. E.2 and E.3, which suggest how these concentration differences appear in the melt, and how they may be approximated as indicated on the metastable eutectic phase diagram. This result and the following approximations taken from the equilibrium phase diagram remain valid provided that the total interfacial supercooling, ΔT_{eu}, is small, i.e., a few Kelvins or less:

1. $\Delta\hat{C}_\alpha \approx \Delta C_\alpha$
2. $\Delta\hat{C}_\beta \approx \Delta C_\beta$
3. $\Delta\hat{C} = \Delta C_\alpha + \Delta C_\beta$

If the interfacial concentration differences appearing in Eq. (E.13) are replaced by their equilibrium approximations, as just itemized, one can find the values of the amplitudes as,

$$B_n \cong \frac{v_{eu}}{D_\ell} \frac{\Delta C \lambda}{(\pi n)^2} \sin \frac{2\pi n S_\alpha}{\lambda}. \tag{E.14}$$

Equation (E.14) defines every amplitude coefficient in the solute diffusion solution, excepting $n = 0$, for which a singularity would occur because n^2 appears in the denominator. The amplitude B_0, however, is the special Fourier coefficient that adjusts the arbitrary *average* bulk eutectic composition of the melt along the eutectic interface, C_0, to match the equilibrium eutectic composition, C_{eu}. This 'adjustment' of the average melt composition insures that the proper interfacial compositions, $\langle C_\alpha \rangle$ and $\langle C_\beta \rangle$, as reflected in the phase diagram, produce the correct α- and β-lamellae compositions, respectively. Equivalently, B_0 also insures conservation of solute mass during steady-state eutectic growth. Selecting the correct averages of the composition in the melt ahead of each phase insures that the diffusion solutions provides the correct overall average interfacial composition of C_{eu} and individual phase fractions as dictated by the equilibrium phase diagram.

Plane-Wave Term

The average interfacial composition ahead of the α-phase is defined directly from the diffusion solution, Eq. (E.4), by setting $z = 0$. Integrating across a single α-melt interface, $0 \le x \le S_\alpha$, for all the Fourier modes yields the interfacial concentration

$$\langle C_\alpha \rangle \equiv C_0 + B_0 + \frac{1}{S_\alpha} \int_0^{S_\alpha} \sum_{n=1}^{\infty} B_n \cos \frac{2\pi n x}{\lambda} dx. \tag{E.15}$$

A uniform composition shift occurs near the interface caused by formation of a solute boundary layer. This uniform shift is referred to as the 'eutectic plane-wave' term, as it is determined by the coefficient B_0, which is the amplitude of the solute boundary layer. If Eq. (E.14), the expression for the amplitudes, B_n ($n \neq 0$), is substituted into Eq. (E.15) one obtains

$$\langle C_\alpha \rangle \equiv C_0 + B_0 + \frac{v_{eu}}{D_\ell} \frac{\Delta C \lambda}{S_\alpha} \times \left(\sum_{n=1}^{\infty} \frac{1}{(n\pi)^2} \sin \frac{2\pi n S_\alpha}{\lambda} \int_0^{S_\alpha} \cos \frac{2\pi n x}{\lambda} dx \right). \tag{E.16}$$

Carrying out the integration indicated in Eq. (E.16) yields the average composition ahead of the α-phase, namely,

$$\langle C_\alpha \rangle \equiv C_0 + B_0 + \frac{v_{eu}}{D_\ell} \frac{\Delta C \lambda^2}{2S_\alpha} \times \left(\sum_{n=1}^{\infty} \frac{1}{(n\pi)^3} \sin^2 \frac{2\pi n S_\alpha}{\lambda} \right). \tag{E.17}$$

The infinite sum appearing in parentheses on the RHS of Eq. (E.17) can be evaluated numerically, because it depends only on the volume fraction of the α-phase in the eutectic, viz., $V_V^\alpha = 2S_\alpha/\lambda$. Defining this summation as the function $\Psi(V_V^\alpha)$, one may re-write Eq. (E.17) as

$$\langle C_\alpha \rangle \equiv C_0 + B_0 + \frac{v_{eu}}{D_\ell} \frac{\Delta C \lambda}{2S_\alpha/\lambda} \Psi(V_V^\alpha). \tag{E.18}$$

The infinite sum, $\Psi(V_V^\alpha)$, is plotted in Fig. E.4 as a symmetrical function of the α-phase volume fraction in the eutectic. This function varies over a relatively narrow range of values for lamellar eutectics. Inasmuch as the phase volume fractions of lamellar eutectics may vary from a value of about 0.3 α-phase plus 0.7 β-phase, to the other extreme of about 0.7 α-phase plus 0.3 β-phase, the corresponding sum function actually falls within a relatively narrow range, $0.025 \leq \Psi(V_V^\alpha) \leq 0.034$.

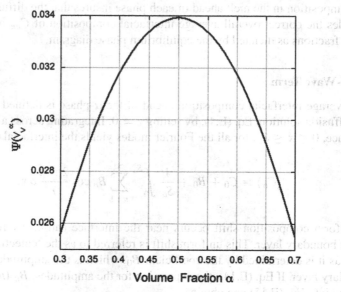

Fig. E.4 The infinite sum, $\Psi(V_V^\alpha)$, needed to determine the composition of the α-phase plotted versus volume fraction of the α-phase. Over the range shown here for volume fractions appropriate to lamellar eutectics, this sum changes about 30% from its peak value for symmetric eutectics, to the limits where the lamellar-rod transitions occurs

Arguments identical to those just presented may be used to determine the average composition ahead of the β-phase lamellae, namely

$$\langle C_\beta \rangle \equiv C_0 + B_0 - \frac{v_{eu}}{D_\ell} \frac{\Delta C \lambda}{(1 - V_V^\alpha)} \Psi(1 - V_V^\alpha). \tag{E.19}$$

Symmetric Eutectics

The special case of a symmetric eutectic, discussed in Section 16.2.6, gives $V_V = 0.5$, and $\Psi_{sym} \approx 0.034$. Symmetry of the phase diagram reduces Eqs. (E.18) and (E.19), respectively, to the simpler forms

$$\langle C_\alpha \rangle \equiv C_0 + B_0 + 2\Psi_{sym} \frac{v_{eu}\Delta C\lambda}{D_\ell}, \tag{E.20}$$

and

$$\langle C_\beta \rangle \equiv C_0 + B_0 - 2\Psi_{sym} \frac{v_{eu}\Delta C\lambda}{D_\ell}. \tag{E.21}$$

The overall average interfacial composition, $\langle C \rangle_\lambda$, for a symmetric eutectic, averaged across a periodic segment of one lamellar spacing, λ, is given by half the sum of Eqs. (E.20) and (E.21). Adding these equations yields

$$\langle C \rangle_\lambda = \frac{\langle C_\alpha \rangle + \langle C_\beta \rangle}{2} = C_0 + B_0. \tag{E.22}$$

Now a symmetric eutectic *must* have an overall composition of $C_0 = C_{eu}$. As the amplitude coefficient, B_0, acts to provide a 'plane wave' solute boundary layer that buffers the interface composition, C_0, to match the required equilibrium eutectic composition, C_{eu}, it becomes evident that for symmetric eutectics the amplitude coefficient $B_0 = 0$. That is, a plane-wave solute boundary layer is not needed when a symmetric eutectic grows from a melt, for which the composition is exactly at C_{eu}. The individual melt compositions ahead of every phase lamellae, $\langle C_i \rangle$, ($i = \alpha, \beta$), are therefore shifted slightly above and below C_{eu},

$$\langle C_i \rangle = C_{eu} \pm 0.068 \left(\frac{\lambda}{D_\ell / v_{eu}} \right) \Delta C, \ i = \alpha, \beta. \tag{E.23}$$

The interpretation of Eq. (E.23) is clear: the average steady-state solute concentrations just ahead of either phase equals the eutectic composition, C_{eu}, plus or minus an extremely small correction from the coupled diffusion field. See again Fig. E.3. This correction is about 7% of the total equilibrium tie-line concentration difference, ΔC, multiplied by the dimensionless ratio of the lamellar spacing to the diffusion distance, the latter of which is already a small number much less than unity.

Non-symmetric Eutectics

When eutectics have non-symmetric phase diagrams the zeroth amplitude coefficient, B_0 that provides the composition buffer, or solute boundary layer, can be determined, in general, from an interfacial solute mass balance. This mass balance is expressed as piece-wise continuous integrals spanning a pair of lamellae, at $z = 0$. These integrals represent the Stefan solute mass balances given previously in Eqs. (E.7) and (E.8). The integral flux condition for any Fourier mode, $n = 0, 1, 2, \ldots$, may be written as

$$v_{eu} B_0 \int_0^{\lambda/2} dx + D_\ell \int_0^{\lambda/2} \sum_{n=1}^{\infty} B_n \frac{2\pi n}{\lambda} \cos \frac{2\pi n x}{\lambda} dx = \quad \text{(E.24)}$$

$$v_{eu} \Delta \hat{C}_\alpha \int_0^{S_\alpha} dx + v_{eu} \Delta \hat{C}_\beta \int_{S_\alpha}^{\lambda/2} dx.$$

The second integral on the LHS of Eq. (E.24) that contains the infinite sum of B_n's vanishes over the periodic interval, $\lambda/2$. The integral solute mass balance may be solved for B_0 as

$$B_0 = \frac{\Delta \hat{C}_\alpha S_\alpha + \Delta \hat{C}_\beta (\lambda/2 - S_\alpha)}{\lambda/2}, \quad \text{(E.25)}$$

or, alternatively, as

$$B_0 = \Delta \hat{C}_\alpha V_V^\alpha + \Delta \hat{C}_\beta \left(1 - V_V^\alpha\right). \quad \text{(E.26)}$$

Equations (E.25) and (E.26) show that the zeroth amplitude coefficient, B_0, now represents the volume-fraction weighted departure of the melt's concentration at the eutectic interface from the eutectic composition. In contrast to the special case of symmetric eutectics, non-symmetric eutectics may have their overall composition $C_0 \neq C_{eu}$. Equation (E.26) suggests that in order for the average interfacial solute concentration to remain at C_{eu} during steady-state growth, a non-symmetric eutectic can alter the relative volume fractions of its conjugate solid phases, and, thereby, compensate for the off-eutectic overall composition. Thus, we conclude that for general (non-symmetric) lamellar eutectics, B_0 is a small quantity given by

$$B_0 = C_{eu} - C_0. \quad \text{(E.27)}$$

This general result helps explain why the overall alloy composition of a lamellar eutectic need not be precisely set at C_{eu} and yet still avoids the appearance of any pro-eutectic primary phase. This surprising aspect of off-eutectic solidification, uncovered through this analysis, expands our understanding of microstructure selection in eutectic alloys, which is indeed a complex one. The ability to induce coupled lamellar or rod eutectic solidification in directionally solidified off-eutectic alloy compositions is now well established experimentally. Processing off-eutectic compositions, yet producing coupled in-situ composites, adds greatly to one's ability to process well-designed eutectic microstructures with unique properties [2].

Capillary Effects

Hunt and Jackson next considered the influence of interfacial curvature at the tips of the lamellae, noting that on a sufficiently small scale the interface is, in fact, non-planar. Figure E.5 shows a sketch of the general situation at the eutectic triple points, where α, β, and melt are in simultaneous contact, and in local equilibrium. These fine-scale ripples on the eutectic interface were previously ignored in Section 4.1 when seeking the solute diffusion solution. Figure E.5 indicates that, θ_α and θ_β are the (complementary) dihedral angles of the eutectic phases, which form a distinct groove, or cusp, at the interface. These contact angles are, in fact, exceedingly difficult to measure accurately during eutectic growth because of the high resolution needed and the magnified motion of the interface.

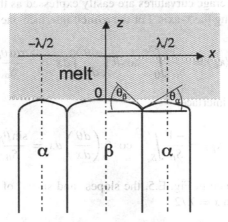

Fig. E.5 Sketch of the profiles of the phases in an unsymmetric lamellar eutectic, showing the individual contact angles and phase curvatures against the melt

Interfacial Curvatures

The in-plane curvatures, κ_i, of the α-melt and β-melt interfaces may be described in the z, x coordinate system shown in Fig. E.5. The standard Cartesian curvature formula of the two interface profiles, $\zeta_i(x)$, is

$$\kappa_i(x) = -\frac{\frac{d^2\zeta_i(x)}{dx^2}}{\left[1 + \left(\frac{d\zeta_i(x)}{dx}\right)^2\right]^{3/2}}, \quad (i = \alpha, \beta). \tag{E.28}$$

Equation (E.28) may be transformed to a more useful form by recognizing that the slopes of the interface profiles $d\zeta_i(x)/dx = \tan\theta_i$. If this equality is substituted into Eq. (E.28) one obtains

$$\kappa_i(\theta) = -\frac{\frac{d}{dx}\tan\theta_i}{\left(1 + \tan^2\theta_i\right)^{3/2}}.$$ (E.29)

Carrying out the indicated differentiation in Eq. (E.29) yields the result

$$\kappa_i(\theta) = -\cos\theta_i\,\frac{d\theta_i}{dx}.$$ (E.30)

Equation (E.30) relates the local curvatures of the eutectic interface to the dihedral angles at the eutectic triple points. The *average* curvatures developed along the α-melt interface, $\langle\kappa_\alpha\rangle$, or along the β-melt interface, $\langle\kappa_\beta\rangle$, can each be determined from the relationship between the interfacial curvatures and the associated dihedral angles, θ_i. These average curvatures are easily expressed as the following integrals taken piecewise along the x-axis. For the α/melt interface one obtains

$$\langle\kappa_\alpha\rangle = \frac{-1}{S_\alpha}\int_0^{S_\alpha}\cos\theta\left(\frac{d\theta}{dx}\right)dx = -\frac{\sin\theta_\alpha}{S_\alpha},$$ (E.31)

and for the β/*melt* interface,

$$\langle\kappa_\beta\rangle = \frac{-1}{S_\beta}\int_{S_\alpha}^{\lambda/2}\cos\theta\left(\frac{d\theta}{dx}\right)dx = \frac{\sin\theta_\beta}{S_\beta}.$$ (E.32)

Note that, as indicated in Fig. E.5, the slopes (and $\sin\theta$s) of the eutectic lamellae vanish at $x = 0$, and $x = \lambda/2$.

Gibbs–Thomson Effect

The temperature depression at the α-phase lamellae tips associated with their average curvature, $\langle\kappa_\alpha\rangle$ is given by the Gibbs–Thomson relationship, developed earlier in Section 8.3.

$$\Delta T_\kappa^\alpha = \Gamma^\alpha\frac{\sin\theta_\alpha}{S_\alpha},$$ (E.33)

where the capillary coefficient, Γ^α, for the α-phase appearing in Eq. (E.33) is defined in terms of the α/melt interfacial energy, $\gamma_{\alpha\ell}$, and the entropy of melting, ΔS_f^α, for $\alpha \rightleftharpoons \ell$,

$$\Gamma^\alpha \equiv \frac{\gamma_{\alpha\ell}\Omega}{\Delta S_f^\alpha}.$$ (E.34)

Similar arguments applied to Eq. (E.33) for the β/melt interface yields the Gibbs–Thomson temperature depression at the tips of the β-lamellae,

$$\Delta T_\kappa^\beta = \Gamma^\beta \frac{\sin \theta_\beta}{S_\beta}, \tag{E.35}$$

where the capillary coefficient, Γ^β, is defined as

$$\Gamma^\beta \equiv \frac{\gamma_{\beta\ell}\Omega}{\Delta S_f^\beta}. \tag{E.36}$$

Interfacial Isothermality

A requirement for the coupled growth of a type-I eutectic growing from a nominally planar interface is that the interface must remain *isothermal*. This requirement
suggests that the supercooling associated with the curvature of the lamellae, $\Delta \hat{T}_{\alpha\beta}$,
and the supercooling required to drive the diffusion, $\Delta \hat{T}_{diff}$ must sum to the *same*
total supercooling, $\Delta \hat{T}_{tot}$ for each phase. Figure E.6 schematically indicates this
condition of interfacial isothermality for a lamellar eutectic. The individual effects
on the α/melt interfacial supercooling arising from the processes of solute diffusion
and capillarity (the latter of which is responsible for energy storage among the α-β
interfaces) can be expressed, respectively, as

$$\Delta T_{diff}^\alpha = -m_\ell^\alpha \left[C_\ell(x, 0) - C_{eu} \right], \tag{E.37}$$

and

$$\Delta T_\kappa^\alpha = \Gamma^\alpha \langle \kappa_\alpha \rangle. \tag{E.38}$$

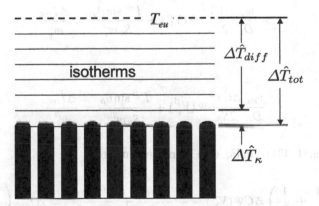

Fig. E.6 Thermal requirements in the melt near an isothermal eutectic interface. As suggested by
Hunt and Jackson [1] the supercooling required for solute diffusion, $\Delta \hat{T}_{diff}$, and the supercooling
required by the Gibbs–Thomson effect at the curved tips of the lamellae, $\Delta \hat{T}_\kappa$, must sum to the
same total supercooling for each phase, $\Delta \hat{T}_{tot}$, below their common equilibrium eutectic temperature, T_{eu}

The total interfacial supercooling developed at the tips of the α-lamellae is given by the sum of Eqs. (E.37) and (E.38),

$$\Delta \hat{T}_{tot}^{\alpha} \equiv T_{eu} - \hat{T}_{\alpha} = \Delta T_{diff}^{\alpha} + \Delta T_{\kappa}^{\alpha}. \qquad (E.39)$$

In order for the eutectic interface to be isothermal, Hunt and Jackson suggested that $\hat{T}_{\alpha} = \hat{T}_{\beta}$.

If Eqs. (E.18), (E.27), (E.33), and (E.37) are substituted into the RHS of Eqs. (E.39), and then for the β-phase lamellae similarly are substituted Eqs. (E.19), (E.27), (E.36), and (E.37), one obtains the following isothermal conditions based on total supercooling: For the α-lamellae,

$$-\frac{v_{eu}}{D_{\ell}} \frac{\Delta C \lambda^2}{2 S_{\alpha}} \Psi(V_V^{\alpha}) m_{\ell}^{\alpha} + \frac{\Gamma^{\alpha} \sin \theta_{\alpha}}{S_{\alpha}} = \Delta \hat{T}_{tot}, \qquad (E.40)$$

and for the β-lamellae,

$$\frac{v_{eu}}{D_{\ell}} \frac{\Delta C \lambda^2}{2 S_{\beta}} \Psi(V_V^{\beta}) m_{\ell}^{\beta} + \frac{\Gamma^{\beta} \sin \theta_{\beta}}{S_{\beta}} = \Delta \hat{T}_{tot}. \qquad (E.41)$$

Divide Eq. (E.40) through by the liquidus slope for the α-phase, and similarly divide Eq. (E.41) through by the liquidus slope for the β-phase. The sum functions for either phase are the same, because this function is symmetrical about $V_V^{\alpha} = 0.5$. Thus, at any volume fraction, $\Psi(V_V^{\alpha}) = \Psi(V_V^{\beta})$. See again Fig. E.4. Carrying out these steps and substitutions yields

$$-\frac{v_{eu}}{D_{\ell}} \frac{\Delta C \lambda^2}{2 S_{\alpha}} \Psi(V_V^{\alpha}) + \frac{\Gamma^{\alpha} \sin \theta_{\alpha}}{S_{\alpha} m_{\ell}^{\alpha}} = \frac{\Delta \hat{T}_{tot}}{m_{\ell}^{\alpha}}, \qquad (E.42)$$

and

$$\frac{v_{eu}}{D_{\ell}} \frac{\Delta C \lambda^2}{2 S_{\beta}} \Psi(V_V^{\beta}) + \frac{\Gamma^{\beta} \sin \theta_{\beta}}{S_{\beta} m_{\ell}^{\beta}} = \frac{\Delta \hat{T}_{tot}}{m_{\ell}^{\beta}}. \qquad (E.43)$$

Subtract Eq. (E.42) from Eq. (E.43) and one finds that

$$\frac{v_{eu}}{D_{\ell}} \frac{\lambda^2}{2} \left(\frac{1}{S_{\alpha}} + \frac{1}{S_{\beta}} \right) \Delta C \Psi(V_V^{\alpha}) - \frac{\Gamma^{\alpha} \sin \theta_{\alpha}}{S_{\alpha} m_{\ell}^{\alpha}} + \frac{\Gamma^{\beta} \sin \theta_{\beta}}{S_{\beta} m_{\ell}^{\beta}} = \Delta \hat{T}_{tot} \left(\frac{1}{m_{\ell}^{\beta}} - \frac{1}{m_{\ell}^{\alpha}} \right). \qquad (E.44)$$

The last term on the RHS of Eq. (E.44) defines the system's average (reciprocal) liquidus slope, $\langle 1/m_{\ell} \rangle$, which is a negative quantity,

$$\langle 1/m_\ell \rangle \equiv -\frac{m_\ell^\alpha - m_\ell^\beta}{m_\ell^\alpha m_\ell^\beta} \quad\quad\quad\quad\quad (E.45)$$

Substitution of this definition provided by Eq. (E.45) into Eq. (E.44), and recalling that $S_\alpha = \lambda V_V^\alpha/2$, and $S_\beta = \lambda(1 - V_V^\alpha)/2$, gives after several steps of algebra

$$-\frac{v_{eu}}{4D_\ell}\frac{\Delta C \Psi(V_V^\alpha)}{(1 - V_V^\alpha)V_V^\alpha}\lambda + \left(\frac{\Gamma^\alpha \sin\theta_\alpha}{V_V^\alpha m_\ell^\alpha} - \frac{\Gamma^\beta \sin\theta_\beta}{V_V^\beta m_\ell^\beta}\right)\frac{2}{\lambda} = \Delta\hat{T}_{tot}\langle 1/m_\ell \rangle. \quad (E.46)$$

The functional dependences of Eq. (E.46) on the lamellar spacing, λ, makes it clear that the general dependence of the total supercooling on the eutectic spacing is the same as that found using Tiller's analysis. Cf. Eq. (16.18). Note, for example, how the two terms on the left-hand side of Eq. (E.46) depend on λ and $1/\lambda$. The procedure to determine the optimal eutectic spacing, λ^*, which provides the minimum interface supercooling is also identical to that described in Tiller's theory, Section 16.2.7. The scaling law found by Hunt and Jackson relating the eutectic spacing and the interface speed is also $\lambda^{*2}v_{eu} =$ constant, but is cast in the more explicit and testable form,

$$\lambda^{*2}v_{eu} = \frac{8D_\ell}{\Delta C\Psi(V_V^\alpha)} \times \left(\frac{\Gamma^\beta \sin\theta_\beta V_V^\alpha}{m_\ell^\beta} - \frac{\Gamma^\alpha \sin\theta_\alpha(1 - V_V^\alpha)}{m_\ell^\alpha}\right). \quad\quad (E.47)$$

Finally, the scaling law for the minimum total supercooling, $\Delta T_{tot}^\star$, is found by substituting Eq. (E.47) back into Eq. (E.44). That substitution plus some addition algebra yields Hunt and Jackson's eutectic scaling law

$$\frac{\Delta T_{tot}^{\star 2}}{v_{eu}} = 2\frac{\Delta C \Psi(V_V^\alpha)}{\langle 1/m_\ell \rangle^2 D_\ell(1 - V_V^\alpha)V_V^\alpha} \times \left(\frac{\Gamma^\beta \sin\theta_\beta}{(1 - V_V^\alpha)m_\ell^\beta} - \frac{\Gamma^\alpha \sin\theta_\alpha}{V_V^\alpha m_\ell^\alpha}\right). \quad (E.48)$$

Results for Symmetric Eutectics

The best comparisons between Tiller's theory and the more general treatment of lamellar eutectics by Hunt and Jackson should be made on the basis of predictions for symmetric eutectics. The following simplifications apply in the symmetric case:

- Interfacial contact angles are equal, so $\theta_\alpha = \theta_\beta = \theta^\star$.
- Liquidus slopes are equal and opposite, so therefore $m_\ell^\alpha = -m_\ell^\beta = m_\ell$.
- The average reciprocal liquidus slope is $\langle 1/m_\ell \rangle = 2/m_\ell$.
- Volume fractions of the conjugate phases are equal, so $V_V^\alpha = V_V^\beta = 0.5$.
- As a consequence of equal volume fractions, the sum function, $\Psi(0.5) \approx 0.034$.
- The solid-melt interfacial energies are equal, so $\gamma_{\alpha\ell} = \gamma_{\beta\ell} = \gamma_{s\ell}$.
- In a symmetric eutectic, the tie-line composition difference between the conjugate solid phases is $\Delta C = 2(1 - k_0)C_{eu}$.

Inserting all of the special conditions listed above into Eq. (E.48) gives the scaling law for symmetric lamellar eutectics,

$$\frac{\Delta T_{tot}^{*2}}{v_{eu}} \cong -1.08 \frac{m_\ell(1 - k_0)C_{eu}\Omega 2\gamma_{s\ell}\sin\theta^*}{D_\ell \Delta S_f}. \tag{E.49}$$

Equation (E.49) can be simplified further using the surface tension balance at the eutectic triple points, indicated in Fig. E.7,

$$\gamma_{\alpha\beta} = 2\gamma_{s\ell}\sin\theta^*. \tag{E.50}$$

Substituting Eq. (E.50) into Eq. (E.49) yields the scaling law for symmetric eutectics that connects the interface supercooling, $\Delta\hat{T}$, to the rate of eutectic freezing, v_{eu}.

$$\frac{\Delta T_{tot}^{*2}}{v_{eu}} \cong -1.08 \frac{m_\ell(1 - k_0)C_{eu}\Omega\gamma_{\alpha\beta}}{D_\ell \Delta S_f}. \tag{E.51}$$

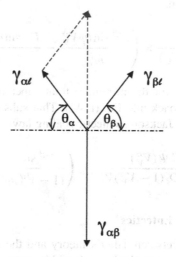

Fig. E.7 Surface tension vectors in equilibrium at an eutectic triple point. Contact angles and vector magnitudes between the crystalline phases and the melt are sketched for the case of a symmetric eutectic. See also Fig. 16.8

A comparison between Tiller's result, Eq. (16.25), and Eq. (E.51), derived from Hunt and Jackson's theory, shows that the scaling laws are consistent. The linear approximation for the solute gradients at the eutectic interface, used in Tiller's theory, introduced an unknown geometric factor, ξ. Comparison of Eqs. (16.25) with (E.51) shows that $\xi \approx 0.5$. Considering the elegant simplicity of Tiller's model of directionally solidified lamellar eutectics, his approximate results are indeed impressive.

References

1. J.D. Hunt and K.A. Jackson, *Trans. Metall. Soc.*, **236** (1966) 1129.
2. R. Trivedi and W. Kurz, *Solidification Processing of Eutectic Alloys*, D.M. Stefanescu, G.J. Abbaschian and R.J. Bayuzick, Eds. The Metallurgical Society, Warrendale, OH, 1989, p. 4.

Index